Evolution Education Around the Globe

Hasan Deniz · Lisa A. Borgerding
Editors

Evolution Education Around the Globe

Editors
Hasan Deniz
College of Education
University of Nevada Las Vegas
Las Vegas, NV
USA

Lisa A. Borgerding
College of Education, Health,
 and Human Services
Kent State University
Kent, OH
USA

ISBN 978-3-319-90938-7   ISBN 978-3-319-90939-4 (eBook)
https://doi.org/10.1007/978-3-319-90939-4

Library of Congress Control Number: 2018940410

© Springer International Publishing AG, part of Springer Nature 2018
This work is subject to copyright. All rights are reserved by the Publisher, whether the whole or part of the material is concerned, specifically the rights of translation, reprinting, reuse of illustrations, recitation, broadcasting, reproduction on microfilms or in any other physical way, and transmission or information storage and retrieval, electronic adaptation, computer software, or by similar or dissimilar methodology now known or hereafter developed.
The use of general descriptive names, registered names, trademarks, service marks, etc. in this publication does not imply, even in the absence of a specific statement, that such names are exempt from the relevant protective laws and regulations and therefore free for general use.
The publisher, the authors and the editors are safe to assume that the advice and information in this book are believed to be true and accurate at the date of publication. Neither the publisher nor the authors or the editors give a warranty, express or implied, with respect to the material contained herein or for any errors or omissions that may have been made. The publisher remains neutral with regard to jurisdictional claims in published maps and institutional affiliations.

Printed on acid-free paper

This Springer imprint is published by the registered company Springer International Publishing AG part of Springer Nature
The registered company address is: Gewerbestrasse 11, 6330 Cham, Switzerland

Preface

Is the teaching of evolution only controversial in a few places or for members of a few conservative religious sects? Our travels and communication with our evolution education research colleagues around the world made us anecdotally answer no to this question, but we realized that we lacked a systematic way of making such comparative assertions. *Evolution Education Around the Globe* begins to answer this question and provides what we hope is an internationally informed conversation about evolution education.

The origin of this book goes back to our years together at Indiana University Science Education Doctoral Program. During our time together in the doctoral program, we realized our common interest in evolution education and evolution education research. We collaborated over several research projects exploring evolution acceptance and understanding among Turkish and American preservice biology teachers. Our manuscripts from these projects appeared in *Journal of Research in Science Teaching* and *Reports of the National Center for Science Education*. We both graduated in 2007 and took Assistant Professor of Science Education positions at University of Nevada, Las Vegas (Hasan Deniz) and Kent State University (Lisa A. Borgerding). Our collaborative work had a pause as we struggled with the responsibilities of our new positions and life events until we resumed our collaboration when we met in Washington, D.C., to attend a National Science Foundation event in 2015. At this point, we both secured our tenure and were promoted to the rank of Associate Professor of Science Education. We realized that our passion for evolution education persisted, and we discussed possible ways to resume our collaboration. We quickly agreed on a need for a book providing a global view on evolution education and evolution education research. Immediately after our meeting at Washington, D.C., we secured a book proposal from Springer. When we mentioned our intention for such a book to our colleagues in American Education Research Association (AERA) and National Association for Research in Science Teaching (NARST) conferences, we were encouraged that our colleagues from around the world praised our efforts, and some of them were eager to submit a chapter describing status of evolution education in their respective countries or regions. With this encouragement, we sent a call for chapters to e-mail listservs or contact persons of international science education organizations. In the call for chapters, we expressed the need for a book and invited colleagues around the world to provide a chapter systematically summarizing evolution education literature in their country or geographical region and address the following topics:

- Public acceptance of evolutionary theory within the social and cultural context of the country;
- Whether there are anti-evolution movements in the country;
- Place of evolutionary theory in the curriculum;
- Emphasis given to evolutionary theory in biology teacher education programs;
- Biology teachers' attitudes toward teaching evolutionary theory;
- Suggestions to improve evolution education in the country.

This book differs from other books about evolution education in at least three distinct ways. First, and most importantly, the proposed book has an international focus. The vast majority of evolution education books are mono-national, and almost all exclusively focus on the evolution education controversy in the USA. Second, the individual chapter contributions for the proposed book include common elements that facilitate a cross-cultural meta-analysis. This meta-analysis will serve as the culmination of this international inquiry. Finally, this book is written for a primarily academic audience in an effort to provide a much-needed common background for future evolution education research across the globe.

We are indebted to each of the authors for their willingness to provide an overview of evolution education and evolution education research in their respective countries or regions. These authors excavated and synthesized research and policy documents in these different regions, and many even gathered and analyzed new or previously unreported data in this context. Very selfishly, we have had the pleasure of working with these excellent scholars from around the world and engaging in conversations we have long sought to have. We have learned from the chapters in this volume and hope that evolution education will benefit from this international perspective.

We are also grateful to our shared doctoral advisor, Dr. Valarie Akerson, who has long supported our careers and provided thoughtful advice as we first endeavored to initiate this book project.

Finally, we are grateful to our families for supporting us throughout the production of this book.

Las Vegas, NV, USA Hasan Deniz
Kent, OH, USA Lisa A. Borgerding

Contents

Part I Introduction

1 **Evolutionary Theory as a Controversial Topic in Science Curriculum Around the Globe** 3
Hasan Deniz and Lisa A. Borgerding

Part II North and South America

2 **Pedagogical Implications of American Muslims' Views on Evolution** 15
Khadija E. Fouad

3 **Project Teach Evolution: Preparing Biology Pre-service Teachers to Teach Evolution in Missouri, U.S.A.** 41
Patricia J. Friedrichsen, Larry G. Brown and Johannes Schul

4 **Controversial Before Entering My Classroom: Exploring Pre-service Teacher Experiences with Evolution Teaching and Learning in the Southeastern United States** 59
Amanda L. Glaze and M. Jenice "Dee" Goldston

5 **Case Studies in Teaching Evolution in the Southwestern U.S.: The Intersection of Dilemmas in Practice** 81
Rachel J. Fisher

6 **Evolution Education in Mexico, Considering Cultural Diversity** 101
Alma Adrianna Gómez Galindo, Alejandra García Franco, María Teresa Guerra Ramos, Eréndira Alvarez Pérez and José de la Cruz Torres Frías

7 **Evolution Education and the Rise of the Creationist Movement in Brazil** 119
Alandeom W. Oliveira and Kristin L. Cook

8   Evolution Education in Galápagos: What Do Biology Teachers
    Know and Think About Evolution? 137
    Sehoya Cotner and Randy Moore

Part III   Europe

9   Evolution Education in England 155
    Michael J. Reiss

10 Evolution Education and Evolution Denial in Scotland.......... 169
    J. Roger Downie, Ronan Southcott, Paul S. Braterman
    and N. J. Barron

11 Teaching Evolution in Greece 195
    Panagiotis K. Stasinakis and Kostas Kampourakis

12 Evolution Education in France: Evolution Is Widely Taught
    and Accepted 213
    Marie-Pierre Quessada and Pierre Clément

13 Evolution Education in the German-Speaking Countries 235
    Erich Eder, Victoria Seidl, Joshua Lange and Dittmar Graf

Part IV   Middle East

14 An Insight into Evolution Education in Turkey 263
    Ebru Z. Muğaloğlu

15 Evolution Education in Iran: Shattering Myths About Teaching
    Evolution in an Islamic State 281
    Mahsa Kazempour and Aidin Amirshokoohi

16 Evolution Education in the Arab States: Context, History,
    Stakeholders' Positions and Future Prospects 297
    Saouma BouJaoude

Part V   Asia

17 Evolution Education in Hong Kong (1991–2016): A Content
    Analysis of the Biology Textbooks for Secondary School
    Graduates 315
    Ka Lok Cheng and Kam Ho Chan

18 Evolution Education in Indonesia: Pre-service Biology Teachers'
    Knowledge, Reasoning Models, and Acceptance of Evolution 335
    Arif Rachmatullah, Ross H. Nehm, Fenny Roshayanti and Minsu Ha

19 A Glimpse of Evolution Education in the Malaysian Context 357
    Yoon Fah Lay, Eng Tek Ong, Crispina Gregory K. Han
    and Sane Hwui Chan

20	**Biological Evolution Education in Malaysia; Where We Are Now** .... Kamisah Osman, Rezzuana Razali and Nurnadiah Mohamed Bahri	375
21	**Evolution Education in the Philippines: A Preliminary Investigation** .... Jocelyn D. Partosa	391

Part VI Africa

22	**The Unusual Case of Evolution Education in South Africa** ...... Martie Sanders	409

Part VII New Zealand

23	**Evolution Education in New Zealand** .... Alison Campbell	431

Part VIII Conclusion

24	**Evolution Education Around the Globe: Conclusions and Future Directions** .... Lisa A. Borgerding and Hasan Deniz	449

About the Editors

Hasan Deniz is an Associate Professor of Science Education at University of Nevada Las Vegas (UNLV). He teaches undergraduate, masters, and doctoral level courses in science education program at UNLV. His research agenda includes students' and teachers' epistemological beliefs about science (nature of science) and evolution education. He is recently engaged in professional development activities supported by several grants targeting to increase elementary teachers' knowledge and skills to integrate science, language arts, and engineering education within the context of Next Generation Science Standards.

Lisa A. Borgerding is an Associate Professor of Science Education at Kent State University. She teaches undergraduate, masters, and doctoral level courses in science education and research methodology. Her research centers upon the teaching and learning of biological evolution from early childhood through college, nature of science instruction, preservice and inservice teacher development, and service learning in teacher education.

Part I
Introduction

Chapter 1
Evolutionary Theory as a Controversial Topic in Science Curriculum Around the Globe

Hasan Deniz and Lisa A. Borgerding

Abstract Evolutionary theory is considered as one of the greatest scientific achievements in history of science on par with the theory of heliocentricism, general and specific relativity, and the theory of plate tectonics. However, public controversy over teaching evolutionary theory urges science educators to consider conceptual, epistemic, worldview/religious, and social/cultural factors simultaneously when teaching about evolutionary theory. In this book, we aimed to explore the influence of social and cultural domain on evolution education.

Evolutionary theory is considered as one of the greatest scientific achievements in history of science on par with the theory of heliocentricism, general and specific relativity, and the theory of plate tectonics. Heliocentricism challenged earth-centered Ptolemaic system and replaced it with sun-centered view of the universe. Relativity changed our concept of time. The theory of plate tectonics changed our view of unmoving continents and replaced it with the view that each continent was part of a single continent that broke apart. Similarly, the theory of evolution changed the concept of fixed species and replaced it the view that new species can arise from the old species. None of these aforementioned scientific theories are controversial within the scientific community and there is no debate among scientists whether these theories meet the standards of a scientific theory. However, evolutionary theory stands out from other scientific theories in that it tends to create a public controversy. The controversy over teaching evolutionary theory is a global phenomenon not merely confined to a single country or region. The controversial nature of evolutionary theory makes teaching evolution a difficult task for biology teachers, thereby creating a very interesting and unique research agenda for science educators. Many science education researchers around the world

---

H. Deniz (✉)
College of Education, University of Nevada Las Vegas, Las Vegas, NV, USA
e-mail: hasan.deniz@unlv.edu

L. A. Borgerding
College of Education, Health, and Human Services, Kent State University, Kent, OH, USA
e-mail: ldonnell@kent.edu

are interested in investigating issues that are salient to teaching and learning of evolutionary theory.

Science educators simultaneously need to consider various domains when teaching about evolutionary theory:

- Conceptual domain
- Epistemic domain
- Worldview/religious domain
- Social and cultural domain.

1.1 Conceptual Domain

The conceptual domain includes both scientifically accepted evolutionary concepts and students' nonscientific conceptions related to evolutionary theory. A collection of students' current concepts including scientifically accepted evolutionary concepts and alternative conceptions about evolution called conceptual ecology (Toulmin, 1972) influences how students learn about evolutionary theory. Students' views on evolutionary theory are probably reflective of general population's views on the theory, which we know from survey data is quite skeptical of evolutionary theory in many countries (Miller, Scott, & Okamoto, 2006). Therefore, it is quite reasonable to expect students to enter into classroom with a number of misconceptions about evolutionary theory. Common student misconceptions about evolutionary theory includes (a) all evolutionary change is adaptive, (b) evolutionary change is progressive, (c) evolutionary change is teleological (goal-directed), (d) evolutionary theory is a form of atheism, and (e) evolutionary process in general, and natural selection in particular are equated with event-like ontology rather than equilibration type ontology (Ferrari & Chi, 1998). Many influential authors writing about conceptual change warned about how students' prior conceptions (the current conceptual ecology) might interfere with the learning process (e.g. Bransford, Brown, & Cocking, 1999; Chin & Brever, 1993; Pintrich, Marx, & Boyle, 1993). Students' current conceptual ecology can either facilitate or impede the learning process (Pintrich et al., 1993). Therefore, science teachers need to assess their students' current conceptual ecology about evolutionary theory before they start teaching about the theory. Ascertaining students' prior conceptions about evolutionary theory is in line with the practice of a teacher who adopts constructivist teaching and learning principles.

1.2 Epistemic Domain

The epistemic domain can be examined at two levels (a) personal epistemology of students and (b) students' epistemological beliefs about science, i.e., students' nature of science views.

Personal epistemology shapes one's sense of what constitutes reality and how one comes to know something. Personal epistemology serves as a standard against which one judges his or her understanding during learning (Hewson, 1985; Hofer, 1997). Perry (1970) described a person's epistemological development in nine stages. Perry's original scheme contained nine stages, but it was convenient for most researchers to organize these nine stages into four: dualism, multiplicity, relativism and commitment to relativism. According to Perry (1970), many students come to college at the dualism stage. In dualism stage, students see the things as "right or wrong" or "black or white." They think that knowledge is objective and the instructor is the representative of authority. As the students are exposed to conflicting views of different authorities on the same issue, they question the dichotomous view of the world. They think that there are some issues that cannot be definitively known. Students who are thinking at this level are in multiplicity stage. Within this stage, students believe that there is truth, but that there is room for uncertainty. In relativity stage, students come to think that there are few issues that can be known for sure. This stage is much different from other stages because there is a major departure from dualistic way of thinking. Authority becomes open to debate and criticism in this stage. In the commitment to relativism position, students find relativism disorienting. Students seek to develop commitments to do away with disorienting while they continue to acknowledge other peoples' positions.

Nature of science (NOS) refers to epistemology of science, i.e., values and beliefs specific to the scientific knowledge and its development (Lederman, 2007). There is no single definition of NOS among philosophers of science, historians of science, scientists, and science educators, but certain NOS ideas are uncontroversial and promoted by most science educators. These NOS ideas include but are not limited to conceptions that scientific knowledge is empirically-based, tentative, subjective, inferential, socially and culturally embedded, and depends upon human imagination and creativity. In addition to these NOS ideas, three additional NOS ideas are relevant to evolutionary theory:

(1) The functions of and relationships between scientific theories and laws: There is no hierarchical relationship between theories and laws. Theories do not turn into laws. Everyday usage of the word theory is problematic for students' science learning. The everyday usage refers to some sort of wild idea, which may or may not be empirically supported. In science, theories are extremely well-supported web of hypotheses that are constructed to explain natural phenomena.
(2) The notion that experiments are not the only way to perform scientific research: Experimentation is a useful scientific method in science, but is not the only method to conduct scientific research. Scientific research in astronomy and evolutionary biology are based on extensive observations rather than experiments.
(3) The demarcation criteria for scientific knowledge: Science is a limited way of knowing. Science cannot answer all questions. For example, science cannot answer moral and ethical questions. Scientists do not invoke supernatural explanations when conducting scientific research.

It can be conceived that students' overall personal epistemology and their epistemological beliefs about science, i.e., nature of science views are commensurate with each other. It can be thought that students' overall epistemological sophistication can influence to what extent one can improve their nature of science views (Akerson, Morrison, & McDuffie, 2006) and in turn, improved nature of science views can facilitate one's transition from a lower epistemological stage to a higher epistemological stage (Deniz, 2011). It is quite possible to conceive that students' overall epistemic beliefs and nature of science beliefs together shape students' learning about evolutionary theory. In other words, students will learn about evolutionary theory from the lens of their overall epistemic beliefs in general and nature of science views in particular.

1.3 Worldview/Religious Domain

Philosophical materialism (atheism) and theism are the two major worldviews that are relevant to teaching evolutionary theory (Anderson, 2007). According to philosophical materialism, matter exists and that is all; there is no acknowledgement of God or gods; and ethics are constructed by humans. According to theism, God exists, God created the universe and the living things; there is life after death; and ethics originate from God. A person with a theistic worldview may not be necessarily religious, however, all religious people subscribe to the theistic worldview.

Epistemologically, religion and science can be considered as "nonoverlapping magisteria" (Gould, 1997) but pedagogically these two magisterias can potentially overlap with each other in a student's mind. This potential overlap specifically fuels the opposition to teaching the theory of evolution and antievolution movements in various countries around the globe. The driving force behind antievolution movements is not just about students' learning about science content (evolution), but the implications of evolutionary theory for students' worldviews and religious beliefs. Confining evolutionary theory within philosophical materialist worldview provides additional fuel and fervor for antievolution movements that are motivated by their strong commitment to the preservation of their theistic worldview. We need to convey to our students that scientists are methodological materialists but they are not necessarily philosophical materialists. In other words, scientists do not use supernatural explanations while conducting and publishing scientific research, but they can interpret their scientific understanding and research findings from the perspective of their theistic worldview. In fact, Easterbrook (1997) reported that about 40% of working scientists have serious religious beliefs, based on survey items including explicit statements including believing in a personal God and praying.

1.4 Social and Cultural Domain

Miller et al. (2006) compared the acceptance of evolutionary theory among adults in 34 countries. Most Western European countries such as Iceland, Denmark, Sweden, France, and United Kingdom have 75 percent or more acceptance rates of evolutionary theory. Countries such as United States and Turkey have acceptance rates of 40% and 25% respectively. If a concept has little leverage within a cultural milieu, it will not be readily acceptable and it will be difficult for that concept to be included in the school curriculum. Costa (1995) stated that successful transition of students from their own world to school science depends on the compatibility of family and school cultures. Deniz, Donnelly and Yilmaz (2008) found that Turkish preservice biology teachers with well-educated parents are more likely to accept evolution as a scientifically valid theory. This makes sense considering the fact Turkish education system is historically modeled based on Western education principles especially after the declaration of Republic of Turkey in 1923.

According to Aikenhead and Jegede (1999) when the culture of school science is compatible with a students' social and cultural values science instruction tend to happen smoothly. However, if there is a conflict between the culture of science and a student's socio-cultural values science instruction tends to damage students' socio-cultural values by forcing students to abandon their indigenous values. For this reason, Aikenhead and Jegede (1999) called for developing culturally sensitive curricula and teaching methods to be able to avoid the clash between students' cultural values and the culture of Western science.

1.5 Evolution Education Research Around the Globe

Many researchers from different countries around the world conduct research on evolution education. A critical of review of the evolution education research from different parts of the world allows us to have a global view of the issues that are salient to teaching evolution. This critical review of the literature from different countries also enables us to appreciate the influence of social and cultural context on evolution education topics under investigation. Science education researchers have investigated student and teacher evolution understanding in Greece (Athanasiou & Mavrikaki, 2015), Canada (Nieswandt & Bellomo, 2009); England (Tenenbaum, To, Wormald, & Pegram, 2015), and the Netherlands (Geraedts & Boersma, 2006). Several studies have investigated evolution acceptance in countries such as Turkey (Akyol, Tekkaya, Sungur, & Traynor, 2012; Deniz et al., 2008; Peker, Comert, & Kence, 2010), Greece (Athanasiou & Papadopoulou, 2012), Korea (Ha, Haury, & Nehm, 2012), Lebanon (BouJaoude, Wiles, Asghar, & Alters, 2011b), Belize (Nunez, Pringler, & Showalter, 2012), and Jordan (DeBaz & El-Weher, 2012). Other non-U.S. studies have investigated evolution as it pertains to the nature of science in Lebanon and Egypt (BouJaoude et al., 2011a; Dagher & BouJaoude, 2005) and

Korea (Kim & Nehm, 2011) and religiosity in Greece (Athanasiou & Papadopoulou, 2012), Lebanon (BouJaoude et al., 2011a, 2011b; Hokayem & BouJaoude, 2008), Israel (Dodick, Dayan, & Orion, 2010); Scotland (Downie & Barron, 2000), Austria (Eder, Turic, Milasowszky, Van Adzin, & Hergovich, 2011); Australia (Ferguson & Kameniar, 2014); England (Hanley, Bennett, & Ratcliffe, 2014), Singapore (Seoh, Subramaniam, & Hoh, 2016), and Thailand (Yasri & Mancy, 2014). There are international concerns about how teachers approach and actually teach evolution in Brazil (Marcelos & Nagem, 2012), South Africa (Abrie, 2010), Israel (Dodick, Dayan, & Orion, 2010); the Netherlands (Schilders, Sloep, Peled, & Boersma, 2009), and Turkey (Akyol et al., 2012). Finally, science educators have investigated the presentation and minimization of evolution in national curricula in New Zealand (Campbell & Otrel-Cass, 2011), Belize (Nunez, Pringle, & Showalter, 2012), and France (Quessada & Clement, 2007).

In this book, we aimed to present a global view of evolution education by asking science educators around the world to address the following topics in their own country or region:

- Public acceptance of evolutionary theory within the social and cultural context of the country or region
- Whether there are anti-evolution movements in the country or region
- Place of evolutionary theory in the curriculum
- Emphasis given to evolutionary theory in biology teacher education programs
- Biology teachers' attitudes toward teaching evolutionary theory
- Suggestions to improve evolution education in the country or region.

The book includes a total of 24 chapters: the introductory and conclusion chapters; seven chapters from North and South America (Brazil, Galápagos, Mexico and four chapters from the United States-US Muslims, Missouri, Southwestern US, and Southern US); five chapters from Europe (England, France, German speaking countries, Greece, and Scotland); three chapters from Middle East (Arab States, Iran, and Turkey); five chapters from East Asia (Hong Kong, Indonesia, Philippines, and two chapters from Malaysia); one chapter from South Africa; and one chapter from New Zealand.

In this book, we aimed to explore the influence of social and cultural domains on evolution education. Therefore, cognitive and epistemic aspects of evolution education are not targeted in this volume. Even though we asked our authors to systematically address the above points in each chapter, we also allowed some chapters to include empirical studies related to evolution education while addressing as many common points as possible.

1.6 Conclusion

We believe that this volume will contribute to the evolution education research by providing a global view on the status of evolution education and evolution education research. The work done in evolution education is substantial and it is time to coordinate the efforts of educators and researchers interested in evolution education. We believe that this volume will underscore evolution education research as a significant area of study within international science education research and bring the challenges of teaching evolutionary theory to the attention of international public opinion. The contributors to this volume addressed the above bulleted points by including the relevant evolution education research conducted in the designated country or region. The systematic treatment of topics and the inclusion of relevant literature across different countries or regions allowed us to assess the state of evolution education and evolution education research in each country or region, thereby providing a global view. Evolution education will benefit from the work of the contributors to this volume and from those who draw on our contributors' insightful suggestions to improve evolution education.

References

Abrie, A. L. (2010). Student teachers' attitudes towards and willingness to teach evolution in a changing South African environment. *Journal of Biological Education, 44*, 102–107.

Aikenhead, G. S., & Jegede, O. J. (1999). Cross-cultural science education: A cognitive explanation of a cultural phenomenon. *Journal of Research in Science Teaching, 36*, 269–287.

Akerson, V. L., Morrison, J. A., & McDuffie, A. R. (2006). One course is not enough: Preservice elementary teachers' retention of improved views of nature of science. *Journal of Research in Science Teaching, 43*(2), 194–213.

Akyol, G., Tekkaya, C., Sungur, S., & Traynor, A. (2012). Modeling the interrelationships among pre-service science teachers' understanding and acceptance of evolution, their views on nature of science and self-efficacy beliefs regarding teaching evolution. *Journal of Science Teacher Education, 23*(8), 937–957.

Anderson, R. D. (2007). Teaching the theory of evolution in social, intellectual, and pedagogical context. *Journal of Research in Science Teaching, 91*(4), 664–677.

Athanasiou, K., & Papadopoulou, P. (2012). Conceptual ecology of the evolution acceptance among Greek education students: Knowledge, religious practices and social influences. *International Journal of Science Education, 34*(6), 903–924.

Athanasiou, K., & Mavrikaki, E. (2015). Conceptual inventory of natural selection as a tool for measuring Greek university students' evolution knowledge: Differences between novice and advanced students. *International Journal of Science Education, 36*, 1262–1285.

BouJaoude, S., Asghar, A., Wiles, J. R., Jaber, L., Sarieddine, D., & Alters, B. (2011a). Biology professors' and teachers' positions regarding biological evolution and evolution education in a Middle Eastern society. *International Journal of Science Education, 33*(7), 979–1000.

BouJaoude, S., Wiles, J. R., Asghar, A., & Alters, B. (2011b). Muslim Egyptian and Lebanese students' conceptions of biological evolution. *Science & Education, 20*, 895–915.

Bransford, J., Brown, A., & Cocking, R. (Eds.). (1999). *How people learn: Brain, mind, experience, and school*. Washington, DC: National Academy Press.

Campbell, A., & Otrel-Cass, K. (2011). Teaching evolution in New Zealand's schools—Reviewing changes in the New Zealand science curriculum. *Research in Science Education, 41*, 441–451.

Chinn, C. A., & Brewer, W. F. (1993). The role of anomalous data in knowledge acquisition: A theoretical framework and implications for science instruction. *Review of Educational Research, 63*, 1–49.

Costa, V. B. (1995). When science is "another world": Relationships between worlds of family, friends, school, and science. *Science Education, 79*(3), 313–333.

Dagher, Z. R., & Boujaoude, S. (2005). Students' perceptions of the nature of evolutionary theory. *Science Education, 89*(3), 378–391.

De Baz, T., & El-Weher, M. (2012). The effect of contextual material on evolution in the Jordanian secondary-school curriculum on students' acceptance of the theory of evolution. *Journal of Biological Education, 46,* 20–28.

Deniz, H. (2011). Examination of changes in prospective elementary teachers' epistemological beliefs in science and exploration of factors meditating that change. *Journal of Science Education and Technology, 20*(6), 750–760.

Deniz, H., Donnelly, L., & Yilmaz, I. (2008). Exploring the factors related to acceptance of evolutionary theory among Turkish preservice biology teachers: Toward a more informative conceptual ecology for biological evolution. *Journal of Research in Science Teaching, 45*(4), 420–443.

Dodick, J., Dayan, A., & Orion, N. (2010). Philosophical approaches of religious Jewish science teachers toward the teaching of 'controversial' topics in science. *International Journal Of Science Education, 32*(11), 1521–1548.

Downie, J. R., & Barron, N. J. (2000). Evolution and religion: Attitudes of Scottish first year biology and medical students to the teaching of evolutionary biology. *Journal of Biological Education, 34,* 139–146.

Easterbrook, G. (1997). Science and God: A warming trend? *Science, 277,* 890–893.

Eder, E., Turic, K., Milasowszky, N., Van Adzin, K., & Hergovich, A. (2011). The relationships between paranormal belief, creationism, intelligent design, and evolution at secondary schools in Vienna (Austria). *Science & Education, 20,* 517–534.

Ferguson, J. P., & Kameniar, B. (2014). Is 'learning' science enough?—A cultural model of religious students of science in an Australian government school. *International Journal of Science Education, 36,* 2554–2579.

Ferrari, M., & Chi, M. T. H. (1998). The nature of naïve explanations of natural selection. *International Journal of Science Education, 20*(10), 1231–1256.

Geraedts, C. L., & Boersma, K. T. (2006). Reinventing natural selection. *International Journal of Science Education, 28*(8), 843–870.

Gould, S. J. (1997). Nonoverlapping magisteria. *Natural History, 106,* 16–22.

Ha, M., Haury, D. L., & Nehm, R. H. (2012). Feeling of certainty: Uncovering a missing link between knowledge and acceptance of evolution. *Journal of Research in Science Teaching, 49*(1), 95–121.

Hanley, P., Bennett, J., & Ratcliffe, M. (2014). The interrelationships of science and religion: A typology of engagement. *International Journal of Science Education, 36,* 1210–1229.

Hewson, P. W. (1985). Epistemological commitment in the learning of science: Examples from dynamics. *European Journal of Science Education, 7,* 163–172.

Hofer, B. K. (1997). The development of personal epistemology: Dimensions, disciplinary differences, and instructional practices. Doctoral dissertation. University of Michigan, Ann Arbor.

Hokayem, H., & BouJaoude, S. (2008). College students' perceptions of the theory of evolution. *Journal of Research in Science Teaching, 45*(4), 395–419.

Kim, S. Y., & Nehm, R. H. (2011). A Cross-cultural comparison of Korean and American science teachers' views of evolution and the nature of science. *International Journal of Science Education, 33,* 197–227.

Lederman, N. G. (2007). Nature of science: Past, present, and future. In S. K. Abel & N. G. Lederman (Eds.), *Handbook of research on science education* (pp. 831–879). Mahwah, NJ: Erlbaum.

Marcelos, M. F., & Nagem, R. L. (2012). Use of the "Tree" analogy in evolution teaching by biology teachers. *Science & Education, 21,* 507–541.

Miller, J. D., Scott, E. C., & Okamoto, S. (2006). Public acceptance of evolution. *Science, 313,* 765–766.

Nieswandt, M., & Bellomo, K. (2009). Written extended-response questions as classroom assessment tools for meaningful understanding of evolutionary theory. *Journal of Research in Science Teaching, 46*(3), 333–356.

Nunez, E. E., Pringle, R. M., & Showalter, K. T. (2012). Evolution in the Caribbean classroom: A critical analysis of the role of biology teachers and science standards in shaping evolution instruction in Belize. *International Journal of Science Education, 34*, 2421–2453.

Peker, D., Comert, G. G., & Kence, A. (2010). Three decades of anti-evolution campaign and its results: Turkish undergraduates' acceptance and understanding of the biological evolution theory. *Science & Education, 19*, 739–755.

Perry, W. G. (1970). *Intellectual and ethical development in the college years: A scheme.* Cambridge, Mass: Harvard University Press.

Pintrich, P. R., Marx, R. W., & Boyle, R. A. (1993). Beyond cold conceptual change: The role of motivational beliefs and classroom contextual factors in the process of conceptual change. *Review of Educational Research, 63*(2), 167–199.

Quessada, M. P., & Clement, P. (2007). An epistemological approach to French syllabi on human origins during the 19th and 20th centuries. *Science & Education, 16*, 991–1006.

Schilders, M., Sloep, P., Peled, E., & Boersma, K. (2009). Worldviews and evolution in the biology classroom. *Journal of Biological Education, 43*, 115–120.

Seoh, K. H. R., Subramaniam, R., & Hoh, Y. K. (2016). How humans evolved according to grade 12 students in Singapore. *Journal of Research in Science Teaching, 53*, 291–323.

Tenenbaum, H. R., To, C., Wormald, D., & Pegram, E. (2015). Changes and stability in reasoning after a field trip to a natural history museum. *Science Education, 99*, 1073–1091.

Toulmin, S. (1972). *Human understanding: An inquiry into the aims of science.* Princeton, NJ: Princeton University Press.

Yasri, P., & Mancy, R. (2014). Understanding student approaches to learning evolution in the context of their perceptions of the relationship between science and religion. *International Journal of Science Education, 36*(1), 24–45.

Hasan Deniz is an Associate Professor of Science Education at University of Nevada Las Vegas (UNLV). He teaches undergraduate, masters, and doctoral level courses in science education program at UNLV. His research agenda includes students' and teachers' epistemological beliefs about science (nature of science) and evolution education. He is recently engaged in professional development activities supported by several grants targeting to increase elementary teachers' knowledge and skills to integrate science, language arts, and engineering education within the context of Next Generation Science Standards.

Lisa A. Borgerding is an Associate Professor of Science Education at Kent State University. She teaches undergraduate, masters, and doctoral level courses in science education and research methodology. Her research centers upon the teaching and learning of biological evolution from early childhood through college, nature of science instruction, preservice and inservice teacher development, and service learning in teacher education.

Part II
North and South America

Chapter 2
Pedagogical Implications of American Muslims' Views on Evolution

Khadija E. Fouad

Abstract American Muslims' rates of acceptance of evolution and those of the population as a whole are similar, because they form three groups: those who accept both macroevolution and microevolution for all species, those who accept macroevolution for all species except humans, and those who reject macroevolution for all species, and because people who have one way of negotiating the relationship between science and religion may be resistant to adopting another method of negotiating this relationship. A difference is that American Muslims generally accept an old age for the Earth, whether or not they accept evolution. Pedagogical implications of these views for Muslims are that curricula could be sequenced to teach microevolution before macroevolution, and that a robust treatment of both the science supporting evolutionary theory and important NOS concepts could help students avoid common misconceptions promoted by American creationists. Introducing students to different methods of negotiating the relationship between religion and science, and to practicing Muslim evolutionary biologists and Muslims from the past who developed proto-evolutionary theories, might help them to view acceptance of evolution in a more favorable light.

2.1 Introduction

2.1.1 The American Context for Islam

American Muslims constitute a small minority in the United States of about 1%. Muslims have been a part of the United States since its inception, mainly coming involuntarily due to the slave trade, but also some voluntarily even from early on (GhaneaBassiri, 2010). There have been successive waves of Muslim immigration to the U.S., with 40% of the current American Muslim population having arrived after 1960 due to changes in immigration laws (GhaneaBassiri, 2010; Pew, 2007).

K. E. Fouad (✉)
Appalachian State University, Boone, NC, USA
e-mail: fouadkd@appstate.edu

Converts to Islam and their children constitute more than a third of American Muslims, a feature that is unique to the United States compared to Muslim populations in other countries (Gallup, 2009).

American Muslims include groups seen elsewhere in the world, such as Sunni, Shia, Sufi, and Ahmadi Muslims (GhaneaBassiri, 2010; Pew, 2007). In addition, there are many American Muslims who self-identify simply as Muslim. That is, when they are asked about their affiliation, they will reply that they are "just Muslim" (Pew, 2007). There are some from indigenous, uniquely American forms of Islam, such as the Nation of Islam, as well. Half of American Muslims identify as Sunni, 16% as Shi'a, 22% as "just Muslim," with the remaining 12% containing Muslims from other groups, such as the Nation of Islam and the Ahmadiyya Movement in Islam.

Brief history of Islam. The religion of Islam was founded in the seventh century in Mecca in present-day Saudi Arabia when Muhammad ibn Abdullah began having experiences that he interpreted as divine revelations starting around 610 CE and continuing until his death in 632 (Aslan, 2006). These revelations were collected to form the Quran, or the Recitation, the scripture of the Muslims. The main teaching of Islam is that God is One and that He alone is worthy of worship. Muslims engage in various practices to attain nearness to God, such as prayer, charity, fasting, and performing the pilgrimage to Mecca.

The early Muslim community faced severe persecution in Mecca, including ridicule, torture, boycott, and death (Aslan, 2006). In response to this harsh treatment, Muhammad and some of his followers migrated to present-day Medina in 622 CE. In Medina, Muhammad became a political as well as a spiritual leader. After his death, his followers passed on many of his sayings and actions by oral tradition. These were collected in later centuries and written down to become known as the hadith collections.

The formation of groups in Islam. After Muhammad's death in 632 CE, there was disagreement among his companions as to who should succeed him (Aslan, 2006). One party supported his longtime friend and father-in-law Abu Bakr, while others supported his cousin and son-in-law, 'Ali ibn abu Talib. These two groups gave rise to the Sunni and Shi'a Muslims, respectively. Sunni Muslims hold that leadership of the Muslim community could rest in any pious, knowledgeable man. Shi'a Muslims hold that leadership of the Muslims should be by divine appointment only, and that this divine office of leadership in Islam was bestowed on descendants of Prophet Muhammad through his daughter, Fatimah, and 'Ali, because they believe these people to be wiser and more pious than others (Tabataba'i, 1971). Currently in the United States Sunnis have a diffuse, decentralized leadership, although umbrella organizations, such as the Islamic Society of North America, provide cohesion and structure for Islamic activities (GhaneaBassiri, 2010). Shi'a Muslims in the United States have religious scholars who provide them with guidance and leadership, as well as umbrella organizations, such as the Muslim Students' Association—Persian Speaking Group.

The Ahmadiyya Movement in Islam was founded in 1889 by Mirza Ghulam Ahmad of Qadian, India, who maintained that he was the long-awaited reformer of

Islam, the Imam Mahdi, as well as the Promised Messiah and metaphorical second coming of Jesus anticipated by Christians and Muslims alike, and the reincarnation of Krishna that the Hindus expected (GhaneaBassiri, 2010). The Ahmadiyya Movement sent missionaries to the United States in the 1920s and 1930s, successfully winning a number of converts. For this reason, they claim to be the oldest Muslim organization in the United States. Currently Ahmadi leaders are chosen on a local level under regional and national leadership, with separate organizations for women and men. These report directly to the Khalifah, the spiritual head of the community, headquartered in London, UK (Saliha Malik, personal communication, 2010).

The Nation of Islam (NOI) is a distinctly American form of Islam that originated in the early part of the 20th century (GhaneaBassiri, 2010). It was brought into national prominence under the leadership of Elijah Muhammad, who began leadership of the community in 1934. When he died in 1976, his son Warith Deen Muhammad took over leadership of the organization and later renamed it the Muslim American Society. He led his followers to an American version of Islam rooted in the Quran and mainstream Islamic practices. A couple of years after Warith Deen Muhammad took over the leadership of the NOI, Louis Farrakhan formed a splinter group that broke off from the main body of the organization and retained the original name. He resisted Warith Deen Muhammad's guidance toward a more mainstream version of Islam and instead retained the beliefs and practices promulgated by Elijah Muhammad.

Those American Muslims who say they are "just Muslim" without claiming membership in any specific group are a diverse group, and have different approaches to Islam. Some rely on the Quran alone for religious guidance, while others may rely on the hadith traditions as well. Among the reasons that they identify as just a Muslim are that they do not identify with ancient animosities or foreign cultural traditions that they view as intrinsic parts of Muslim groups, or they may have a desire to avoid sectarian arguments. In practice, many of these Muslims attend Sunni, Shi'a, or other mosques.

2.2 Public Acceptance of Evolutionary Theory Within the Social, Political, and Cultural Context of the United States

2.2.1 American Muslims' Views on Evolution

American Muslims' acceptance rate for evolution is 45%, similar to the acceptance rate for American Christians, but lower than the 53% acceptance rate for Muslims worldwide (Pew, 2013).

Everhart and Hameed (2013) conducted a mixed methods study of the views of 23 Pakistani-American medical doctors on evolution. They found four positions on evolution when they asked the physicians to choose a statement that was closest to their beliefs, theistic evolution, "all species, including humans, have evolved over millions of years, but Allah guided the process," naturalistic evolution, "all species, including humans, have evolved over millions of years, and Allah played no part," the special creation of humans, "Allah created humans, but all other species have evolved over millions of years," and the special creation of all species, "Allah created humans and all other species in the form they exist today." A qualitative study was conducted to examine the relationship between 60 American Muslim undergraduates' views on evolution, their understandings of nature of science, their understanding of natural selection, and the manner in which they negotiate the relationship between science and religion (Fouad, 2016a). Respondents in this study all believed that God was responsible for creation, whether or not they believed He used evolution as a mechanism for these changes. They generally accepted the idea that natural selection is responsible for microevolutionary changes in all organisms, including humans, but differed over whether all organisms, all organisms except humans, or no organisms are the product of macroevolutionary changes. These positions corresponded to theistic evolution, belief in the special creation of humans, and belief in the special creation of all species, respectively. None of the undergraduates chose the naturalistic evolution position. These positions are not unique to American Muslims, and similar positions can be found among American Christians (Legare, Evans, Rosengren, & Harris, 2012). Table 2.1 lists examples of people articulating these positions taken from an unpublished data set consisting of qualitative research interviews of 63 American Muslim undergraduates (Fouad, 2016b).

Factors affecting American Muslims' views on evolution. Although the evolution acceptance rate among American Muslims is similar to the country as a whole, there are some distinctive features about the manner in which American Muslims view evolutionary theory. We will examine these features in more detail.

The relationship between science and religion. Most U.S. Muslims do not believe there is any conflict between science and religion. The manner in which people negotiate the relationship between science and religion can be classified as conflict, independence, dialog, and integration (Barbour, 2000). Those with a conflict view see science and religion as competing methods of making sense of the world. Those who take an independence view see science and religion as having different, independent functions so that both can be used to make sense of the world, although each explains different aspects. Those who take a dialog view use metaphors from one to explain the other, or view religion as providing answers to questions that science cannot answer. Those who view the relationship between religion and science as being integrated use both together to formulate their understandings of natural phenomena. These categories can generally be a useful way to think about American Muslims' views on the relationship between science and religion, although some do not fit into these categories, either because they are disengaged from this question or because they are in the process of sorting out this relationship for themselves (Fouad, 2016a). Examples of American Muslims articulating each of these positions are presented in Table 2.2.

Table 2.1 Examples of American Muslims' views on evolution

Stance	Example
Theistic Evolution (Both macroevolution and microevolution for all species)	Abbas: There's a lot of evidence scientifically that proves evolution, but being Muslim, we believe that the source of all life or all matter in the universe comes from a Supreme Being, Allah, and it just makes sense this way without conflicting with my religious beliefs Angela: I feel the evolution debate is null and void, considering the scientific evidence we have. As Muslims we are required to read and understand science, and be exemplary in learning. So, for me it's like the judgment of how basing Allah's creation on human understanding is a little faulty, so I really just don't see how evolution can't co-exist with a belief in Allah and His creation of Earth, because we don't, we can't even have any understanding of Allah's mercifulness. How can we have understanding of something as complex as how He decided to create the world? Habib: If as a Muslim you take it that Allah, along with His 99 names, if He's capable of anything, then He would be capable of implementing such a system as evolution
Special Creation of Humans (Microevolution for humans and both microevolution and macroevolution for all other species)	Rafiq: I believe that we did evolve from previous ancestors, but when you tell me actually that when we first evolved from the very first human being, that's kind of, you know, that they're come from another species, we're not, we didn't come from monkeys.... Because religiously, obviously Adam and Eve were the first human beings on earth, correct? ...So, that's why I'm telling you that we, the very first human beings did not evolve from previous species, but we did evolve from our ancestors, such as Adam and Eve. That's my view on it. I do agree that we did evolve, but not from other animals, from our own species Salahuddin: It makes sense to me, because if you look at the Quran and also the Bible, God says that He blew His soul into Adam, but it also says that the heavens and the Earth were as one unit of creation, and also, "We created from water every living thing." So, I don't see them as being apart. The fact that God blew His spirit into Adam can be taken symbolically, but I think that might be stretching it, although I wouldn't be surprised if we did evolve with the other species....I mean, I wouldn't be surprised

(continued)

Table 2.1 (continued)

Stance	Example
	if it is more evolution, but just from the way the verse is, it sounds to me like that Adam and the jinn and the angels were all created separately from that process, but at the same time, one of God's names is al-Bari, which has been translated as the Evolver
Special Creation of All Species (Microevolution for all species with macroevolution for no species)	Akilah: I believe that Allah created everything, and nothing evolved by itself. Everything's from Allah so you know how people say, oh, from evolution, the dinosaurs and all this kind of stuff? But I believe like Allah created everything on the planet. He created the world and everything Hadiyah: Well, I know that I've seen different types of animals: birds and reptiles and different things like crocodiles and alligators. I'm sure that over time that their environment changed, and they changed with their environment. So, to me, this is a just another thing to marvel at. When you think about Allah and His creation, everything changes over time, but how does it change? Well, of course, as a Muslim, I believe it changes with the will of Allah, with the power of Allah, so I do believe that even the land, not just the animals, every creation, the trees, the plants, everything has changed over time, so of course it's only logical for the things that live in the environment to change with it, and I think that is something that, you know, it shows us the power of Allah, like how He can adapt the things over time, and things change with their environment

Note All names are pseudonyms. Data taken from a sample of 63 American Muslim undergraduates (Fouad, 2016b)

American Muslims' stances on evolution can be seen in the light of the manner in which they negotiate the relationships between science and religion. Hadiyah's response given in Table 2.1 is an example. She uses integration to incorporate both the scientific evidence and her religious beliefs to form a coherent view of biological evolution. She can accommodate the strong scientific evidence she learned in her biology and anthropology classes by allowing for microevolution of all species, but her literal interpretation of the Quran precludes her from accepting the idea that evolution was responsible for their emergence.

Religious texts. American Muslims' stances on evolution can also be seen as a response to both the scientific evidence and their religious scriptures as well as the

Table 2.2 Examples of American Muslims' views on the relationship between science and religion

Stance	Example
Conflict	Brittany: Either you believe what your religious book says, or you believe what this theory says Lubna: With religion it's, everything is written. With science, it's everything is to be proven....Very religious people, they don't necessarily think science is correct, because they think that everything has already been written, and that it doesn't have to be proved Nabila: I think the border is crossed when one decides to specifically focus on scientific points of view, one is trying to understand the world and completely disregard any religious aspects like forgetting to acknowledge the fact that, okay, these discoveries aren't human discoveries really. We have to acknowledge that apart from the scientific understanding and the scientific explanations for these phenomenas, at the end of the day, really everything can be explained by Allah, and everything was created by Allah
Independence	Haroon: Religion is different from science, because science is the study of how things work in the universe. Religion is the study of how you should live in this universe Carlene: Religion and philosophy, it seems that those fields, they function to tell us why things happen, and science and physics and all the rest, they tell us how things happened Nafisa: I think science tends to explain what's going on in the world whereas religion kind of gives it a purpose
Dialog	Adam: There is a big gap in science. How did something come from nothing? It's a gap they try to fill up with reason, but it's God, not science Nadira: General umbrella of science .... Some parts are incomplete without religion. There is not a conflict because one is a tool to explain the other. Science cannot stand alone, because it is a tool to explain what is written in the Quran, to gain an appreciation of what Allah says in the Quran, because Allah is al-Malik, King of Everything
Integration	Jason: Science and religion, they go hand in hand Latifa: I don't think you have to separate science and religion, because if we're talking about religion in terms of what God has a part in and we assume that God has a part in everything, it doesn't really make sense to separate them
Disengagement	Nusaybah: I really couldn't take a side, honestly, I really don't take sides

Note All names are pseudonyms. Data taken from a sample of 63 American Muslim undergraduates (Fouad, 2016b)

weights and interpretations they give to each of these types of explanations. The two main textual sources used by Muslims are the Quran, which Muslims hold to be the word of God as revealed to the Prophet Muhammad, and the hadith, which are traditions attributed to Prophet Muhammad (Aslan, 2006). All Muslims consider the Quran as authoritative, but there are disagreements over which hadith are considered authentic among the different groups of Muslims (Aslan, 2006). For example, Sunni Muslims use traditions that were collected from the Prophet's companions and retold by later generations. Shi'a Muslims use traditions transmitted by the imams, descendants of Prophet Muhammad whom they believe to be his pious successors. Some scriptures used by American Muslims in formulating their stances on evolution are considered in detail here.

Muslims consider God to be the Creator of the universe and to be responsible for its care and maintenance in response to verses such as the following. "And We have not created the heavens and the earth and whatever is between both of them as one who indulges in idle play"[1] (21:16). Here, creation is described as teleological in its essence, as everything has been created for a set purpose determined by God.

Not only did God create the universe, but He is responsible for maintaining it, and encompasses it with His knowledge, as described in the following verse:

> God – there is nothing worthy of worship but He, the Living, the Self-subsisting, Eternal. Neither drowsiness nor sleep can seize Him. For Him is whatever is in the heavens and whatever is in the earth. Who is there who can intercede with Him except with His permission? He knows whatever is in front of them and whatever is behind them, and they will not encompass anything from His knowledge except what He wills. His authority extends over the heavens and the earth, and He does not weary of guarding and preserving them both, for He is the Most High, the Always Most Magnificent. (2:255)

Here God is depicted as being continually necessary for the perpetuation of the creation. If He were to shift His attention from it for only a moment, it would cease to exist. However, He is constantly awake and alert, preserving the universe and everything in it.

Most Muslims do not have any problem accepting an old age for the Earth. Although creation is described in the Quran as taking place in six days (سِتَّةِ أَيَّام), "days" is generally understood to mean periods of time, and not necessarily 24-hour "days." For example, "God is He Who created the heavens and the Earth and whatever is between both of them in six eons" (32:4).

Noah's flood is mentioned in the Quran, but it engulfs only Noah's people, and not the entire Earth, for example, the following verse.

> And We helped him against the nation who belied Our miraculous signs. Indeed they were an evil nation, so We drowned them all together (21:77).

This verse does not pose any problem to Muslims who wish to accept evolution, as verses in the Bible concerning the flood do for some Christians. Christians who believe in a literal interpretation of the Biblical version of the flood must somehow explain the evolutionary bottleneck that would have occurred on the ark. Muslims, on the other hand, believe that only Noah's people were flooded, so plants and animals could have easily survived outside of the flood zone. Even a literal interpretation of the version in the Quran would not be incompatible with acceptance of evolution.

There are many verses in the Quran that could be interpreted as specifying how Adam was created, but it does not give a similar treatment to the creation of other

---

[1] All translations of the Quran from the Arabic are my own unless otherwise noted.

organisms. Therefore Muslims consider the creation of Adam differently in formulating their stances on evolution than they do the creation of other organisms.

One example of a verse mentioning the creation of plants and animals is the following:

> He created the heavens without any visible pillars and He cast in the Earth anchors (firm mountains) lest it shake with you, and He spread on it every living, crawling creature, and We sent down water from the sky and germinated on it every noble pair. (31:10)

Verses such as this one do not specify exactly how animals and plants were created, and therefore leave open the possibility that they could have evolved as part of the creative process.

Evolution of human beings is problematic for some Muslims because of verses that could be interpreted to specify how human beings were created. The following is one such verse.

> Indeed the example of Jesus with God is like the example of Adam. He created him from dust, then He said to him, "Be!" so, he became. (3:59)

This verse is not problematic in itself, but traditional interpretations of the verse based on hadith can be seen as presenting a barrier to the idea that human beings were not specially created. According to the traditional exegesis, a delegation of Christians came to Prophet Muhammad in Medina and claimed divinity for Jesus because he was born without a father (Ibn Kathir, n.d.). This verse was revealed to counter this argument by claiming that, although Jesus was born without a father, Adam was born without a father or a mother, so if Adam has no claim to divinity because he was born without any parents, then Jesus would not have a claim to divinity by being born from only one parent. According to this interpretation, neither Adam nor Jesus came from normal births, but were instead specially created, and therefore Adam could not have come into being as the result of natural evolutionary processes.

Another verse that describes the creation of Adam is the following.

> And when your Sustainer said to the angels, "Indeed, I am One Who creates a human being from clay dried from stinking dark mud. So, when I have proportioned him and I have breathed into him from My Spirit, then all of you fall down in prostration to him." 15:28–15:29.

Many Muslims interpret this verse to signify that God created Adam at a specific point in time and in a specific manner. From this, they infer that Adam was specially created, and that therefore he could not have evolved.

There are some Muslims who not only accept evolution, but claim that verses in the Quran are consistent with the idea that human beings evolved, such as the following verses.

> And when your Sustainer said to the angels, "Indeed I am One Who Makes a *khalifah* (خَلِيفَة) on the Earth." They said, "Will you make on it one who will cause corruption in it and shed blood, while we glorify with Your praise and purify for You?" He said, "Indeed I am the most knowledgeable of whatever you all do not know." And He taught Adam the names, all of them. Then, He presented him to the angels. So, He said, "Inform Me of these names if

you are truthful." They said, "Your glory! We have no knowledge, except whatever You taught us. Indeed, You are the Always All-Knowing, the Always All-Wise." (2:30–2:32)

The term "*khalifah*" in the preceding passage can be translated as "successor." In this interpretation, Adam would be a successor to someone who came before him. Therefore, he would not be the first human being. In this passage, angels are depicted as saying that human beings will cause corruption and shed blood on Earth. However, the succeeding passages could be interpreted to suggest that their knowledge is limited. Therefore, their statements that people would shed blood and cause corruption would have to be based on prior observation. If they had an opportunity to observe human behavior before the creation of Adam, then he would not have been the first human being. From this, these Muslims conclude that there must have been people on Earth before Adam. If Adam were not the first human being, then these verses could be interpreted to argue against special creation of human beings, and could further be interpreted as not precluding the idea that human beings evolved.

In a more traditional exegesis of this passage the term *khalifah* is interpreted to mean "vicegerent" or "steward," rather than "successor." According to this interpretation, the angels had not observed humans before Adam, but instead had observed the jinn, or unseen beings, before the creation of Adam. According to this interpretation, the angels' assessment of human beings was based on their observations of unseen beings and not on observations of humans who lived prior to Adam. When interpreted in this manner, this passage does not have any bearing on the evolution of humans.

The following hadith from Sunni sources describing the creation of Adam can be interpreted to support microevolution of human beings, because it seems to suggest that people have decreased in average height since the time of their creation.

> Allah created Adam, making him 60 cubits tall. …People have been decreasing in stature since Adam's creation.[2]

The purported decrease in stature of people since the time of the creation of Adam could be considered a microevolutionary change if interpreted in biological terms. Some American Muslims use this hadith to justify the idea that humans are subject to microevolution, even though they do not accept the idea that humans evolved from non-human ancestors. By accepting microevolution for humans, they can incorporate both their interpretations of the special creation of Adam and scientific evidence supporting the idea of evolution of human beings into their schema.

Islamic scholars and organizations. American Muslims' views on evolution are influenced by popular scholars whose speeches they hear in person at a mosque or conference, or on online formats, such as You Tube. What follows is a brief examination of views on evolution expressed by scholars from the three main

---

[2] From Sahih al Bukhari Vol. 4, Book #55, Hadith #543 retrieved from http://sunnah.com/bukhari/60.

groups of American Muslims, Sunni, Shi'a, and "just Muslim." The Ahmadiyya Movement in Islam and the Nation of Islam are included as representing two of the earliest American Islamic organizations, and to give a flavor of the diversity of the American Muslim community. In addition, the views of a Turkish creationist organization that has widespread influence among American Muslims are examined. These represent differing positions on evolution that are representative of those found among American Muslims by people who have widespread influence in their respective Muslim communities.

Yusuf Estes. One popular internet preacher is Yusuf Estes, a former evangelical Christian who holds a doctorate in theology. He identifies as "just a Muslim" because he interprets verses of the Quran that warn against dividing into sects as precluding him from joining any of the groups of Muslims that exist today. He has been listed as one of the 500 most influential Muslims, has traveled the world to lecture on Islam for popular audiences, and has a large internet presence, including a website that had accumulated more than 13 million unique hits as of 2011 (Schleifer, 2011).

Estes (2009) takes a strictly creationist stance, claiming that the theory of evolution "lacks any real, testable evidence. The most we can come up with is not even a possibility, more or less like a dream that they're trying to use evidences, mix them together, stack the deck, as we say, to come up with something" (Estes, 2009). He raises issues that he feels disprove the idea of evolution, such as, "If we evolved from monkeys, how come we still have monkeys?" (Estes, 2009). Such arguments are quite similar to those raised by Christian creationists. Perhaps Yusuf Estes finds them attractive in part because of his background as a former evangelical Christian. Estes sees evolution as part of a strategy used by atheistic scientists to turn believers away from God. Estes (2006) even goes on to suggest that since evolution is so nonsensical, scientists must have some sort of ulterior motive for promoting it. He suggests their desires to publish papers in academic journals and to obtain academic appointments as possible ulterior motives.

Harun Yahya. Yusuf Estes cites Harun Yahya as one source of his ideas on evolution. Harun Yahya is a pseudonym used for a popular form of Islamic creationism originating from Turkey and propagated worldwide using both print and electronic media (Edis, 2009). Harun Yayha's arguments are taken from American creationists and other sources to produce a form of old earth creationism. An example of a typical argument against evolution from the Harun Yahya corpus is, "A 450-million-year-old fossil horseshoe crab, no different from those crabs of our day" (Yahya, 2008, p. 32).

Yasir Qadhi. Yasir Qadhi, the son of parents who immigrated from Pakistan in the 1960s, is a popular Sunni theologian who teaches Islamic studies at Rhodes College in Memphis, TN, and is Dean of Academic Affairs and instructor for the Maghrib Institute. He has been named as one of the 500 most influential Muslims (Schleifer, 2017). He is well-known among Sunni Muslims in the U.S. and serves as a speaker at the Islamic Society of North America's conventions, which draw over 30,000 participants annually. The so-called Islamic State called for his assassination because he was one of 126 Muslim scholars who served as signatories of a letter condemning their actions as contrary to Islam (Schleifer, 2017). The video referenced below where he discusses his views on evolution has nearly 25,000 views on You Tube (Qadhi, 2013).

Qadhi (2013) integrates his understandings of both the Islamic faith and the science behind the theory of evolution. In light of the scientific evidence for evolution he states the following.

> So, what the theory of evolution does, it takes these facts – these are undeniable facts – and then proposes a system that takes into account all these facts.... To say that the theory of evolution is only a theory ignores the whole point.... The theory of evolution from a purely scientific standpoint, in my humble opinion, makes a lot of sense.

He adheres to scriptural literalism, which he claims is not a problem for Muslims because "the Quran is the divine, uncorrupted speech of Allah; it is the literal word of Allah" (Qadhi, 2013). He reconciles his understanding of the Quranic teachings with the theory of evolution by making an exception for human beings. He uses a metaphor to explain this exception.

> Imagine if you like, a series of dominoes tumbling, and they're all going, as we've seen on You Tube clips and what not, going in different directions, having been caused by one beginning domino, and eventually, if these dominoes continue, one line of that domino will lead to that domino which is a final domino known as man, because we know that nothing has been evolved *from* us. We are the final domino....All of these dominoes came about, all of these species came about, and right when it was our turn, right when the next domino should have been our domino, Allah, *subḥanahu wa ta'ala* [God, Glorified and Most High], inserted that domino directly, and that's *Banu Adam* [Adam's descendants]. And, of course, that domino, which is us, fits in perfectly with all the other dominoes, because, why would it not fit in perfectly? Allah is perfect in His creation, and all of the other species are evolving the way that they are supposed to, and when it was the right time at the right place, Allah, *subḥanahu wa ta'ala*, placed us where we were supposed to be such that a neutral observer, who doesn't believe in Allah, quote unquote a *kāfir* [non-believing] observer, would automatically say, "Obviously, this domino comes from the one before it," and he has every right to make that claim.

Qadhi (2013) argues that Muslims should not consider scientists as part of some conspiracy. Instead, they should understand that scientists are operating under a different paradigm.

In Qadhi's view all of evolution can be accepted, except human evolution. In this manner, he can accept the scientific evidence for non-human species without reservation. By claiming that although human beings are an exception to evolution, they were created as if they evolved, he can accommodate scientific evidence for human evolution. He has sophisticated understandings of both nature of science and nature of religion, so he is able to formulate his position without compromising his beliefs in either sphere.

Hassanain Rajabali. Hassanain Rajabali is a popular speaker among Shi'a Muslims, who holds a master's degree in molecular biology and a degree in psychology from the University of Colorado (Qul, 2014). He is well-known in American Shia circles, and has traveled the country to give lectures on Islam to both Shi'a and popular audiences. Videos of these lectures are widely available on Shi'a websites. The video referenced below where he discusses his views on evolution has more than 25,000 views on You Tube (Rajabali, 2008).

He does not think that acceptance of evolution is necessarily contradictory to having a belief in God (Rajabali, 2008). He explains, "There is no verse in the Quran where Allah forbids it, and therefore, we have to be silent about it and say maybe it's possible." He reiterates that science and religion are indeed compatible, because science and religion take different approaches. According to him, science is basically a tool that people can use to advance knowledge, while religion presupposes belief in God, but there is no reason that a person who believes in God cannot use the tool of science.

> From an Islamic perspective, and this is very important for us to understand, we must not think that science [is a bad thing]. No, science is one of the greatest gifts God has given us. It's one of the greatest tools we have been given, and in my opinion, thank God for science! (Rajabali, 2008).

According to Rajabali (2008), "Evolution is a process; it's a methodology; it's a system." He claims that although the Quran categorically states that God created everything, it does not explicitly state the method of creation. Therefore, it is possible that evolution was one of the methodologies He used.

For Rajabali (2008), the creation of Adam is a sticking point. "The Quran is very clear on this issue, that Adam was created and placed on Earth" (Rajabali, 2008). However, a scientist would argue that everything has to be within the system, and must have come from some branch of some tree, from some predecessor. "I said that is *a* system, but it is not the *only* system," counters Rajabali (2008). He claims that one cannot take evolution back to infinity, because it must have started at some point. Therefore if species were created at some point in the distant past, then it is not a stretch to say that God created Adam without a predecessor.

According to Rajabali (2008), to reject God outright is to be dogmatic. He argues that there is no evidence that God does not exist, so, at the most, one could be agnostic without going beyond the bounds of reason. On the other hand, he thinks that rejecting the scientific viewpoint outright without examining the arguments in its favor, on the basis of religion is also being too dogmatic. He believes both the religious and scientific arguments should be scrutinized to see if they stand up to the light of reason.

> [A]ll these realities have to be met with a clear understanding of a holistic human being who lives within the spectrum of science, ethics, ideologies, etc., etc., which brings about the completion of who we are....[I]n reality, it's not us vs. them, or this vs. that. I think at the end of the day, they both have a position, and we need to reconcile them. (Rajabali, 2008)

Mirza Tahir Ahmad. Mirza Tahir Ahmad (1928–2003) was the fourth *khalifat ul-masih*, or successor to the founder of the Ahmadiyya Movement in Islam (AMI), Mirza Ghulam Ahmad (1835–1908). He served as a homeopathic physician prior to his election to the office of khalifa in 1982. Although his views on evolution are widely known within the AMI, most other Muslims would not be familiar with them. Ahmad (1998) wrote a book, *Revelation, Rationality, Knowledge, and Truth*, which is widely read and referred to by scholars and speakers within the AMI. In it, he explains how his position in favor of evolution of all species is compatible with his interpretation of the Quran.

Ahmad (1998) believed that evolution, like all other aspects of the natural world, was under the control of God and that He purposefully directed it. He began his discussion with the following verses of the Quran:

> Blessed is He in Whose hand is the kingdom, and He has power over all things;
>
> *It is He* Who has created death and life that He might try you - which of you is best in deeds; and He is the Mighty, the Most Forgiving, *The Same* Who has created seven heavens in stages (Tibaqan). No incongruity can you see in the creation of the Gracious God. Then look again: Do you see any flaw? Aye, look again, and yet again, your sight will *only* return to you tired and fatigued. (67:2–4 of Mawlawi Sher Ali translation)

He claims these verses demonstrate that there is no contradiction in creation, because they describe it as not flawed, and also that God creates things via stage by stage development, as exemplified by the mention of His creation of the heavens in stages. He connects this to human evolution by stating that this stage by stage development applies to humans by linking the previous passage to the verse, "That you [human beings] shall assuredly pass on from one stage [Tibaqan] to another" (84:20). Ahmad (1998) interprets these and other verses of the Quran to mean that the selection processes that went into the creation of human beings were by the choice and design of the All-Knowing and All-Powerful Creator, and not by random chance or blind necessity.

According to Ahmad (1998), although the Quran was revealed more than 1400 years ago, it contains verses that could not be properly interpreted until the modern age. Among these are verses that describe the origins of life and the creation of human beings. It should be noted that although the idea that the Quran contains verses that somehow presage modern scientific discoveries is common among Muslims in the West, not all of them would include the theory of evolution under this umbrella (Guessom, 2011).

Human kind is described in the Quran as having been created from dust, clay, pottery clay, and dark, fermenting mud. Ahmad (1998) interprets these verses as referring to early stages in the creation of primordial organic molecules on Earth by inorganic processes. He contends that these verses refer to the creation of human beings, because they were the ultimate result of these processes. These processes would have been reversible in the oceans due to hydrolysis of the resulting molecules. Consequently, some scientists propose a wet beginning with dry intermediate stages and others propose that the initial stages must have been dry. Ahmad (1998) goes on to explain that clay has been proposed as a surface that would be amenable for

> an initial or intermediary dry stage. This stage was reached when the oceanic prebiotic soup was concentrated and dried in the form of laminated micro-thin layers of clay. The Quran is evidently on the side of those who support a wet beginning with an intermediary stage of dryness where concentrated primordial soup was moulded into plates like dry ringing clay, such as broken pieces of earthenware. (Ahmad, 1998, p. 373)

Ahmad (1998) scoffs at the idea from literalist readings of the scripture that Adam's creation from clay signifies that God molded him out of clay and then suddenly created a human being from that as being as absurd as the idea scientists hold that human beings were created from a process that proceeds by blind chance. Rather, he believes it was a slow and deliberate process, under God's direction, guidance, and care.

> The scenario of natural selection as against the scenario of purposeful design, would require hundreds of thousands of variant atmospheres, accidentally created by the interplay of billions of chances over millions of earths, of which only one could be rightly proportioned to support life on earth....There are many ... verses in the Quran to the ... effect that life has to be protected by God, every moment of its existence, or it will cease to be. (Ahmad, 1998, pp. 400–402)

According to Ahmad (1998), God is the Creator, but uses the process of evolution to bring living things into existence. He is involved in every step; nothing proceeds by blind chance. Ahmad (1998) claims that this is evident in the fine-tuning of such structures as transport proteins in cell membranes and also of the universe as a whole, configured precisely so that it could produce a planet that would support life.

Nation of Islam. Although the Nation of Islam is a minority group with only a few tens of thousands of the more than two million U.S. Muslims, their charismatic leader, Louis Farrakhan, has an influence that extends beyond his religious community to African-Americans in general. The video referenced below where Farrakhan discusses his views on evolution has well over a million views on You Tube.

The position of the NOI is that Darwin's theory of evolution was concocted to cover up the true origins of human beings. According to Farrakhan (2013), White people "would rather say that they are the descendants of apes rather than *admit* that the Black man and woman is their father and mother."

> I understand by God's grace the teachings of the honorable Elijah Muhammad and why these teachings must be spoken to White people, to yellow and brown people, to every human being on the earth. Everyone must know the Black Man, because to know the Black Man is to know something of yourself. You cannot know the tree as well if you just study the fruit. You must also study the root. Now, we said ... historically speaking, anthropologically speaking, genetically and biologically speaking, there is no human being on the earth that predates the Black man and the Black woman. Now, you may wish to argue, but there is no argument. The honorable Elijah Muhammad asked us the question, who is the original man? And he gave us the answer. The original man is the Asiatic Black man, the owner, the maker, the cream of the planet earth, the God of the universe....

> ... Notice in the answer, the word "Africa" never is mentioned. The original man is not the African Black man. The original man is the Asiatic Black man.... In the lessons given to us by the Honorable Elijah Muhammad, I repeat, Africa is not mentioned....The question is asked, why does the devil call our people Africans? Now, he didn't say why do *we* call *ourselves* Africans. Um mm. He said why does the *devil* call our people Africans? Now, by devil we mean the Caucasian people, nobody under the ground, getting ready to burn you after you are dead, the White man on top of the ground burning you while you are alive.... Why does the devil call our people Africans?... To make our people of North America believe that the people on that continent are the only people that we have, and that they are all savage. Every time they show Africa, they attempt to show you our people in a savage

condition. They want you and me to focus our minds on that continent and that continent alone.... They do this to try to divide us. We have Black people that have been all over this Earth and have settled everywhere on the Earth. You may not know it, but there are Black people in China, Black people in Japan, Black people in Korea, Black people in India, ... in Fiji, in new Zealand, in Australia, Black people in Indonesia, ..., in the Hawaiian islands, Blacks there. When you come to North America, we came here before Columbus. There is a sign that Blacks were here in the Americas long before Christopher Columbus was even a thought in the mind of his father. ...

So to understand that it was a White man that named the continent of Africa Africa, and we predated the White man, then what was it called before the White man named it Africa? The honorable Elijah Muhammad said the original people called the *planet* Asia. The whole *planet* was once called Asia, not just that one part over there that is called Asia today, but all of it was Asia. The part that you call Europe was called Asia. Some of the old maps called it Eurasia. ...

So now if we are the original inhabitants of the earth, and we are, and our color as the first creatures of almighty God coming up out of darkness, the honorable Elijah Muhammad said we take our color from the darkness out of which we originated, so we are Black, symbolizing that we are the first human beings, and from us came all other human beings. That is the teaching of the honorable Elijah Muhammad, and you, Black man, and you, Black woman, if there were no people before you and you were the first of God's creation, then you are a direct descendant of the originator of the heavens and the earth. Therefore the nature of God is your nature, and if you are left alone and fed properly, spiritually, mentally, morally, you will grow up into God Himself. So, the Bible in the book of Pslams said, Ye [you] are all gods, children of the most high God. (Farrakhan, 2013)

According to Farrakhan (2013) the Asiatic Black man, a direct descendant of God, was the original human being. White people were descended from the Asiatic Black man.

Timothy Muhammad (2013), writing for the Nation of Islam Research Group, explains the origin of White people from "the Aboriginal People of the Earth; the Dark People of the Earth—The Black Man and Woman of the Earth from which every species of human being has come." According to Muhammad (2013), it is these aboriginal people that are referred to as "Us" in the Bible when it says, "Let Us make man in our image and after our likeness." That White people were derived from them is supported by recent scientific evidence that the White race was born when "a major genetic alteration occurred exactly 6,600 years ago.... [T]he White race is a young race—a 'new man' who, as the Honorable Elijah Muhammad has said, 'came from us, but he is different from us.'" He continues that people had civilization and advanced scientific knowledge long before the White race came on the scene.

Muhammad claims that Darwin's theory of evolution was devised to cover up the fact that the White race was "selectively bred into existence" and to place "doubt in the minds of the Black professional class ... about the true reality of the Original Man, Who is God." Muhammad (2013) concludes that, "the theory of evolution is not an empirical science, but a "false knowledge," made up of racist doctrines whose aim and purpose is to deny and cover up the reality of the original people, who are God." He then goes on to question the logic of believing "a people who called 'Us' three fifths of a human being. We cannot and should not believe and follow the white supremacist model of education that our former slave masters and their children have foisted upon us."

In NOI thought Darwin's theory of biological evolution is antithetical to belief in God and does not tell the true story of the history of human beings, but is instead being taught to cover it up. They contend that Black people were not descended from apes, but, rather, had noble origins. They claim that White people, on the other hand, had ignoble origins, as they were selectively bred into existence, and had to be taught and civilized by Black people before they could make any advancements or achievements or develop a civilization.

2.3 Suggestions to Improve Evolution Education in the United States

2.3.1 Pedagogical Implications for Evolution Education of American Muslims

In light of the foregoing discussion of American Muslims' views on evolution, some pedagogical implications of these views for both K-12 and post-secondary education are examined here. Research into specific pedagogical strategies for Muslim students in the American context is currently lacking, so the intent of this discussion is to start a conversation and to suggest areas for further research.

Place of evolutionary theory in the curriculum. In the U.S., K-12 state and national science curricula are typically spiraled, so that concepts are introduced in elementary school, and then successively elaborated on in middle and high school. An example of a widely-used set of standards on which to base curricula is the Next Generation Science Standards (NGSS) (NGSS Lead States, 2013). Although they were intended to serve as national standards, fewer than 20 states have adopted them so far. Even so, state standards on evolutionary biology generally follow a similar sequence. The NGSS recommend that on the elementary level, biodiversity is introduced in second grade and differential survival is introduced in third grade. In middle school, students learn about biological evolution by studying the fossil record and how this can be used to infer common ancestry. They also examine evidence for evolution from embryonic development and selective breeding. In high school, students infer common ancestry through macromolecular evidence, and study the mechanisms of natural selection and how it leads to adaptation of organisms to their respective environments.

Treatment of microevolution and macroevolution. Regardless of the position that American Muslims take on macroevolution, in the main they accept microevolution. For this reason, it might be beneficial to start with microevolution when teaching evolution. Once students have a grasp of the role of natural selection in producing microevolutionary changes, then macroevolution could be introduced.

This sequence might be difficult to implement for K-12 education in the United States, however. The sequence that is commonly taught, starting with evidence for macroevolution and then teaching microevolution, is the reverse of what I am suggesting here.

However, some have suggested that natural selection deals with abstract concepts, such as genes, while macroevolution can be deduced from the fossil record, which is more concrete. Therefore the sequence of dealing with macroevolutionary changes in middle school and microevolutionary changes in high school is perhaps best suited to students' cognitive abilities at these levels (Jackson, 2007). As this sequencing by grade level in national and state standards is unlikely to change, perhaps high school teachers, who would normally be tasked with teaching microevolutionary changes to their students, could begin their units on evolution with this material, and then move on to the macroevolutionary topics, which are harder for students to accept, after they have mastered microevolution.

At the post-secondary level where macroevolution and microevolution are taught together, it would be easier to sequence the course to start with microevolutionary changes before dealing with macroevolutionary ones. One of my colleagues has successfully used this approach with religious Christian students (S. W. Seagle, personal communication, March 1, 2017). He reported that in the past he frequently had some of his religious students express their concern to him in response to learning about evolution by coming to his office hours and offering to pray for him. He changed the sequencing of the evolution unit by introducing his students to the more easily accepted microevolutionary concepts before delving into macroevolution. He reported that after this change his students no longer feel the need to express their concerns to him in response to this unit. As this tactic has been successful with religious Christian students in the American context, it is a promising line of inquiry to pursue with Muslim undergraduates as well.

It would also be important to help students understand the distinction between microevolution and macroevolution, rather than simply using the more ambiguous term "evolution" as a catch-all. The terms "macroevolution" and "microevolution" are not generally introduced until high school in the U.S. For example, a popular middle school life sciences textbook, Prentice Hall's *Life Science*, deals with biological evolution without mentioning these terms (Padilla et al., 2009), while Holt Mc Dougal's high school textbook, *Biology*, uses the term "microevolution" in a discussion of natural selection (Nowicki, 2010). At the college level, the terms are used extensively. For example, Campbell's *Biology*, the most popular college level general biology textbook, uses the terms "macroevolution" and "microevolution" repeatedly in its treatment of evolution (Urry et al., 2017). Raven and Johnson's (2002) *Biology* uses these terms in its discussion of evolution as well, and Brooker and colleague's (2011) *Biology* uses them in section heads as well as in the text. Therefore, it is reasonable to expect that K-12 teachers would be familiar with these terms from their college biology courses. Since these terms are common in both high school and college level biology textbooks, making this distinction could be easily implemented at both levels.

From a pragmatic standpoint many of the important practical applications of evolution, such as preventing antibiotic resistance in human pathogens or formulating flu vaccines, rely on understanding of microevolutionary changes, so stressing microevolution would probably not have serious negative practical consequences for people who go on to study further in biology.

Countering creationism. Addressing evidence that directly refutes Christian creationist arguments and their old-Earth variants promulgated by Harun Yahya could prevent some students from being swayed by these types of arguments. For example, explaining how some ancestral forms, such as lemurs, co-exist with descendent forms, such as monkeys, in the present day could counteract arguments such as, "If humans are descended from apes, why are there still apes?" Teaching amendments to evolutionary theory, such as the idea of punctuated equilibrium, could counteract arguments that evolution does not happen because there are some extant species that do not appear to have changed appreciably in hundreds of millions of years when compared with their fossil counterparts. Helping students to understand theory-laden NOS could help counteract the idea the Charles Darwin had an "agenda" in a way that other scientists do not. Helping students understand other NOS concepts, such as the nature of scientific theories, the logic of testing scientific theories, the validity of observationally based theories and disciplines, and the use of inference and theoretical entities in science, might help counteract other creationist arguments on weaknesses in Darwin's theory (Clough, 1994; Smith, 2010). Teaching the history of the development of evolutionary theory and the manner in which it has been critiqued from within the scientific community and how these criticisms have been dealt with based on scientific evidence could also be useful in countering these "holes in the theory" arguments. This need not entail even mentioning the creationist counterparts to these arguments, and I do not suggest bringing these into the science classroom. However, the teacher could have these in mind when designing lessons to arm students with information that could counteract these arguments when students encounter them outside of science class. The foregoing is intended as a brief suggestion of possible strategies that could be employed in the classroom, rather than as an exhaustive list of possible creationist arguments and methods to counter them. The intention here is to start a dialog on the usefulness of these strategies and to suggest avenues for future research.

Modeling how to negotiate the relationship between science and religion for students. U.S. textbooks at both secondary and post-secondary levels commonly recommend teaching an independence view of the relationship between science and religion, and this view is commonly expressed in the biology departments of American colleges and universities. This is due in part to the influence of Stephen J. Gould (1997) who espoused the independence view by claiming that science and religion have "non-overlapping magisteria." He explains, "The lack of conflict between science and religion arises from a lack of overlap between their respective domains of professional expertise—science in the empirical constitution of the universe, and religion in the search for proper ethical values and the spiritual meaning of our lives." This viewpoint is recommended to counteract the conflict view to help religious people to accept the theory of evolution.

However, there are some problems with this approach. Many Muslims think of science and religion as integrated rather than as independent, for example, the influential Muslim scholar Yasir Qadhi discussed above. The majority of the theistic evolutionists who took part in a qualitative study on American Muslim undergraduates had an integrated view of the relationship between religion and science, while only a small minority of all respondents used independence to negotiate this relationship (Fouad, 2016a). A couple of the respondents who used integration expressed their opposition to using independence instead, at the urging of a teacher or a parent, because this simply made no sense to them.

Similar difficulties exist for non-Muslim theistic evolutionists. For example the noted geneticist Francis Collins stated the following in response to Gould's position.

> That doesn't work for me. To me, being a scientist who is also a believer is a wonderful, comforting, harmonious experience, so that as a scientific discovery looms into view (and we scientists have the chance to do that from time to time), it is both a remarkable moment of realizing that you've discovered something that no human knew before, but God knew it, and so you are both experiencing discovery, and also a chance to glimpse just a little bit of God's mind. For me, that is just a privilege and a wonderful experience not to be missed." (Flato, 2006)

For these reasons, it might be preferable to give students examples of different ways of thinking about the relationship between science and religion rather than insisting that everyone take the independence view. Presenting more than one way of negotiating this relationship would make it more likely that students would find a method that is suitable for them.

Smith (2010) advocates a related approach in his review of evolution pedagogy. He suggests explicitly introducing students to Barbour's (2000) typology and inviting them to reflect on how their personal positions relate to these categories. Smith (2010) states, "at least in classrooms with substantial numbers of students from religiously conservative backgrounds, it is my opinion that the largest barriers to studying and learning about evolution are the philosophical and religious issues involved." Therefore he advocates an explicit, reflective examination of nature of science as well as a discussion of the ways in which religious people can negotiate the relationship between science and religion.

Muslim scientists as role models for accepting evolution. Muslim scientists and anthropologists who are currently working to push the boundaries of our knowledge in the field of evolution could potentially serve as role models for Muslim students (Hameed, 2013). As people who have found successful strategies for negotiating the relationship between science and religion, they can serve as examples of how to accept evolution by natural selection as a mechanism for the production of biological diversity in general and of human beings in particular while still maintaining an active faith.

Ehab Abouheif. One such researcher is Ehab Abouheif who holds the Canada research chair in evolutionary biology at McGill University (Verdone-Smith, 2015). His collaborative research group focuses on the evolution of ants. He has authored numerous publications in prestigious journals, including *Science* (Abouheif &

Wray, 2002) and *Proceedings of the National Academy of Sciences* (Smith, et al., 2011). He discusses his position on the scientific evidence for evolution.

> There's a lot at stake here, because it's well beyond evolution. If it's not about the evidence, if you reject science, if you reject evolution as a science and you're not willing to listen to evidence, then that means that for all of science, when it comes into contact with sociological, political conflicts, then you won't believe it either. (Farell, 2012, para 7)

He stressed the importance of Muslims studying evolution so that they could be innovators of science and technology and not just consumers.

Fatimah Jackson. Fatimah Jackson (2015) conducts research at Howard University on microevolutionary changes that lead to human diversity and on human-plant co-evolution. She has published in *Science* (Jackson, Lee, & Taylor, 2014), and other scientific journals. On accepting evolution she stated, "I studied evolution before I accepted Islam. It was no hindrance for me to become Muslim" (thedeeninstitute, 2013). She negotiates the relationship between science and religion by seeing them as independent.

> Remember, science, especially evolutionary science, is designed to tell you how things change, not why. Why comes from our Islam. You know, when we want to know why something happened we go to the Islam. (thedeeninstitute, 2013)

She uses a metaphor to describe her position as a theistic evolutionist.

> Look at the similarities, the genetic similarities among all of the life that has been created. That is a sign of the signature of a single artist… you would never confuse a Monet painting with a VanGough. You would never confuse it, because every artist has a signature, has a style of presenting their creativity, and the style that we see is in the unity of the genetic message across all living species on this planet. (thedeeninstitute, 2013)

Researchers such as Fatimah Jackson and Ehab Abouheif could serve as role models for Muslim students on how to successfully negotiate the relationship between religion and science to accept biological evolution. The role models for negotiating this relationship would not necessarily have to be Muslims themselves. People from other faith traditions who have successfully negotiated this relationship, such as Francis Collins as quoted above or Theodosus Dobzhansky in his seminal 1973 article "Nothing in biology makes sense except in the light of evolution" could also potentially serve as role models for Muslim students.

Abu Uthman al-Jahiz. Historical figures from the Golden Age of Islam, such as Abu Uthman al-Jahiz (781–869) are another possible source of role models for Muslim students. He was a prolific writer on many subjects, including animals adapting to their environments. His work was known to European scientists, including Lamarck. Such scientists who contributed their proto-evolutionary theories to the discourse on evolution are often overlooked in science textbooks. Since their ideas were foundational to modern Western science and some history of evolutionary thought is normally presented in lessons on evolution in textbooks and in the classroom in the U.S., it would be fairly easy to include them in discussions on evolution.

Further justification. In the United States, proponents of creationism attempt to undermine evolution education using three tactics (Berkman & Plutzer, 2015). One is to exploit common misconceptions in NOS understandings by suggesting that there is some controversy surrounding evolutionary theory in scientific circles. Another is to suggest that since a controversy exists, it is only "fair" to teach both sides. A third is to promote the idea that religion and science are incompatible.

Some American high school biology teachers have been susceptible to these tactics (Berkman & Plutzer, 2015). They may attempt to avoid controversy in their classrooms by concentrating on microevolution without mentioning macroevolution, by discussing evolution of microbes while avoiding that of humans, or by using terms such as "adaptation" or "change over time" in place of evolution. They may discuss creationist views in their classrooms in the interests of "fairness." Some tell students that they must learn about evolutionary theory because it is included in standardized tests, but without advocating for it on the basis of the scientific evidence that supports it.

It is important to note here that the pedagogical strategies mentioned above could potentially counteract these three creationist tactics. Therefore, they should be implemented in the context of a scientifically robust evolution unit.

The suggestion to begin the evolution unit with microevolution and then follow that with macroevolution once students have mastered natural selection is *not* meant to suggest that macroevolution should be de-emphasized in the treatment of evolution in either the high school or university biology classroom. Rather, it is meant to suggest that since most American students, whether Muslim or not, are willing to accept microevolution, they may be more inclined to learn about evolution if this is used as the gateway to the unit. Beginning the unit with those aspects of evolution that they are more likely to reject may turn them off of the subject entirely and prevent them from learning even those aspects that they might otherwise accept. The suggestion to stress to students the distinction between microevolution and macroevolution is meant to introduce proper terminology to students.

In the United States the courts have ruled that it is unlawful for public schools to promote religious views or to teach creationism or its variants, such as intelligent design, in the classroom (NRC, 2008). This is one reason that it is important to avoid mentioning creationist arguments in the science classroom, even while teaching material that could serve to counter these arguments. Another is that mentioning creationist arguments in the classroom could confuse students by making it appear that there is indeed a controversy about the science behind evolutionary theory (Clough, 1994). These are reasons to include both the scientific evidence and informed NOS views that would help students to counter these arguments should they encounter them, but not to include the creationist arguments themselves in the science classroom.

Although advocating for a particular religious viewpoint is not allowed in American public schools, teaching students about religion is not prohibited. Introducing students to the views of people who have used varying strategies to negotiate the relationship between science and religion would be allowable as long as the teacher refrained from promoting or advocating for one of these positions. In addition to the benefit mentioned above of giving students examples of these strategies to help them find one that may work for them, this serves to counteract the creationist strategy of promoting the false idea that religion and science are necessarily incompatible.

2.4 Conclusions

Regarding acceptance of evolution, there are some ways in which American Muslims are similar to other Americans and other ways in which they differ. Rates of acceptance are similar. Also similar is the way that American Muslims differ in their views on evolution, forming three groups: those who accept both macroevolution and microevolution for all species, those who accept macroevolution for all species except humans, and those who reject macroevolution for all species, but could accept microevolution for all species. Another similarity is that people who have one way of negotiating the relationship between science and religion may be resistant to adopting another method of negotiating this relationship.

American Muslims differ from their compatriots in some important ways. They are far more likely to accept an old age for the Earth, even if they do not accept evolution as the best explanation for the appearance of new species. A related concern, that Noah's ark would have served as a bottleneck for species, with their subsequent development from kinds, is mostly absent for American Muslims.

There are several pedagogical implications of these views for Muslims. One is that curricula at the secondary and post-secondary levels could be sequenced to teach microevolution before macroevolution in order to accommodate those students who accept the former, but not the latter. This would benefit non-Muslim students who reject macroevolution as well.

A robust treatment of important NOS concepts, including theory-laden NOS, the nature and logic of testing scientific theories, the validity of observationally based theories and disciplines, and the use of inference and theoretical entities in science, could help both Muslim and non-Muslim students avoid common misconceptions about evolutionary theory that are often exploited by creationists in formulating their arguments against it. Helping students understand how evolutionary theory has been modified over time to enhance its explanatory power, and providing more robust explanations of the nature of lineages could potentially counteract other common creationist arguments against evolution.

It could be useful for both Muslim and non-Muslim students to introduce them to different methods of negotiating the relationship between religion and science, rather than expecting that only one method will work for all students, since there are multiple ways that people have successfully negotiated this relationship in order to avoid conflict. Introducing Muslim students to practicing Muslim evolutionary biologists and to Muslims from the past who developed proto-evolutionary theories might help them to view acceptance of evolution in a more favorable light.

References

Abouheif, E., & Wray, G. A. (2002). Evolution of the gene network underlying wing polyphenism in ants. *Science, 297*(5579), 249–252.

Ahmad, M. T. (1998). *Revelation, rationality, knowledge, and truth.* Surrey, UK: Islam International Publications Ltd.

Aslan, R. (2006). *No god but God: The origins, evolution, and future of Islam.* New York, NY: Random House.

Barbour, I. G. (2000). *When science meets religion.* New York, NY: Harper Collins.

Berkman, M. B., & Plutzer, E. (2015). Enablers of doubt: How future teachers learn to negotiate the evolution wars in their classrooms. *The Annals of the American Academy of Political and Social Science, 658,* 253–270.

Brooker, R. J., Widmaier, E. P., Graham, L. E., & Stilling, P. D. (2011). *Biology.* New York: McGraw Hill.

Clough, M. (1994). Diminish students' resistance to biological evolution. *The American Biology Teacher, 56,* 409–415.

Dobzhansky, T. (1973). Nothing in biology makes sense except in the light of evolution. *The American Biology Teacher, 35,* 125–129.

Edis, T. (2009). Modern science and conservative Islam: An uneasy relationship. *Science & Education, 18,* 885–903. https://doi.org/10.1007/s11191-008-9165-3.

Estes, Y. (2006). *No Brainer.* Retrieved from http://www.scienceislam.com/audio/no_brainer.html.

Estes, Y. (2009). *Faith Science & Common Sense, 10th (Final) Lecture, The Malaysian Tour 2008.* Retrieved from https://www.youtube.com/watch?v=COzRpbiIGF4.

Everhart, D. and Hameed, S. (2013). Muslims and evolution: A study of Pakistani physicians in the United States. *Evolution: Education and Outreach, 6.* Retrieved from: http://www.evolution-outreach.com/content/6/1/2.

Farell, J. (2012, November 30). God and evolution: Easier for Muslims than Christians? *Forbes.*

Farrakhan, L. (2013) *Origin of the White Man—Part 1 (a).* Retrieved from http://www.youtube.com/watch?v=vwmMbOgadTs.

Flato, I (host). (August 4, 2006). Francis Collins interview [radio program] in Bishop L. & Goodwin, S. (producers). *Talk of the Nation.* National Public Radio.

Fouad, K. E. (2016a). *American Muslim undergraduates' views on evolution.* Doctoral dissertation. IU ScholarWorks. http://hdl.handle.net/2022/20879.

Fouad, K. E. (2016b). *American Muslim undergraduates' views on evolution.* Unpublished raw data.

Gallup. (2009). *Muslim Americans: A national portrait, an in-depth analysis of America's most diverse religious community.* Gallup, Inc.

GhaneaBassiri, K. (2010). *A history of Islam in America.* New York: NY: Cambridge University Press.

Gould, S. J. (1997). Nonoverlapping magesteria. *Natural History, 106,* 16–22.

Guessoum, N. (2011). *Islam's quantum question.* London: I. B. Tauris.

Hameed, S. (January 11, 2013). Muslim thought on evolution takes a step forward. The Guardian. Retrieved from https://www.theguardian.com/commentisfree/belief/2013/jan/11/muslim-thought-on-evolution-debate.

Ibn Kathir. (n.d.) Quran Tafisr Ibn Kathir. Retrieved from http://www.qtafsir.com/index.php?option=com_content&task=view&id=531&Itemid=46.

Jackson, D. F. (2007). The personal and the professional in the teaching of evolution. In L. S. Jones & M. J. Reiss (Eds.), *Teaching about scientific origins: Taking account of creationism.* New York: Peter Lang Publishing Inc.

Jackson, F. L. C. (2015). *Department of biology faculty profile: Fatimah Jackson.* Retrieved from http://www.biology.howard.edu/faculty/jackson/jackson.html.

Jackson, F. L. C., Lee, C. M., & Taylor, S. (2014). Let minority-serving institutions lead. *Science, 345*(6199), 885.

Legare, C. H., Evans, D. M., Rosengren, K. S., & Harris, P. L. (2012). The coexistence of natural and supernatural explanations across cultures and development. *Child Development, 83*(3), 779–793.

Muhammad, T. (2013). *The false doctrine of evolution*. Nation of Islam Research Group. Retrieved from http://noirg.org/the-false-doctrine-of-evolution-2/.

National Research Council. (2008). *Science, evolution, and creationism*. Washington, DC: The National Academies Press.

NGSS Lead States. (2013). *Next Generation Science Standards: For States, By States*. Washington, DC: The National Academies Press.

Nowicki, S. (2010). *Biology*. Holt McDougal.

Padilla, M. J., Miaoulis, I, Cyr, M., Coolidge-Stolz, E., Cronkite, D., Janner, J., … Lisowski, M. (2009). *Science explorer: Life science*. Boston, MA: Pearson Education, Inc.

Pew Forum on Religion and Public Life. (2013). *The world's Muslims: Religion, politics, and society*. Washington, DC: Pew Research Center.

Pew Forum on Religion and Public Life. (2007). *Muslim Americans: Middle class and mostly mainstream*. Washington, DC: Pew Research Center.

Qadhi, Y. (2013). *The Quran and evolution: Thoughts from a believing, rational Muslim*. Retrieved from https://www.youtube.com/watch?v=Ydlrg7zFP6w.

Qul. (2014). Hassanain Rajabali, in Qul, the library for all your needs. Retrieved from http://www.qul.org.au/audio-library/lectures-majalis/1385-hassanain-rajabali.

Rajabali, H. H. (2008). *Evolution and God in Islam*. Retrieved from https://www.youtube.com/watch?v=yH1dwXEV51E.

Raven, P. H., & Johnson, G. B. (2002). *Biology*. Boston, MA: McGraw-Hill.

Schleifer, S. A. (Ed.). (2011). *The Muslim 500: The 500 most influential Muslims 2011*. Amman, Jordan: The Royal Islamic Strategic Studies Center.

Schleifer, S. A. (Ed.). (2017). *The Muslim 500: The 500 most influential Muslims 2017*. Amman, Jordan: The Royal Islamic Strategic Studies Center.

Smith, C. R., Smith, C. D., Robertson, H. M., Helmkampf, M., Zimin, A., Yandell, M., … Cash, E. (2011). Draft genome of the red harvester ant Pogonomyrmex barbatus. *Proceedings of the National Academy of Sciences, 108*(14), 5667–5672.

Smith, M. U. (2010). Current status of research in teaching and learning evolution: I. Philosophical/epistemological issues. *Science & Education, 19*, 523–538.

Tabataba'i, M. H. (1971). *Shi'ite Islam*. Houston: Free Islamic Literatures Inc.

thedeeninstitute. (2013). Have Muslims misunderstood evolution? [You Tube video]. Retrieved from https://www.youtube.com/watch?v=FbynBJVTWKI.

Urry, L. A., Cain, M. L., Wasserman, S. A., Minorsky, P. V., & Reece, J. B. (2017). *Campbell Biology*. New York, NY: Pearson.

Verdone-Smith, C. (2015). *The Abouheif lab: Canada research chair in evolutionary developmental biology*. Retrieved from http://biology.mcgill.ca/faculty/abouheif/publications.html.

Yahya, H. (2008). *Atlas of Creation* (Vol. 1). Istanbul, Turkey: Global Publishing.

Khadija Fouad is a Visiting Assistant Professor of Science Education in the Biology Department at Appalachian State University in Boone, North Carolina. She teaches secondary science methods and history and philosophy of science to preservice science teachers, as well as microbiology laboratories for the biology department. Her current research focuses on the interactions between students' religious beliefs and their understandings of science content and nature of science (NOS). In addition, she is investigating methods of improving students' NOS understandings in both K-12 and university settings, including incorporating history of science into explicit, reflective consideration of NOS aspects. Another line of inquiry involves including science from non-European cultures to provide culturally relevant pedagogy to non-mainstream students studying in Western contexts and to challenge the misconception that the rise of Western modern science happened without input from non-European cultures.

Chapter 3
Project Teach Evolution: Preparing Biology Pre-service Teachers to Teach Evolution in Missouri, U.S.A.

Patricia J. Friedrichsen, Larry G. Brown and Johannes Schul

Abstract We highlight our evolution education efforts in the state of Missouri, United States of America. Acceptance of evolution among Missourians is compared to results from a national survey; the religiousness, education, and age of Missourians help explain state and national differences. To further examine regional influences in the state, a brief history of the Ozarks region and its culture are included. Anti-evolution efforts in the state are examined through the frequency of anti-evolution legislative bills and the state science standards. The authors describe their evolution education efforts, focusing primarily on a hybrid evolution content and pedagogy undergraduate course for pre-service biology education students. Course curriculum, assignments, and assessments are described. Challenges teaching the hybrid course include differing science teaching orientations of the two instructors, as well as a tension between the emphases given to content versus pedagogy.

3.1 Introduction

"More than four in 10 Americans continue to believe that God created humans in their present form 10,000 years ago, a view that has changed little over the past three decades" (Newport, 2014, p. 1). Beginning in 1982, Gallup has conducted the Values and Beliefs survey every two years in the United States (Newport, 2014). Over this time span, the percentage of individuals holding a creationist position has stayed fairly stable, varying only in the range of 40–47%. The remainder of Americans believe human evolution occurred, but they are divided as to whether God was involved in guiding the process. The theist evolution position has dropped from 38% (1982) to 31% (2014), while the secular evolution position has risen from nine percent (1982) to 19% (2014). "Historically, Americans' views on the origin of humans have been related to their religiousness, education, and age" (Newport,

---

P. J. Friedrichsen (✉) · L. G. Brown · J. Schul
University of Missouri, Columbia, MO, USA
e-mail: friedrichsenp@missouri.edu

© Springer International Publishing AG, part of Springer Nature 2018
H. Deniz and L. A. Borgerding (eds.), *Evolution Education Around the Globe*, https://doi.org/10.1007/978-3-319-90939-4_3

2014, p. 1). Younger Americans, who tend to be less religious, and Americans with college degrees are more likely to have an evolutionary viewpoint on the origin of humans (Newport, 2014).

3.2 Public Acceptance of Evolution in the State of Missouri

The United States of America is the third largest country in the world, based on land mass, encompassing over 3.8 million square miles (NationMaster, n.d.). Regional and state differences exist within this large, diverse country. Therefore, in this chapter, we focus on one state, Missouri, located in the Midwestern region of the country. In 2015, Missouri's population was approximately 6 million people, comprised of 80% White, 12% Black or African American, 4% Hispanic or Latino, 2% Asian, and 2% identified as two or more races (United States Census Bureau, n.d.).

How does the state of Missouri compare to the country as a whole in regard to acceptance of evolution? According to the 2014 U.S. Religious Landscape Study, 38% of Missourians indicated humans and other living things have existed in their present form since the beginning of time in comparison to 34% nationally (Pew Research Center, 2017). In Missouri, 25% indicated humans and other living things evolved due to God's design, equal to the percentage held by all Americans. Twenty-nine percent of Missourians surveyed held a secular evolution position in comparison to 33% nationally. Of the remainder of the Missourians surveyed, 4% indicated life evolved, but do not know how, and 3% did not know or refused to answer this question (Pew Research Center, 2017). According to the results of this survey, the percentage of Missourians holding creationist views is slightly higher (4%) than the national percentage.

The religiousness, education, and age of Missourians help explain this trend (Newport, 2014). According to the 2014 U.S. Religious Landscape Study, the religious composition of Missouri adults includes: 77% Christian, 20% unaffiliated, and 3% non-Christian faiths. Within the Christian category, the largest sub-groups are: evangelical Protestant (36%), mainline Protestant (16%), and Catholic (16%). Eighty-two percent of Missourians say that religion is either very important (56%) or somewhat important (26%) (Pew Research Center, 2017). In regard to college education, Missouri is slightly below the national average. In 2011, 36.4% of Missourians attained at least an associate degree (two-year college degree) while the national average was 38.7% (Lumina Foundation, 2013). In 2010, the median age of Missourians was 37.9 years. The state population is aging; in 2000, the 45–64 age group comprised 22% of the total population, and, in 2010, this age group increased to 27% (Missouri Economic Research and Information Center, March 2012). These statistics support Newport's findings that older, religious individuals without a college degree are more likely to hold creationist views. In the next section, to further understand these statistics, we explore the culture of an influential region in the state, the Ozarks.

3.3 Ozarks History and Culture

"The Ozarks is one of the America's great regions, set apart physically by rugged terrain and sociologically by inhabitants that profess political conservatism, religious conservatism and sectarianism, and strong belief in the value of rural living" (Rafferty, 1988, p. 1). The Ozarks region covers a large portion of southern Missouri and northern Arkansas. By 1830, this region was settled by Scots-Irish immigrants who moved westward, extending Appalachia to the Ozarks. Experiencing the similar rocky and thin soils as Appalachia, and the difficulty of farming, they continued their "slash and burn" subsistence agriculture and their itinerant ways, living in relatively isolated small groups. The culture developed with an attitude for low taxes, few schools and libraries, and less literacy. They considered themselves honest farmers, as opposed to wealthy land-holding aristocrats, keeping a certain distrust of political parties and governance they perceived were used to control morality. Hence they maintained social distance from most national institutions, centering their life in family and the local community (Woodward, 2011).

Religious belief and behavior significantly contributed to the independent, emotional, and locally authoritarian aspects of Ozark culture. These patterns have roots in Scots-Irish Calvinism and the revivalism of 17th century British Isles where plain worship, individual moral behavior, and a more effective role of laity was emphasized. In Appalachia and the Ozarks, this faith was expressed in an oral folk/traditional religion that rejected previous institutional patterns. Revival meetings began in North Carolina and Kentucky, and spread westward, including Methodist, Presbyterian, and Baptist leaders, which in turn influenced the beginnings of the Christian Church, the Shakers, and the Cumberland Presbyterian Church. The public ritual of immersion baptism was advocated, in which adults made the decision to define and express their own faith. This new "mountain religion" became a status movement, a multiplicity of equals (McCauley, 1995).

Baptist, Cumberland Presbyterian, Methodist, and Christian (Disciples of Christ) denominations became the voice of righteousness and morality. Congregations developed flexible, decentralized patterns, primarily under lay leadership and itinerant clergy, and perpetuated camp meeting revival evangelism. Today one can still observe camp meetings, fellowship gatherings, springtime baptisms, and other social occasions centered in local congregations (Blevins, 2002).

During the 19th century, other religious communities formed out of the Manifest Destiny narrative, which called for pioneers to develop the new lands of the West, as they perceived God intended. The Stephenites (later Missouri Synod Lutherans), Mennonites, Mormons, and others forged new identities out of new inspirations in a new land. This strain of separatist, idealist communities nurtured in the freedom of the frontier was another powerful influence on the religious formation of the Ozarks in the subsequent century (Cherry, 1971).

Greater Appalachia and Ozark religiosity also served to separate it from Yankee, Midland, or Deep South expressions of faith. However, Ozark faith did share central theological tenets with the Deep South from late 19th century, through the 1920s and 30s, into the present. Southern Evangelical Christianity emphasized the private dimensions over the larger community/national dimensions of faith expression with such beliefs as: personal salvation from a sinful world and redemption from the oppression of the present era. The Ozarks joined the Deep South in its opposition to modernism, standing on a platform that included Biblical inerrancy and the teaching of religion, not science (Woodward, 2011).

The Ozarks embraced the Fundamentalist trend in the early 20th century, with anti-evolution campaigns, organizing Bible colleges and Bible Fellowships, and supporting Fundamentalist and Evangelical radio preachers and their organizations. Into the 1990s and to the present, there is significant support for creationism, prayer in school, abstinence-only sex education, bans on abortion, and state and local rights. Religion became the last best refuge of family, community, and traditional ways (Woodward, 2011).

The Ozarks Region is also shaped by the persistent belief that this region has a sacred quality. Harold Bell Wright's (1907) book, *The Shepherd of the Hills*, contributed to this perception of sacredness. Wright was part of the Country Life Movement that envisioned rural locations as the best of all worlds for fostering healthy living and democratic values. Soon thereafter Chautauqua and YMCA camps located near Branson, Missouri, and Eureka Springs, Arkansas. Various Christian denominations also built camps for conferences, retreats, and educational events. Springfield, Missouri, became a center of revival faith, as represented by the world headquarters of The Assemblies of God, a Pentecostal denomination founded in the Ozarks (Morrow & Myers-Phinney, 1999).

After the rising popularity of Wright's *Shepherd of the Hills* (Wright, 1907) and his other books, tourists came to the Branson area to see the sites and the people that inspired his work. Religious spectacles were established at Branson and Eureka Springs in the form of outdoor pageants and dramas. Caves, springs, clear mountain streams, and mountain vistas all provided settings for inspiration and spirituality to those seeking it. Tourism developed by melding conservative Christian values with musical entertainment, all cast in the reimagined country culture of the Ozarks. "Christian" entertainment venues came into being during the boom years of the 1980s and 1990s, giving such places as Branson the iconic landscape of religious nationalism. Branson hosts the country's largest Veteran's Day celebration, has centers for the Trinity Broadcasting Network and Focus on the Family, as well as Camp Kanakuk, the largest Christian athletic camp in the United States. Branson continues to represent the Ozarks as a sanctuary for religious pilgrims who seek a largely protestant, Anglo-Saxon, working and middle class, rural-imaged faith that encourages evangelical economic prosperity. This quest is often framed as a backlash against progressive secular culture (Ketchell, 2007) which is often represented in anti-evolution bills in the state legislature.

3.4 Influence of Anti-evolution Movements in Missouri: Legislative Bills

From 2004 to 2016, 14 anti-evolution bills were introduced in the Missouri House of Representatives; although none of these bills became law (http://www.house.mo.gov/billcentral.aspx). Over time, the strategy and wording of the bills have evolved. In 2004, House Bill (HB) 1722 called for equal treatment of science instruction regarding evolution and intelligent design. Beginning in 2008, anti-evolution bills were labelled as "Teacher Academic Freedom" bills, and, in 2012, a critical analysis of the evidence of biological and chemical evolution became the focus.

3.4.1 Place of Evolutionary Theory in the Curriculum: Missouri Standards and Teacher Practice

In the United States, there is no mandated national K-12 curriculum; each state sets its own educational standards, as well as selects and supervises standardized testing. The Missouri Science Standards, at the high school level (grades 9–12), are assessed by a state-mandated exam only in biology, and not in the other science disciplines. Students typically enroll in biology courses in 10th grade (16 years old). Beyond the state standards and the state-mandated biology assessment, each of the 550 school districts in Missouri write their own curriculum, determine instructional approaches, and select textbooks and other instructional materials for the school district.

Prior to 2017, the Missouri Science Standards were referred as Course Level Expectations (Missouri Department of Elementary & Secondary Education [DESE], 2008). Table 3.1 shows the section related to evolution at the high school level.

The Fordham Report *State of State Science Standards 2012* reviewed individual states' science standards and identified the "undermining of evolution" as the number one problem across all states' standards (Lerner, Goodenough, Lynch, Schwartz, & Schwartz, 2012, p. 9). In this report, Missouri's state standards received a grade of "C" with a score of four out of seven points for content and rigor and a score of two out of three points for clarity and specificity, resulting in an overall score of six out of 10. The life science section received a score of six out of seven; however, the authors note that many of the individual learning objectives related to evolution are marked with asterisks. An asterisk "indicates the item is essential to the curriculum of the Course but will not be assessed at the State level. The indicated expectation should be taught and assessed locally" (DESE, p. 1). So, although the state does a better than average job including evolution in the biology standards, the accountability in state testing is missing.

In 2012, the first author conducted three focus groups with teachers [N = 6] working in rural schools in the state, although more were originally planned. It was

Table 3.1 Missouri course level expectations for evolution

Missouri course level expectations Biology standards for natural selection
Genetic variation sorted by the natural selection process explains evidence of biological evolution
A. Evidence for the nature and rates of evolution can be found in anatomical and molecular characteristics of organisms and in the fossil record (a) *Interpret fossil evidence to explain the relatedness of organisms using the principles of superposition and fossil correlation (b) *Evaluate the evidence that supports the theory of biological evolution (e.g., fossil records, similarities between DNA and protein structures, similarities between developmental stages of organisms, homologous and vestigial structures)
B. Reproduction is essential to the continuation of every species (a) *Define a species of terms of the ability to mate and produce fertile offspring (b) Explain the importance of reproduction to the survival of a species (i.e., the failure of a species to reproduce will lead to the extinction of that species)
C. Natural selection is the process of sorting individuals based on their ability to survive and reproduce within their ecosystem (a) Identify examples of adaptations that may have resulted from variations favored by natural selection (e.g., long-necked giraffes, long-eared jack rabbits) and describe how that variation may have provided populations an advantage for survival (b) *Explain how genetic homogeneity may cause a population to be more susceptible to extinction (e.g., succumbing to a disease for which there is no natural resistance) (c) Explain how environmental factors (e.g., habitat loss, climate change, pollution, introduction of non-native species) can be agents of natural selection (d) *Given a scenario describing an environmental change, hypothesize why a given species was unable to survive
Note * indicates the item is essential to the curriculum of the Course but will not be assessed at the State level. The indicated expectation should be taught and assessed locally *Source* Missouri Department of Elementary and Secondary (2008). Biology Course Level Expectations. Retrieved from https://dese.mo.gov/sites/default/files/cle-biology-science.pdf

challenging to find teachers in rural schools who were willing to meet and discuss teaching evolution. In the focus groups, beginning teachers often shared stories of being confronted by students, saying they did not want evolution taught in their school. In contrast, more experienced teachers shared information about how, over time, they had carefully built trust and respect in the community. The experienced teachers said they taught natural selection but avoided the "E" word in their classes, or they waited to teach evolution in upper-level, elective biology courses. In each focus group, the teachers reminded the first author that evolution was a locally assessed state standard, indicating they felt no accountability pressure from the state to teach evolution.

In 2016, a closely-aligned version of the national science education standards, *Next Generation Science Standards* (*NGSS*) (NGSS Lead States, 2013), was adopted in Missouri. In *NGSS*, evolution is identified as one of the four disciplinary core ideas in Life Science. In Missouri, implementation of the new standards will begin in the 2017–2018 school year, with state assessments scheduled for the 2018–2019 school year. This new set of state standards has a stronger emphasis on evolution (introducing adaptation and differential survival in Grade 3, the fossil record and natural selection in middle school, and common ancestry, evidence for evolution, natural selection, and speciation in high school), and it remains to be seen how this will implemented in local school districts in the state. Across the U.S., 28% of biology teachers are advocates for evolution, 13% advocate for creationism, while the remaining fall into the "cautious 60%," who advocate for neither evolution or creationism (Berkman & Plutzer, 2011). We could find no published studies of the attitudes of Missouri biology teachers toward teaching evolution; however, the first author conducted a survey of Missouri biology teachers' professional development needs and teaching practices related to evolution (Friedrichsen, Linke, & Barnett, 2016).

In the survey, Missouri biology teachers who taught evolution (N = 276) self-assessed their understanding of specific evolution topics, estimated the amount of class time they spent teaching individual evolution topics, and identified challenges in teaching evolution (Friedrichsen et al., 2016). Eighty percent reported having adequate or an in-depth understanding of all the listed evolution topics. In regard to the most often taught topics, 100% of the teachers reported teaching natural selection with 93.5% reported spending at least one class period on it. Sixty-seven percent of the teachers reported spending at least one class period or more teaching nature of science. The least taught topics and the percentages of teachers reporting teaching these topics were: human evolution (26.8%), cladograms/phylogenetic trees (23.6%), origin of life (21.7%), microevolution (18.8%), and geological timelines (17.4%). The teachers who were teaching evolution reported the two biggest challenges were a lack of good labs and supplemental instructional materials. Teachers were also asked to rate their familiarity with a list of evolution education resources (e.g., *Understanding Evolution* website and various NSTA publications); the majority of teachers were unfamiliar with these available resources.

3.5 University of Missouri Science Teacher Education Program

In 2002, when the first author, Pat Friedrichsen, joined the faculty at University of Missouri, she taught the third science methods course in a three-course sequence. To demonstrate innovative ways to teach natural selection and help pre-service teachers (PSTs) understand argumentation, she engaged them with the software *Beak of the Finch* (http://bguile.northwestern.edu). She also included readings and discussions about the nature of science (i.e., scientific laws versus scientific theories) and the controversy surrounding the teaching of evolution in public schools. She became aware that some of the biology PSTs held creationist viewpoints and were conflicted about teaching evolution. Pat was sympathetic to the students' dilemma, because of their late realization that their personal beliefs conflicted with high school biology teaching expectations. When Pat explored this issue, she found that biology education majors were not required to take an evolution course; PSTs could choose between an evolution course or another course, Community Ecology, in which the emphasis on evolution varied by the instructor, from little or no evolution to half of the semester. Berkman and Plutzer (2011) reported, "teachers who are advocates for evolutionary biology are more likely to have completed a course in evolution than teachers who are ambivalent about evolution or who teach creationism" (p. 405). Consequently, the undergraduate biology education and the post-baccalaureate certification program entry requirements were changed to require a full semester evolution course.

Pat's research focuses on science teacher learning with a focus on pedagogical content knowledge and skill (PCK&S) development. PCK&S is defined as the knowledge, reasoning, planning, and teaching of "a particular *topic*, in a particular *way* for a particular *purpose* to particular *students* for enhanced *student outcomes*" (Gess-Newsome, 2015, p. 36). This line of research is predicated on the understanding that content knowledge alone is not enough, and teachers need to develop a specialized knowledge base in which they transform their content knowledge to make it comprehensible for learners (Shulman, 1986). Pat began increasing the number of evolution readings and class discussions in her science methods course, focusing on common student misconceptions and strategies for teaching evolution. Berkman and Plutzer (2011) recommend that the best way to influence the "cautious 60%" of biology teachers is to focus on pre-service teacher education. We agree with this recommendation; however, within the context of our secondary science teacher education program, it became challenging to meet the diverse needs of all PSTs. Our science methods courses include PSTs seeking certification in physics, earth science, chemistry, and biology. Increasing the emphasis on evolution education in the methods courses resulted in too much emphasis for the PSTs in the physical sciences and not enough to adequately prepare biology PSTs.

In a review of the literature focused on K-12 teachers and evolution education, Sickel and Friedrichsen (2013) proposed four goals for biology teacher preparation. The first goal is to improve PSTs' evolution content knowledge and includes a list of specific evolution concepts found in science standards. The second goal is improving PSTs' understanding of the nature of science, with emphasis on the following tenets: nature of scientific questions, the empirical nature of scientific knowledge, nature of scientific theories, and the tentativeness of scientific knowledge. The third goal is PSTs' acceptance of evolution as a valid scientific theory (not to be conflated with personal acceptance of evolution). The fourth goal has received little attention in the research literature. It seeks to develop PCK for teaching evolution, including knowledge of evolution curricula resources; commonly held misconceptions and student difficulties in learning evolution; instructional strategies, including way to challenge students' misconceptions; and strategies for assessing student understanding of evolution. To work towards achieving these four goals and to develop a network of evolution educators in the state, Pat collaborated with the third author, Johannes Schul, an evolutionary biologist, in Project Teach Evolution. As part of Project Teach Evolution, Pat and Johannes co-designed and co-taught a hybrid evolution content and pedagogy course for biology PSTs.

3.6 Project Teach Evolution: Hybrid Evolution Content and Pedagogy Course

In this section, we describe the design of the hybrid course, student feedback, challenges, and future directions. Our course design was informed by the results of the survey conducted of Missouri biology teachers (Friedrichsen et al., 2016). We included human evolution, phylogenetic trees, and geological timelines, as these were some of the least taught topics identified in the survey. We also focused on including labs appropriate for high school use, and we incorporated evolution education resources, such as the *Understanding Evolution* website (http://evolution.berkeley.edu). We offered this new course as an additional section of the existing *Evolution* lecture course, a 3-credit biology course. The course was co-taught in two evening sessions a week to minimize scheduling issues for biology education majors. The goals of the hybrid course were for students to develop evolution content knowledge, develop emerging PCK for teaching evolution, have an understanding of various anti-evolution strategies (e.g., critical analysis) and criticisms of evolution, and be able to articulate a strong rationale for teaching evolution in high school biology courses. In Year 1, 15 students were enrolled in the course, and in Year 2, 11 students were enrolled.

3.6.1 Course Overview

In the first year, we struggled to meld the evolution content, taught by Johannes, and the evolution education pedagogy taught by Pat. Initially, Pat tended to teach pedagogy or discuss the public controversy surrounding the teaching of evolution during the one-hour Monday evening sessions, while the three-hour block focused on evolution content (similar to the other sections of the course). Table 3.2 gives an overview of the course topics and activities, and the separate columns are indicative of our struggle. Over the course of the semester, we gradually started to find ways to overlap the content and pedagogy. The required course materials were: *Evolutionary Analysis* (Freeman & Herron, 2007), *Not in our Classrooms* (Scott & Branch, 2006) and SimBio Virtual Labs *Darwinian Snails* and *Mendelian Pigs* (SimBio.com).

Table 3.2 Course overview year

Week	Evolution content	Evolution education content
1	• HIV introduction lecture	• Discussion of Berkman and Plutzer (2011, p. 106) reading • Overview of misconceptions • Introduction to *Understanding Evolution* website (http://evolution.berkeley.edu/evolibrary/home.php) • Activity using Natural Selection Concept Cartoons (Anderson, 2012)
2	• Selection on HIV treatment • HIV trade-off multi-level selection • Weekly content test	• Administered Conceptual Inventory of Natural Selection (CINS) (Anderson, Fisher, & Norman, 2002) • Plant FastPlants for AP Biology Artificial Selection Lab (The College Board, 2017)
3	• Evidence for evolution: Dog breeds • Ring species, archaeopteryx • Homology, atavisms, geology • Wallace and Huxley • Weekly content test	• Nature of Science: Law versus Hypothesis versus Theory • Introduction to state and national high school biology standards • Activity: Create posters comparing state and national high school evolution standards. Galley walk of posters
4	• History of life • Weekly content test	• Discussion of *How Science Works* website (http://undsci.berkeley.edu/index.php) • Nature of Science and Theory of Evolution
5	• Weekly content test	• Introduce Earth Calendar Assignment • Natural Selection Simulation (Pasta activity) • Introduction to AP Biology Artificial Selection Lab 1 (The College Board, 2017).

(continued)

Table 3.2 (continued)

Week	Evolution content	Evolution education content
6	• Natural selection: Four Postulates, katydid research, eye evolution • Weekly content test	• Reviewing flower structure and pollination • FastPlant Artificial Selection Lab: Pollinate FastPlants • Student presentations of Earth Calendar using a second analogy of choice (Ex: football field, map, dictionary) • Explore free online natural selection simulations, students present their critique of simulations.
7	• Weekly content test	• Fastplants • Discuss creationist objections to teaching evolution
8	• Alleles, Mendelian Genetics • Hardy-Weinberg • Fitness and Selection • Patterns: Mutation • Migration: Snakes • Genetic drift, using PopGen Fishbowl 1 (Jones, 2008) • Weekly content test	• Fastplants • Review correct responses to CINS • Introduce Milestone Project • Discussion: Missouri anti-evolution bills • Introduce Evolution Teaching Rationale Paper assignment
9	• Nonrandom mating • Tree Lab HIV • Primate hemoglobin • Weekly content test	• FastPlants • Discussion: *Not in our Classrooms*, Chap. 1 (Scott, 2006) • Discussion: *Understanding evolutionary trees* (Gregory, 2008)
10	• Tree lab choice: HIV and primates • Speciation • Mechanisms of divergence: Sexual selection • Origin of life: Eukaryotes • Mammalian evolution: Ear and color vision in primates • Weekly content test	• Student Milestone Presentation: Cambrian Explosion • Discussion: *Not in our Classrooms* Chap. 2 (Matzke & Gross, 2006) • Introduce Evolution Teaching Position Paper assignment
11	• Weekly content test	• FastPlants: Plant F_1 generation • Student Milestone Presentations: Bony Skeleton & Jaws, Tetrapods
12	• Weekly content test	• Guest speaker on phylogenies
13	• Weekly content test	• Activity: Guppy Sexual Selection (Sampson & Schleigh, 2013) • Discuss *Not in our classrooms* Chap. 3 (Hewlett & Peters, 2006)
14	• Field trip to rock quarry to collect fossils	• Student Milestone Presentation: Dinosaur Radiation
15	• Human evolution • Chimp versus Bonobo evolution • Weekly content test	• Student Milestone Presentation: Feathers and Flight, K-T Extinction, Whale Evolution

In the second year, we modified the course structure to organize it around five evolution stories: HIV, Evolution of Sex, the Dover Trial, Mammalian Evolution, and Human Evolution. The Dover Trial story focused on the public controversy surrounding the teaching of evolution at Dover Area High School in Pennsylvania. We retained many of the projects and assignments from Year 1, which we describe briefly below.

3.6.2 Use of Existing Evolution Resources

In response to the survey findings (Friedrichsen et al., 2016), we were deliberate in incorporating existing evolution education resources and labs. Next, we describe several of the resources we used in more detail along with our rationale for their selection.

FastPlant Artificial Selection Investigation 1. For this investigation, we used the Advanced Placement Biology Artificial Selection Lab (The College Board, 2017). PSTs grew FastPlants and observed the trichome (hair) number. Using the class data, they selected the top 10% hairiest FastPlants to pollinate. Later, PSTs collected the seeds and grew the F_1 generation. The investigation showed the dramatic effect of artificial selection within two generations. With this lab, PSTs became familiar with an instructional resource for teaching artificial selection and gained experience growing FastPlants, which can be used to demonstrate a wide range of biological concepts.

SimBio virtual labs. We used two SimBio virtual labs, *Darwinian Snails*, and *Mendelian Pigs* (SimBio.com) as homework assignments. *Darwinian Snails* emphasizes experimental design, genetic variation, heritability, and natural selection; *Mendelian Pigs* emphasizes Hardy-Weinberg equilibrium, Mendelian genetics, mutation, and population genetics. The SimBio Virtual Labs use an inquiry-oriented environment, refer to actual biological organisms (as opposed to fictional creatures), and are based on data from published scientific studies. These characteristics were appealing to the instructors because the virtual labs were more authentic than typical natural simulations used by many high school teachers (e.g., colored pasta representing individual organisms). High school students may fail to learn the intended concept (i.e., natural selection) when pasta, candy, or toothpicks are used because this simulation is far removed from authentic scientific investigations (Sickel & Friedrichsen, 2012).

NetLogo PopGen Fishbowl. We incorporated this modeling software to teach genetic drift. Jones (2008) designed the simulation to allow students to conduct virtual experiments, allowing students to violate each of the assumptions of Hardy-Weinberg to see the effect. We incorporated this modeling software to highlight modeling as an *NGSS* practice.

3.6.3 Instructor-Designed Projects and Position Paper

To further support the development of PCK for teaching evolution, we designed two additional projects: Earth Calendar assignment and the Milestone Project. To help students synthesize course readings and discussions related to the public controversy, we required the PSTs to write a position paper.

Earth calendar assignment. Students, using a spreadsheet, mapped the age of the Earth to a 12-month calendar. They were given a list of biological milestones (e.g., evolution of photosynthesis) to place on their calendar. As part of that assignment, students created a second analogy of their choice. Students chose a variety of analogies, including mapping the age of the earth to yards on a football field, mile markers on an interstate highway, and pages and word entries in a dictionary. This assignment helped students visualize deep time and developed their PCK for representations for deep time.

Milestone project. Pairs of students were given a different evolutionary milestone (e.g., land plants, internal skeleton, feathers) to research and present to the class. They had to address a list of questions, such as: When did the milestone occur? What was the environment in which the milestone took place? What were the effects of this innovation on the environment and other organisms? In this assignment, PSTs were placed in the role of teacher, as they considered the best way to represent and share their information with the class, further developing their PCK for evolution.

Position paper. PSTs were asked to write a position paper articulating the science education field's position on teaching evolution. Students were required to reference supporting evidence from the following categories: national science education standards, education professional organizations, and nature of the biology discipline. The purpose of the paper was to have students synthesize their understanding of the science education field's position on teaching evolution, and gain confidence in their ability to defend the teaching of evolution to administrators and parents.

3.6.4 Student Feedback

At the beginning and end of the course, PSTs were asked to rate their understanding of evolution content, their preparedness to defend the teaching of evolution, and their preparedness to teach evolution, using a 4-point scale: 1 = weak and 4 = strong. In both Year 1 and Year 2, there were significant gains for each dimension on the post-tests as determined by a paired t-test (see Table 3.3). Students were also given the opportunity to write comments on their course evaluations. Table 3.4 contains representative student comments.

Table 3.3 PST pre and post self-evaluations of understanding and preparedness

Items	Year 1 N = 15		Year 2 N = 11	
	Pre mean (SD)	Post mean (SD)	Pre mean (SD)	Post mean (SD)
Your understanding of evolution	2.03 (0.59)	3.53* (0.50)	2.56 (0.53)	3.56* (0.53)
Your preparedness to participate in the social controversy surrounding evolution teaching and defend the teaching of evolution	1.47 (0.62)	3.40* (0.61)	1.67 (0.71)	3.44* (0.73)
Your preparedness to teach evolution	1.60 (0.71)	3.40* (0.71)	1.67 (0.50)	2.89* (0.78)

Note $*p < 0.05$, two-tailed

Table 3.4 Sample student feedback

Sample student feedback
We honestly need more upper level classes for ed students. This is the first upper level content class I didn't feel was geared towards pre-med students. I've felt a little like an after thought in some science classes and have a challenging content course specifically geared toward helping me as a professional has been an invaluable experience
Excellent class!!! I learned so much and everything was relevant. Great balance + connections between evolution teaching and teaching component
This class was very helpful in preparing me to teach evolution to high school students. I had some knowledge beforehand, but being exposed to the "controversy" I feel more prepared to confront it
It was a really cool concept to combine the content with pedagogy. Especially the pedagogy of teaching evolution since it's such a controversial topic
It's important to be able to defend your stance on why teaching evo is so important, and having something to reference if this situation were to arise
I really was able to understand effective ways to teach evolution to high school students and felt really supported that I'll be able to do it well in the future

3.6.5 Challenges

As we co-designed and co-taught the hybrid course, we experienced several challenges, including tension created from different science teaching orientations, and an on-going tension between depth versus breath of content, as well as content versus pedagogy.

Differing science teaching orientations. Both instructors shared a common vision of better preparing future teachers to teach evolution. Prior to Project Teach Evolution, we had interacted in meetings, and Pat had attended a full semester of Johannes' *Evolution* lecture course. As we planned and co-taught the course, we continuously negotiated how to engage students in the course. In retrospect, these challenges could be attributed to our differing science teaching orientations or conceptions of teaching. Pat's science teacher orientation might best be described as project-based (Magnusson, Krajcik, & Borko, 1999). She designed her science methods course around a series of projects (e.g., designing a curriculum unit) to help students meet the course objectives. As a science teacher educator, her courses focused on helping students learn the processes of teaching (e.g., how to plan a lesson, how to assess student learning), and consequently, her teaching was more process-oriented. As a faculty member in a science department, Johannes' conception of teaching was more content-oriented, and might be described as transmitting structured content knowledge (Kember, 1997). Our differing conceptions of teaching are reflective of the cultures of two different departments (Biology and Learning, Teaching, & Curriculum). Over time, we negotiated and experimented with different ways to engage students; for example, we found the use of NetLogo PopGen Fishbowl allowed us to teach content while modeling scientific practices called for in *NGSS*. Because of differing academic cultures and teaching orientations reflective of those cultures, co-teaching hybrid courses can be challenging work that requires time and negotiation to find common ground.

Negotiating a balance between content (depth vs. breadth) and pedagogy. In the hybrid course, students earned the same number of biology credits as students in the regular section of the *Evolution* course. Johannes was also teaching a regular section of the *Evolution* course at the same time, and the hybrid course was a teaching overload for him. These factors created constraints on our collaboration. In Year 1, by keeping the content similar in both sections of the course, Johannes' teaching load was more manageable. However, this created a tension because the PSTs needed less depth but a greater breadth of content knowledge to teach evolution in high school. For example, macro-evolution, the geological time scale, and human evolution were not originally included in the regular *Evolution* course, but were added to the hybrid course. In Year 2, the content was re-structured around five evolution themes to better address the needs of the PSTs and the course content varied more between Johannes' sections of the course.

The hybrid course met for one additional hour a week to accommodate the pedagogy components of the course, although students did not earn education credit. In the beginning of the collaboration, Pat and Johannes viewed themselves as the education expert and the content expert, respectively, so in Year 1, Pat tended to teach one hour of pedagogy and Johannes taught 3 hours of evolution content. Over time, through negotiation and experimentation, our roles and teaching began to overlap and we found ways to merge content and pedagogy (e.g., designing the Milestone Project and the Earth Calendar assignment). Co-teaching a hybrid course requires time to negotiate and develop new teaching practices.

3.7 Next Steps and Conclusion

After two years of teaching the hybrid course, a decision was made to discontinue it. Several factors contributed to this decision, including the extensive planning time required to effectively co-teach the course. Pat now teaches a new course, *Biology Methods*. This new course has an emphasis on developing PCK for teaching evolution, but also addresses all of the *NGSS* Disciplinary Core Ideas (DCI) for Life Science. Within each life science DCI, PSTs research common misconceptions and ways to challenge specific misconceptions, unpack the *NGSS* Performance Expectations to identify daily learning targets, identify and critique instructional resources, and design instruction to meet specific Performance Expectations. Within the evolution portion of the course, we discuss articles about the public controversy surrounding evolution and PSTs write a rationale paper articulating why evolution should be included in the high school biology curriculum. Our Science Teacher Education Program continues to require an evolution content course while adding the requirement of the *Biology Methods* course.

In conclusion, public school biology teachers are at the frontlines of the public controversy surrounding the teaching of evolution. We have chosen to focus our efforts on pre-service teacher education to better prepare biology teachers to teach evolution and to address the public controversy surrounding its teaching. Pat's efforts have evolved over the years, from emphasizing evolution teaching in secondary science methods courses that include all disciplines, to adding an evolution course as a requirement of the biology education degree program, to co-teaching a hybrid evolution content and pedagogy course, to now teaching a specialized biology methods course that addresses teaching evolution. Requiring an evolution content course for pre-service biology teachers is only part of the solution. PSTs also need to develop PCK for teaching evolution, and have a thorough understanding of creationist arguments and the controversy surrounding the teaching of evolution. By better preparing future biology teachers to teach evolution, we can improve the biological literacy of all citizens.

Acknowledgements This work is based on work supported by the National Science Foundation Transforming Undergraduate Education in Science (TUES) program under Grant 1140462. Any opinions, findings, and conclusions or recommendations expressed in this material are those of the authors(s) and do not necessarily reflect the views of the National Science Foundation.

References

Anderson, D. L. (2012). Natural selection concept cartoons. Retrieved from http://www.pointloma.edu/experience/academics/schools-departments/department-biology/faculty-staff/dianne-anderson-phd/concept-cartoons/natural-selection.

Anderson, D. L., Fisher, K. M., & Norman, G. J. (2002). Development and evaluation of the conceptual inventory of natural selection. *Journal of Research in Science Teaching, 39*(10), 952–978.

Berkman, M. B., & Plutzer, E. (2011). Defeating creationism in the courtroom, but not in the classroom. *Science, 331*(6016), 404–405.

Blevins, B. (2002). *Hill folks: A history of Arkansas Ozarkers and their image*. Chapel Hill, N.C.: University of North Carolina Press.

Cherry, C. (1971). *God's new Israel: Religious interpretations of American destiny*. Englewood Cliffs, N.J.: Prentice-Hall Inc.

Freeman, S., & Herron, J. C. (2007). *Evolutionary analysis* (4th ed.). Upper Saddle River, New Jersey: Prentice Hall.

Friedrichsen, P. J., Linke, N., & Barnett, E. (2016). Biology teachers' professional development needs for teaching evolution. *Science Educator, 25*(1), 51–61.

Gess-Newsome, J. (2015). A model of teacher professional knowledge and skill including PCK. In A. Berry, P. Friedrichsen, & J. Loughran (Eds.), *Re-examining pedagogical content knowledge in science education* (pp. 28–42). New York: Routledge.

Gregory, T. R. (2008). Understanding evolutionary trees. *Evolution: Education and Outreach, 1* (2), 121.

Hewlett, M., & Peters, T. (2006). Theology, religion, and intelligent design. In E. C. Scott & G. Branch (Eds.), *Not in our classrooms: Why Intelligent Design is wrong for our schools* (pp. 57–82). Boston, MA: Beacon Press.

Jones, T. C. (2008). PopGen Fishbowl 1. Retrieved from http://ccl.northwestern.edu/netlogo/models/community/PopGen_Fishbowl_1.

Kember, D. (1997). A reconceptualisation of the research into university academics' conceptions of teaching. *Learning and instruction, 7*(3), 255–275.

Ketchell, A. (2007). *Holy hills of the Ozarks: Religion and tourism in Branson, Missouri*. Baltimore, MD: The Johns Hopkins University Press.

Lerner, L. S., Goodenough, U., Lynch, J., Schwartz, M., & Schwartz, R. (2012). *State of state science standards 2012*. Retrieved from https://edex.s3-us-west-2.amazonaws.com/publication/pdfs/2012-State-of-State-Science-Standards-FINAL_2.pdf.

Lumina Foundation. (2013). *A stronger Missouri through higher education*. Retrieved from https://www.missourieconomy.org/pdfs/lumina_missouri_report.pdf.

Magnusson, S., Krajcik, J., & Borko, H. (1999). Nature, sources, and development of pedagogical content knowledge for science teaching. In J. Gess-Newsome & N. G. Ledermann (Eds.), *Examining pedagogical content knowledge: The construct and its implications for science education* (pp. 95–132). Dordrecht, NL: Kluwer Academic Publishers.

Matzke, N. J., & Gross, P. R. (2006). Analyzing critical analysis: The fallback antievolutionist strategy. In E. C. Scott & G. Branch (Eds.), *Not in our classrooms: Why Intelligent Design is wrong for our schools* (pp. 28–56). Boston, MA: Beacon Press.

McCauley, D. V. (1995). *Applachian mountain religion: A history*. Urbana, IL: University of Illinois.

Missouri Department of Elementary & Secondary Education [DESE]. (2008). Biology course level expectations. Retrieved from https://dese.mo.gov/sites/default/files/cle-biology-science.pdf.

Missouri Economic Research and Information Center. (2012, March). Missouri population data series: Age and gender demographics, 2010 Census. Retrieved from https://www.missourieconomy.org/pdfs/age_and_gender_demographics.pdf.

Morrow, L., & Myers-Phinney, L. (1999). *Shepherd of the hills country: Tourism transforms the Ozarks, 1880–1930s*. Fayetteville, AR: The University of Arkansas Press.

NationMaster. (n.d.). Countries Compared by Geography > Total area > Sq. km. International Statistics at NationMaster.com. Retrieved from http://www.nationmaster.com/country-info/stats/Geography/Total-area/Sq.-km.

Newport, F. (2014). In U.S., 42% believe creationist view of human origins. Retrieved from http://www.gallup.com/poll/170822/believe-creationist-view-human-origins.aspx.

NGSS Lead States. (2013). Next Generation Science Standards: For states, by states. Retrieved from http://www.nextgenscience.org/.

Pew Research Center. (2017). The 2014 Religious Landscape Study. Retrieved from http://www.pewforum.org/about-the-religious-landscape-study/.

Rafferty, M. (1988). The Ozarks as a region: A geographer's description. *OzarksWatch Magazine, 1*.

Sampson, V., & Schleigh, S. (2013). *Scientific argumentation in biology: 30 classroom activities*. Arlington, VA: NSTA Press.

Scott, E. C. (2006). The once and future Intelligent Design. In E. C. Scott & G. Branch (Eds.), *Not in our classrooms: Why Intelligent Design is wrong for our schools* (pp. 1–27). Boston, MA: Beacon Press.

Scott, E. C., & Branch, G. (Eds.). (2006). *Not in our classrooms: Why intelligent design is wrong for our schools*. Boston, MA: Beacon Press.

Shulman, L. S. (1986). Those who understand: Knowledge growth in teaching. *Educational Researcher*, 4–14.

Sickel, A. J., & Friedrichsen, P. (2013). Examining the evolution education literature with a focus on teachers: Major findings, goals for teacher preparation, and directions for future research. *Evolution: Education and Outreach, 6*(23). https://doi.org/10.1186/1936-6434-6-23.

Sickel, A. J., & Friedrichsen, P. J. (2012). Using the FAR guide to teach simulations: An example with natural selection. *The American Biology Teacher, 74*(1), 47–51.

The College Board. (2017). Investigation 1 Artificial Selection. Retrieved from http://media.collegeboard.com/digitalServices/pdf/ap/bio-manual/Bio_Lab1-ArtificialSelection.pdf.

United States Census Bureau. (n.d.). Quick Facts Missouri. Retrieved from https://www.census.gov/quickfacts/table/PST045216/29.

Woodward, C. (2011). *American nations: A history of the eleven rival regional cultures of North America*. New York City, NY: Peguin Group.

Wright, H. B. (1907). *The shepherd of the hills*. New York City, NY: Lasso Press.

Patricia Friedrichsen is a Professor of Science Education in the Department of Learning, Teaching, and Curriculum at the University of Missouri. She teaches in both the undergraduate science teacher education program and the doctoral program. Her research agenda focuses on secondary science teacher learning across the professional continuum, using a variety of perspectives including pedagogical content knowledge, beliefs, communities of practice, and core practices. She is currently researching teacher learning in the context of teaching science using a socio-scientific issues-based pedagogical approach. Pat is a former high school biology teacher who taught evolution to her students.

Larry G. Brown is a retired MU Assistant Professor of Geography where he taught courses in Human Geography, including the Geography of Missouri, from 1990 through 2014; having earned a PhD in Policy Studies, an MA in Geography, a Masters of Divinity, and a BA in Sociology. He has researched White Nationalism, and published articles on the Christian Identity Movement. He continues to be a public speaker on topics related to White Nationalism, Domestic Terrorism, and Ozark Culture. Larry is a professional storyteller, and past-president of Missouri Storytelling, Inc. (MO-TELL). Larry is an ordained minister with standing in the Christian Church (Disciples of Christ), having served as pastor in NE, IN, and MO for over 30 years. He is currently a regular instructor in MU Extension's Osher Lifelong Learning Institute.

Johannes Schul is Professor of Biological Sciences at the University of Missouri. He teaches undergraduate and graduate level courses in evolution, neuroscience, and general biology. His research program studies function and evolution of acoustic communication in insects with an integrative approach. Questions and methods range from molecular to behavioral and evolutionary levels. He has been engaged in several grant funded projects developing and implementing integrative content and inclusive teaching practices for general biology and evolution courses.

Chapter 4
Controversial Before Entering My Classroom: Exploring Pre-service Teacher Experiences with Evolution Teaching and Learning in the Southeastern United States

Amanda L. Glaze and M. Jenice "Dee" Goldston

Abstract Evolution continues to be a polarizing topic amongst the public as well as in K-12 and post-secondary classrooms. One issue that contributes to the polarization is the absence of accurate and meaningful instruction on evolution. The divide is especially pronounced in regions such as "The South"—Alabama, Georgia, Arkansas, Louisiana, Mississippi, South Carolina, North Carolina, and Tennessee—where cultural underpinnings strongly align against scientific topics dealing with human origins and change. Research shows that acceptance or rejection of evolution provides a reference for teachers' choice whether to teach controversial topics such as evolution as well as the depth, breadth and duration of instruction. In this chapter we take a deeper look at the lived experiences of pre-service science teachers at a teaching college in the Southeastern United States in an effort to frame a context within the region by which later choices regarding teaching are made. Furthermore, we provide suggestions for improvements to teaching and learning that have implications beyond this critical region. Although public controversy surrounding evolution is widely regarded as being defining of the United States, the implications of studies here have translational value to teaching and learning evolution around the world.

4.1 Introduction

The United States holds an anomalous position within the ranks of nations when it comes to the teaching and learning of evolution. Whereas most industrialized, or as they are often described "first-tier nations," have demonstrated little or no

---

A. L. Glaze (✉)
Georgia Southern University, Statesboro, GA, USA
e-mail: aglaze@georgiasouthern.edu

M. Jenice "Dee" Goldston
The University of Alabama, Tuscaloosa, AL, USA

controversy surrounding evolutionary concepts, the United States consistently ranks below other nations in scores relative to acceptance and understanding of evolution due to high levels of conflict between the concepts set forth by scientific explanations and public opinion (Miller, Scott, & Okamoto, 2006). "Cultural clashes between students' life-worlds and the world of western science challenge science educators who embrace science for all, and the clashes define an emerging priority for the 21st century" (Aikenhead & Jegede, 1999, p. 269). Nowhere are clashes between culture and science more prominent than in the Southeastern United States and more specifically the culturally connected sub-region of "The South"—Alabama, Georgia, Arkansas, Louisiana, Mississippi, South Carolina, North Carolina, and Tennessee—where educational board decisions, state laws, and legal cases demonstrate a very public showcase of anti-evolution, and often anti-science, sentiment (Price, 2013; Rissler, Duncan, & Caruso, 2014; Wilson, 1996).

To highlight some of the demographics of the area, we will focus on one state at the center of the region. Known as the literal and figurative "Heart of Dixie", explorations in the state of Alabama highlight the conflict and controversy that often surrounds evolution teaching and learning in the region (Glaze, 2013; Glaze, Goldston, & Dantzler, 2015; Goldston & Kyzer, 2009). The state of Alabama has a population of approximately 4.78 million people that includes a variety of cultures, socioeconomic levels, and backgrounds, although this cumulative variety represents a small minority in the state (U.S. Census Bureau, 2016). According to state records, licensure tests in the state are offered in 13 languages, however English is the primary language spoken by 3.99 million of the state's residents, followed by Spanish (89,000), Indo-European languages such as French and German (43,800), Asian languages (22,000), and other Native American, Africa, and Arabian languages (6,800) (Echevarria, 2013). Approximately 60% of the state population identifies as religious (Alabama State Religion, n.d.). In terms of type, Christianity accounts for 58% of the state population, with 46% of the population identifying as Protestant (36% Baptist), 8% "other" Christian, and only 4% Catholic (Alabama State Religion, n.d.). Non-Christian religions represent less than 2% of the state population (Alabama State Religion, n.d.). While the United States has a constitutional focus on the separation of church and state, in Alabama, and the South as a whole, there is a greater inclusion of religion as a part of culture, and that underpinning is mirrored in legislation and government action. There is a heavy focus on the importance of state choice in matters of government and education. Education decision-making in Alabama falls to elected local school boards that operate under the shared oversight of an elected state Department of Education. Local education control lends itself to greater autonomy in what is taught in the classroom, despite what is written in standards and widely expected at the national, or even at the state, level (Urban, 1992).

Education in Alabama is free to all students through age 21 and compulsory between the ages of six and sixteen, with some age exceptions based on individual considerations such as health. Children of this age range can attend public schools or private schools, secular or parochial, that consist of twelve grade levels plus the availability of additional early training in pre-kindergarten and kindergarten classes for those not yet six. Grades are typically divided into three groups: primary/elementary (K-4), intermediary/middle/junior high (5–8), and secondary/high schools (9–12). However, the housing of these grade levels may vary based on individual school board decisions and the size of the local population. Alabama is mostly classified as rural, therefore, it is not uncommon to find several school structures: grades K-12 housed in a single school; grades K-6 as elementary and 6–12 as the high school; or, in larger systems, grades K-4 or 5 in elementary, 5 or 6–8 as middle or "junior" high schools, and 9–12 high schools. The goal of these groupings is separation of age groups based on learner maturity, departmentalization of subject areas, and community need.

Over the last decade, nationally mandated standardized testing has impacted science teaching in Alabama. Testing in the state has focused primarily on reading and math causing greater time to be put into development in these subjects. Only these subjects are assessed each year from grades 3–8. Science is assessed by state examination only in grades 5 and 7, then later at the national level on the American College Testing College Readiness Examination (ACT) in grade 10. As a result, it is not uncommon for students to have little or no formal science classroom experiences until reaching grade 5. Testing in the state is largely done for the purpose of tracking student progress and measurement for benchmarks set by federal or state legislation. Students who fail to meet proficiency in these tests are not withheld from the next grade level, as that determination is based on in-class performance. In the post-*No Child Left Behind* era, it will be interesting to see how, or whether, these tests continue to be utilized. What is known is that the atmosphere of the South—the culture, the beliefs, and the conflict—is as prevalent in the schools as it is among the public.

4.2 Public Acceptance of Evolutionary Theory Within the Social, Political, and Cultural Context of the Southeastern United States

The Southeastern United States provides a unique venue to study the perspectives of students, teachers, and the public regarding the perceived controversy surrounding elements of evolution. In 2008, Kristi Bowman brought attention to the evolution struggles in the geographical Southeast, noting that students therein were 84% less likely than students elsewhere in the United States to receive accurate instruction regarding evolution and ten times more likely not to have any evolution instruction in their primary or secondary experiences (Bowman, 2008). The uncommon history of the South, the depth of the Southern identity, and the highly evident and influential "power of place as a category of social and personal experience" make it all the more important to understand the dynamics of evolution teaching and learning in the region (Kincheloe & Pinar, 1991, p. 167).

For instance,

> Not only does the South find itself inhabited by the living presence of a unique history, a peculiar literary tradition, and an unusual set of social relationships but Southerners might also be said to possess a distinctive way of knowing, an epistemology of place. (Kincheloe & Pinar, 1991, p. 10)

According to Kincheloe and Pinar (1991), the concept of "place" is an element of social and cultural influence that guides each individual's learning and development. Religion is but one facet of the sense of place. However, when "place" is viewed from a perspective responsive to the nature of religiosity and evangelical literalist traditions found in the South, religion cannot be removed from consideration due to the impact it has demonstrated on evolution acceptance and decision making (Glaze et al., 2015; Kincheloe & Pinar, 1991; Nadelson & Sinatra, 2008; Whitaker, 2010). In essence, shared cultural norms and social expectations surrounding the very word "evolution" point to elements of a shared worldview as a part of a sense of place in the South as a region; one that makes all conversations on evolution more complex and delicate.

4.3 Existence and Extent of Influence of Anti-evolutionary Movements in the Southeastern United States

Beginning in the 1920s, states including Kentucky, Tennessee, Arkansas, Mississippi, and Florida had laws or bills in place to prevent or circumvent teaching of evolution in public schools while other states, including Louisiana and Texas, had their state board of education restrict evolution instruction and strike the mention of evolution from textbooks (Elsberry, 2001). Anti-evolution legislation and local actions began facing public challenge in the science-driven decades after World War II, when a number of key court decisions were passed down to counter earlier anti-evolution efforts. Key cases in supporting the teaching of evolution and restricting the teaching of non-scientific alternatives originated from states around the South. Table 4.1 summarizes key court decisions at the federal level that were integral in striking down anti-evolution or alternative evolution education laws and actions across the South.

While great strides have been made in legally supporting evolution and drawing lines as to what is and is not acceptable in science classrooms, there are still challenges to the teaching and learning of evolution, including "Academic Freedom Laws" (Glaze & Goldston, 2015; Pobiner, 2016; Smith 2010a, 2010b). These laws follow guidelines from the Discovery Institute—a group largely focused on promotion of the blending of science and religion called Intelligent Design—and utilize carefully structured language to avoid raising flags based on existing court decisions (National Center for Science Education [NCSE], 2009b, March 20). These laws circumnavigate the rulings of prior cases on evolution and creationism in classrooms under the guise of providing protections to teachers

Table 4.1 Federal Legal Cases from the Southeastern United States

Year	Case	Summary ruling/Impact
1968	*Epperson v. Arkansas*	Arkansas's anti-evolution legislation was unconstitutional
1975	*Daniel v. Waters*	Tennessee law requiring equal teaching of creationism and evolution is in violation of the Establishment Clause
1982	*McLean v. Arkansas Board of Education*	Arkansas laws requiring the teaching of creation science are in violation of the Establishment Clause, provided legal definition of science
1987	*Edwards v. Aguillard*	Louisiana legislation that allowed evolution teaching only when taught with creationism is unconstitutional, violates the Establishment Clause, and undermines science education
1997	*Freiler v. Tangipahoa Parish Board of Education*	Louisiana Board of Education policy requiring a disclaimer against evolution for religious purposes is unconstitutional, Intelligent Design identified as creation science
2005	*Selman v. Cobb County School District*	Cobb County, Georgia, School District requirement of an anti-evolution disclaimer in textbooks was in violation of the Establishment Clause

"for presenting scientific information pertaining to the full range of scientific views regarding biological and chemical evolution" and protection for students "concerning their positions on views regarding biological and chemical evolution" (NCSE, 2009b, March 20, p. 1). Such laws have been successfully passed in Southern states (e.g., Louisiana Science Education Act [Act 473, SB733 2008], Tennessee's Teacher Protection and Academic Freedom Act [SB0893/HB0368, 2012]. Following the success of early attempts, similar laws have been brought to the floor in Alabama, Florida, and Kentucky but have yet to pass as of this writing (NCSE, 2009a, March 11).

4.4 Place of Evolutionary Theory in the Curriculum

The approach to evolution in the curriculum nationally has improved by leaps and bounds as evidenced by the increasing coverage and focus on evolution as a unifying concept in life science in the *Next Generation Science Standards* (NGSS Lead States, 2013). However, the South has a reputation, both historically and presently, as a hotbed of division when it comes to evolution in the classroom (Goldston & Kyzer, 2009). In addition to legislative action, Southern states maintain state-written standards, with many incorporating elements from the *NGSS* but crafting their own adjusted version of the standards to avoid conflict given the existing sociopolitical cultures in each state. The standards adopted by the state of

Alabama, wherein the word evolution is noticeably missing, provide evidence of this divergence from the *NGSS* (Alabama State Department of Education, 2015). While there is some inclusion of evolutionary concepts that align with unity and diversity of life in what is commonly referred to as micro-evolution, in comparison to the coverage of evolution in the *NGSS*, it is minimal and demonstrates the extent to which people will go to avoid the "e-word" if at all possible. There remains a state-required disclaimer in the front of all biology textbooks that continues to draw criticism for the threat it poses to science literacy in the South (Branch, 2017; Glaze, 2016; Goldston & Kyzer, 2009; Rissler et al., 2014). One such criticism is that it provides support for students and teachers to ignore the topic of evolution on the grounds that it is not scientifically supported and is being called into question by the state board. The mixed-signals sent by actions and disclaimers alike, do little to improve the experiences or attitudes of classroom teachers toward evolution.

4.5 Biology Teachers' Attitudes Toward Teaching Evolutionary Theory

Science teachers in the South often find evolution to be "frustrating and challenging" as well as a source of criticism from family or community (Goldston & Kyzer, 2009). Teachers often feel added pressure to avoid evolution due to "persistent and publicly sanctioned hostility" regarding the teaching of evolution (Shankar & Skoog, 1993). As shown in three teachers in Alabama, when teachers "perceive the topic of evolution to be in direct conflict with their own or their students' personal beliefs" teaching evolution becomes even more problematic (Goldston & Kyzer, 2009). Similarly, when teacher beliefs are in opposition to the curriculum, internal conflict occurs that is likely to influence whether a teacher and/or their students are open to accepting evolution (Chinn & Samarapungavan, 2001; Davson-Galle, 2004; Jones & Carter, 2007; Meadows et al., 2000). While studies in the region are rare, the South tends to represent the extreme of what is found among teachers in other regions, mirroring the strongest of impacts that are seen in many areas (Glaze et al., 2015; Rissler et al., 2014).

A number of additional factors that influence the teaching of evolution have been identified that include misconceptions regarding evolution, comfort with the content, and conflict with religious beliefs in the Southeastern United States (Aguillard, 1998; Bowman, 2008; Glaze et al., 2015). Among Louisiana classroom teachers, Aguillard (1998) identified educational background as important and "subjects were often critical of their college biology training" explaining that they had less than three classes in biology where they specifically address evolution (Aguillard, 1998, p. 172). Religious beliefs have emerged in several studies as the most important factor in teacher or pre-service teacher acceptance or rejection of evolution in the South as well (Glaze et al., 2015). In fact, having the ability to reconcile religious beliefs and scientific ways of explaining the world is key to an individual's ability to overcome conflict between religion and beliefs (Meadows et al., 2000; Shipman, Brickhouse, Dagher, & Letts, 2002; Wiles, 2008).

Other research suggests that the most influential factor outside of religious beliefs is rooted in the understanding of the nature of science itself, specifically as it relates to how scientific knowledge is generated and the practice of science by those in the field (Jorstad, 2002; Nadelson, 2007; Wiles, 2008; Woods & Scharmann, 2001). Thus, understanding of the nature of science, science content background, and open-mindedness to religion and scientific issues represent key factors identified as influential in the acceptance or rejection of evolution and thereby impact choices surrounding the teaching of evolution in the classroom (Berkman & Plutzer, 2010; Fowler & Meisels, 2010; Glaze et al., 2015; Trani, 2004).

Exploring external socio-cultural relationships between those we interact with regularly, such as parents, friends, church and community members impact ideas and choices about evolution teaching (Demastes, Good, & Peebles, 1995; Goldston & Kyzer, 2009; Woods & Scharmann, 2001). Support within the system and the school itself, district and school guidelines for teaching of evolution, knowledge of legal cases regarding the teaching of evolution, membership in professional organizations, and the textbooks used to teach biology all serve as positive external forces for teaching evolution in the South (Aguillard, 1998). Similarly, parent attitudes and perceptions of support or discord from others related to their stance on evolution are also powerful influences on teacher choices surrounding teaching evolution in the South (Aguillard, 1998). Further compounding teacher attitudes toward evolution and choices regarding the teaching of evolution in their classrooms is the fact that little is done in teacher education to prepare them for the barrage of intersecting factors that will impact their teaching.

4.6 Emphasis Given to Evolutionary Theory in Biology Teacher Education Programs

Evolution is as much an issue with pre-service science teachers and science majors as it is with students in K-12 classrooms and classroom teachers (Glaze et al., 2015; Ha, Haury, & Nehm, 2012; Rutledge & Warden, 2000). Pre-service teacher preparation programs represent an area of focus in debates about evolution and how to maximize its impact on classroom teaching and learning (Deniz, Cetin, & Yilmaz, 2011; Deniz, Donnelly, & Yilmaz, 2008; Glaze et al., 2015). Historically, there have been questions surrounding the level of content expertise acquired by teachers of science and whether science courses for education majors are too specific to translate into the classroom (Adams & Krockover, 1997; Rice, 2003; Wenner, 1993). In the teacher education program associated with the narratives in this chapter, science coursework was housed in the science departments and consisted of a core of required classes, including two levels of introductory biology, genetics, cell biology, and ecology. For those seeking general science certification, additional courses were required in chemistry, physics, and geography/geology. Apart from those courses, teacher candidates took a seminar in research and had a choice of two additional courses each from two groups of upper-level biology courses that included a course on evolutionary adaptation. Most of these were specialized courses and there were no options for science undergraduate courses for teachers that would address topics targeted for the classes they would teach once certified. While evolution is frequently wound into the subjects of these other courses in science, the limited background in evolution leading to university study coupled with the broad range of topics covered in university science courses, leaves little room for depth of understanding. Furthermore, the nature of the evolution course as elective allows those who are not accepting of evolution to avoid the course and the content that is needed for deeper comprehension and teaching.

Pre-service teachers' pedagogical content knowledge—the understandings of how students learn and ability to teach the content within the context of that understanding—with respect to evolution has been questionable due to the variance in approaches found in teacher education programs as well as a lack of training in specific strategies to approach sociocultural concerns and classroom management (Aguillard, 1998; Berkman & Plutzer, 2010; Bloom 2007; Griffith & Brem, 2004; Veal & Kubasko, 2003). In the program associated with the narratives of this chapter, evolution was not specifically addressed as a topic in science methods coursework, which consisted of a semester-long secondary science methods course that met 3 h, once a week, for fifteen weeks of the term. The course was designed for pre-service teachers who would later teach in grades 6–12 and be departmentalized as science teachers. Largely, the content was determined by the professor tasked with teaching the methods courses and focused less on specific content topics and more on the greater issues of science education and pedagogy, such as inquiry learning, conceptions of science, and lesson/unit planning.

4.7 Pre-service Teachers as a Lens for Understanding Evolution Conflict

Studies exploring views of evolution by teachers are wide in scope and provide us snapshots of thinking, teaching, and navigation of conflict. However, there have been few studies conducted in the Southeastern United States (Aguillard, 1998; Glaze & Goldston, 2015; Meadows et al., 2000; Rissler et al., 2014) and even fewer done with the pre-service teacher sub-group in the South (Glaze, 2013; Glaze et al., 2015). The role that pre-service teachers play in the big picture of understanding teaching and learning of evolution is important because they represent the transition, if you will, between student and teacher. In some ways they represent a sort of "missing link" in our understanding of the experiences that frame how, and whether, evolution is taught. The lived experiences of pre-service science teachers create a highly detailed narrative of the intersections and divergence of the two roles in which they operate—teacher and student—and demonstrate the intensity of thought in which they engage as they reconcile their worldviews with the scientific ways of knowing they encounter in their training and later in the classroom.

World view ("worldview" in contemporary research) refers to the deeply personal collection of understandings, beliefs, and explanations an individual develops about the world around them (Cobern, 1994a, 1994b). This part of our personal identity is crafted over the course of a lifetime, serves as the lens through which we view and evaluate our experiences, and holds sway over our decisions to accept or reject all new information (Branch & Scott, 2008; Johnson, Hill, & Cohen, 2011). When worldview interacts with experiences and conflict arises, research suggests that the path of least resistance is taken. As a result, individuals opt to accept that which most closely aligns to their lens (Branch & Scott, 2008; Jakobi, 2010; Nehm & Reilly, 2007). Conversations with secondary preservice science teachers about evolution indicated that they are already thinking about evolution and how the perceived conflict with evolution will impact their classrooms long before they take on their first teaching position (Glaze, 2013; Nehm & Schonfeld, 2007). A snapshot of the lived experiences of these pre-service science teachers can be seen in conversations with a small group of individuals enrolled in a science teacher education program at one teaching college in the Southeastern United States.

A qualitative personal narrative approach enabled participants to describe, in their own words, their lived experiences surrounding evolutionary theory in and out of the teacher preparation settings. Participants in this study, upper-level undergraduate students, had completed their core courses and had begun study in the college of education. Participants completed an initial online survey and were given the opportunity to be chosen for interview if they met the following criteria: (1) they were an undergraduate, formally admitted to the college of education, (2) their content major was either general science or biology education, both of which teach biology courses in the state of study, (3) they had completed the first block of education courses but were pre-internship at the time of the interview, and (4) they had completed the following science courses-BIO 101/103 (Introduction to Biology I), 102/104 (Introduction to Biology II), 322 (Genetics) or 332 (Ecology), and 373 (Cell Biology).

Of the 79 students who received the request to complete the survey, 37 students completed the Measure of Acceptance of Theories of Evolution (MATE) survey that was used to ascertain their levels of acceptance of evolution (Rutledge & Warden, 2000). The MATE measure has been highly utilized as a possible measure of acceptance of evolution and has demonstrated both validity and reliability when taken by both students and teachers (Rutledge & Warden, 2000). The participants were representative of the program in which they participated, meaning they were mostly white and balanced between male and female, however they do not match the overall state teacher demographic, which is slightly more racially diverse, even after steps were taken in an attempt to diversify the pool. Based on scores, participants were grouped by their level of acceptance with six as *very low (20–52)*, seven as *low (53–64)*, nine as *moderate (65–75)*, ten scoring as *high (77–88)*, and five scoring *very high (89–100)* (Rutledge & Warden, 2000). Two individuals were randomly selected from each group for the initial set of interviews for a total of ten participants with an additional participant added from the mid-line (moderate) group for more balanced representation of clusters (high/moderate/low). The final group was nearly equal between male (6) and female participants (5) and representative of white (9), black (1), and mixed racial backgrounds (1). Participants selected for interviews participated in on-campus individual interviews with the researchers for no less than one hour each and no more than three hours each, including follow up interviews that were conducted.

In the tradition of narrative storytelling, we utilized open-ended questions to elicit participants to share their experiences in their own words, allowing them to be as detailed as they preferred and allowing them to speak freely in whatever direction their thinking took them. In order to compare the narratives of our participants, we transcribed all of the interviews, adding follow up questions where there were areas in need of greater elaboration or where phrasing was unclear. From those stories, common threads emerged, including shared experiences, obstacles, and ways of thinking about teaching that would frame future experiences for participants and their students. Analysis consisted of a basic synthesis of the narratives relative to the common experiences shared in the group in response to the questions posed. What follows is a discussion of the experiences of our participants surrounding evolution under those common threads that were present in their stories.

Content knowledge is inadequate and misconceptions prevail, even in teacher education. Our starting point in exploring participant experiences relative to evolution was to ask them to define evolution in their own words and then explain it to us. In exploring their ideas about evolution content, participants described evolution as a process of changes occurring in a population over a period of time but, beyond that, the definitions were limited, incomplete, and replete with misconceptions regarding when, to whom, and how evolution occurs. There were frequent mentions of Darwin, but few other scientists that were connected with evolution. Participants mostly described natural and artificial selection, but with superficial nods to changes in the environment and mutations rather than explicit focus on reproduction, variation, and selection. When probed regarding their knowledge and beliefs about evolution, several misconceptions specific to evolutionary theory were elicited. These included misconceptions with respect to ancestry with particular emphasis on humans, monkeys and apes; adaptation as an individual response to changes to the environment; and evolution as an unsupported and untestable idea.

The human ancestry issue was addressed by Charles who said, "Do I believe in that little picture that they show you in middle school about how man evolved from other species? I don't. I don't believe that, but I do believe in natural selection as a means to produce a fit species." Still other participants held ideas that are reminiscent of Lamarckian explanations of change. As one participant noted, "Evolution is about organisms adapting to their environment and changing their physical appearances in order to adapt over time." Declan found evolution to encompass "the measureable changes that an animal makes due to environmental stimuli and the changes that the animal passes on to its offspring." Another participant challenged evolution as unsupported, claiming "they say that evolution has been proven, when it hasn't and they just, you know, try to bring up arguments to do with that." Though not discussed here, these quotes were rife with incomplete understandings or misconceptions about evolutionary theory as well as the nature of science itself. One thing agreed upon by all participants was the concept of evolution as change over time; however, there were a range of negative feelings expressed regarding human common ancestry and whether humans were also changing as a part of evolution.

Charlotte   I'm not convinced people eventually evolved from amoebas.
Declan   My feelings about evolution are that, besides the fossil record, there is no absolute concrete evidence that cross species evolution has occurred, or at least not to my current knowledge.
Catherine   The only part of evolution that conflicts with my personal beliefs is that all species of life descends from common ancestry.
Lane   That (evolution) applies to all living things according to science books but I was taught that believing evolution means the Bible is not real.

The statements above illustrate some of those concerns expressed regarding evolution at different levels. While participants were generally open to the idea that things change, a very simplified definition for evolution, they expressed specific concerns with the idea that those small changes could result in the creation of new species and were most resistant to the idea that humans were susceptible to the same changes as all other living things. For better understanding of their definitions of these concepts in evolution, we asked them about their school experiences relative to evolution, including when they first learned about evolution and what they were taught.

Learning experiences in evolution reflect contention in the public realm. When participants were asked about their experiences with science prior to attending college, all reported having several science courses in high school with most reporting having at least one positive experience. For some participants, positive experiences their teachers were leading influences in their choice to become a science teacher. When asked about their early experiences with evolution, participants displayed a wide variety of exposure both in and out of school. Three participants recalled hearing about evolution in elementary school or other settings, but were not sure of the details other than it was negative when it occurred outside of school. Five participants experienced evolution first in life science classes in middle school, including one who explained, "I was taught about evolution in 8th grade and I went into it thinking that this is not what I was supposed to believe, that I wasn't supposed to believe in what my teacher was telling me."

Four participants were in high school before they were exposed to evolution in the classroom, reporting that evolution was barely covered with limited explanations of change over time. They further described vague definitions of adaptation and fitness being applied to evolution in their experiences as well as the absence of any discussion of human origins. For instance,

Jacob   In high school, I learned that evolution is just when organisms or populations change over time. Evolution is essentially heritable changes that occur in species and populations over a period of time. Since the beginning time, species have died off due to lack of adapting to climate change, adapted to climate changes to survive, or evolution to produce offspring better equipped for the region in which the population lives. In other words, we focused in on micro-evolution subjects.

Lane   Evolution is when organisms start from more basic cellular makeup and go to more advanced cellular makeup.

None of the participants in this study reported a thorough study of evolution in their middle or high school years, nor had any taken an evolution-specific science course in their university study. It should be noted that the participants could have taken an upper level course on evolutionary biology, but did not. The evolution course was not required for education majors.

As noted earlier, the participants recalled negative connotations in what they were taught in school or heard outside of school regarding evolution. One participant recalled a sticker being placed in the beginning of the biology textbook as a disclaimer regarding the teaching of evolution that is still present in adopted biology textbooks in Alabama (Branch, 2017; Goldston & Kyzer, 2009; Rissler et al., 2014). As one participant explained, "I first heard about the idea of evolution in high school. It was when my teacher said she refused to teach it. I was formally introduced to evolution in college." Furthermore, the participants stated that evolution was often directly addressed as it related to church or church-related groups, each time with negative implications. One preservice teacher reported that church classes were offered that addressed evolution specifically in regards to their religious beliefs and expectations. Robert explained,

> Every time I ever heard of evolution, it was referred to in a negative way. You know, like, this is Satan's plans to you know, discredit the Bible and stuff like that. So I really didn't have a good understanding of it till later on when I started reading.

Charles reflected,

> I know it has been talked about in church. I don't know if that was the first time that I've heard about evolution though. But it was looked upon in church like it was a ridiculous idea. Like it was, you, you were unintelligent if you actually could possibly believe that human beings came from monkeys.

While the misconception of human descent from monkeys was not novel, the use of this position by community leaders as the foundation for their need to counter teaching of evolution was demonstrative of the lack of acceptance of evolutionary theory prevalent in the public (Gallop, 2014). Jacob added that evolution was discussed in his church but stopped, saying "I would rather not talk about it." In keeping with the nature of evolution as taboo, three of the participants responded that they would rather not discuss the events any further due to the discomfort they felt.

Still, eight participants recalled having discussions with family or friends regarding evolution at varying times in their lives. Most were in the context of their parents teaching their own closely-held beliefs and expectations, with few parents having accurate knowledge of the concepts of evolutionary theory. Others were more clandestine discussions, perhaps due to the shadow of taboo over the topic of evolution, held with close friends regarding questions and conflict between beliefs and readings, or teachings of the individual. Charlotte shared, "I'm pretty sure [I heard about evolution] from my parents or church, so it wasn't an accepted idea at all. It did come up in church, but I really remember my mom, who is very religious, talking about it—always negatively. William also reflected a negative intergenerational view of evolution,

> Yeah, she (my mom), she told me this story about when she was in school and how they taught evolution…she came home from school and told her mother what she had learned in school and her mother, my grandmother, chastised her and told her that she shouldn't talk about that and that that was wrong. And so, my mother told me that story. And my mother isn't as closed off as my grandmother is, but she did say that evolution is not right and that it'll, um, it's not what we believe and, yeah, that's the first thing I ever heard about evolution. I didn't know what evolution was until she told me that because I hadn't learned about it in school yet. That was in elementary school, in like 5th grade.

Three participants specifically addressed difficulty or unwillingness to discuss evolution with peers, family, and others due to the controversial or taboo nature of the topic or a desire to avoid conflict with those they know oppose evolution. Catherine stated, "I have never talked with my family about evolution. It would be a difficult conversation because they would have such narrow thoughts on it." Charles held an enlightening view of the taboo nature of asking hard to answer questions that challenged the status quo, seen in the following:

> In Sunday school, when I would ask the hard questions, I can't think of anything specifically about evolution but just questions in general, I was actually pulled into my pastor's office once when I was a kid and was told I need to stop asking questions like that in Sunday school because I'm being a disturbance.

Within Southern culture, asking questions of things like beliefs or Biblical explanations, especially in literalist traditions, often resulted in tangible repercussions such as being outcast from loved ones and removed from the social and cultural foundation of the community, which was traditionally been built on the cornerstones of church and shared beliefs. Shared experiences, whether cultural or social were also evident in participant narratives.

Shared backgrounds inform expectations and define experiences. Although the eleven participants interviewed herein were representative of the full spectrum of levels of acceptance there was far less diversity in their backgrounds, culturally and otherwise. Among the eleven, all but two described growing up in areas within 100 miles of where they attended college. Three were from self-described urban settings. However, urban in the Southeastern United States often meant a population of 20,000–50,000 and very rarely a "large" city, like Atlanta or Birmingham—both of which are relatively small in comparison to the super cities of the eastern or western coasts. More frequently these preservice science teachers came from small towns of a few thousand that typify the South, places where everyone knew everyone and where many would return to teach following graduation. Growing up in small Southern communities was viewed both as positive and negative by the participants. On one hand, they noted a positive feeling of closeness and support from those around them, while on the other, a lack of anonymity based on the small-town dynamic was problematic. When asked what they felt were the strongest influences on the person they had become, participants mentioned their families and how any decision they faced was largely affected by the consequences that it could have on their loved ones. Furthermore, religion was a major influence in their daily lives and the choices made regarding what they learned or thought about science and their eventual role as science teachers.

The interactions and connections between the participants and influences such as family, religion, peers, and community provided the framework for understanding —the worldview—by which each evaluated all other experiences and knowledge. For instance, where there was agreement between major influences, such as parents and church, peers and family, there tended to be a higher occurrence of rejection of evolution. Rejection was elected because the family, church community, and peers were all in agreement that the acceptance or discussion of evolution was taboo. When there was agreement between church and family but disagreement among peers, school, or mentors, the participant tended to side with that of church and parents—the earliest contributors to their worldview. As a result, the participant addressed the conflict by managing it, ignoring it, or trying to find reconciliation (Griffith & Brem, 2004; Hermann, 2013). In other cases, participant stance on evolution resulted from a critical incident that strongly impacted their worldview, such as abuse that was not addressed by leaders in their church or being ostracized for questioning literal interpretation of Biblical events.

Church plays a role in social and cultural context. Shared background rooted in attendance in church and participation in church-sponsored or related events was common to all but one of the participants. In their experiences, it was common to be in church each time there was a service or activity, which often meant three or more times a week. In fact, many noted the church as the cornerstone of their childhood experiences. When asked about their experiences with religion, many were currently or had been personally involved at a high level. As William explained,

> I'm a pretty devout Christian. I'm a deacon in my church and I also sing and travel around singing at different churches and stuff like that. So I would say that most of my community involvement is through my church and working with the people of my community with my church.

Charles, another preservice teacher, shared similar experiences growing up where church was, as he described, a "major social venue" like school. He further noted, "My particular church was very large for the area. It was and still is the largest building in the town. There were no outlets that I was aware of for kids interested in science." This social connection was also shared by Robert, whose childhood and youth involved moving often and attending many different schools, some for very short periods of time. As he explained,

> The interactions I had with my community were actually kind of shallow, I made no long-term friends that I kept in contact with until I was a teen. Most of my friends were not from school. They were from church, especially in the rural areas. No matter where we lived there was always a church and I found myself involved there more.

As to their current religious background, nine participants identified, at varying levels of intensity, with being either religious or spiritual. Participants generally viewed being religious as a practice and tradition with expected behaviors, as opposed to feelings or beliefs they held which were considered much more personal and spiritual in nature. Others saw religion as an all-encompassing term to mean the beliefs and practices of one who believes in a certain deity and the principles included therein. In many cases, the term Christian, born-again Christian, or Believer were specifically used by participants to self-identify not only as religious or spiritual, but as a certain type of believer, namely that of literalist evangelical Protestant Christianity attributed to the Southeastern region of the United States.

In each case, the participants articulated strong personal stances regarding the role of religious beliefs or religion in their lives as a part of who they were, rather than a superficial trait. Their self-defined religious beliefs centrally served as a moral compass that directed their decisions and actions. For some, this included a literal interpretation of religious text as a foundation of his or her belief system. Literal interpretation of events in the Bible, namely Genesis and the six days of creation, posed the greatest perceived conflict with evolution in the views of the participants, one of whom stated that "the Bible obviously tells us something different from evolution." The evolution concept being referred to in conflict with the creationist view was human evolution, specifically that human beings have evolved from other life forms as opposed to being specially created by the God of the Christian Bible.

Rejection of evolution is highly associated with conflicts between concepts and religious beliefs. The greatest perceived obstacle to accepting evolution found among participants was that of biblical creation and its direct contradiction to accounts of creation, particularly, human origins, as seen in the following:

Jacob   Macro-evolution conflicts with my personal beliefs because I believe that God created this universe and also created us and the other organisms that live here…I'm sticking to my belief that evolution occurs for survival to take place among the populations and species that God placed here on this earth. Evolution in the sense of organisms coming into existence from a common ancestor is false and God created the earth and its creatures.

William   You know, in Christianity the reason that there is death, the reason that people die is because of sin, because of Adam and Eve's first sin. It was God's original plan to never have death and… the Bible says because of sin there is death. So, if all of that is true, and there were millions of years before man even came and dinosaurs or you know, whoever was before man that would mean that there was a death before there was ever sin and I don't, I don't believe that there was death before there was sin.

The preceding statements were demonstrative of the level of conflict that surrounded evolution on a very personal basis among participants. Some examples of conflicting ideas the participants held were such a fundamental part of their closely-held religious belief system that acceptance of evolution would mean no choice but to reject their faith and personal beliefs. For these participants it was not simply a matter of misunderstanding or lacking scientific evidence supporting evolution so much as that consideration of any evidence for evolution was seen as questioning what they had accepted as a literal fact in their religious beliefs. As such, a large part of changing attitudes and openness to teaching and learning of evolution fell upon finding ways to mitigate the perceived conflict between science and religion. It was easily noted through the stories where participants would fall on the scale of acceptance based solely on their discussion of navigation of beliefs relative to evolutionary content. Those who could not reconcile, or who were unwilling to even consider evolution were on the low and very low ends. Those who were able to move beyond the conflicts they originally perceived between their beliefs and understanding were on the high and very high ends of the acceptance scale. Those in the middle were just that, those able to, at the least, compartmentalize their thinking about evolution and religion in a way that allowed them to consider both. As much as science can be rejected based on conflict with religion, science can also become a refuge on the opposite end of the spectrum from those who are turned away from their beliefs. One thing the narratives point to is that the struggle with evolution goes well beyond simple understanding and knowledge, although those two elements are also in need of address.

4.8 Conclusions

The dynamics of scientific and other worldviews, as seen here with Southern preservice teachers, are key to understanding the controversy surrounding evolution in the public sphere and the impact worldviews have on teaching and learning in classrooms today (Anderson, 2007; Glaze et al., 2015; Hansson & Lindahl, 2010; Hokayem & BouJaoude, 2008). Teachers, as well as students, "bring with them ideas and values about the natural world that they have formulated based on their own socio-cultural environment or from previous educational experiences" (Cobern, 1999, p. 1). Worldviews vary among individuals, however, certain patterns or themes can be found among those who are situated similarly in background, education, location, and religion as in the Southeastern United States (Kincheloe & Pinar, 1991). These worldviews may diverge from scientific worldviews and other ways of knowing and make conflict resolution among existing and new information very difficult (Griffith & Brem, 2004; Hermann, 2012; Sinatra, Brem, & Evans, 2008; Hermann, 2012).

As Cobern stated, "nowhere in science is the overlap between scientific ideas and other ideas in society more clear than with the theory of evolution. Evolution has acceptance problems because it is hard for students to accommodate the concepts of this theory within their cognitive culture" (1994b, p. 584). As a result, confrontation with de facto information that clashes with these ideas and values results in rejection and revolt simply because acceptance of conflicting ideas involves the restructuring of deeply held beliefs and even personal identity (Aikenhead, 2006; Hanson & Lindahl, 2010; Kahan, 2010, 2014; Shuumba, 1999; Wood, Douglas, & Sutton, 2012). The navigation of these conflicts throughout the lives of preservice teachers paints their choices regarding what and how they teach in their classroom (Glaze, 2013).

Making headway in improving evolution instruction—and by extension public scientific literacy—is much more complex than simply increasing training in content and pedagogy. When a teacher's own beliefs are contradictory to the science, even when it is in the standards, they often avoid or wash over the topic (Berkman & Plutzer, 2010; Borgerding, et al, 2015; Griffith & Brem, 2004; Meadows et al., 2000; Scharmann & Harris, 1992). For those who elect to teach so-called "controversial" topics, they often do so at their own risk—risk of job, risk of conflict with administration, and risk of social effects (Bloom, 2007; Borgerding, et al., 2015; Bramschreiber, 2013; Hermann, 2013; Moore & Kraemer, 2005). As such, teacher perceptions of evolution and how it will impact or be received by those they teach largely inform their choices whether and how to approach it in the classroom (Bowman, 2008; Catley, 2006; Veal & Kubasko, 2003).

4.9 Suggestions to Improve Evolution Education in the Southeastern United States

The goal for science education in the 21st century is science for all, to build a society that is scientifically literate and able to share in the logical decision making needed for society to thrive as we move forward (Natural Research Council, 2011). If pilot approaches are effective in the most difficult of audiences, then there is potential for the resulting strategies and interventions to be successful in regions where there is less controversy or in where there are similar underpinnings that inform teaching and learning. To improve the teaching and learning of evolution and, by extension, scientific literacy, scientists and science teacher educators alike must reevaluate the ways we are reaching the public, preparing teachers, and providing support that extends beyond the classroom. Decisions regarding the teaching of evolution have wide implications in that each teacher will, over the course of their career, impact thousands of students. Many of those students will receive their only exposure to science during those primary and secondary education years. For that reason, preservice education provides the most logical place to focus our attentions and have the widest reach for impacting science education.

In order to impact the teaching and learning of evolution on a wide scale in the future, it is imperative that pre-service science teacher training focus on multiple facets of the problem to address misconceptions, improve understandings, and develop approaches for responsive teaching in k-12 classrooms. First, it is important that the nature of science be included explicitly in both content courses and methods courses for science teachers. Misunderstandings of the nature and processes of science represent some of the most commonly used excuses for rejection of evolution and the only way to correct these misconceptions is to address them directly. Second, to counter known deficiencies in content when it comes to topics such as evolution that are frequently skimmed or skipped, evolutionary theory must play a larger role in the content training of pre-service science teachers. If we wish for evolution to be taught as the unifying theory that it is, pre-service science teachers must have a conceptually sound understanding of the content in that context. Third, worldviews must be considered and addressed in pre-service science teacher education. Specifically, pre-service science teachers should, in their education coursework, have opportunities to explore and define their worldviews, develop an understanding of the roles that worldviews play in their learning and in teaching, and develop strategies that are culturally and socially responsive to address some of those challenges. The goal is not taking away their worldview, but rather to make pre-service science teachers aware of their worldviews and those of others with whom they are engaged. The result of this more profound personal connection is a concerted effort to leverage knowledge and experiences, adapt instruction, and enhance explorations based on understandings of the dynamic nature of learning and the learner.

References

Adams, P. E., & Krockover, G. H. (1997). Concerns and perceptions of beginning secondary science and mathematics teachers. *Science Education, 81,* 29–50. https://doi.org/10.1002/(SICI)1098-237X(199701)81:1<29:AID-SCE2>3.0.CO;2-3.

Aguillard, D. W. (1998). *An analysis of factors influencing the teaching of biological evolution in Louisiana public secondary schools* (Doctoral dissertation or master's thesis). Available from ProQuest Dissertations & Theses database (UMI No. 9922044).

Aikenhead, G. S. (2006). *Science education for everyday life. Evidence-based practice.* New York, NY: Teachers College Press.

Aikenhead, G. S., & Jegede, O. J. (1999). Cross-cultural science education: A cognitive explanation of a cultural phenomenon. *Journal of Research in Science Teaching, 36*(3), 269–287.

Alabama State Religion. (n.d.). Retrieved April 10, 2017, from http://www.bestplaces.net/religion/state/alabama

Anderson, R. (2007). Teaching the theory of evolution in social, intellectual, and pedagogical context. *Science Education, 91,* 665–677.

Berkman, M., & Plutzer, E. (2010). *Evolution, creationism, and the battle to control America's classrooms.* Cambridge University Press.

Bloom, J. W. (2007). Preservice elementary teachers' conceptions of science: Science theories, and evolution. *International Journal of Science Education, 4,* 401–415.

Borgerding, L. A., Deniz, H., & Anderson, E. S. (2015). Evolution acceptance and epistemological beliefs of college biology students. *Journal of Research in Science Teaching.* https://doi.org/10.1002/tea.21374.

Bowman, K. L. (2008). The evolution battles in high school science classes: Who is teaching what? *Frontiers in Ecology and the Environment, 6*(2), 69–74.

Bramschreiber, T. L. (2013). Teaching evolution: Strategies for conservative school communities. *Race Equality Teaching, 32*(10), 10–14.

Branch, G. (2017). "The Cadillac of disclaimers": Twenty years of antievolutionism in Alabama. In C. Lynn, A. Glaze, L. Reed, & W. Evans (Eds.), *Evolution education in the American South: Culture, politics, and resources in and around Alabama* (pp. 103–119). New York, USA: Palgrave McMillian Publishers.

Branch, G., & Scott, E. (2008). Overcoming obstacles to evolution education: In the beginning. *Evolution: Education and Outreach, 1*(1) 53–55.

Catley, K. M. (2006). Darwin's missing link-a novel paradigm for evolution education. *Science Education,* 767–783. https://doi.org/10.1002/sce.20152.

Chinn, C. A., & Samarapungavan, A. (2001). Distinguishing between understanding and belief. *Theory into Practice, 40*(4), 235–241.

Cobern, W. W. (1994a) *Worldview theory and conceptual change in science education.* A Paper Presented at the 1994 Annual Meeting of the National Association for Research in Science Teaching. Retrieved on August 25, 2011, from http://www.wmich.edu/slcsp/SLCSP124/SLCSP-124.pdf

Cobern, W. W. (1994b). Belief, understanding, and the teaching of evolution. *Journal of Research in Science Teaching, 31,* 583–590.

Cobern, W. W. (1999). The nature of science and the role of knowledge and belief. *Science & Education, 9,* 219–246.

Daniel v Waters, 515 F.2d 485 (6th Cir. 1975)

Davson-Galle, P. (2004). Understanding: "Knowledge", "belief", and "understanding". *Science & Education, 13*(6), 591–598.

Demastes, S. S., Good, R. G., & Peebles, P. (1995). Patterns of conceptual change in evolution. *Journal of Research in Science Teaching, 33*(4), 407–431.

Deniz, H., Cetin, F., & Yilmaz, I. (2011). Examining the relationships among acceptance of evolution, religiosity, and teaching preference for evolution in Turkish pre-service biology teachers. *Reports of the National Center for Science Education, 31*(4), 1.1–1.9.

Deniz, H., Donnelly, L. A., & Yilmaz, I. (2008). Exploring the factors related to acceptance of evolutionary theory among Turkish pre-service biology teachers: Toward a more informative conceptual ecology for biological evolution. *Journal of Research in Science Teaching, 45*(4), 420–443.

Echevarria, D. (2013). *Speaking other languages in Alabama.* Retrieved April 10, 2017, from https://www.altalang.com/beyond-words/2010/05/06/speaking-other-languages-in-alabama.

Edwards v. Aguillard, 482 U.S. 578 (1987).

Elsberry, W. R. (2001). *Anti-evolution and the law.* Retrieved from http://www.antievolution.org/topics/law/.

Epperson v. Arkansas, 393 U.S. 97, 37 U.S. Law Week 4017, 89 S. Ct. 266, 21 L. Ed 228 (1968).

Fowler, S. R., & Meisels, G. G. (2010). Florida teachers' attitudes toward teaching evolution. *The American Biology Teacher, 72*(2), 96–99.

Freiler v. Tangipahoa Board of Education, No. 94-3577 (E.D. La. Aug. 8, 1997).

Gallop. (2014). *In U.S. 42% believe creationist view of human origins,* June 2014. [Data set]. Retrieved from http://www.gallup.com/poll/170822/believe-creationist-view-human-origins.aspx.

Glaze, A. M. L. (2013). Evolution and pre-service science teachers: Investigating acceptance and rejection. *ProQuest Dissertations and Theses.*

Glaze, A. L. (2016, March). Textbook evolution sticker hurts children's understanding of science but also their faith. Invited Op-Ed for Alabama.com. Retrieved from http://www.al.com/opinion/index.ssf/2016/03/textbook_evolution_sticker_hur.html#incart_river_index also available at http://evostudies.org/2016/04/sticker-shock-in-alabama/.

Glaze, A. L., & Goldston, M. J. (2015). Evolution and science teaching and learning in the United States: A critical review of literature 2000–2013. *Science Education, 99*(3), 500–518.

Goldston, M. J., & Kyzer, P. (2009). Teaching evolution: Narratives with a view from three southern biology teachers in the USA. *Journal of Research in Science Teaching, 46*(7), 762–790.

Griffith, J. A., & Brem, S. K. (2004). Teaching evolutionary biology: Pressures, stress, and coping. *Journal of Research in Science Teaching, 41*(8), 791–809.

Ha, M., Haury, D. L., & Nehm, R. H. (2012). Feeling of certainty: Uncovering a missing link between knowledge and acceptance of evolution. *Journal of Research in Science Teaching, 49*(1), 95–121.

Hanson, L., & Lindahl, B. (2010). "I have chosen another way of thinking", students' relations to science with a focus on worldview. *Science & Education, 19,* 895–918. https://doi.org/10.1007/s11191-010-9275-6.

Hermann, R. S. (2012). Cognitive apartheid: On the manner in which high school students understand evolution without believing in evolution. *Evolution: Education and Outreach, 5*(4), 619–628.

Hermann, R. S. (2013). On the legal issues of teaching evolution in public schools. *The American Biology Teacher, 75*(8), 539–543.

Hokayem, H., & BouJaoude, S. (2008). College students' perceptions of the theory of evolution. *Journal of Research in Science Teaching, 45*(4), 395–419.

Jakobi, S. R. (2010). "Little monkeys on the grass..." how people for and against evolution fail to understand the theory of evolution. *Evolution Education Outreach, 3,* 416–419.

Johnson, K. A., Hill, E. D., & Cohen, A. B. (2011). Integrating the study of culture and religion: Toward a psychology of worldview. *Social and Personality Psychology Compass, 5,* 137–152. https://doi.org/10.1111/j.1751-9004.2010.00339.x.

Jones, M. G., & Carter, G. (2007). Science teacher attitudes and beliefs. In S. K. Abell & N. G. Lederman (Eds.), *Handbook of research on science education* (pp. 1067–1104). Mahwah, NJ: Erlbaum.

Jorstad, S. (2002). *An analysis of factors influencing the teaching of evolution and creation by Arizona high school biology teachers* (Doctoral dissertation). Available through ProQuest Dissertations and Theses database (UMI No. 3053890).

Kahan, D. (2010). Fixing the communications failure. *Nature, 463*(7279), 296–297.

Kahan, D. M. (2014). Ordinary Science Intelligence: A science comprehension measure for use in the study of risk perception and science communication. In *The Cultural Cognition Project Working Paper, 112th ed. Yale Law and Economics Research Paper* (504).

Kincheloe, J. L., & Pinar, W. F. (Eds.). (1991). *Curriculum as social psychoanalysis: The significance of place*. Albany, NY: State University of New York Press.

Louisiana Science Education Act, Act 473, SB733 (2008).

Meadows, L., Doster, E., & Jackson, D. F. (2000). Managing the conflict between evolution and religion. *The American Biology Teacher, 62,* 102–107.

McLean v. Arkansas Board of Education, 529 F. Supp. 1255, 1258–1264 (ED Ark. 1982).

Miller, J., Scott, E., & Okamoto, S. (2006). Public acceptance of evolution. *Science, 313*(5788), 765–766.

Moore, R., & Kraemer, K. (2005). The teaching of evolution and creationism in Minnesota. *The American Biology Teacher, 67*(8), 457–466.

Nadelson, L. S. (2007). *Pre-service teachers' understanding of evolution, the nature of science, and situations of chance* (Doctoral Dissertation, University of Nevada, Las Vegas, 2007). Dissertations & Theses: Full Text database, AAT 3261083.

Nadelson, L. S., & Sinatra, G. E. (2008). Education professionals' knowledge and acceptance of evolution. *Evolutionary Psychology, 7*(4), 490–516.

National Center for Science Education. (2009a, March 11). "Academic freedom" legislation. Retrieved January 11, 2017, from https://ncse.com/creationism/general/academic-freedom-legislation.

National Center for Science Education. (2009b, March 20). Discovery Institute's "model academic freedom statute on evolution". Retrieved January 11, 2017, from https://ncse.com/node/11929.

National Research Council. (2011). *A framework for k-12 science education*. Washington, DC: The National Academies Press.

Nehm, R. H., & Reilly, L. (2007). Biology majors' knowledge and misconceptions of natural selection. *BioScience, 57*(3), 263–272.

Nehm, R. H., & Schonfeld, I. S. (2007). Does increasing biology teacher knowledge of evolution and the nature of science lead to greater preference for the teaching of evolution in schools? *Journal of Science Teacher Education, 18*(5), 699–723.

NGSS Lead States. (2013). *Next Generation Science Standards: For States, By States*. Washington, DC: The National Academies Press.

Pobiner, B. (2016). Accepting, understanding, teaching and learning (human) evolution: Obstacles and opportunities. *American Journal of Physical Anthropology, 159*(S61), 232–274.

Price, T. (2013, March 22). Science and religion. *CQ Researcher, 23*, 281–304. Retrieved from http://library.cqpress.com/.

Rice, J. K. (2003). *Teacher quality: Understanding the effectiveness of teacher attributes*. Washington, D.C.: Economic Policy Institute.

Rissler, L., Duncan, S. I., & Caruso, N. M. (2014). The relative importance of religion and education on university students' views of evolution in the Deep South and state science standards across the United States. *Evolution Education & Outreach, 7*(24). https://doi.org/10.1186/s12052-014-0024-1.

Rutledge, M. L., & Warden, M. A. (2000). Evolutionary theory, the nature of science and high school biology teachers: Critical relationships. *The American Biology Teacher, 62*(1), 23–31.

Selman v. Cobb County School District, 449 F.3d 1320 (11th Cir. 2006).

Scharmann, L. C., & Harris, W. M. (1992). Teaching evolution: Understanding and applying the nature of science. *Journal of Research in Science Teaching, 29*, 375–388.

Shankar, G., & Skoog, G. (1993). Emphasis given evolution and creationism by Texas high school biology teachers. *Science Education, 77*(2), 221–233.

Shipman, H. L., Brickhouse, N. W., Dagher, Z., & Letts, W. J., IV. (2002). Changes in student views of religion and science in a college astronomy course. *Science Education, 86*, 526–547.

Shuumba, O. (1999). Relationship between secondary science teachers' orientation to traditional culture and beliefs concerning science instructional ideology. *Journal of Research in Science Teaching, 36*(3), 333–335.

Sinatra, G., Brem, S., & Evans, M. (2008). Changing minds? Implications of conceptual change for teaching and learning about biology. *Evolution Education and Outreach, 1*(2), 189–195.

Smith, M. U. (2010a). Current status of research in teaching and learning evolution: I. Philosophical/epistemological issues. *Science & Education, 19*, 523–538. https://doi.org/10.1007/s11191-009-9215-5.

Smith, M. U. (2010b). Current status of research in teaching and learning evolution: II. Pedagogical issues. *Science & Education, 19*, 539–571. https://doi.org/10.1007/s11191-009-9216-4.

Teacher protection and academic freedom act, SB0893/HB0368 (2012).

Trani, R. (2004). I won't teach evolution, it's against my religion: And now for the rest of the story. *American Biology Teacher, 66*(6), 419–427.

Urban, W. J. (1992). A curriculum for the south. *Curriculum Inquiry, 22*(4), 433–441.

U.S. Census Bureau. (2016). *Alabama quick facts*. Retrieved from https://www.census.gov/quickfacts/table/PST045216/01.

Veal, W. R., & Kubasko, D. S. (2003). Biology and geology teachers' domain-specific pedagogical content knowledge of evolution. *Journal of Curriculum and Supervision, 18*(4), 334–352.

Wenner, G. (1993). Relationship between science knowledge levels and beliefs toward science instruction held by preservice elementary teachers. *Journal of Science Education and Technology, 2*(3), 461–468.

Whitaker, W. (2010). All things pointing home. *Journal of Curriculum Theorizing, 26*(2), 123–128.

Wiles, J. R. (2008). *Factors potentially influencing student acceptance of biological evolution* (Doctoral dissertation or master's thesis). Available from ProQuest Dissertations & Theses database.

Wilson, C. R. (1996). The south, religion, and the scopes trial. *Letters Archive, 4*(2). Retrieved from http://www.vanderbilt.edu/rpw_center/_archived-letters/scope.htm.

Wood, M. J., Douglas, K. M., & Sutton, R. M. (2012). Dead and alive: Beliefs in contradictory conspiracy theories. *Social Psychological and Personality Science, 3*(6), 767–773.

Woods, C. S., & Scharmann, L. C. (2001). High school students' perceptions of evolutionary theory. *Electronic Journal of Science Education, 6*(2). Retrieved on September 13, 2011 from E:\EvoResearch\Electronic Journal of Science Education V6, N2, December 2001, Woods and Scharmann.mht.

Amanda L. Glaze is an Assistant Professor of Middle Grades & Secondary Science Education at Georgia Southern University. She teaches undergraduate and graduate level courses in science education and leadership at GSU and is currently involved in a National Science Foundation DRK12 grant seeking to improve the teaching of evolution in the Southeastern United States. Her research agenda targets the teaching and learning of evolution where there is conflict with cultural beliefs and public opinion as well as culturally responsive teaching practices and high-quality teacher education practices for science education.

M. Jenice Goldston is a Professor Emerita at the University of Alabama (UA). She is currently involved in a National Science Foundation MSP grant and book writing. She has taught science education courses from undergraduates to doctoral levels. Her research has centered on the preparation of high quality science teachers and their on-going professional development; sociocultural-political influences (diversity and equity) affecting teaching and learning science; reform in teacher education and undergraduate science teaching in higher education; university-school partnerships and reform; and currently she is involved with research in teaching and learning evolution.

Chapter 5
Case Studies in Teaching Evolution in the Southwestern U.S.: The Intersection of Dilemmas in Practice

Rachel J. Fisher

Abstract Over the last century, anti-evolution sentiment has been reflected in major legislative initiatives across the United States, including Arizona. Despite recent science education reform documents citing evolution as a core concept to be taught in grades K-12 in the U.S., research shows problems with how it is currently taught. Evolution is often avoided, teachers minimize its importance within biology, infuse misconceptions, and/or interject non-scientific ideologies into lessons. The current study focused on how teachers in two geographically and culturally distinct school districts in Arizona negotiated dilemmas during an evolution unit. One district was rural with a large population of Mormon students, while the other was urban, with a majority of Mexican/Mexican-American students. Using a case study approach, I observed three biology teachers during their evolution lessons, interviewed them throughout the unit, co-planned lessons with them, and collected artifacts, including anonymous student work. I also included data from genetics lessons for each teacher to determine if the issues that arose during the evolution unit were a result of that teacher's practices, or if they were unique to evolution. Findings showed teachers' backgrounds and comfort levels with evolution, in addition to their perceptions of community context, affected how they negotiated pedagogical, conceptual, political, and cultural dilemmas. This study's findings inform in-service teachers' future practice and professional development tools to aid their teaching; this includes methods to negotiate some of the political (e.g. state standards) or cultural (e.g. religious resistance) issues inherent to teaching evolution.

R. J. Fisher (✉)
Amplify Education, Inc., Atlanta, GA, USA
e-mail: racheljfisher2@gmail.com

5.1 Introduction

Evolution is the unifying concept in biology, and as Dobzhansky (1973, p. 125), a distinguished evolutionary geneticist noted, "Nothing in biology makes sense except in the light of evolution." Despite the overwhelming *lack* of controversy among scientists regarding the scientific validity of evolutionary theory, the social controversy that accompanies evolutionary theory affects if and how this topic is currently taught in many science classrooms in the United States. Public skepticism of evolution is alive and well today. In a Gallup poll from 2014, 42% of Americans held a creationist view that humans were created in their present form during one event within the last 10,000 years, and one-third of all Americans held the view that God guides evolutionary processes ('theistic evolution'). Nineteen percent held a more 'secular' view in that humans evolved and God had no role in this process (Gallup, 2014).

5.2 Anti-evolution Legislation in the Southwest

This skepticism of evolution, and even clear anti-evolution sentiment by a segment of the public, has been reflected in major legislative documents across the U.S. over the last century. The history of anti-evolution legislation in Arizona, more specifically, mirrors that which took place in the rest of the country from the 1920s up to today. Arizona's political conservatism, more generally, has provided fertile ground for heated discussion over the teaching of evolution in public schools (Webb, 1981). For example, in 1927, Reverend R.S. Beal of Tucson's First Baptist Church, proposed a law in Arizona that was similar to the Tennessee Butler Act of 1925, attempting to outlaw the teaching of evolution. He rejected evolution as a science and categorized it as a religion that, "...humanizes God and defies humanity," (Webb, 1981, p. 139). This attempt at legislation never gained much ground, and similar to the rest of the country, Arizona lost interest in this issue for the next few decades (Webb, 1981).

As a result of the Soviet Union's 1957 launch of Sputnik, education experts in the U.S. re-examined the state of science education (Bleckman, 2006; Scott, 2009). Consequently, the National Science Foundation (NSF) established the Biological Sciences Curriculum Study (BSCS) that created three science textbooks, all of which emphasized the role of evolution as a major concept in the biology curriculum. Evolution was now prominent in public science education, and once again back in the public eye (Scott, 2009). Similar to the rest of the U.S., the 1960s saw a resurgence of the evolution debate in Arizona public schools. Phoenix area schools were one of the testing grounds for the BSCS biology curricula. Some parents spoke out against the emphasis of evolution in science classes. For example, in 1963, a Phoenix parent asked the state board of education if his student could leave class during the evolution unit—the board shifted responsibility to the local school district, and ultimately, the parent was allowed to excuse his student from this unit (Webb, 1981).

Some of the fiercest opposition in Arizona during the 1960s stemmed from three Mormon Stake Presidents, including Junius E. Driggs. They wrote letters opposing the teaching of evolution to local school superintendents, and to the *Arizona Republic* newspaper. In the latter, Driggs stated that the teaching of evolution "is a very dangerous situation and in view of the fact that the theory has not been established as a fact, we think it should not be taught in schools," (Webb, 1981, pp. 141–142). The school superintendent responded to this opposition by stating that students were not required to take the course where evolution was taught, and that they were not expected to "believe" it if they did enroll (Webb, 1981). Additionally, in 1964 and 1965, the Arizona State Legislature saw the introduction of two bills similar to one another (House Bill 301 and Senate Bill 172, respectively), both requiring equal time for the teaching of evolution and divine creation, neither of which passed (Webb, 1981; Wilhelm, 1978).

From the 1970s until today, resistance to teaching evolution continues to be alive and well in Arizona. In 1976, a Republican Congressman, John Conlan, sponsored an anti-evolutionary amendment to the National Defense Education Act. This amendment would "...prohibit federal funding of any curriculum project with evolutionary content or implications," (Moore, Decker, & Cotner, 2009, p. 277). This passed the House, but was defeated in the Senate (Moore et al., 2009). Later, in 2004, around the same time as the *Kitzmiller* intelligent design case in Dover, Pennsylvania, Arizona's state board of education was lobbied (unsuccessfully) to include an order for science teachers to discuss intelligent design in the state science standards (National Center for Science Education, 2013). As recently as 2013, Arizona introduced an antiscience bill (Senate Bill 1213), with similar academic freedom type prose as was introduced in other bills in the U.S.; however, the bill died and was never enacted (National Center for Science Education, 2013).

5.3 Educational Problem

Despite several legislative attempts to thwart the teaching of evolution in public schools in Arizona and the rest of the U.S., educational reform documents highlighted the importance of teaching this topic. The *National Science Education Standards* (NSES) (National Research Council, 1996), and the *Benchmarks for Science Literacy*, developed by the American Association for the Advancement of Science (AAAS) in 1993, both underscored evolution. More recently, the *Next Generation Science Standards* (NGSS)—which are based on the *Framework for K-12 Science Education* and replaced the NSES—highlight the need for change in science education due to advances in scientific research and knowledge of how students learn science (National Research Council, 2012). The *NGSS* stress evolution as a core concept throughout grades K-12 (National Research Council, 2011).

Although national science education reform documents exist that emphasize the importance of teaching evolution in U.S. biology classrooms, recent research nationwide has shown that many teachers avoid the topic altogether, only teach the tested concepts within evolution, minimize the importance of this unifying theme of biology (Berkman & Plutzer, 2010), and/or infuse misconceptions into lessons (Smith, 2010). If evolution is taught at all in science classes, the topics are limited in scope, focusing more on microevolutionary processes (at the species level—e.g. only teaching natural selection), rather than emphasizing macroevolution (e.g. speciation). When educators (e.g. teachers, curriculum developers, etc.) leave out major components of evolutionary theory, this results in an incomplete and/or incorrect understanding of the evolutionary process (Catley, 2006).

5.3.1 The State of Arizona

Within the state of Arizona more specifically, the treatment of evolution is given marginal attention in the state's educational standards. Although a superficial standard exists in 8th grade for coverage of 'factors that allow for survival of organisms,' (Arizona State Board of Education, 2005), evolution is not explicitly stated in the standards until high school. And within the high school standards, 'biological evolution' is one of four major concepts of the life sciences, with an emphasis on natural selection and adaptations (Arizona Department of Education, 2005). Fewer than 10% of the questions on the high school state science exam include questions on evolution (Arizona Department of Education, 2009). As of 2017, the state had not yet adopted the *NGSS*, and is undergoing a revision process of its state science standards.

Additionally, there is a dearth of research on how teachers approach evolution in Arizona. An exception is Griffith and Brem's (2004) work with secondary biology teachers' implementation of their evolution units in Arizona schools. These teachers discussed the sources of pressure, the stress that results, and the coping strategies teachers use (in the form of pedagogical strategies) to deal with stressors when teaching evolution. Results showed that teachers' responses placed them into three different categories—'conflicted' teachers, who struggle with their own beliefs and how their teaching of evolution will impact students; 'scientist' teachers, who have no internal stressors and who see no place for social issues in their classroom; and 'selective' teachers, who avoid difficult topics and situations.

5.3.2 Cultural Issues in Evolution

Most of the culturally-based studies on the teaching of evolution do not address issues other than religion. For example, the focus of a paper may be on the fundamentalist Christian religiosity of teachers and/or students, but there is very little consideration of ethnicity in context as well. One of the contexts of the current study was in a community with a large population of Mexican/Mexican-American students. With this population comes not only different language, but also different discourse about scientific concepts than students who identify as White and American born. In studies that approach the issue of religiosity, most focus on conservative fundamentalist Christian communities due to their history of resistance to evolution (an exception are the few studies on Islamic beliefs and the teaching of evolution, which have taken place in Lebanon (Dagher & BouJaoude, 1997), Pakistan (Asghar, Wiles, & Alters, 2010), and Turkey (Deniz, Donnelly, & Yilmaz, 2008). However, a second context for the current study is in a community in the southwest with a large population of Latter-day Saints (LDS, or Mormon) students and families. Although this faith has a history of conservative values, there is a lack of research on how the LDS community (teachers and/or students) approaches/ responds to evolution in the classroom. Arizona's proximity to Utah (where a large population of LDS individuals resides) makes this second context unique as well.

5.4 Arizona: Demographics and Education System

In 2016, Arizona's population was estimated at 6.9 million (U.S. Census Bureau, 2016). The primary language spoken is English, with 75% of the population five years and over speaking this language. The second most prevalent language spoken in Arizona is Spanish (Modern Language Association, 2010). Approximately two-thirds of adults in the state (67%) identify as Christian (primarily Evangelical Protestant and Catholic), and 'unaffiliated' (atheist or agnostic) adults comprise 27% of the population. Only 5% of the adult population in Arizona identifies as LDS (Pew Research Forum, 2014c).

Most of the public schools in Arizona are divided by elementary (Kindergarten through fifth grade—kindergarteners can enroll at age 5), middle (sixth through eighth grades), and high school (ninth through twelfth grades). However, this will vary with school district, albeit slightly; for example, some districts include sixth grade in their elementary school, and then middle school will only comprise seventh and eighth grades. Children ages six through sixteen (or 10th grade) are covered under Arizona's compulsory education laws and therefore must attend a public school. Students that are homeschooled are required to receive an education that is equal to what they would receive at a school (Education Commission of the States, 2010).

At the end of the year in grades 3–8, students take the state AzMERIT exam for reading and math. In high school, the AzMERIT End-of-Course tests are English Language Arts (grades 9, 10, and 11), Algebra, Geometry and Algebra II. Arizona's Instrument to Measure Standards (AIMS) is currently the only state science exam, and is given at the end of fourth and eighth grades, and once in high school (Arizona Department of Education, 2016). The grade in which the AIMS science test is given in high school varies by district. Neither the AzMERIT nor the AIMS science exams are requirements for graduation.

5.4.1 The Cultural Context of the Study

The current study focused on how high school biology teachers in two geographically and culturally distinct school districts in Arizona negotiate dilemmas during an evolution unit. As noted above, this area of the country has a history of both political conservatism and anti-evolution legislative attempts that mirrors what took place in the U.S. from the 1920s until today. One school district in the current study was rural and had a large population of students affiliating with the Mormon faith, while the other district was urban, with a large majority of Mexican/Mexican-American students. Neither of these populations had been studied in-depth in the context of evolution education prior to this project.

Although the specific religious demographics of the Latino population are not currently known, a recent Pew Survey (2014a) cited a shift away from Catholicism for a large portion of this country's Latino population, moving towards Protestantism. Almost one-fourth of Latinos in the U.S. consider themselves Protestant, while a good portion of those identified as such considers themselves evangelical. Historically, Protestant evangelicals have been one of the strongest opponents of evolution, citing the literal interpretation of the Bible instead (Scott, 2009).

Mormons have historically espoused both conservative values along with conservative leaders, similar to conservative Protestants; however, documents show a range of reactions by LDS individuals to evolution (Eddington, 2006). For example, in both 1909 and 1925, Presidents from the Church of Jesus Christ of Latter-day Saints released public statements on evolution. During both of these declarations, they stated, "Man is the child of God, formed in the divine image and endowed with divine attributes" (Pew Research on Religion and Public Life Project, 2014b). Both of these declarations carried with them an anti-evolution rhetoric. A recent Pew Survey (2016) showed that 52% of Mormons nationwide reject evolution, and believe that humans have always existed in their present form. However, a Biology Professor from Brigham Young University, Duane Jeffery, and a Physics Professor from Utah Valley State College, William Evenson, both agree that the church does not espouse a firm position on evolution (Eddington, 2006). Currently, leaders of the Church have not adopted a firm position on evolution either.

5.5 Case Studies: Ben, Jane, and Diane

In the current study, I used a case study approach (Yin, 2014) to examine the conceptual, pedagogical, cultural, and political dilemmas of practice of three teachers during their evolution units. These dilemmas were described by Windschitl

(2002) in the context of teaching in a constructivist manner; however, in the present study, I apply these terms more broadly with respect to teaching evolution. Case studies have increasingly been used as a research tool and in a variety of situations; they have helped researchers better understand the knowledge of individual, group, political, or socially related phenomena. This approach allows the researcher to retain meaningful characteristics of real-life events (e.g. how a neighborhood changes, or in the current research, how teachers approach evolution). Case studies should also be used when contextual conditions are pertinent to a study (Yin, 2003) (e.g. community in which evolution is taught), and when looking for interactions within a context (Stake, 1995).

I observed three biology teachers during all of their evolution lessons, interviewed them throughout the unit, co-planned lessons with them, and collected artifacts from this unit, including anonymous student work. I also observed four genetics lessons per teacher to determine if they changed their pedagogy when they taught evolution (all teachers taught genetics just prior to evolution). I used Lemke's (2001) sociocultural theory in science education as a theoretical lens—this includes thinking about how the subculture of science education fits into the cultural ecology of a larger community, and with other subcultural systems in which it is aligned or in conflict. This lens also considers how science education is, as a community, dependent on economic and political forces outside it, and how it resists/accommodates to this dependence. These two areas of Lemke's (2001) work are interesting as they relate directly to issues of this study on teaching evolution. When considering teaching this unit, it is important to take into account the community in which it is taught (and the religiosity therein, more specifically) and the potential conflicts that may arise. The practice of teaching evolution is also dependent on political forces, and how each teacher/school deals with this is also unique. Finally, Lemke (2001) argues about the importance of including sociocultural theory as a lens in research since science education affects not only individuals, but it has political, economic, and cultural implications as well.

In the sections that follow, I discuss the findings from each teacher, and then present a brief summary. I show how each teacher dealt with intertwined dilemmas, which resulted in evolution units unique to each teacher and school context. Table one below briefly describes the background and community/school context for each teacher—Jane, Ben, and Diane. These three teachers were chosen based on convenience, as they were the sole respondents to inquiries for participation in this study.

5.5.1 The Case of Jane

Jane taught for ten years in a rural community where she also lived. Despite her experience teaching various subjects during this time, only recently had she started teaching biology.

Pedagogical issues. Jane's general pedagogy did not shift much between units (evolution and genetics), as she emphasized the use of videos and textbooks/associated worksheets as tools throughout. However, she employed controversy avoidance strategies only during her evolution unit. In an interview, Jane specifically stated that she kept evolution "paced and factual" to prevent any student from arguing—this was supported by my observations of her unit. She gave students very little time to ask questions during any of the evolution lessons. Jane was clearly uncomfortable including human evolution in her unit, which she thought might cause a 'stir' among her LDS students who were "likely creationists." These pedagogical strategies were a result of the following issues—Jane's perceptions of her students; her lack of comfort with evolution specifically (this unit was one she felt "least comfortable teaching"); and her general unfamiliarity with the biology content (evidenced by her lack of educational background in this content, and inexperience teaching it-see Table 5.1).

Conceptual issues. Jane's evolution unit focused primarily on microevolution, as she emphasized Darwin and natural selection throughout most of the unit. So much of the content within the videos and worksheets focused on Darwin's journey, that it felt repetitive at times. All the while, she excluded any concepts relating to deep time or the history of Earth. Jane's lack of biological sciences background may have at least partially prompted the omission of these latter concepts. The content Jane chose to exclude from her unit was another example of how she avoided potential controversy. As mentioned previously, she taught in an environment where many of the students and their families were "likely creationists." The potential controversy that could arise as a result of introducing ideas relating to speciation affected her content choice during her unit. Additionally, Jane's own limited knowledge and lack of comfort with evolution (including inadequate comprehension of more complex topics such as speciation, per several discussions and interviews), influenced her content as well.

Political issues. A significant component of the dilemmas Jane faced during her teaching of biology, in general, and more specifically during the evolution unit, related to political issues. Although vocal parental resistance to evolution was not much of a problem for Jane, the administration had a significant influence on many aspects of her teaching. It was clear from my interactions with her that the administrative directives were unwelcomed since they reduced her autonomy as a teacher. These included required weekly 'data' meetings, the development of a curriculum map, the daily documentation of standards, and thus, the tests that were ultimately aligned with these standards. As she noted, the "standards and tests definitely drive content taught."

Prior to the beginning of the school year, Jane and other biology teachers district-wide identified standards from those developed by the state they deemed as 'most important' to teach. These eventually became the 'Essential Questions' and 'Big Ideas' for her lessons, as mandated by the district. Although I would argue that Jane was somewhat in control of the standards, the initial requirement of reducing these standards was not her choice. Jane also discussed how in evolution and other units the standards limited the content she could teach. Ultimately, though, Jane

Table 5.1 Background of teachers and school/community context

	Jane	Ben	Diane
Background	– BS Community Health – MS Elementary Education – Certified in Special Education, Secondary Education, History, Geography, General Science and Biology – Social worker	– BS Molecular Biology (with Honors) – MS Ecology & Evolutionary Biology – Certified in Biology	– BS Biology/Certification to Teach Biology; BA in Painting
Teaching experience	– 10 years (4 Years Biology)	– 2 Years (General and AP Biology)	– 3 Years (2 Years Biology)
School/student context	– Rural Public School – Title I – 62% White; 30% Hispanic – Latter-day Saints (LDS or Mormon) Religion Prevalent Among Students	– Urban Public School – Title I – 57% Hispanic; 37% White	– Urban Public School – Title I – 57% Hispanic; 37% White

used the standards to justify teaching different content in biology, including evolution, to the administration. The administration required Jane to provide a list of these standards, Essential Questions, and Big Ideas at the beginning of the year—she posted the latter two daily on her front board solely for administration purposes.

As noted above, several mandatory tests were an important part of Jane's curriculum. She administered both an evolution mid-term and the Performance Based Pay (PBP) test (both of which were district specific) to her students during the current research project. She was careful to focus her evolution lessons on topics that would not only be in the standards, but also on the tests—the latter test (PBP) was especially important to Jane since it would result in extra pay if her students did well. Her careful exclusion of particular topics from in-class lessons (such as cladograms and Hardy-Weinberg) were largely a result of the content in the standards and thus, the test questions.

Cultural issues. In this case study, Jane showed evidence of how the community where she lived and worked for over a decade, which had a large population of LDS students, affected several decisions relating to her evolution unit. Jane's perception of her students as "pretty heavy into creationism" was confirmed as accurate by an LDS expert in my study. As the expert noted, despite lack of an official church position on evolution (Evenson & Jeffery, 2005), LDS students in more rural areas were likely to be creationists. Jane also noted that her LDS students were some of her best academically—they did well in class, and rarely voiced resistance to evolution. However, according to Jane, students' lack of vocal resistance in class, despite potentially disagreeing with the evolution content, was due to their upbringing and teachings to respect authority.

Jane made several pedagogical and conceptual decisions that were at least partially a result of her perceptions of her students' beliefs as creationists. Her conceptual focus on microevolution, with a clear emphasis on Darwin and natural selection, prevented her from diving into ideas that could be viewed as more personal to her students. Jane chose not to teach human evolution, and even chose not to review a video on this topic for use in her classroom (despite its presence on her desk). When I asked how she would decide whether or not to include this video in the future, she wanted to ensure it was "not anti-creationism" because she did not want to "offend her students." However, at no time did she engage in conversations with students regarding their beliefs, as she clearly understood that it was "illegal to teach creationism."

Jane's overall approach to evolution was one that clearly took into account community religiosity. On multiple occasions, she acknowledged that many individuals in her community believe that "teaching evolution interferes with their religious beliefs," and cited the possibility of a community member becoming her boss one day. These contextual factors, along with her own upbringing, (which included being taught creationism at a Lutheran Missouri Synod school for several years,) affected her overall views on issues of religion and science. When she discussed how she dealt with a resistant student outside of class, Jane's response to the student highlighted her own view of evolution and religion as complementary. She mentioned evolution as "marrying" with creationism and how they can be "unified."

5.5.2 The Case of Ben

Ben was in his second year of teaching during this study—unlike Jane, he did not live in the community in which he taught. He came to teaching with a Master's in evolutionary biology, which included experience in this field of research.

Pedagogical issues. Ben's pedagogy did not shift between units, as he included interactive PowerPoint lectures and bell work, along with labs, during both the evolution and genetics units. Despite this, interviews and observations clearly showed he was *more* passionate about evolution than other topics he covered in class, including genetics. This was evident in his pedagogy from day one of the unit —he told the students how excited he was to start the topic, and emphasized his goal for them—"to view the natural world through the eyes of evolution." This goal stemmed from his sincere passion and in-depth knowledge of the subject. Ben took this goal seriously and sincerely wanted his students to reach this objective. Ben used bell work (at the beginning of class) as a time for students to discuss their ideas about previously learned evolutionary concepts. He created a safe space for students to engage in discussions, and provided the means for them to demonstrate if they achieved the goals he set for them. Ben allowed all students to answer questions, and when students' responses were incorrect, he used a positive scaffolding technique rather than rejecting their responses outright and making them feel as if they could not be wrong. However, he clearly struggled internally to manage his in-depth knowledge of and passion for evolution (due to his graduate level work in the field), along with expectations of his first and second year students in an introductory level biology course. As a result, this unique internal struggle made the teaching of evolution difficult for him.

Conceptual issues. During his evolution unit, Ben included a diversity of topics that were based both in micro- and macroevolution—he clearly placed importance in understanding a range of concepts in order to understand the entirety of evolution. Ben seemed apprehensive about including too many ideas relating to microevolution since it was very "tedious" and it "requires some math" (this was in reference to gene frequency calculations); his perceptions were that his students generally had low math abilities. As such, Ben shied away from teaching too much microevolution because it required particular skills that he felt many of his students lacked. Ben's ample scientific background and the confidence he exhibited with the content throughout the evolution unit are evidence that his decisions to choose particular concepts to teach was unrelated to any potential conceptual misunderstandings.

Within the diversity of topics Ben covered during his evolution unit, he stressed ideas relating to macroevolution. For example, he deemed the 'types of evolution' a "keystone" lesson for the unit since it shows students how organisms become different from one another. This idea, along with the infusion of other macroevolution concepts, such as the lesson on Hominid evolution, and discussions around the idea of common ancestry, showed Ben's comfort with these topics. He even discussed the 'types of evolution' lesson as one topic within evolution he felt most comfortable teaching. One concept Ben excluded related to geologic time—he felt it did not "fit" in with the evolution unit. At no time did he state any apprehension teaching this concept because of a lack of background knowledge.

Political issues. The political issues observed at Ben's urban school were in direct contrast to those at Jane's rural school. I argue that political issues had little effect on Ben's teaching, compared to other dilemmas (other than the administration of the AzMERIT state test, which caused a brief interruption to the evolution unit, and will be discussed below). Ultimately, this lack of administrative micromanagement provided Ben with a great degree of autonomy. Additionally, Ben did not regularly used standards in this biology course—he did not have any standards written on the board, nor were they written on any artifacts for either genetics or evolution units. I probed his views on standards by asking specific questions in interviews since he did not mention them in class or during informal discussions. It was clear that Ben felt that the Arizona State Standards were obsolete; although the newer *Next Generation Science Standards* (NGSS) were moving in the direction of how he teaches, Ben did not directly incorporate them into his lessons. Spending time aligning standards with lessons was not something Ben wanted to do, and since he had a lot of autonomy at his school, he chose not to do so.

There was also a lack of administrative pressure at Ben's school to teach content that was on a biology standardized test. He rarely spoke about tests since his students did not have to pass the state science test (AIMS), as the test results did not affect the students' grades or graduation, and had no impact on teacher evaluations (per the administration at his school). The only administrative directive that impacted Ben's unit, albeit in a slight way, was the last-minute decision by the administration to issue a new state standardized test called AzMERIT. Although he took time away from his 'official' evolution unit during testing (since students alternated between testing while others were present in the class), Ben used it as a time to apply evolution concepts with review activities. Ben's comfort and background in the topic may have resulted in the lack of pressure he felt regarding this last-minute administrative directive.

Cultural issues. Although Ben taught at a Title I urban school with primarily Mexican/Mexican-American students, he never outwardly acknowledged much about his own religiosity/cultural upbringing or practices, or that of his students. As such, there was no way to know whether or not he was Mexican-American himself. Alternatively, Ben's lessons primarily included connections to his perceptions of students' knowledge of biology, and their lives/interests as teenagers more generally (e.g. including video games, musicians, etc.). Despite the majority of his students being Latino/a, when I asked him to describe his students, never once did he mention this cultural identity. Rather, when he initially discussed student demographics, he described them based on the two groups at his school—"One is a population that comes from a little further north. And they are coming from a more affluent area. And one is a population that is coming basically from a 2.5 mile radius around the school… from an area a little bit closer to the school that is much, much less affluent… but we have students whose parents are going to check their grades and call them out if their grades don't [sic], if they aren't in the right place.

And then we have students where that's not the case." Most of the students (~70%) in his regular biology class live in the less affluent area near the school.

Ben never discussed his own religious upbringing, and only once did he reference his perception of his students' religion. He speculated that his students were taught not to believe in evolution, but he provided little evidence of this being the case. However, Ben was clearly aware of the social controversy surrounding evolution, as was evident in his bell work on the first day of the evolution unit. At this time, he asked the students what they heard, good or bad, about evolution, and then discussed it. Ben's reason for discussing the science and religion issues upfront was to 'get them out of the way.' He was aware of the social controversy, and indirectly made it clear in class that the students should not discuss these ideas at any other time during the unit. Ben's discussion with students included his awareness of a potential conflict between "strongly held beliefs" and evolution. However, he clearly delineated between science and religion, and mentioned how students had to learn evolution "for the test." However, I would argue that this latter statement invoking the test was not attempting to undermine science—most of the themes from this study relate back to Ben's passion for evolution, and the importance he placed upon his students' viewing all things in the natural world through the eyes of evolution. As a result, the purpose of this brief discussion was to keep the focus on the science during the unit, as Ben appeared uncomfortable discussing the science/religion issue on this day. The remainder of his unit focused solely on the scientific content, and the central role of evolution as a powerful scientific explanation.

5.5.3 The Case of Diane

Diane taught the same biology course as Ben, only a few doors down at the same urban high school. Despite being in her third year of teaching, this was her second year of teaching the biology course.

Pedagogical and conceptual issues. Because Diane's pedagogical and conceptual issues were highly intertwined, I will discuss them in context of one another. Other than Diane and Ben's shared passion for and comfort with evolution, *how* Diane approached the evolution unit was quite distinct from Ben. Diane's pedagogy was quite similar between her genetics and evolution units. For example, she started all of her classes with bell work where she allowed for ample student discussion, and therefore scaffolding of student knowledge in both units. She was clearly comfortable with the topics and allowed for co-construction of knowledge with her students during bell work and the interactive lectures and/or activities that normally followed. Additionally, there was nothing apparent during evolution that caused discomfort or reason for her to change how she taught during this unit, as compared to genetics. It was clear from my interactions with Diane that she thoroughly enjoyed the topic of evolution, referencing her focus on evolution and ecology during her undergraduate work in biology.

Diane's main goal for the evolution unit was to help her students 'dispel myths,' and correct and prevent the perpetuation of misconceptions about the topic. She thought it was important for students to be mindful of the general public's misuse of scientific terms, and overall distrust of science. To this end, helping the students identify and 'debunk' myths would enable them to handle any incorrect knowledge they heard outside of class. Diane corrected students throughout the unit if they used language that inferred any type of misconception, such as assigning purpose to organisms. Her approach was to stop the students in a kind way, ask them questions about what they said, and help scaffold their misconception into a correct response. She regularly revisited the meaning of the terms 'adaptation' and 'fitness' throughout the evolution unit, primarily in context of their differing colloquial and scientific uses.

The conceptual focus of Diane's unit was primarily on microevolution concepts, with an emphasis on genetic (allele) changes in species over generations (which was evident in the labs, the bell work and other in-class work). However, one reason she focused on microevolution stemmed from her inexperience as a teacher, coupled with the fact that she only taught two classes of biology, and four classes of environmental science. Thus, biology was not at the forefront of her curriculum development focus. Although phylogenetics was a topic she mentioned as most comfortable teaching within evolution, she did not include it in her unit solely because of time restrictions. Thus, the exclusion of such macroevolutionary principles was not due to a lack of comfort with or inadequate conceptual background in the content area.

Because Diane only taught two biology classes, and considered it 'secondary' to her environmental science class, she relied heavily on an informal mentor teacher at her school for the content and pedagogy of biology, including evolution. From my in-depth discussions with Diane, it was clear that her decision to rely upon this mentor did not originate from lack of confidence or comfort with teaching any areas within biology. This mentor was an experienced biology teacher whom Diane trusted—she regularly spoke highly about her, both with regard to her content knowledge and types of activities she provided to Diane. As she stated in an interview, "I know that she understands the breadth of the whole year and the scope of what's happening throughout the year. So I feel very confident using her resources and just knowing it's all going to work out."

Political issues. As aforementioned with the case of Ben, the urban school where he and Diane taught put few, if any, administrative pressures on either of these teachers. Biology teachers, such as Diane and Ben, were not under pressure for their students to pass a state science test, as the results did not affect the students or teachers. Ultimately, this lack of pressure provided both teachers with autonomy to teach what they want and how they wanted to teach it, including the evolution unit. Not being micromanaged is "wonderful" according to Diane. Similar to Ben, Diane rarely used standards (neither state nor national) in her biology course. Only when I probed her thinking about standards, more generally, during an interview did her viewpoint on them arise—Diane clearly thought the *Next Generation Science Standards* represented the direction in which science teaching should move, and her administration supported this idea; but again, she chose not to use them to create lessons. The administration supported the teaching of evolution, and if the teachers did not teach it (according to Diane), *that* decision would cause an issue.

With Ben's case, I had discussed the last-minute decision by the administration to issue a new state standardized test (the AzMERIT). Since all biology teachers were affected by this decision, Diane was impacted as well. However, she dealt with this brief time away from her evolution unit to administer the test much differently than Ben. She appeared more frustrated by the situation, and leaned on her mentor, (as she did at many other times during her teaching), to help devise a plan and implement activities during this testing window. She even mentioned how this issue had the "most negative impact" on her evolution unit.

Cultural issues. Similar to Ben's students, a majority of Diane's students were Mexican/Mexican American. Despite this, Diane only occasionally discussed her own or her students' cultural backgrounds. As such, there was no way to definitively conclude her cultural heritage vis-à-vis her students. When I asked about her students' backgrounds, she mentioned there was a "really big Hispanic population" at her school. Furthermore, she talked about the students in context of their difficult home lives, and the perception of their parents as potentially having 'issues.' Despite this, Diane did not connect to the cultural background of the students, whether it related to their Latino/a culture, or religiosity, during any of her evolution lessons. By excluding any connection to the everyday lives of students in any way (e.g. through family, popular culture, or other knowledge, as Ben did), Diane did not leverage their funds of knowledge during class. Additionally, in context of her focus on addressing misconceptions throughout her evolution unit, Diane mentioned how *all* students, despite their socioeconomic status, come to class with misconceptions. This blanket statement, however, does not individualize students from different cultural upbringings, and therefore, varying beliefs, which could result in different misconceptions/views of evolution.

5.5.4 Summary

The current research focused on teachers in schools that are culturally and geographically distinct. Results of the study showed that teachers do not always make pedagogical shifts during evolution, as compared to other units in biology. For example, Jane's use of videos and textbooks was a routine she practiced across units, as was Ben's regular use of bell work and laboratory-based activities. However, an individual teacher's perceptions of students' ideas, whether relating to content or culture (and whether accurate or not), can affect his/her teaching. These highly individualized cultural and pedagogical dilemmas were situated in context of political pressures (or lack thereof) which result in differing degrees of teacher autonomy. All of these intertwined dilemmas, and how a teacher negotiates them, resulted in evolution units unique to the teacher. I essentially highlighted these complexities as a result of my in-depth case study approach, where the teachers' voices became crystalized through multiple lines of evidence. As Goldston and Kyzer (2009, p. 783) noted, "…if we interpret teachers' practices by reducing them to only their beliefs and acceptance regarding evolution, we minimize the rich contexts and their entitlement as the voice of their situated practices."

5.6 Suggestions for Improving Evolution Education

The findings of this study show areas where in-service professional development (PD) of biology teachers should focus at the local and national level, with the ultimate goal of supporting all teachers in becoming effective educators of evolution. Many scholars have supported the increase in PD programs due to the widespread and complex nature of teaching evolution (Berkman & Plutzer, 2015; Pobiner, 2016; Wei, Beardsley, & Labov, 2012). It is crucial for these programs to be accessible to all teachers (Wei et al., 2012), including those in lower income districts, and those with a student body with varied cultural and religious views. There is a dearth of PD programs more generally for this purpose, and especially those that are sustaining. Ample evidence shows the effectiveness of PD that is not just one day long, but rather, is longer term and involves teacher participation follow-up (Freeman, Marx, & Cimellaro, 2004; Smith, 2013).

More specifically, future PD programs in evolution should include components that help the teacher become more comfortable and confident with teaching this content. In the current study, Jane did not have a science degree, nor did she feel comfortable teaching biology content in-depth, or providing time for student questions. If a teacher is more comfortable, he or she may be more likely to engage students in discussions and/or allow for students' questions on the subject. Including content-based information is important, but it should not be the entirety of the PD. The current research showed that there are multiple interrelated factors that affect how evolution is taught in all three classrooms. Coupling content (at all levels—from micro- to macroevolution) in addition to pedagogical tools to engage students, is helpful for teachers that do not have as much experience teaching or lack a background in the content. However, an important part of PD that is not a component of most current programs, includes the discussion about the broader social and religious contexts where educators teach, and the impact on the evolution unit. Although these latter ideas have been recommended for PD on evolution (Glaze & Goldston, 2015; Pobiner, 2016), they have not been widely implemented. This type of PD would be particularly helpful for teachers similar to Ben and Diane who both lived outside of their school districts, and rarely acknowledged their students' culture. Such a program could address the following types of teachers: those who know about their students' backgrounds but intentionally ignore them; and those who are sincerely unaware of their students' cultures.

My research supports the need for PD to broaden its base to include community context, which allows a space for teachers to engage in discussions with colleagues about their own backgrounds, and that of the students they teach. These programs should also include discussions on how teachers can learn to engage with their communities outside of school time, so they can better understand the students they teach. This could ultimately result in pedagogical approaches that consider students' cultural views and/or knowledge. These suggestions about PD for in-service teachers should be applied to pre-service teachers (PST) as well. During science methods courses, it is also crucial for the PST to become not only scientifically literate, but also to become knowledgeable about the community context in which they plan to teach or are currently student teaching. Engaging PSTs in discussions about their particular cultural context and/or the implications of including certain content in evolution (ex. religious ideas, etc.) is crucial for the development of a successful biology teacher. Within these discussions, methods instructors should include the history, legal and otherwise, of the teaching context (both locally and nationally), to help aspiring teachers better understand the totality of issues they will be up against as new biology teachers of evolution.

This study also showed how a mentor teacher greatly influenced another teacher's curricular decisions during her biology units, not just evolution. However, having this experienced teacher for Diane (the novice teacher) to rely on clearly helped her approach a subject she had little experience teaching and not much time to plan, given her focus on her other course. School districts nationwide should implement mentor programs for biology teachers at all experience levels, including novice and experienced in-service teachers, in addition to student teachers, especially during units that may be more difficult to teach, such as evolution. The current research showed how even an experienced teacher (Jane), (but not necessarily in the content area), could have potentially benefited from a 'mentor' teacher with greater conceptual understanding of the content. As such, I argue that mentor teachers can range in specialty from understanding content, pedagogy, and/or student and community context issues.

Providing mentors, in addition to PD outside of school time (as discussed above), will help teachers become effective educators of evolution, whether they are novice or experienced. This support at multiple levels—both in local districts and nationwide—can provide educators with the tools they need to teach the science behind evolution, and engage their students in meaningful discussions about the importance of evolution in their everyday lives. As a former President of the National Association for Biology Teachers (NABT), a large organization that firmly supports teaching evolution, once stated, "one strong teacher…who knows how to address the teaching of evolution…can impact change at a local level, and we do not want to underestimate the impacts that those teachers can have," (Wei et al., 2012, p. 14).

References

Arizona Department of Education. (2005, March). *Science Standards Articulated by Grade Level: High School*. Retrieved from https://cms.azed.gov/home/GetDocumentFile?id=550c512eaadebe15d072a957.

Arizona Department of Education. (2009, November). *AIMS Science High School Test Blueprint*. Retrieved from https://cms.azed.gov/home/GetDocumentFile?id=584ede25aadebe050c573eff.

Arizona Department of Education. (2016). *Assessment*. Retrieved from http://www.azed.gov/assessment/.

Arizona State Board of Education. (2005, March). *Arizona Academic Content Standards: Science*. Retrieved from https://cms.azed.gov/home/GetDocumentFile?id=550c5129aadebe15d072a8d1.

Asghar, A., Wiles, J. R., & Alters, B. (2010). The origin and evolution of life in Pakistani high school biology. *Journal of Biological Education, 44*(2), 65–71.

Berkman, M., & Plutzer, E. (2010). *Evolution, creationism, and the battle to control America's classroom*. New York, NY: Cambridge University Press.

Berkman, M., & Plutzer, E. (2015). Enablers of doubt: How future teachers learn to negotiate the evolution wars in their classrooms. *The Annals of the American Academy of Political and Social Sciences, 658*(1), 253–270.

Bleckman, C. A. (2006). Evolution and creationism in *Science*: 1880–2000. *BioScience, 56*(2), 151–158.

Catley, K. M. (2006). Darwin's missing ink—A novel paradigm for evolution education. *Science Education, 90*(5), 767–783.

Dagher, Z. R., & BouJaoude, S. (1997). Scientific views and religious beliefs of college students: The case of biological evolution. *Journal of Research in Science Teaching, 34*(5), 429–445.

Deniz, H., Donnelly, L., & Yilmaz, L. (2008). Exploring the factors related to acceptance of evolutionary theory among Turkish pre-service biology teachers: Toward a more informative conceptual ecology for biological evolution. *Journal of Research in Science Teaching, 45*(4), 420–443.

Dobzhansky, T. (1973). Nothing in biology makes sense except in the light of evolution. *The American Biology Teacher, 35*(3), 125–129.

Eddington, M. (2006, March 1). *LDS book: Evolution is not incompatible with religion*. Retrieved from http://www.sltrib.com/utah/ci_3557150.

Education Commission of the States. (2010). *Compulsory School Age Requirements*. Retrieved from https://www.ncsl.org/documents/educ/ECSCompulsoryAge.pdf.

Evenson, W. E., & Jeffery, D. E. (2005). *Mormonism and evolution: The authoritative statements*. Salt Lake City, UT: Greg Kofford Books Inc.

Freeman, J. G., Marx, R. W., & Cimerrallo, L. (2004). Emerging consideration for professional development institutes for science teachers. *Journal of Science Teacher Education, 15*(2), 111–131.

Gallup. (2014). *Evolution, creationism, intelligent design*. Retrieved from http://www.gallup.com/poll/21814/evolution-creationism-intelligent-design.aspx.

Glaze, A. L., & Goldston, M. J. (2015). U.S. science teaching and learning of evolution: A critical review of the literature 2000-2014. *Science Education, 99*, 500–518. https://doi.org/10.1002/sce.21158.

Goldston, M. J., & Kyzer, P. (2009). Teaching evolution: Narratives with a view from three southern biology teachers in the USA. *Journal of Research in Science Teaching, 46*(7), 762–790.

Griffith, J. A., & Brem, S. K. (2004). Teaching evolutionary biology: Pressures, stress, and coping. *Journal of Research in Science Teaching, 41*(8), 791–809. https://doi.org/10.1002/tea.20027.

Lemke, J. L. (2001). Articulating communities: Sociocultural perspective on science education. *Journal of Research in Science Teaching, 38*(3), 296–316.

Modern Language Association. (2010). *Most spoken languages in Arizona*. Retrieved from https://apps.mla.org/cgi-shl/docstudio/docs.pl?map_data_results.

Moore, R., Decker, M., & Cotner, S. (2009). *Chronology of the evolution-creationism controversy*. Santa Barbara, CA: Greenwood Publishing Group.
National Center for Science Education. (2013). Antiscience bill dies in Arizona. Retrieved from http://ncse.com/news/2013/02/antiscience-bill-dies-arizona-0014725.
National Research Council. (1996). *National Science Education Standards*. Washington, DC: National Academies Press. ISBN-10: 0-309-21742-3.
National Research Council. (2011). *A Framework for K-12 Science Education: Practices, crosscutting concepts, and core ideas*. Washington, DC: The National Academies Press.
National Research Council. (2012). *The Next Generation of Science Standards*. Washington, DC: The National Academies Press.
Pew Forum on Religion and Public Life. (2014a). *The shifting religious identity of Latinos in the U.S.* Retrieved from http://www.pewforum.org/2014/05/07/the-shifting-religious-identity-of-latinos-in-the-united-states/.
Pew Forum on Religion and Public Life. (2014b). *Religious groups' views on evolution*. Retrieved from http://www.pewforum.org/2009/02/04/religious-groups-views-on-evolution/.
Pew Forum on Religion and Public Life. (2014c). *Religious landscape study: Adults in Arizona*. Retrieved from http://www.pewforum.org/religious-landscape-study/state/arizona/.
Pew Forum on Religion and Public Life. (2016). *On Darwin Day, 5 facts about the evolution debate*. Retrieved from http://www.pewresearch.org/fact-tank/2016/02/12/darwin-day/.
Pobiner, B. (2016). Accepting, understanding, teaching, and learning (human) evolution: Obstacles and opportunities. *The American Journal of Physical Anthropology, 159*(S61), 232–274.
Scott, E. C. (2009). *Evolution vs. creationism: An introduction*. Berkeley, CA: University of California Press.
Smith, M. U. (2010). Current status of research in teaching and learning evolution: II. *Pedagogical issues. Science and Education, 19*, 539–571. https://doi.org/10.1007/s11191-00909216-4.
Smith, S. M. (2013). Professional development for science teachers. *Science, 340*, 310–313.
Stake, R. E. (1995). *The art of case study research*. Thousand Oaks, CA: Sage.
U.S. Census Bureau. (2016). Quick Facts: Arizona. Retrieved from https://www.census.gov/quickfacts/table/PST045215/04.
Webb, G. E. (1981). The evolution controversy in Arizona and California: From the 1920s to the 1980s. *Journal of the Southwest, 33*(2), 133–150.
Wei, C. A., Beardsley, P. M., & Labov, J. B. (2012). Evolution education across the life sciences: Making biology education make sense. *CBE Life Sciences Education, 11*(1), 10–16. https://doi.org/10.1187/cbe.11-12-0111.
Wilhelm, R. D. (1978). *A chronology and analysis of regulatory actions relating to the teaching of evolution in public schools* (Doctoral dissertation). Retrieved from ProQuest (7817731).
Windschitl, M. (2002). Framing constructivism in practice as the negotiation of dilemmas: An analysis of the conceptual, pedagogical, cultural, and political challenges facing teachers. *Review of Educational Research, 72*(2), 131–175.
Yin, R. K. (2003). *Case study research: Design and methods* (3rd ed.). Thousand Oaks, CA: Sage.
Yin, R. K. (2014). *Case study research: Design and methods* (5th ed.). Thousand Oaks, CA: Sage.

Rachel J. Fisher is a Science Pedagogical Analyst with Amplify Education. She provides conceptual and pedagogical support to K-8 science teachers nationwide as they implement the Amplify Science curriculum. As a science teacher educator, Rachel has also designed and implemented professional development activities (based on Next Generation Science Standards) for these science teachers. She is also engaged in outside work relating to the development of educators' knowledge and design of their evolution units.

Chapter 6
Evolution Education in Mexico, Considering Cultural Diversity

Alma Adrianna Gómez Galindo, Alejandra García Franco,
María Teresa Guerra Ramos, Eréndira Alvarez Pérez
and José de la Cruz Torres Frías

Abstract Mexico is a megadiverse country with great biological and cultural diversity. In this chapter, we address the analysis of the evolution education considering the enormous challenge related with these diversity, specially the presence of indigenous groups, which speak more than 365 varieties of 65 languages. To exemplify this challenge the comparison of two regions in Mexico are presented: Monterrey city, in the Northwestern state of Nuevo Leon, characterized by a development based on industrial growth and the Mayan Highlands in the Southeastern state of Chiapas, which is one of the most culturally diverse places in the country with over seventy percent of the population being indigenous. In our analysis, two main issues emerge that require attention to improve the evolution education in Mexico. The first one is evolution is not considered as a transversal approach to biology curriculum rather it is presented as a list of concepts that would need to be covered. The second one is the presence of a national curriculum in which cultural diversity is not explicitly addressed and the diverse contexts of the students are ignored. The dimension of this challenge to promote real evolution education in Mexico and some suggestions to consider an intercultural perspective are discussed in this chapter.

A. A. Gómez Galindo (✉) · M. T. Guerra Ramos
Cinvestav Monterrey, Vía del conocimiento 201, Km. 9.5 Carretera nueva al aeropuerto,
Parque PIIT, 66600 Apodaca, Nuevo León, México
e-mail: agomez@cinvestav.mx

A. García Franco
División de Ciencias Naturales e Ingeniería, Departamento de Procesos y Tecnología,
Universidad Autónoma Metropolitana—Cuajimalpa, Cuajimalpa, México

E. Alvarez Pérez
Facultad de Ciencias, Departamento de Biología Evolutiva, UNAM,
Ciudad de México, México

J. de la Cruz Torres Frías
UdeG, Guadalajara, México

© Springer International Publishing AG, part of Springer Nature 2018
H. Deniz and L. A. Borgerding (eds.), *Evolution Education Around the Globe*, https://doi.org/10.1007/978-3-319-90939-4_6

6.1 Introduction

The United States of Mexico, hereinafter Mexico, is one of the five countries in the world considered as megadiverse. Within its territory 12% of the terrestrial biodiversity is represented (CONABIO). This biological diversity can be related to cultural diversity (Maffi & Woodley, 2010) conforming what is known as biocultural diversity. Mexico is also home of more than sixty different indigenous groups, which speak more than 365 varieties of 65 languages. This multicultural composition was recognized in 1992 in the Constitution whose second article claims that the nation has a multicultural composition originally based on its indigenous peoples (CIESAS, CGEIB-SEP, et al., 2014).

The recognition of cultural diversity and the need to incorporate the knowledge and language of indigenous cultures has been considered in formal education only in recent times. For the most part, the goal of formal education was to integrate and assimilate those who spoke a language other than Spanish (Ferreiro, 1994). The 2006 educational reform noted the importance of considering "the diversity of ways of interpreting the world and how, in some cases, they [indigenous people] have contributed to scientific development (for instance, herbalism), or indigenous technological development which is beneficial for communities' relationship with the environment," (Barahona et al., 2014, p. 2261). Although textbooks have been published in different indigenous languages and indicative texts for teachers have been produced, there are very few materials that incorporate indigenous knowledge or that propose concrete ways in which teachers could introduce indigenous knowledge in the classroom (Ramírez Castañeda, 2006; García Franco, 2015). Recognition of cultural diversity in science teaching is almost non-existent in everyday practices in secondary education (Lazos Ramírez, 2015).

Diversity poses an enormous complexity when trying to characterize the state of teaching evolution in México. In this chapter, a panorama of the state of teaching evolution in the country will be presented. However, the need for considering an intercultural dialogic education for the teaching of evolution will also be discussed and reflected upon. This approach is currently missing from the discussion of education in the country.

Besides presenting general considerations about teaching evolution in Mexico, when possible, two contrasting regions in the country will be analyzed in order to exemplify similarities and differences that could be relevant for teaching and learning about evolution. One is in the Northwestern part of the country: Monterrey City, in Nuevo León state, is characterized by a development based on industrial growth. The other is in Southeastern Mexico, the Mayan Highlands in the state of Chiapas, which is one of the most culturally diverse places in the country with a 72% of the population being indigenous, particularly Tzeltal and Tsotsil. Most people in this region are involved in subsistence agriculture and the region remains fundamentally rural and non-industrialized.

Finally, it will be argued how teaching evolution in Mexico has not acknowledged the diversity of cultural and socioeconomic contexts of the Mexican population. Teaching evolution has failed to articulate the fundamental ideas of evolution with the different contexts in which they can become significant. Some ways in which biocultural diversity could be contemplated to the benefit of teachers and students will be considered.

6.1.1 Country Context

Mexico is a country of North America, with 1,960,668 km^2 of surface and more than 119 million inhabitants (INEGI, 2015). The density of population is 61 inhabitants per km^2, but the distribution of the population in the country is diverse. There are 3 main cities with very high population density, for example Mexico City considered among the second most populated city in the world, with 5900 persons per square kilometer. The main language in Mexico is Spanish. Currently Mexico is the country with the largest number of Spanish speakers in the world. However, there are also 65 indigenous languages with 365 different variants, and it is considered the seventh country with the largest linguistic diversity in the world. Over seven million people in Mexico speak an indigenous language (6.5% of the population). However, 24% consider themselves indigenous (INEGI, 2016).

The diversity in Mexico is recognized in the Constitution, whose second article (reformed in 1992) states "The Nation has a pluricultural composition based originally in its indigenous people who descent from populations that inhabited the current territory when colonization started, and that conserve their own social, economic, cultural, and political institutions".

In Mexico, the religious composition is mainly Catholic, however those who profess a religion other than the majority or do not have a belief are almost fifteen percent of the population and, for their classification, more than 250 religious categories are needed (INEGI, 2010).

Mexico, is a secular country that contemplates a constitutional separation between the State and the churches since the 19th century, but a strong link between nationalism and Catholicism prevails (De la Torre and Gutiérrez, 2013).

The right to education is granted in the Mexican Constitution (article 3) that states: "Every individual has the right to receive education. The State - Federal, State, Federal District and Municipalities - will provide preschool, primary, secondary and upper secondary education; ... shall be compulsory." The general Law of education establishes three levels of education: basic, upper secondary and higher education. The basic level is comprised of preschool, primary and secondary, serves children from 3 to 15 years old, and is certified by official certificate. By law, education in Mexico is secular and should exclude any religious doctrine; it should be oriented by scientific progress and will fight against ignorance, fanaticism and prejudices.

6.2 Public Acceptance of Evolutionary Theory Within the Country's Social, Political, and Cultural Context

Darwin's ideas about evolution were introduced in education in Mexico in 1875 by Justo Sierra, who was a prominent educator and jurist (Moreno, 1989). From that moment, these ideas were the subject of debate amongst biologists and philosophers and were disseminated in the population. They also encountered opposition from religious groups who tried to prevent such ideas from getting into education (Comas, 2010).

However, the separation between the State and the Church that had been legislated in the "Laws of Reform" in 1860 was fundamental to prevent that religious ideas won the battle. Justo Sierra argued that scientific ideas should be taught despite their differences with common sense, ideological or theological positions. This debate was also important to disseminate evolutionary ideas for the public (Barahona & Bonilla, 2009). In 1902 in the school dedicated to teacher preparation (*Normal de Maestros*), the lecture of General Biology was created with an evolutionary approach. A couple of years later the book 'Notions of Biology' was edited and widely used by generations of biology teachers in the country.

Despite its early introduction in the country, when compared to other regions in Latin America (Comas, 2010), the current situation is not very optimistic. In the National Survey about Perception of Science and Technology (INEGI, 2013) in which more than 40 million people were surveyed, a large number of Mexicans (forty percent) did not agree with the statement "Human beings are the product of evolution from other animal species". The proportion of respondents is different according to education and gender. For example, seventy three percent of women with higher education agreed with the statement and twenty eight percent of women without any education agreed with the statement.

In this same survey two statements were presented and people were asked to determine which was valid: (1) "Every living being, including human beings, plants and animals have evolved through a process of natural selection" and (2) "All living beings were created by a supreme being (God)". Forty percent of the population answered that both were valid which speaks of the relation people find between evolution and a supreme being. In this same question, thirty four percent of respondents (close to 14 million people) answered that the only valid statement was that "All living beings were created by a supreme being (God)". Even though twenty three percent of respondents consider evolutionary theories as valid; this percentage is less than those who consider the creation of living beings by God.

In a different study undertaken by the National Autonomous University of Mexico (UNAM) called 'Mexicans viewed by themselves', over fifty percent of respondents adhere to creationists ideas about the origin of life and the Universe, whereas only forty were convinced that living beings have evolved over time. Even between those who adhere to evolutionary theories, almost fifty percent said that evolutionary process is guided by a supreme being, whereas thirty percent responded that biological evolution is explained by natural processes such as natural selection. More relevant for the present chapter, forty-seven percent considered that creationism should be taught in the schools, and only thirty percent agreed that schools should teach evolutionary ideas (Ruiz, 2016).

When confronted with this panorama, one could think that the public acceptance of evolutionary theory is not very high, and that such acceptance is related to the acceptance of evolution being taught in schools. There is, however, a need to undertake more systematic research and profound analysis of the reasons behind the trends in survey responses.

The relevance of religious beliefs and their influence on science learning is very complex. It is necessary to acknowledge that science and religion are two different realms of people's lives. Within a pluralistic epistemology (Olivé, 2009), different sets of beliefs can coexist for any individual. However, the assumption that they do not influence one another is an oversimplification and a comfortable position that ignores the complex ways in which religious beliefs and science interact in daily life and in school (Taber, 2017a).

Diversity of beliefs in the classroom is something that should be considered when teaching evolution. Just as an example of such diversity we will present a brief analysis of the diversity of religious beliefs in Monterrey and in the Chiapas highlands. In a country with 119.5 million inhabitants; more than 92 million are self-reported as Catholic (77%) and only 5 million reports having no religion (INEGI, 2010). But, as has been stated, the country is very diverse and this diversity is also present in the ways in which religious ideas permeate society, culture and even politics. In the State of Chiapas, the proportion of Catholics is lower than in the whole country and only 58% of the population is reported as such (De la Torre & Gutiérrez, 2013). In the Chiapas Highlands, where more than 70% of the population is indigenous, there is an ample diversity of religions including Pentecostal, Protestant, Evangelical and Islam. Mexico, like other Latin American countries, has experienced a steady decline in the percentage of Catholic population. There are reforms of territorial hegemony at municipal scales that are significant for understanding contemporary religious change. An example of this is that the 2010 census detected 70 municipalities where Catholicism has come to occupy a place of religious minority; most of them are concentrated in the southeastern of the country, predominantly in the state of Chiapas (43 out of 70) (De la Torre & Gutiérrez, 2013).

This change of religion has brought about new identities and ways of interacting in society (Robledo Hernández & Cruz Burguete, 2005). It is worth mentioning that there are many different versions of Catholicism and, in indigenous communities, traditional (prehispanic) beliefs are highly intertwined with religious beliefs. Lisbona Guillen (2013) has shown how even in large cities in Chiapas, the indigenous presence is very relevant in form of last names, food and an intricate array of festivities and social responsibilities related to religious creeds (particularly Catholic). On the other hand, Monterrey is an industrial and business city with more than 20 daily flights to the capital of the country and to the United States of America and Europe. The state of Nuevo León reported 4.6 million inhabitants and 4 million as Catholic (87%).

It should be expected that having such a different set of beliefs within one classroom could impact the way in which evolutionary theory is learnt. These differences, however, are not considered in the national curricula and teachers are not prepared to deal or even consider how this diversity could be relevant for the learning of their students (Lazos Ramírez, 2015).

6.3 The Existence and Extent of Influence of Anti-evolution Movements in the Country

The presence of anti-evolution movements in Mexico has not been documented even though some researchers (Comas, 2010) state that groups of activists are proposing creationist or intelligent design ideas as an alternative to explain origin, diversity and adaptation of organisms and are looking for their introduction in general education. However there is no formal registry of any demand of excluding contents related to evolution.

Conservative movements of families (such as the National Union of Parents) have paid more attention to sexual education and how it is incorporated into the curricula than to evolutionary education. But even if there is no formal movement against teaching evolution or in favor of excluding certain contents or including others, the widely held religious beliefs could play an important role in the actual possibility of teaching evolution to all students despite their religious creeds (Taber, 2017b).

Biocultural research has extensively probed the inextricable links between biological, linguistic and cultural diversity (Terralingua, 2014). There are significant correlations between regions of high biodiversity and areas or concentration of human diversity (Oviedo, Maffi & Larsen, 2000). Therefore, there is a need to consider cultural differences as well as the different beliefs present in the classroom in order to teach evolution in a relevant way.

6.4 Place of Evolutionary Theory in the Curriculum

Until 2013, only basic education (i.e. preschool, primary, and secondary) was compulsory in Mexico. In 2013 a bill declaring upper secondary education (high school) as compulsory, was passed by the Congress. However it has hardly become a reality.

By 2015, practically all children aged 5–12 years were registered in school, an age range that covers the final year of preschool, primary and the first year of secondary education. From the age of 13, the enrolment rate begins to decline (from ninety seven at 13 years of age to seventy three at 15, falling to eight at age 24) (INEE, 2015). This suggests that, for more than a third of the population, knowledge of evolutionary biology is limited to what is learnt until secondary school.

This section focuses principally on biological evolution content in the curriculum for secondary education. The secondary curriculum is the same for the whole country. The complete cycle lasts three years and is normally studied by students between the ages of 12 and 15 (SEP, 2011).

The way in which evolutionary content has been considered in the curricula has changed over time. As a historical reference, in the 1970 reform, content dealing with evolution in primary education changed from being a list of topics to being one of the foci in the free textbook underpinned by the 1993 curriculum (Barahona & Bonilla, 2009). This guideline was also, though not very successfully, included in the 2006 curriculum. Currently, the primary and secondary school curricular content is a long way from being integrated into the perspective of evolutionary biology.

The list of topics related to evolution in the current curriculum would seem to respond more to what has been branded official pedagogical rhetoric. This is also applicable to the intercultural approach, which is neither part of nor related to the evolutionary topics in this curriculum, but is to be found in the official discourse. In the curriculum, there is no connection between the basic ideas of evolution and the distinct cultural contexts in which they can become meaningful.

There are good reasons for believing that evolutionary theory rarely reaches classrooms in an appropriate form even when, to a greater or lesser extent, it has been taken into account in curricular designs. This is partially due to the inconsistency between curricular changes and early and in-service teacher training. This is in addition to the quantity and complexity of content in science subjects, which have a scientificist focus that can lead to encyclopedism and rote memorization (Candela et al., 2012).

While this section deals with the place of evolutionary theories in the curriculum, suffice it to say that research efforts in Mexico point in essentially the same direction as many others throughout the world as regards the difficulties of learning evolutionary concepts. Exactly the same results were obtained in two studies carried out twelve years apart using similar instruments (Sánchez, 2000; Alvarez, 2015), that is, poor learning on the part of secondary students in these subjects.

In both studies, the instruments used were multiple choice questionnaires that covered problems relating to the origin of variation, its randomness with regard to the needs of organisms, population changes over time resulting from natural selection and the result they lead to (adaptation). The Sánchez study (op. cit.) aimed at identifying `alternative conceptions` (in a sample of 90 students aged 12–15) and the Alvarez study (op. cit.) was guided by the concept of epistemological obstacles that González Galli (2011) (in a sample of 194 students aged 12–14). The results of both authors show that, even after teaching, students predominantly opt for teleological thinking (that is, the assumption that things, including variation and evolution, happen for a reason). Students should instead learn that evolutionary phenomena do not revolve around predetermined purposes and that variation arises at random (independently of the needs of the organisms and of the selective pressures). Furthermore, in parallel with González Galli, Alvarez reported students' persistence in ideas centered around the individual (described in other studies as 'not thinking in terms of populations'). As is well known, thinking in terms of populations, rather than in terms of individuals, is essential for understanding evolution. Finally, these studies also found that linear causal thinking is predominant in students (namely the assumption that all phenomena have a single cause which operates in one single direction, for example, supposing that if an organism needs certain traits, it obtains them). Here, students need to be helped to understand that evolutionary phenomena are complex, usually have multiple causes, and involve, moreover, probability and chance. Associated with this, biological evolution frequently presents situations where there are reciprocal rather than unidirectional effects, for example, when the environment influences the organisms, and in turn these influence the environment.

In the two studies carried out at normal state schools in Mexico cited above, students' average grade was under 5 in the first case, and 5 in the second (both on a scale of 10). This suggests that the curricular changes implemented between 2000 and 2012 did not lead to improved learning of biological evolution.

The current curriculum includes one natural science subject per year in secondary school: biology in the first year, physics in the second and chemistry in the third. This means that, for those who do not go beyond secondary education, this is the only opportunity they will have to learn about biological evolution. Additionally, this knowledge, which has transformed our way of seeing the world, life, and humankind, will not be studied by high school students for more than another two years.

In science I (emphasizing Biology), there is specific evolutionary biology content and the Teacher's Guide stresses its importance. However, in at least one of the frequently used free textbooks endorsed by the SEP (Limón et al., 2016) what is emphasized is the description of adaptive traits, which is a long way from including the evolutionary approach, lacking from the course content (Barahona et al., 2014). To convey an evolutionary approach, course content would have to be directed towards explaining how characteristics are acquired by species, and clarify the historical process of interaction between the inheritable variation and natural selection that produced them. In other words, it would be necessary to understand the scientific model that explains adaptation and teach the cases that illustrate it in the classroom.

In the secondary school curriculum, there are notable inconsistencies between stated aims and evolutionary biology content. The model of evolution by inheritable variation and natural selection is, of course, fundamental knowledge and is related to events in daily life, as recognized in the Teacher's Guide (1, 2010). The problem is that no guidance is given on how it should be taught. What is needed when explaining adaptation is a rigorous, sufficient and coherent selection of knowledge. Similarly, it is necessary to teach the updated, theoretically contextualized Darwinian model, which implies awareness of, its scope and its limits, as well as pointing out that further models exist to explain other evolutionary phenomena (Alvarez, 2015; Alvarez & Ruiz, 2015).

Despite the importance of variation as a universal characteristic of living beings and crucial for understanding evolution, it is scarcely mentioned in the content.

Without acquiring an understanding of individual differences, there is little chance of understanding evolution. The concept of variation is particularly relevant in a country as megadiverse as Mexico and is linked to the immediate environment. It should be important to recognize that humans, as well as all other living beings have the property of being different in each one of their traits, one of which is culture. As long as the human species is part of biodiversity, cultural diversity (with its emergent characteristics) is part as well. This relation, which is absent in the current curriculum is particularly relevant in Mexico as well as in other countries such as Mexico, Indonesia, India, Australia, Zaire and Brazil where there is a strong correlation between cultural and biological diversity (Loa et al., 1998; Oviedo et al., 2000).

In short, what the secondary education curriculum lacks is an evolutionary focus; the content includes evolutionary biology concepts, but these have no structure. There is a lack of linking elements in the curriculum from one school year to the next creating obstacles for the consolidation of the scant content related to evolutionary biology at primary school (children from 6 to 12). If this content were suitably selected and focused, it could provide the foundation for acquiring fundamental knowledge of evolutionary biology in secondary school. The subject-based curricular approach in these two school cycles hinders the progression of learning, a deficiency that extends into high school.

In addition to the above, the curricular content related to biological evolution does not take into account the contextual mosaic that epitomizes Mexico, pointing in the opposite direction to the intercultural dialogical educational approach that such a biologically and culturally diverse country requires.

As mentioned at the beginning of this section, secondary education has a national curriculum. At the same time, as Mexico is a megadiverse country in biological and cultural terms, the universality of scientific knowledge and every citizen's right to learn about this portion of humankind's inheritance goes hand in hand with the right to recognize, value and ponder the traditional knowledge of the indigenous peoples. This diversity could be considered as an asset and could be used to explain evolution and to contextualize its relevance in terms of phenomena that are familiar and relevant for students, particularly for those whose culture is alien to science. But this is yet to be achieved.

6.5 Emphasis Given to Evolutionary Theory in Biology Teacher Education Program

In Mexico, initial teacher training (for pre-school, primary and secondary school) is undertaken in *Escuelas Normales* that can be public or private. These schools are oriented by the National Program of Teacher Education of the Education Ministry (SEP).

In secondary school, besides the teachers graduated from *Escuelas Normales*, anyone who holds a bachelor degree can become teacher. Science teachers can be biologist, chemists, physicists, doctors, engineers, amongst others. The National Institute for the Evaluation of Education (INEE) estimates that forty of the total population of teachers (139,366) was educated in a university. These teachers do not have any kind of initial teacher training.

The curriculum for teachers' education is national and is related to the study programs that have been described previously in the chapter. This curriculum does not consider cultural and socioeconomic diversity present in the country.

In this national curriculum, in preschool level there are two courses that cover some elements for teaching evolution: "Living beings" and "Biodiversity as proof for evolution" (SEP, 2012a).

Something similar happens for primary teachers. In the study programs, there are two courses that cover elements for teaching evolution, including the subjects: "Environment and Ecosystems", "Recognition of ecosystems", and "Ecology and Biodiversity" (SEP, 2012b). Contents related to evolution are introduced as anecdotes doing very little to contribute to the construction of biological knowledge articulated by theoretical-evolutive knowledge (Taber, 2017a).

In primary and secondary teacher education programs, evolutionary theory and its teaching are approached superficially with a 'hands-on' approach centered only on phenomena and with little relation to theoretical underpinnings (for example, environment and ecosystems issues are exemplified but not related with evolutionary ideas, something similar happens with ecology and biodiversity). In secondary teachers' education, there is a dense theoretical conceptual approach that does not leave room to contextualize examples or to introduce students' and local knowledge (SEP, 2012c). Even in the specialization for biology teaching, evolution is considered very simplified as an opening theme, as an historical anecdote as has been found in other countries (Taber, 2017b). This does not allow that knowledge about evolutionary biology is integrated as a focus to explain species' transformation through time.

In teacher education, evolution does not have a transversal integrative structure that allows to comprehend evolution as a perspective to teach biology in every educational level. In the curricular content, the description of the proximal causes (physiological, morphological, etc.) do not incorporate distal causes, such as evolutionary ones, related to variability and natural selection, adaptation, phylogeny, etc. (Mayr, 1998).

Teacher education in Mexico does not have a diversified proposal that considers cultural diversity and different socioeconomic contexts. There is not even a mention for teachers to articulate disciplinary knowledge and students' every day and communitarian lives. Indigenous or local knowledge is left out of the school, and practices such as artificial selection of maize and of different vegetables are not even considered. This knowledge could be readily related to teaching evolution that would add more significant content since it is associated to the daily life of indigenous populations (García Franco & Gómez Galindo, 2015).

Currently in all the country, there is only one *Escuela Normal* which is indigenous, intercultural and bilingual. It is located in the Zinacantán municipality in the state of Chiapas. This *Escuela Normal* was the result of a struggle of a group of teachers convinced that it was necessary to strengthen and reassess the linguistic and cultural diversity of the native peoples in order to construct a national citizenship that considers the contributions of the indigenous people (Baronet, 2008). But even in this kind of teachers' school, there is no clear articulation between local and scientific knowledge.

Teaching evolution in a different way would require an articulation of efforts in order to explain evolutionary processes, how these were developed and how can they explain phenomena relevant for students' daily life.

6.6 Biology Teachers' Attitudes Towards Teaching Evolutionary Theory

Science teachers' attitudes, in general, have received scarce attention in empirical studies in Mexico. Specifically, Mexican biology teachers' attitudes towards teaching the theory of evolution remains as a theme to be investigated despite the introduction of the evolutionary perspective in the science curriculum for elementary education since 1993 (Barahona & Bonilla, 2009).

This section offers the preliminary findings of our own effort to start exploring teachers' acceptance of the theory of evolution and their attitudes towards teaching evolution. Assuming that the acceptance of the theory of evolution is the basis for a positive attitude towards teaching evolution, the aim was to explore both aspects and look for relationships between them. In consequence, the MATE instrument (Rutledge & Warden, 1999) was chosen as a research instrument. It consists of 20 Likert scale items which explore acceptance of (a) process of evolution, (b) scientific validity of evolutionary theory, (c) evolution of humans, (d) evidence of evolution, (e) scientific community's views of evolution and (f) age of the Earth. The MATE instrument was translated into Spanish to be used with Mexican population and this version validity and reliability is investigated in a larger ongoing study. Additionally, 12 items more were developed to specifically address attitudes towards teaching evolution. Consequently, the questionnaire ended up with 32 items.

This survey was taken by 43 secondary Biology teachers in the context of a diploma course on science teaching competencies in which one of the authors of this chapter participated during November 2016 in Monterrey, Nuevo León. In Chiapas, there was not a similar opportunity because occasions for teacher preparation are scarcer and the work that two of the authors were doing was with a handful of teachers. This situation is also representative of the diversity of conditions in the country.

Participant teachers were practicing teachers from Monterrey, Nuevo Leon. They were 18 males and 25 females, aged 23–57, all working in state secondary schools. They had between 1 and 35 years of teaching experience. Concerning their academic background, 16 teachers held a first degree in education from teachers' colleges, 14 hold a first degree from a university, 12 teachers held a master's degree and one had a doctorate.

In this pilot study of the 32-item instrument, data analysis sought statistical evidence on the discriminative power of the items, which could tell us about the validity of the questionnaire as a research instrument. For this purpose, t tests were performed with SPSS v.14. Data were also processed statistically to identify patterns and tendencies in teachers' responses. The frequencies of teachers' responses to the MATE items and the ones regarding acceptance of the teaching of evolution were obtained.

In our adaptation of the MATE instrument, the original Likert scale responses (strongly agree, agree, undecided, disagree and strongly disagree) were conserved. The items designed to explore attitudes towards teaching evolution followed the same response format and were the following:

21. I include examples and ideas related to evolution in my classes.

22. Evolution is a complex theme for students, only natural selection should be taught.

23. It is convenient to teach about evolution in pre-school education.

24. Studying evolution helps my students to understand natural processes and phenomena.

25. I avoid examples and ideas related to evolution in my classes

26. It is convenient to teach about evolution in primary education.

27. Evolution is an accessible theme for students, any aspect of it can be taught

28. Teaching evolution and my religious beliefs enter in contradiction.

29. Teaching evolution should be excluded in education to children and teenagers.

30. Studying evolution does not help my students to understand natural processes and phenomena.

31. Learning evolution and my students' religious beliefs enter in contradiction

32. It is convenient to teach about evolution in secondary education.

The item analysis indicated that 17 of 20 items of MATE in their Spanish version had adequate discrimination power (t test, $p < 0.05$), this suggest that teachers who obtain the higher scores and the lowest scores respond differently to the items. Concerning the additional items about attitudes towards teaching evolution, only one item obtained a non-significant t-test. Therefore, our comments on the tendencies in teachers' responses should be taken as preliminary.

Teachers' scores in the MATE items were grouped into categories of acceptance as suggested by Rutledge and Sadler (2007). Teachers tended to obtain high scores indicating that 32 out of 43 showed high or very high acceptance of the theory of evolution (Fig. 6.1).

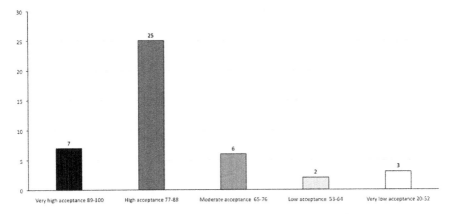

Fig. 6.1 Frequencies in categories of acceptance corresponding to responses to 20 items of the MATE instrument

When teachers' scores in items regarding teaching evolution were grouped, it was also noticed that most teachers (27 out of 43) reported to hold positive attitudes towards teaching this theme (Fig. 6.2).

These preliminary findings indicate that most teachers tended to show moderate, high or very high acceptance of the theory of evolution. Similarly, most of them hold from moderate to high acceptance of teaching evolution. Therefore, at least in these preliminary findings, teachers in the sample tended to accept evolution theory and its teaching showing no major conflict in these two aspects.

Teachers' attitudes towards evolution theory and its teaching may be considered of high relevance from a research perspective. However, it must be acknowledged that for Mexican teachers' evolution is one curriculum theme among so many

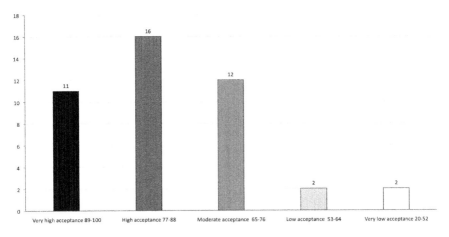

Fig. 6.2 Frequencies in categories of acceptance corresponding to responses to 12 items regarding teaching of evolution

others. Much more controversy has arisen in the Mexican educational context from themes related to human sexual reproduction and the use of contraceptives. Despite this, it can be argued that teachers' attitudes towards the theory of evolution and its teaching deserves more research since evolution is intended to be a central axis of biology curriculum.

6.7 Suggestions to Improve Evolution Education in Mexico

There is a clear need to encourage research on the public acceptance of evolutionary theory and especially of teachers as has been made evident throughout this chapter. The exploratory study that has been presented here shows that teachers have a high acceptance of the theory and seem to be willing to teach it in their classrooms. However, we need larger efforts to make sense of the reasons behind teachers' responses as well as to extend the inquiry to include teachers from other states and diverse educational contexts.

The analysis of the national curriculum (for teachers and for students) identifies two large areas that would need to be reformed to improve the teaching of evolution in the country. On one hand is the way in which content related to the theories of evolution is presented. Our analysis shows that both to prepare students and to prepare teachers evolution is presented as a list of concepts that would need to be covered and these concepts are just some amongst a list of many others. To improve understanding and application of evolutionary ideas, these should be incorporated more articulately in teacher preparation. There should be a consideration of learning progressions that acknowledges the previous knowledge required to understand and construct theoretical models used to explain phenomena. The curricula for teacher education should also establish the relation between these models and the way they are introduced and constructed in the classroom. There is no need to include more concepts into the list, but a need to elaborate a more integrated approach that is consistent with the biologist's view of evolution being in the center of understanding every phenomenon, as well as integrate different school levels.

On the other hand, there is a concern with the idea of using a national curriculum that has no correspondence with the cultural diversity that characterizes this country. This monolithic view of the curriculum has been nuanced in the political discourse with the introduction of an intercultural approach. However, even when textbooks have been translated to different indigenous languages and orientative texts have been created, the objectives and ideals of an intercultural education are yet to be realized (García Segura, 2004), an education that considers people knowledge and concerns, and establish relations with the curriculum. This does not mean there are not individual efforts in different parts of the country, but science teaching and teaching evolution have not been the focus of research.

The way in which evolution should be taught to indigenous students, using what examples, and from what perspectives is inscribed in a pluralist epistemology (Olivé, 2009), according to which the diversity of ways of understanding the world should be understood in every context. A diverse country such as Mexico has a number of contexts and examples in which evolution is relevant and could be used to understand biological and cultural diversity. In order to construct a truly intercultural education, we would need to incorporate voices other than scientists' and academics' that could have a perspective on what is relevant in the classrooms, the problems that need to be understood, and to what ends. In this case, the theoretical proposal cannot precede experience; rather it should be co-constructed incorporating indigenous voices both regarding educational policies as well as ways in which these policies get enacted in the classrooms.

Teachers' in Mexico seem to have relatively high acceptance of evolutionary theory and its teaching, which could mean that there is fertile ground to propose and enact ways of teaching evolution more consistent with the current understanding of evolution and that consider the diversity of contexts. There is, however, a need to undertake studies that recognize differences between teachers in different parts of the country.

Incorporating different voices into the discussion and considering cultural differences could not be done within the framework of a national curriculum that dictates that every student regardless of their interests and the place they inhabit should know exactly the same. The reform of national curriculum in progress in Mexico is a great opportunity to consider these ideas. Evolution is one of the tenets of humankind, a theory that explains the current diversity and some ways in which it could be preserved, but in order to make it relevant for the diverse lives of teachers and students, the contextualization of ideas needs to become a reality.

Acknowledgements This work was supported by grants SEP/SEB 2013, No. 231425 and SEP/SEB 2014-01, No. 240192 by Conacyt, Mexico.

References

Alvarez, P. E. (2015). *Conocimientos fundamentales de biología evolutiva: propuesta didáctica para educación secundaria*. Doctoral Thesis in Biological Sciences, Mexico: Facultad de Ciencias, Universidad Nacional Autónoma de México.

Alvarez P, E., & Ruiz, R. (2015). Proposal for teaching evolutionary biology: A bridge between research and educational practice. *Journal of Biological Education*. https://doi.org/10.1080/00219266.2015.1007887.

Barahona, A., & Bonilla, E. (2009). Teaching evolution. Challenges for Mexican Primary Schools. *ReVista (Harvard Review of Latin America), III*(3), 16–17.

Barahona, A., Chamizo, J. A., Garritz, A., & Slisko, J. (2014). The history and philosophy of science and their relationship to the teaching of science in Mexico. In M. Matthews (Ed.), *International handbook of research in history, philosophy and science teaching* (Vol. III, pp. 2247–2270). New York: Springer International Publishing.

Baronet, B. (2008) La Escuela Normal Indígena Intercultural Bilingüe "Jacinto Canek". *Travaux et Recherches dans les Amériques du Centre, 53.*
Candela, A., Sánchez, A., & Alvarado, C. (2012). Las ciencias naturales en las reformas curriculares. In F. Camacho (Ed.), *La enseñanza de la ciencia en la educación básica en México*. México: Instituto Nacional para la Evaluación Educativa. Retrieved April 21, 2017, from http://publicaciones.inee.edu.mx/buscadorPub/P1/C/227/P1C227.pdf.
CIESAS, CGEIB/SEP, et al. (2014). Campaña nacional por la diversidad cultural de México. Retrieved December 15, 2017, from http://www.diversidadcultural.mx/index.php/Conoce/conoce-portada.html.
Comas Rodríguez, O. (2010). Darwin en las aulas. Retrieved December 26, 2016, https://www.researchgate.net/publication/275214333_Darwin_en_las_aulas.
CONABIO. Portal Biodiversidad Mexicana de la Comisión Nacional para el Conocimiento y Uso de la Biodiversidad [National Commission for knowledge and use of biodiversity]. Retrieved June 16, 2016, from http://www.biodiversidad.gob.mx/usos/maices/maiz.html.
De la Torre, R., & Gutiérrez, Ch. (2013). New landscapes of religious diversity in Mexico. In G. Giordan & W. H. Swatos (Eds.), *Testing pluralism. Globalizing belief, localizing gods* (pp. 125–148). Jr. Leiden-Boston: BRILL.
Ferreiro, E. (1994). Diversidad y proceso de alfabetización: de la celebración a la toma de conciencia. *Lectura y Vida, Revista Latinoamericana de Lectura, 15*(3), 6–11.
García Franco, A. (2015). La enseñanza de las ciencias en escuelas indígenas en México: caminos en la sociedad del conocimiento. *Revista Internacional de Tecnología, Conocimiento y Sociedad, 4*(1), 11–18.
García Franco, A., & Gómez Galindo, A. A. (2015). An intercultural approach to teach evolution using maize selection and harvest. In: J. Lavonen, K. Juuti, J. Lampiselkä, A. Uitto, & K. Hahl (Eds.), *Science education research: Engaging learners for a sustainable future* (pp. 1880–1885).
García Segura, S. (2004). De la educación indígena a la educación bilingüe intercultural. La comunidad purépecha, Michoacán, México. *Revista Mexicana de Investigación Educativa, 9*(20), 61–81.
González Galli, L. M. (2011). *Obstáculos para el aprendizaje del modelo de evolución por selección natural*. Doctoral Thesis, Buenos Aires: Facultad de Ciencias, área de Ciencias Biológicas, Universidad de Buenos Aires.
INEGI. (2010). *Panorama de las religiones en México*. México: INEGI.
INEGI. (2013). *Encuesta sobre la Percepción Pública de la Ciencia y la Tecnología en México (2011)*. México: INEGI.
INEGI. (2015). *Encuesta Intercensal 2015*. México: INEGI. Retrieved April 16, 2107, from http://www.beta.inegi.org.mx/proyectos/enchogares/especiales/intercensal/.
INEGI (2016). Estadísticas a propósito del día mundial de los pueblos indígenas. México: INEGI. Retrieved April 22, 2017, from http://www.inegi.org.mx/saladeprensa/aproposito/2016/indigenas2016_0.pdf.
INEE [Instituto Nacional para la Evaluación de la Educación]. (2015). *Los docentes en México. Informe 2015*. México, D.F.: INEE.
Lazos Ramírez, L. (2015). La enseñanza de las ciencias y la diversidad cultural en México: un estudio en la educación básica secundaria. *Revista Internacional de Tecnología, Conocimiento y Sociedad, 4*(1), 1–10.
Limón, S., Mejía J., & Aguilera J. E. (2016). Ciencias 1 Biología. México: Castillo.
Lisbona Guillén, M. (2013). Un carnaval inventado. El disfraz de lo Zoque en Chiapas contemporáneo. *Revista de Museología Kóot, 3*(4), 103–115.
Loa Loza, E., Cervantes, M., Durand, L., & Peña, A. (1998). Uso de la biodiversidad. En CONABIO, *La diversidad biológica de México: Estudio de País 1998*. México: CONABIO. Retrieved April 21, 2017, from http://www.biodiversidad.gob.mx/publicaciones/librosDig/pdf/divBiolMexEPais1.pdf.
Mayr, E. (1998). *This is biology. The science of the living world*. Cambridge: Harvard University Press.

Moreno, A. R. (1989). *La polémica del Darwinismo en México siglo XIX*. México: UNAM.
Maffi, L., & Woodley, E. (2010). *Biodiversity cultural conservation*. London: Earthscan.
Olivé, L. (2009). *Por una auténtica interculturalidad basada en el reconocimiento de la pluralidad epistemológica* (pp. 19–30). CLACSO, Muela del Diablo Editores, Comunas, CIDES-UMSA: Pluralismo epistemológico. La Paz.
Oviedo, G., Maffi, L., & Larsen, P. (2000). *Indigenous and traditional peoples of the world and ecoregion conservation*. Switzerland: World Wild Fund.
Ramírez Castañeda, E. (2006). *La educación indígena en México*. México: UNAM.
Robledo Gutiérrez, G. P., & Cruz Burguete, J. L. (2005). Religión y dinámica familiar en Los Altos de Chiapas. La construcción de nuevas identidades de género. *Estudios Sociológicos, XXIII*(68), 515–534.
Ruiz, R. (2016). La evolución: el concepto y su recepción en México. *El Universal*. 16 de enero. Retrieved December 15, 2016, from http://www.eluniversal.com.mx/entrada-de-opinion/articulo/rosaura-ruiz/nacion/2016/01/16/la-evolucion-el-concepto-y-su-recepcion.
Rutledge, M. L., & Sadler, K. C. (2007). Reliability of the Measure of Acceptance of the Theory of Evolution (MATE) instrument with university students. *The American Biology Teacher, 69*(6), 332–335.
Rutledge, M. L., & Warden, M. A. (1999). The development and validation of the measure of acceptance of the theory of evolution instrument. *School Science and Mathematics, 99*(1), 13–18.
Sánchez, C. (2000). *La enseñanza de la teoría de la evolución a partir de las concepciones alternativas de los estudiantes*. Doctoral Thesis in Science (Biology). México: Facultad de Ciencias, Universidad Nacional Autónoma de México.
SEP [Secretaría de Educación Pública]. (2011). *Plan de Estudios 2011*. [Curriculum 2011] México: Talleres del Centro Gráfico Industrial de la Comisión Nacional de Libros de Texto Gratuitos. Gobierno Federal, México: SEP.
SEP [Secretaría de Educación Pública]. (2012a). Secretaría de Educación Pública. Licenciatura en Educación Preescolar, Plan 2012. México: SEP.
SEP [Secretaría de Educación Pública]. (2012b). Secretaría de Educación Pública. Licenciatura en Educación Primaria, Plan 2012. México: SEP.
SEP [Secretaría de Educación Pública]. (2012c). Secretaría de Educación Pública. Licenciatura en Educación Secundaria, Plan 1999, Especialidad: Biología. México: SEP.
Taber, K. S. (2017a). The relationship between science and religion—A contentious and complex issue facing science education. In D. Akpan (Ed.), *Science education: A global perspective* (pp. 45–69). New York: Springer International Publishing.
Taber, K. S. (2017b). Representing evolution in science education: The challenge of teaching about natural selection. In B. Akpan (Ed.), *Science education: A global perspective* (pp. 71–96). New York: Springer International Publishing.
Terralingua. (2014). Biocultural diversity education initiative. Canada: Terralingua. Retrieved April 21, 2017, from http://terralingua.org/wp-content/uploads/2015/07/BCDEI-Overview.pdf.

Alma Adrianna Gómez Galindo is lecturer-researcher in the Center for Research and Advanced Studies of the National Polytechnic Institute (Cinvestav) in Monterrey, México. She has a degree as teacher of early childhood education, is a marine biologist and has a PhD in didactics of science. She performs qualitative research on biology education, focusing in teaching - learning sequences for modeling by using multimodal representations and analogies. Her current projects include the analysis of dialogic perspectives for teaching evolution in cultural diversity contexts and developing of learning progressions for central models in biology to guide learning from kindergarten to middle school.

Alejandra García Franco is a Chemical Engineer and has a PhD in Pedagogy from the National Autonomous University of Mexico (UNAM). She is a teacher-researcher at the Autonomous Metropolitan University – Cuajimalpa. She is interested in intercultural scientific education and has collaborated for education projects with indigenous people in Mexico. She is also interested in chemistry learning and the design of teaching learning sequences for teacher training.

María Teresa Guerra-Ramos is a lecturer-researcher in Educational Psychology and Biology Education at the Center of Research and Advanced Studies (Cinvestav), Monterrey Unit; located in the North-East region of Mexico. She teaches master and doctoral level courses in collaboration with academic programs within Cinvestav around the country. Her research interest is focused on teachers' representations of scientists and scientific activity, the development of teaching competences, features of discourse in Biology teaching-learning interactions and collaboration among teachers and researchers for pedagogical innovation.

Eréndira Alvarez Pérez is a Full Professor in the Science Faculty of the National Autonomous University of Mexico (UNAM). It is biologist, Master of Science (in education and history of biology) and a PhD in biological sciences. It conducts research and teaching in the didactics of evolutionary biology, focusing on the generation of didactic resources in this field. Her current projects include dialogue between disciplines and research approaches to address the teaching and learning of fundamental knowledge of evolutionary biology throughout the educational system.

José de la Cruz Torres Frías has a PhD in Education and is currently a postdoctoral fellow in didactics of science in the Center for Research and Advanced Studies (Cinvestav) of the Polytechnic Institute Nacional-Unidad Monterrey, Mexico. He is a member of the National System of Researchers in Mexico. Current lines of research: 1) training for research in higher education and postgraduate courses, 2) In-service science teachers' training in primary and secondary school.

Chapter 7
Evolution Education and the Rise of the Creationist Movement in Brazil

Alandeom W. Oliveira and Kristin L. Cook

Abstract In this chapter, we analyze current educational policies such as the National Curriculum Parameters for Secondary Science Education, commonly used curricula (high school biology textbooks), and publicly available resources (media reports, organizational websites, previous studies, etc.) to ascertain the characterization of evolution education and the rise of the creationist movement in Brazil. Additionally, we provide a historical account of how larger societal forces such as religion and politics have shaped the Brazilian educational landscape over time. Our ultimate goal is to better understand not only what evolution education in Brazil is like but also how it came to be (i.e., the dynamic sociological processes behind its current state). Findings indicate that a lack of understanding about evolution as a unifying theory coupled with vague messages present in curricula and from the Ministry of Education about defining parameters regarding what should be taught in a science classroom can lead to teachers getting caught in the crosshairs of public pressure and creationist propaganda.

7.1 Country Context

> Once, I was a chimpanzee
> Now, I walk only on my feet... (Antunes, Brown, & Monte, 2002)

In the above quotation from the popular song *Tribalistas*, three very famous Brazilian singers—Arnaldo Antunes, Carlinhos Brown, and Marisa Monte—make a somewhat implicit and poetic allusion to human evolution, portraying themselves as having reached a more evolved or developed state in their lives. Its high popularity as a song that reached the top of Brazilian music charts at the turn of the

---

A. W. Oliveira (✉)
State University of New York, New York, NY, USA
e-mail: aoliveira@albany.edu

K. L. Cook
Bellarmine University, Louisville, KY, USA

© Springer International Publishing AG, part of Springer Nature 2018
H. Deniz and L. A. Borgerding (eds.), *Evolution Education Around the Globe*, https://doi.org/10.1007/978-3-319-90939-4_7

millennium without eliciting any form of controversy or contentious responses from conservative religious groups highlights the cultural significance (or non-significance) of evolution in Brazilian society. Since then, the Brazilian socio-cultural context has changed drastically with the advent of anti-evolution forces in various political and educational spheres. In this chapter, we examine the current state of evolution education and the rise of the young-Earth creationist movement in Brazil in recent years. More specifically, we analyze current educational policies such as the National Curriculum Parameters for Secondary Science Education, commonly used curricula (high school biology textbooks), and publicly available resources (media reports, organizational websites, previous studies, etc.). Additionally, we provide a historical account of how larger societal forces such as religion and politics have overtime shaped the Brazilian educational landscape. Our ultimate goal is to better understand not only what evolution education in Brazil is like but also how it came to be (i.e., the dynamic sociological processes behind its current state).

It will be convenient to begin by providing demographic information that can help readers have a general sense of the scope and main features of the Brazilian national context. With a population of approximately 200.4 million people, Brazil's dominant spoken language is Portuguese and its two main religious groups are Roman Catholics (64.6%), and Protestants and other Christians (24%). Secular and centralized in nature, the country's educational system is governed by the Ministry of Education and comprises three distinct levels, namely pre-school education, basic education (consisting of primary school ages 6–14 and secondary school ages 15–17) and higher education. Basic education is free and mandatory for those between the ages of 6 and 17, although private pre-K-12 schools are also common in Brazil. Likewise, higher education, including graduate degrees, is free at public universities. The official university entrance exam is named the *Exame Nacional do Ensino Médio* (ENEM). Because this entrance exam is very challenging and public federal universities (the best ones in Brazil) have limited seats, admissions are highly competitive.

7.2 Public Acceptance of Evolutionary Theory

International comparisons of evolution education and public acceptance suggest a "global spread" of the creationist movement and raise concerns about potential growth of anti-evolution attitudes within public educational systems worldwide (Blancke & Kjærgaard, 2016; Harmon, 2011; Miller, Scott, & Okamotto, 2006). Closely aligned with this international trend, Brazil has recently witnessed increasing resistance to the teaching of evolution among its populace. A national survey recently showed that, though evolution is accepted by more than half of the population (54%), the overwhelming majority of Brazilians (89%) now believe that creationism should be taught in schools, and that it should replace the theory of evolution in the school curriculum (75%) (Brum, Fonseca, & Cardoso, 2005).

Pointing to the emergence of a new generation of creationists in Brazil, these numbers have been taken as evidence of a public educational system in need of much improvement in terms of both science teacher preparation and biology instruction.

At first sight, these numbers may not seem surprising given the fact that Brazil is the country with the largest number of Catholics in the world. However, it should be noted that the religious climate in Brazil, while historically predominantly Catholic, has been changing considerably. According to the latest national census data from 2000, 15% of the population was protestant (a growth rate that was beyond 100% in the 1990s). There are also a growing number of Brazilians who are associated with Pentecostal and neo-Pentacostal churches, with an increase of nearly 8.9% a year (whereas the traditional Evangelical groups are growing by about 5.2% per year) (Mariano, 2004). When compared with other countries, Brazilians as a whole tend to believe in a Creator—a divine supernatural—who created humans and other organisms (Tidon & Lewontin, 2004). In a comparative study of life scientists in the UK and Brazil, researchers investigated the relationship between scientific training and religious beliefs. Falcão (2008) found that Brazilian life scientists had more adherence to a belief in the supernatural despite advanced scientific training than those in the UK. In her study, Brazilian scientists, even without ascribing to a particular religion, had a firmer attachment to a belief in God, no matter what their level of training at the University. While Falcão's (2008) work gives us a comparative lens on Brazilians' ideologies, we cannot determine from this whether Brazilian scientists ascribe to creationist views. Furthermore, a firm belief in God and possessing Creationist views are certainly not synonymous. However, to understand Brazilian culture, we must understand the strength of the belief in God regardless of ascribed religion or non-religion.

7.3 Anti-evolution Movements in Brazil

Since the military dictatorship from 1964 to 1985, public schools in Brazil have generally experienced a widespread lack of investment. Though the public schools are said to be secular and the Ministry of Education prohibits the teaching of creationism in science classes, there has been increasing pressure on teachers to incorporate creationism in the public school classroom. High profile politicians, such as Marcelo Crivella (the former Minister of Fishing and Aquiculture) and the former environment minister, Marina Silva, have publicly defended the teaching of creationism (Silva & Prado, 2010). Such prominent influences have blurred the idea of secularism in public schools causing teachers to be unclear about what should be taught in the science classroom. Furthermore, a controversial decision was made by the governor of Rio de Janeiro to introduce the teaching of creationism in schools. These pro-creationism upsurges have resulted in discussions around the country about the separation of church and state as well as inspired researchers to take a closer look at what is being taught in schools and the ways in which teachers may or may not be equipped to navigate these complex issues.

As debates rage on in the media and in political spheres about what should or should not be happening in public schools, creationist advocacy/political groups such as the *Brazilian Creationist Society* (http://www.scb.org.br/scb/), *Brazilian Institute of Intelligent Design* (http://www.designinteligentebrasil.com.br/), and the *Brazilian Association of Creation Research* (http://abpc.impacto.org/) have been established. These anti-evolution groups have actively campaigned for the teaching of creationism. The Brazilian Creationist Society and the Brazilian Association of Creation have been espousing their anti-evolution beliefs since the 1970s, while the Brazilian Institute of Intelligent Design has more recently been founded. These groups have dramatically increased their number of publications, pamphlets, and translated books that present anti-evolution rhetoric. Additionally, one of the largest TV broadcasters in Brazil, *Record*, is owned and operated by the Evangelistic Universal Church of the Reign of the God. Other small broadcasting stations are owned by the Catholic Church. Researchers have noted the negligence of media in this sphere, which present unclear messages regarding evolution (Pazza & Kavalco, 2007). Because there has not been a history of anti-evolution propaganda as long as there has been in the United States, which has been steeped in these debates for some time and has many organizations to speak out against the evangelical proselytizations, Brazil does not yet have many of these groups. As Pazza, Penteado, and Kavalco (2010) state, "in Brazil, a Society for the Study of Evolution has neither been created yet nor has a committee engaged in the study of issues related to the teaching of evolution" (p. 112).

The above state of affairs highlights the centrality of institutionalized sites such as associations and organizations to the emergence of social movements and counter-movements. As Polletta and Jasper (2001) write, "such institutions supply the solidary incentives that encourage movement participation" and shared space for members to forge new identities, develop group affinity, mobilize resources, and coordinate collective action. These were precisely the sort of sociological processes that took place subsequent the founding of the creationist societies and institutes in Brazil. These associations tactically made available to the public a variety of institutionalized spaces (physical and virtual) including educational centers, websites, online stores, multimedia, annual conferences, journals (e.g., the *Creationist Magazine*), books (e.g., *Biblical Creationist Cosmovision, Studies in Science and Religion*) for oppositional identity work and adversarial reframing (Benford & Snow, 2000) of evolution education as an eminent threat to Brazilians' cultural and religious values. A conspicuous example of such tactic can be found on the *Brazilian Association of Creation Research*'s website (http://abpc.impacto.org/), whose homepage requires users to choose between two links: "I'm a creationist" or "I'm not a creationist" (Fig. 7.1). To enter the website, users have to first officially select one of two opposing identities and are then directed accordingly. Such tactics seem very effective in promoting the emergence of a collective creationist identity among Brazilians, thus providing the creationist movement in Brazil with considerable momentum and impetus.

Fig. 7.1 Brazilian Association of Creation Research's website

7.4 Evolutionary Theory in the Curriculum

In order to attain a more comprehensive and broad understanding the place of evolutionary theory in the Brazilian school curriculum (a very large and complex educational system), we turned to current data made public by the *National Textbook Program* (FNDE, 2012). Federally funded, this government program purchases and distributes textbooks in all content areas to public schools (primary and secondary) throughout the country. For more than 80 years, this program has served as the main source of school curricula for the Brazilian public education. True to its mission to provide all public school students with free access to textbooks, the program distributed over 87.6 million secondary textbooks nationally only in the year of 2015. Such a wide scope motivated our decision to examine curricular materials distributed by this particular program as part of our efforts to assess the place of evolution theory in the Brazilian biology curriculum more broadly.

Every three years, the National Textbook Program publishes content-specific textbook guides that are designed to help public school teachers select curriculum by providing detailed descriptions and expert evaluations of commercially available textbooks in their subject areas. Once the selection period ends, the program releases the number of textbook copies distributed nationwide as selected by teachers (PNLD, 2015b). The latest *Biology Textbook Guide* at the time of this

Table 7.1 Brazilian biology textbooks and their national distribution

Textbook title[a]	Number of copies[b, c]	Ranking
Biologia Hoje (Linhares & Gewandsznajder, 2013)	547,228	First
Biologia em Contexto (Amabis & Martho, 2013)	427,116	Second
Biologia (Mendonça, 2013)	265,228	Third
BIO (Lopes & Rosso, 2013)	293,107	Fourth
Biologia (Silva, Sasson, & Caldini, 2013)	170,732	Sixth
Total	1,703,411	

[a]Three-volume book sets used sequentially in each of three years of mandatory biology in Brazilian high school
[b]Includes both student and teacher editions
[c]Includes only the volume covering the topic of evolutionary theory

study had been published in 2015 and provided teachers with nine textbook options from which to choose (PNLD, 2015a). These books were officially approved by the Brazilian government for use in public schools between the years of 2015 and 2017. In this chapter, we examine how the topic of evolution is approached in five of the biology textbooks requested by secondary teachers through this program (Table 7.1). Combined, nearly 1.7 million copies of these biology textbooks were distributed nationally, hence constituting some of the most widely used evolution curricula in Brazil.

Our content analysis of these biology textbooks revealed recurrent allusion to creationism. This curricular trend is described and illustrated below.

7.4.1 Allusion to Creationism

One particularly noticeable feature of the biology textbooks was the explicit attention given to creationism and creationist ideas in evolution units and chapters. Four of the textbooks introduced students to creationism, often before even addressing evolutionary theory itself.

In *BIO*, Lopes and Rosso (2013) included a single paragraph about creationism early in evolution unit (third page):

> Before life was understood as being the result of evolution, living beings were considered divine creations, referred as creationism. Created beings do not undergo changes over time, referred as fixism. Species were considered static groups of organisms, similar to an ideal and unchanging type, characterized by its own essence. Variations among members were merely accidental, and not essential. This way of thinking constitutes what it is known as essentialism. Essentialists do not accept variation, because for them variability is accidental and irrelevant (p. 10).

As can be seen above, Lopes and Rosso (2013) make explicit references to creationism, also identifying it as a conception of life that is aligned with a fixist paradigm no longer prevalent in scientific thought. This is was the only reference made to creationism throughout this entire textbook.

Likewise, Linhares and Gewandsznajder (2013) quickly acknowledge creationism at the onset of the evolution chapter in *Biologia Hoje*. Written in as a historical narrative the chapter begins with the follow passage:

> According to prevalent thought until the XVIII century, each species had appeared in an independent fashion and retained the same characteristics. Even the Swedish naturalist Carl von Linné (1707–1778; Lineu in Portuguese), who created in 1735 the first system of biological classification, accepted this idea, known as creationism or fixism. In the beginning of the XIX century the hypothesis of transformation of species started to be defended by some scientists to explain diversity of species and the existence of fossils of organisms different from current organisms (p. 159).

Like the other textbook authors, Linhares and Gewandsznajder (2013) allude explicitly to creationism, describing it as part of fixism—a paradigm to which famous scientists like Linné once subscribed. However, rather than abandoning the topic altogether, creationism resurfaces at the end of the chapter where one finds a large informative box in blue that occupies an entire page. Entitled "Evolution and Religion," it informs students about present day controversy surrounding evolution:

> Some religious groups, however, disagree with the theory of evolution and defend, for example, the idea that living beings were created by God exactly as written on the Bible, that is, they defend creationism (p. 166).

Linhares and Gewandsznajder (2013) also provide students with numerous quotations from science experts (e.g., Francis Collins, Stephen Jay Gould, and Carl Sagan) and a long list of book references to support the stance that science and religion can be compatible (i.e., a religious person can accept both God and evolution), even though they are epistemologically distinct human endeavors.

In César, Sesar and Caldini's (2013) *Biologia*, the chapter on evolution begins with a large informative box entitled "Fixism and Transformism" that takes up the entire first page and top of second page. It begins with the following passage:

> The notion that living beings change over time – called transformism – is for us as familiar as the idea that cells are the unit of life, or that that the DNA is our genetic material. However, for a long time, it was believed that each living species was fixed and unchanging – belief known as fixism – having appeared by divine creation and maintaining the same original characteristics. The fixist ideas, supported by Aristotle even in Antiquity, remained prevalent until the XIX century. This was due to the fact that Bible study reinforced the notion that the Earth had 6 thousand years, and that each life form was constructed, one by one, by the Creator… (p. 202).

Once again, explicit allusion of creationism as an archaic and invalid notion is followed by the identification of famous scientists who held fixist ideas such as Linné and Georges Cuvier. The remainder of the chapter provides a detailed historical account of transformist ideas, particularly the scientific contributions of Lamarck, Darwin, Mayr, and Dobzhansky. No other mention of creationist notions can be found throughout the text, except in the glossary of scientific terms found at the end of the textbook which includes the following two entries:

Divine creation (creationism) Explanation accepted by creationists about the origin of life, according to which life appeared by intervention of a divinity or superior force (p. 374)

Fixism Belief, not supported by science, that living beings are fixed, that is, do not change over time (p. 375)

In comparison to the above textbooks, *Biologia em Contexto* (Amabis & Martho, 2013) stood out due to its unusually strong creationist focus both textually and visually. Inspection of its contents revealed that the chapter on evolution began with a relatively large and colorful reproduction of the *Michelangelo's Creation of Adam* as portrayed in the Sistine Chapel's ceiling. Located just below the chapter title ("Chap. 9: Fundamentals of Biological Evolution"), this image occupied the entire upper half of the first page, being followed by a fairly long section entitled "Myths of Creation" (see Fig. 7.2). In it, the authors provide a detailed account of creationism across Ancient and Medieval civilizations (creationist myths in Greece, Egypt, Inca, Viking, and Christian cultures). The paragraph dedicated to Christian creationism reads:

> One biblical version of the creation of the world states that in the beginning there was only water, and above it the spirit of God. Then, God blew on the liquid surface and there appeared a light, which began to illuminate the darkness. This happened on the first day of creation. On the second day, He created the sky; on the third, fertile soil and plants, including fruit-bearing ones. On the fourth day, God created the Sun, the Moon, and the stars, placing them in their proper places on the cosmos. On the fifth day, fish in the water and birds in the sky were created. On the sixth day, God filled the earth with all species of animals and created man, to his image and resemblance. On the seven day, he rested (p. 206).

This historical account of creationist mythology is then followed by an informative box entitled "Importance of the Topic" in which the authors first make a generalized statement about how religious groups currently interpret creationism ("most Christians and Jews believe the creation myth as described on the Bible a allegoric explanation for the creation of the universe") and then attempt to justify the study of evolution by stating that "knowing what science says about our origins can help us reflect about our history and our relation with the universe, of which we are part (p. 207).

A noticeable exception to the above curricular trend toward incorporation of creationism was Mendonça's (2013) *Biologia*. Unlike all other textbook authors, Mendonça (2013) did not make any mention to creationism or creationist ideas anywhere in the textbook (terms like creationism were completely absent from the student edition). However, in the teacher manual, Mendonça (2013) warned biology instructors about the possibility of controversy to ensue during the unit on evolution:

> The topic [evolution] can lead to conflict with other schools of thought, like creationism, and it is healthy to open up this discussion in class if it occurs naturally as a result of students' interest. It is an opportunity to understand and emphasize that Science has its own objectives and methods, which distinguishes it from other ways of interpreting nature (p. 73; Mendonça, 2013).

Fig. 7.2 First page of evolution chapter in *Biologia em Contexto* (Amabis & Martho, 2013)

Like the student edition, there is complete avoidance of the word "creationism" in the text. Rather than shy away from discussing, teachers are encouraged to treat any disagreement or contention from students as an opportunity to help them better comprehend the nature of science.

In conclusion, our textbook analysis revealed that biology curriculum developers' approaches ranged from complete avoidance (least favored) to explicit and detailed integration of creationism into evolution chapters (most favored). This finding indicates that creationism has indeed become a part of the operationalized biology curriculum in Brazil. Despite its absence from the Brazilian official biology learning standards, it was a prevalent topic introduced to varied extents alongside evolution in the curricular materials of nearly 1.5 million students in Brazilian public schools between the years of 2015 and 2107. There appears to exist a certain degree of misalignment and tension between the *intended curriculum* (what is formally and ideally envisioned by educators as outlined in educational documents) and the *operationalized written curriculum* (how the vision is operationalized as curricular materials to be put into action by teachers) (van den Akker, 2003). Such an integration of creationism as an official topic of instruction in biology textbooks raises questions and concerns about the possibility of creationism being mistaken by teachers and students as part of the scientific cannon. As emphasized by many scholars, science textbooks are typically seen as a written genre wherein authors (disciplinary experts) produce texts meant to initiate students (novice learners) into the specialized world of a scientific field (Hyland, 2002) by providing them with currently accepted facts organized into a coherent and unproblematic body of knowledge (Myers, 1992) or a "current map of the field" (Sutton, 1989). Given this expectation that the textbook will serve as an important source of a factual base for learners, incorporation of creationism into Brazilian biology textbooks comes with the risk of miscommunicating creationism not only as a specialized term that is part of scientific jargon but also as a concept that is part of scientific knowledge itself.

7.4.2 *Evolution Instruction*

While empirical examination of classroom instruction is beyond the scope of this chapter, previous research studies provide evidence that, regardless of which text biology teachers choose to use in the classroom, profiles of evolutionary teaching in Brazil have been generally characterized by a paltry focus on evolutionary theory. The topic is often covered in a few class sessions at the end of secondary education (Tidon & Lewontin, 2004), rather than being presented a central theme that unites areas of science (a "trans-disciplinary subject" that permeates all other content of biology, as recommended by the Brazilian Ministry of Education).

7.4.3 *Evolution in College-Entrance Exams*

The two main examinations used by Brazilian universities for admitting students are the Vestibular (a university-specific qualification test) and the National High School Exam or ENEM (a non-mandatory, standardized national exam) administered by Ministry of Education since 2009. Because the five biology textbooks above included test items from these two examinations, we were provided with a glimpse of how evolution features in Brazilian college-entrance examinations. Unlike the biology texts, test items on the origin of life and biological change focused exclusively on evolutionary theory. Not a single test item alluded to creationism or prompted students to provide non-scientific explanations for these biological phenomena. Instead, students were consistently assessed in terms of their ability to articulate evolutionary explanations and understanding of the scientific content of evolutionary theory. On an epistemic level, these test items were characterized by either dualism (evolution as a collection of absolute facts that were right/wrong) or contextual relativism (evolution as a type of knowledge that is complex, contextual, and open to reevaluation; right or wrong answers exist in specified contexts, and adequacy judgments must be made). Such a trend suggests a degree of epistemic misalignment between the biology curriculum used in public school and college-entrance exams (Table 7.2).

7.4.4 Evolution in Educational Policies

High-school curriculum development in Brazil is regulated by the National Curriculum Parameters for Secondary Science Education (PCNEM). Published by the Ministry of Education, this document outlines the science content learning standards for secondary schools nationally. A major theme emphasized throughout this document is the importance of approaching biology instruction historically. As stated on the PCNEM (PCN, 2002b), "this [historical] stance seeks to overcome the ahistoric vision that many [biology] textbooks have disseminated" (p. 44). A similar recommendation is made for the specific topic of evolution:

> Evolution should be approached historically, showing how distinct periods and schools of thought held different ideas about the origin of life. It is important to relate them to the historical moment in which they were elaborated, recognizing the limits of each in explain the phenomenon (PCN, 2002a; p. 16)

Students are expected, among other things, to develop an evolutionary conception of life, to be introduced to various lines of evidence (e.g., embryologic, geologic, genetic, paleontological, etc.) that support evolutionary claims and to grasp evolutionary concepts such as adaptation and natural selection.

While no explicit allusion is made to creationism or God, science educators are encouraged to promote student comprehension of the philosophical tenets of different explanatory systems as part of their historical approach to biology:

> In the course of the history of humankind there have been several different explanations for the emergence and diversity of life, so scientific models have lived side by side with other explanatory systems, such as, for instance, those of a philosophical or religious inspiration (PCN, 2002b; p. 43).

Table 7.2 Sample test-items on evolution from Brazilian College-entrance exams

---

Test-item 1: non-human evolution

(Enem, MEC) Snakes are among the poisonous animals that cause the largest number of accidents in Brazil, especially in rural areas. Rattlesnakes (*Crotalus spp.*), despite being extremely poisonous, are snakes that, compared to other species, cause few accidents to humans. This is due to the noise of its "rattle", which helps victims perceive its presence and avoid it. These animals only attack human beings to defend themselves and feed on small rodents and birds. However, they have been continuously hunted for they are easily detected. Lately, scientists have observed that these snakes have become quieter, which is problematic since, if people cannot perceive them, the risk of accident increases. The Darwinist explanation for the fact that rattlesnakes have become quieter is that:

(a). the need not to be found and killed changed their behavior;
(b). alterations in their genetic code occurred to enhance them;
(c). successive mutations kept happening so that they could adapt;
(d). quieter varieties were positively selected;
(e). the varieties undergone mutations to adapt themselves to the presence of human beings.

Test-item 2: human evolution

(FGV-SP[a]) It is common for books and the media to represent the evolution of Homo sapiens as a progressive succession of species, like the figure below:

On the extreme left of the figure, it is placed the oldest species, individuals bent over, with long arms and simian faces. The figure is then constructed by continuously adding more recent species to the right: Australopithecus nearly erect, Neanderthals, and it ends with the modern man. This representation is:

(a). adequate. Human evolution occurred along a continuous and progressive line. The fossil of each species discovered is a direct ancestor of more recent and modern species;
(b). adequate. The species represented demonstrate that men are descendants of older and less evolved species of the family: gorilla and chimpanzee;
(c). inadequate. Some species represented in the figure are extinct and did not leave descendants. Human evolution would be better represented by inserting gaps between species, and keeping only existing species;
(d). inadequate. Some species represented in the figure cannot be ancestors to subsequent species. Human evolution would be better represented as tree branches with each species positioned at the tip of each branch;
(e). inadequate. The species represented in the figure are current and, therefore, should not be represented as a line. Human evolution would be better represented with the species placed alongside.

---

[a]FGV-SP is short for Fundação Getúlio Vargas-São Paulo, a highly ranked institution of higher-education renowned for its large amount of academic research

Throughout the text, strong emphasis is also placed on the tentative nature of science as well as the limitations of the scientific endeavor. Science is to be seen by students as a source of explanatory concepts that can be called into question, not definite answers or facts. However, this document is somewhat unclear about the epistemic status of scientific knowledge, at times conveying a certain degree of multiplicity or relativism—evolution is simply conveyed as one of many possible explanations for the origin of life. Furthermore, adoption of a secular pedagogical approach is not explicitly identified as a requirement anywhere in the text.

7.5 Evolutionary Theory in Biology Teacher Education Programs

Universities have also been linked to anti-evolution sentiments, and this certainly has an impact on teachers in training. Indeed, some institutes of higher education, such as Mackenzie Presbyterian Institute and Pueri Domus in São Paulo (founded in 1870, Mackenzie is one of the oldest institutions of higher education in Brazil and is regarded as a center of excellence having graduated numerous important names in Brazilian politics and society), actually incorporate creationism into their science classes (Silva & Prado, 2010). Pazza et al. (2010) study on freshman biology students at the Universidade Estadual do Centro-Oeste do Paraná revealed significant misconceptions and misunderstandings of evolutionary theory. Although most students in the study understood change passes through inheritance, they were unsure how evolution actually occurred (i.e. random effects, variation, and natural selection). They believed evolution was goal-driven toward progress. This is consistent with other research indicating Lamarckian misconceptions showcased by Brazilian biology teachers (Tidon & Lewontin, 2004). Furthermore, the freshman biology students at the Universidade Estadual do Centro-Oeste do Paraná were amenable to teaching creationism as an alternative to evolution in their future classrooms (Penteado, Kavalco, & Pazza, 2012). Some of the university students will eventually become biology teachers, and if their perceptions and understandings of evolution through the teaching/learning process in universities are unchanged, they might teach misinformation to their students.

Berkman et al. (2008) observed that teachers who attended more class hours of evolution during their majoring courses demanded more time for evolution teaching in their biology classes, and well-prepared teachers used up to 60% more time in teaching evolution than the others. This fact suggests that teachers' preparation is the key to providing students a complete and qualified view of the evolutionary process. In a Brazilian context, Clément and Quessada's (2013) work showed that the amount creationist conceptions is significantly lower given the number of years or teacher training at University. Thus, the role of teacher preparation programs is important in educating students about evolutionary theory and the nature of science as well as equipping teachers to navigate the social, political, and religious confluences that surround the teaching of evolution.

7.6 Biology Teachers' Attitudes Toward Teaching Evolutionary Theory

An international research project, termed the Biohead-Citizen Project, involved 18 countries in 2004–2008 and has since been expanded to 30 additional countries. The aim of the project is to determine teachers' conceptions related to evolution and the separation of science and religion (Carvalho, Clément, Bogner, & Caravita,

2008). Overall, results from the large-scale survey showed the level of education in the various counties made little difference in terms of answers to the survey questions, which probes respondents on their beliefs about the origin of life as well as their own practicing of religion and beliefs in God. Additionally, in most counties, there was little difference in the answers of teachers having degrees in biology than those of their colleagues (Clément, 2015). On the whole, researchers found that countries that were less economically developed seemed to have more teachers practicing a religion and espousing Creationist beliefs. Caldeira, Araujo, and Carvalho's (2012) use of the Biohead-Citizen Project in Brazil showed that, compared with the other countries, the Brazilian sample of teachers showed a higher percentage of creationist conceptions, particularly for biology teachers and future teachers. While their work showed a strong influence of religious values on conceptions about the origin of life, researchers did, however, show that this influence is less strong for biology teachers than for other groups—indicating that time spent learning about evolution has lessened their creationist conceptions.

Researchers have also investigated conceptions of protestant biology pre-service teachers in State University of Feira de Santana and the ways students border cross as they consider their religious worldviews in light of their pedagogical content knowledge and responsibility (El-Hani & Sepulveda, 2010). Qualitative analyses of protestant students close to the end of their pre-service teacher preparation program indicted two groups existed: those possessing a more scientifically compatible worldview and were able to synthesize their understanding of science with their religious worldview and those who were inclined to repudiate scientific knowledge. For both groups, it is essential that their worldviews be respected; however, it is equally important that their understanding of evolution and their pedagogical content knowledge be present and accounted for prior to teaching the young minds of their future classrooms.

7.7 Suggestions to Improve Evolution Education in Brazil

Researchers recommend several approaches to improve evolution education in Brazil: (1) providing continuous training of school teachers through workshops for professional advancement; (2) reinforcing the curricular program of the Ministry of Education; and (3) analyzing the books used in secondary education (Tidon & Lewontin, 2004). Not unlike recommendations for teaching evolution in the United States, constructivist approaches that connect evolution to meaningful and relevant topics have been advocated for teaching evolution (Penteado et al., 2012). Because of Brazilians' deeply rooted adherence to a belief in God, it is especially important to take into account students' worldviews using the strategies of Multicultural Science Education. Conceptions of science should be redefined to include other ways of knowing than only those of Western Modern Science (Mazzocchi, 2006). For instance, indigenous knowledge is key to considering perspectives regarding the interactions of humans and nature. Attending to social and epistemological implications for student learning, teachers should carefully and critically consider the most appropriate and effective facilitative approach to evolution discussion for their specific educational context. Teachers should, El-Hani and Sepulveda (2010) argue, focus on students' understanding and reasoning while "offering them ground to be critical, reflexive, and open-minded toward human knowledge in all its variety" (p. 122). However, notwithstanding the pedagogical approaches advocated by evolution educators, at the core teachers need to possess a solid understanding of the theory of evolution and of the nature of science. As well, they need access to solid resources and curricular materials that can support and reinforce these understandings.

7.8 Conclusions

Undoubtedly, it is problematic when biology teachers themselves do not possess or have not garnered through their teacher preparation programs clear understandings of evolution and the nature of science as well as pedagogical approaches to teaching these unifying concepts. Couple a lack of understanding with vague messages from the Ministry of Education about defining parameters regarding what should be taught in a science classroom, and a teacher can get caught in the crosshairs of public pressure and creationist propaganda. Too often in Brazil, we see teachers who marginalize the topic of evolution to the end of the curriculum without underscoring the ways in which this unifying theory supports many biological phenomena. Moreover, the presentation of multiplicity or relativism supported in public school textbooks is epistemically misaligned with the college-entrance exams, which could inadvertently disadvantage public school students when competing with more privileged students from the burgeoning private school system—a potential social justice issue on which more research is needed. Underlying the integration of creationism into evolution chapters in Brazilian biology textbooks was an assumption that explicit allusion to creationism would invariably lead to friendly and cooperative contentiousness (i.e., engaging and epistemically productive dialogue). Teacher manuals simply encouraged instructors to capitalize on such creationist allusions and to treat them as opportunities for students to openly express their own views and to learn about the nature and history of science. In these manuals, the existence of cooperative learners was simply presumed a priori, and the possibility of agonistic, disruptive disagreement to ensue instead not entertained. However, student cooperation is contingent on rapport (i.e., teacher ability to actively create a healthy and positive social atmosphere), and should not simply be taken for granted by curriculum developers. As emphasized by many scholars of academic discourse, the emergence of *agonism* (Tannen, 2002) in intellectual exchanges is a real possibility that too often obfuscates and even hampers the intellectual pursuit of knowledge. More often than not, intellectual disagreement is approached by the layperson simplistically as a polarized "battle" between two warring camps. Therefore, it stands to reason that teacher manuals should not only acknowledge the possibility of evolution instruction to be met with resistance and opposition (from students, parents, local stakeholders, etc.), but also provide Brazilian biology instructors with guidance on how to deal with adversarial or combative situations characterized by vitriolic attacks, sarcastic innuendo, and even name-calling.

Acknowledgements The authors would like to thank Solângela F. Wanderley, a public school teacher in Brazil, for her invaluable guidance and generous assistance with the identification and purchasing of currently used biology textbooks. This study would not have been possible without her.

References

Amabis, J. M., & Martho, G. R. (2013) *Biologia em contexto: Adaptação e continuidade da vida* (1st ed., Vol. 2). São Paulo: Editora Moderna.
Antunes, A., Brown, C., & Monte, M. (2002). *Tribalistas*. Retrieved from http://www.marisa monte.com.br/pt/musica/tribalistas/letra/tribalistas-tribalistas.
Benford, R. D., & Snow, D. A. (2000). Framing process and social movements: An overview and assessment. *Annual Review of Sociology, 26,* 611–639.
Berkman, M., Pacheco, J., & Plutzer, E. (2008). Evolution and creationism in America's classrooms: A national portrait. *PLoS Biology, 6*(5), e124. https://doi.org/10.1371/journal.pbio. 0060124.
Blancke, S., Peter, C., & Kjærgaard, P. C. (2016). Creationism invades Europe. *Scientific American*. Retrieved from https://www.scientificamerican.com/article/eurocreationism/.
Brum, E., Fonseca, C., & Cardoso, N. (2005). E no princípio era o que mesmo? *Época*. Retrieved from http://revistaepoca.globo.com/Revista/Epoca/0,,EDG68197-6014,00-E+NO+PRINCIPIO +ERA+O+QUE+MESMO.html.
Carvalho, G., Clément, P., Bogner, F., & Caravita, S. (2008). *BIOHEAD-Citizen: Biology, Health and Environmental Education for better Citizenship, Final Report*. Brussels: FP6, Priority 7, Project N° CITC-CT-2004-506015.
Caldeira, A. M. A., Araújo, E. S. N. N., & Carvalho, G. S. (2012). Creationism and evolution views of Brazilian teachers and teachers-to-be. *Journal of Life Sciences, 6,* 99–109.
Clément, P. (2015). Creationism, science and religion: A survey of teachers' conceptions in 30 countries. *Procedia - Social and Behavioral Sciences, 167,* 279–287.
Clément P., & Quessada M. P. (2013). Les conceptions sur l'évolution biologique d'enseignants du primaire et du secondaire dans 28 pays variant selon leur pays et selon leur niveau d'étude. *Actes AREF 2013* (en ligne: symposium 188/3, 19 pp.). Retrieved from http://hal.archives-ouvertes.fr/hal-01026095.
El-Hani, C., & Sepulveda, C. (2010). The relationship between science and religion in the education of protestant biology preservice teachers in a Brazilian university. *Cultural Studies of Science Education, 5*(1), 103–125. Retrieved from http://link.springer.com/article/10.1007/ s11422-009-9212-7.
Falcão, E. (2008). Religious beliefs: Their dynamics in two groups of life scientists. *International Journal of Science Education, 30*(9), 1249–1264.
Fundo Nacional de Desenvolvimento da Educação (FNDE). (2012). *Programa Nacional do Livro Didático*. Retrieved from http://www.fnde.gov.br/programas/livro-didatico.
Harmon, K. (2011). Evolution abroad: Creationism evolves in science classrooms around the Globe. *Scientific American*. Retrieved from https://www.scientificamerican.com/article/ evolution-education-abroad/.

Hyland, K. (2002). Directives: Argument and engagement in academic writing. *Applied Linguistics, 23*, 215–239.
Linhares, S. V., & Gewandsznajder, F. (2013). *Biologia hoje: Genética, evolução, ecologia* (2nd ed., Vol. 3). São Paulo: Editora Ática.
Lopes, S. G. B. C., & Rosso, S. (2013). *BIO* (2nd ed., Vol. 3). Brasilia: Editora Saraiva.
Mariano, R. (2004). Expansão pentecostal no Brasil: o caso da Igreja Universal. *Estudos Avançados, 18*(52), 121–138.
Mazzocchi, F. (2006). Western science and traditional knowledge: Despite their variations, different forms of knowledge can learn from each other. *Science and Society, 7*(5), 463–466.
Mendonça, V. L. (2013). *Biologia: O ser humano, genética, evolução* (2nd ed., Vol. 3). São Paulo: Editora AJS.
Miller, J. D., Scott, E. C., & Okamotto, S. (2006). Public acceptance of evolution. *Science, 313*, 765–766.
Myers, G. A. (1992). Textbooks and the sociology of scientific knowledge. *English for Specific Purposes, 11*, 3–17.
Pazza, R., Penteado, P., & Kavalco, K. (2010). Misconceptions about evolution in Brazilian freshmen students. *Evolution: Education and Outreach, 3*, 107–113.
Pazza, R., & Kavalco, K. (2007). Imprecisão e licença científica, o retorno. *Observer Impact, 12*, 461. Retrieved from http://observatorio.ultimosegundo.ig.com.br/artigos.asp?cod=461OFC001.
PCN. (2002a). *Parâmetros curriculares nacionais do ensino médio: Parte III ciências da natureza, matemática e suas tecnologias*. Brasília: Ministério da Educação. Retrieved from http://portal.mec.gov.br/seb/arquivos/pdf/ciencian.pdf.
PCN. (2002b). *National curriculum parameters secondary education*. Brasília: Ministry of Education. Retrieved from http://portal.mec.gov.br/seb/arquivos/pdf/pcning.pdf.
Penteado, P., Kavalco, K., & Pazza, R. (2012). Influence of sociocultural factors and acceptance of creationism in the comprehension of evolutionary biology in freshman Brazilian students. *Evolution: Education and Outreach, 5*, 589–594.
Polletta, F., & Jasper, J. M. (2001). Collective identity and social movements. *Annual Review of Sociology, 27*, 283–305.
Programa Nacional do Livro Didático (PNLD). (2015a). *Guia dos livros didáticos PNLD 2015: Biologia*. Retrieved from http://www.fnde.gov.br/arquivos/category/125-guias?download=8998:pnld-2015-biologia.
Programa Nacional do Livro Didático (PNLD). (2015b). *PNLD 2015: Coleções mais distribuídas por componente curricular*. Retrieved from https://www.fnde.gov.br/arquivos/category/35-dados-estatisticos?download = 9374:pnld-2015-colecoes-mais-distribuidas-por-componente-curricular-ensino-medio.
Silva, C., Jr., Sasson, S., & Caldini, N., Jr. (2013). *Biologia* (11th ed., Vol. 3). Brasilia: Editora Saraiva.
Silva, H., & Prado, I. G. O. (2010). Creationism and intelligent design: Presence in the Brazilian educational policy. *Procedia Social and Behavioral Sciences, 2*, 5260–5264.
Sutton, C. R. (1989). Writing and reading in science: The hidden messages. In R. Millar (Ed.), *Doing science: Images of science in science education* (pp. 137–159). UK: The Falmer Press.
Tannen, D. (2002). Agonism in academic discourse. *Journal of Pragmatics, 34*, 1651–1669.
Tidon, R., & Lewontin, R. C. (2004). Teaching evolutionary biology. *Genetics and Molecular Biology, 27*, 124–131.
Van den Akker, J. (2003). Curriculum perspectives: An introduction. In J. Van den Akker, W. Kuiper, & U. Hameyer (Eds.), *Curriculum landscapes and trends* (pp. 1–10). Dordrecht: Kluwer Academic Publishers.

Alandeom W. Oliveira is an associate professor of science education at the State University of New York at Albany. He earned a Master's degree in science education at Southeast Missouri State University (2002) and a PhD degree in science education at Indiana University Bloomington (2008). He has taught science education courses to teachers in Brazil and the US and has coordinated multiple professional development programs for school teachers, including Science Modeling for Inquiring Teachers Network, and Technology-Enhanced Multimodal Instruction in Science and Math for English Language Learners. His research interests include cooperative science learning, inquiry-based teaching, and classroom discourse.

Dr. Kristin Cook is the Interim Associate Dean of the School of Education at Bellarmine University in Louisville, KY. She received her doctoral degree at Indiana University- Bloomington in Curriculum & Instruction specializing in Science Education and Environmental Sciences. A former high school biology teacher, Kristin has served as a professional developer and consultant for elementary, middle, and high school STE(A)M-focused school reform and project and problem-based learning development. Kristin's research focuses on engaging students and pre-service teachers with the community of science through the exploration of socio-scientific inquiry and integrated STEAM instruction.

Chapter 8
Evolution Education in Galápagos: What Do Biology Teachers Know and Think About Evolution?

Sehoya Cotner and Randy Moore

Abstract In Galápagos, whose economy is based on tourism, the idea of evolution is common throughout society—including in school curricula. Biology teachers in Galápagos love the idea of evolution and are confident that they understand evolution. However, this confidence is not accompanied by a corresponding knowledge of evolution or an acceptance of several evolutionary principles. For example, although all biology teachers in Galápagos are familiar with Charles Darwin and his book *On the Origin of Species*, most favor Lamarckian explanations for life's diversity over those proposed by Darwin. The cognitive dissonance of accepting evolution, often alongside a literal interpretation of Genesis, suggests that biology teachers' ideas about evolution have been decoupled from economic priorities in the archipelago.

8.1 Introduction

Over 180 years after Charles Darwin set foot in the Galápagos archipelago—and put in motion ideas that would culminate in publication of *On the Origin of Species* (1859) almost 25 years later—these islands have been greatly transformed. Once a barren, rocky landscape host to overwhelming endemism and characteristically unafraid fauna, today they house thousands of non-native species, more than 25,000 human residents, and over 170,000 visitors annually (Moore & Cotner, 2013). Four of the archipelago's major islands—Santa Cruz, San Cristóbal, Isabela, and Floreana—have towns or villages (notably Puerto Ayora, Puerto Baquerizo Moreno, Puerto Villamil, and Puerto Velasco Ibarra, respectively) large enough to demand medical facilities, food markets, and schools. These schools are the focus

---

S. Cotner · R. Moore (✉)
Department of Biology Teaching and Learning, University of Minnesota,
420 Washington Avenue SE, Minneapolis, MN 55455, USA
e-mail: rmoore@umn.edu

S. Cotner
e-mail: sehoya@umn.edu

© Springer International Publishing AG, part of Springer Nature 2018
H. Deniz and L. A. Borgerding (eds.), *Evolution Education Around the Globe*, https://doi.org/10.1007/978-3-319-90939-4_8

of the work described below. Here, in the cradle of evolutionary thought, what do biology teachers—who work in these schools—know and think about evolution?

Darwin in Galápagos

HMS *Beagle* visited Galápagos for five weeks in 1835, a late addition to the ship's five-year voyage. The *Beagle*'s primary intent was to map the coastline of South America. However, along the way, Charles Darwin—the ship's 26-year-old unofficial naturalist—took several hikes to study geology and collect plants and animals. While in Galápagos, Darwin visited four islands and collected hundreds of specimens, including several varieties of mockingbirds. These mockingbirds (or "mocking-thrushes") fascinated the young Darwin, who commented on their place-specific features: "In the Galapágos Archipelago, many even of the birds, though so well adapted for flying from island to island, are distinct on each: thus there are three closely-allied species of mocking-thrush, each confined to its own island" (Darwin, 1859, p. 402).

After he returned to England, Darwin gave the specimens he collected in Galápagos to several prominent naturalists (e.g., ornithologist John Gould), whose observations helped Darwin formulate his ideas about evolutionary mechanisms. After decades of deliberation, Darwin published *On the Origin of Species*, among the most—if not *the* most—important books in science (Laddaran, 2015; Moore, Decker, & Cotner, 2009). In *Origin*, Darwin presented his theory of evolution by natural selection, whereby heritable variation and differential reproductive success lead to adaptive evolutionary changes. While Darwin's conclusions were (Bowler, 2003; van Wyhe, 2008), and continue to be (Moore et al., 2009), societally controversial, in the 150-plus years since publication of *Origin*, copious compelling evidence has established natural selection as the only scientifically supported explanation for adaptive change (Larson, 2004).

Today's visitors to Galápagos often feel overwhelmed by Darwin and evolution. For example, vendors sell t-shirts, backpacks, stickers, books, DVDs, and other merchandise adorned with images of evolution, and most of the inhabited islands feature statues and busts of Darwin in prominent places (Fig. 8.1). In Puerto Ayora, the archipelago's largest town, and on Isabela, the archipelago's largest island, branches of the Charles Darwin Research Station lure thousands of visitors per month. Streets, buildings, businesses, and other sites named in honor of Darwin are abundant in Galápagos.

Tourism constitutes roughly 70% of the economy in Galápagos (Honey, 2008), and much of this tourism is linked—via place-names (e.g., Darwin Bay), merchandise (t-shirts featuring Darwin images and quotes from *The Origin*), and tour boats (Yaté Darwin)—with Darwin's visit and subsequent work. Therefore, one might hypothesize that people living in the archipelago would be better versed in evolutionary theory, and more accepting of Darwin's conclusions, than people elsewhere. Indeed, Darwin made the islands famous, and the livelihoods of the islands' residents depend on tourists who come to Galápagos to learn more about Darwin's visit and see the organisms that contributed to Darwin's ideas.

8 Evolution Education in Galápagos: What Do Biology Teachers Know

Fig. 8.1 Darwin and evolution are everywhere in Galápagos. This bust of a young Charles Darwin is at a public plaza along Avenida de Charles Darwin in Puerto Baquerizo Moreno on San Cristóbal. (Sehoya Cotner and Randy Moore)

Evolution education and acceptance, around the world

In many parts of the world, teaching about evolution has been, and continues to be, controversial, in most cases contributing to a populace that is relatively ill-informed about evolutionary biology. Several factors correlate with acceptance of evolution, among them the extent to which one is an "analytical" thinker (Gervais, 2015), and whether one was taught evolution but not creationism in high school (Moore & Cotner, 2009). Other factors correlate with denial of evolution, such as religious beliefs (Gervais, 2015; Moore & Cotner, 2009; National Center for Science Education, 2010), and political conservatism (Cotner, Brooks, & Moore, 2014). At the population level, acceptance of evolution can be predicted somewhat by Gross Domestic Product (the United States being a notable exception; Heddy & Nadelson, 2012). The situation is likely complicated by biology educators who themselves perceive a conflict between the biology they are teaching and religious convictions (Barnes & Brownell, 2016).

Yet rather than being extinguished by an ever-growing and overwhelming body of compelling evidence, creationism is globally pervasive (e.g., Blancke, Hjermistslev, & Kjaergaard, 2014; Clément, 2015; Miller, Scott, & Okamoto, 2006). Creation museums have become commonplace in the United States

(Creation Museums and Learning Centers, 2017), and they are now expanding elsewhere (Visit Creation, 2017). In 2007, the Council of Europe responded to the spread of creationism by passing Resolution 1580, which urges educational authorities in member states to "promote the teaching of evolution as a fundamental scientific theory" and to "oppose the teaching of creationism as a scientific discipline" (Blancke et al., 2014; Council of Europe Parliamentary Assembly, 2007). Similarly, the Interacademy Panel (IAP, a global network of science academies) issued a statement on the teaching of evolution, in which the signatories—from over 60 countries throughout Europe, Africa, Asia, Oceania, and South America—asserted that certain scientific truths have been well established, such as "our Earth formed approximately 4.5 billion years ago" and "Commonalities in the structure of the genetic code of all organisms living today, including humans, clearly indicate their common primordial origin" (IAP, 2006). In the United States, numerous scientific organizations support the teaching of evolution and reject creationism, most states' educational standards mandate the teaching of evolution, and numerous court-decisions have ruled that the teaching of creationism (e.g., as "creation science," "intelligent design") is unconstitutional.

Despite these endorsements, creationism remains surprisingly popular. Moreover, the rejection of evolution is not restricted to a particular region or religious group. For example, in Iceland, a relatively small percentage (20%) are either unsure, or reject outright, the scientific validity of evolution, but in Turkey, 75% reject evolution. The United States lags behind all western European countries, yet ahead of Turkey, with ~40% of adults accepting evolution (Miller, Scott, & Okamoto, 2006).

A recent (2014) PEW study, "Religion in Latin America," reports that throughout Latin America and the Caribbean, the percentage of people who agree that humans and other living things have evolved over time varies from 41% (in the Dominican Republic) to 74% (in Uruguay). Ecuador, which includes Galápagos, falls in the middle of this spectrum. Fifty percent of Ecuadorans surveyed agree that humans and other living things have evolved over time, and 44% think living organisms have existed in their present form since their creation. Similarly, 50% of those surveyed perceive a conflict between science and religion. In Ecuador, as in other countries, education influences these perceptions. For example, among adults with secondary education or higher, 58% agree that humans and other living things have evolved over time; in comparison, 43% of those with less than secondary education agree with this statement.

Evolution education in Galápagos

Given that evolutionary theory is typically taught in biology class, we wanted to learn—from biology teachers in Galápagos—answers to the following questions:

1. What do biology teachers in Galápagos know about the basic aspects of evolutionary thought?
2. To what extent do biology teachers accept the theory of evolution?
3. Charles Darwin is closely associated with Galápagos. What do biology teachers in Galápagos know about this connection?
4. How do biology teachers perceive their role as evolution educators, and do they value the role that Galápagos has played in the history of evolutionary thought?
5. How does the knowledge and acceptance of evolution by biology teachers in Galápagos compare with that of college students in the United States?

8.2 Methods

We addressed these questions by administering a survey to teachers on the three most populated islands in the archipelago: Santa Cruz, San Cristóbal, and Isabela. We did not survey teachers on Floreana Island (population < 300 people) because that island has only an elementary school.

We measured teachers' knowledge of evolution with the Knowledge of Evolution Exam (KEE; Moore & Cotner, 2009; Rissler, Duncan, & Caruso, 2014), a 10-item, multiple-choice quiz that assesses basic concepts of evolution. We measured the teachers' acceptance of evolution with the Measure of Acceptance of the Theory of Evolution (MATE; Rutledge & Warden, 1999), a survey that identifies the extent to which individuals accept or reject key principles of evolution. We also created several novel, Likert-scale response items to help us understand teachers' knowledge and perceptions specific to Darwin in Galápagos (e.g., approximately when Darwin was in Galápagos).

The survey was created in English and translated into Spanish. Three native Spanish-speaking people in Galápagos (and others elsewhere) examined the survey to refine the survey and ensure its accuracy and clarity. The full survey, titled "La Enseñanza de la Evolución en las Islas Galápagos" ("The Teaching of Evolution in the Galápagos Islands"), is available from the authors.

We obtained a list of Galápagos' schools and their directors from the Ministry of Education. At each school, we were given lists of the secondary-school teachers who teach biology. These teachers, who are the teachers we surveyed, are college graduates who had earned their degrees from colleges on the mainland. Although there is no specific requirement for teaching evolution in Galápagos, the topic is common in biology and other courses, exhibits, murals, and conversations in the schools and elsewhere in Galápagos (see below).

At least one of us met with each teacher to explain (and answer questions about) the survey, and we stayed with each teacher until he or she completed the survey. Teachers who completed the survey were offered a $25 honorarium and a certificate of completion. All teachers signed consent forms and were aware that they were free to omit any items on the survey. Christian Bastidas Bustos, the District Analyst for Support, Monitoring, and Regulation in Ecuador's Minister of Education office, approved our survey and granted us permission to visit all of the archipelago's schools and administer the survey. There was no time-limit for teachers taking the survey.

To provide an external comparison-group to help contextualize the teachers' responses, we have also included responses to the KEE and MATE from 535 undergraduate students in introductory biology courses for non-majors at the University of Minnesota. These surveys were given a week before the start of classes so that we could assess the students' knowledge and perceptions before their biology courses began. This survey and its administration were approved by the University of Minnesota's Institutional Review Board.

In addition to scoring the KEE, we calculated the percentage of respondents who agreed or disagreed with Likert-scale items, and the average score for each scaled (1–5) response. We used the Student's t-test to determine significant differences in mean responses between biology teachers in Galápagos, and non-biology students at the University of Minnesota.

8.3 Results

Respondents

The biology teachers in this survey ranged, roughly, between 30 and 50 years of age (we did not specifically as for their ages); 42% were male, and 58% were female. All of the teachers we contacted agreed to take the survey. On Santa Cruz, three teachers were not included in our survey; two of these teachers were on the mainland (i.e., not in Galápagos), and we could not find the other teacher. In San Cristóbal, two teachers were not included in our survey; both of these teachers were on the mainland. Thus, 38 of 43 (88.4%) of all of the targeted teachers (i.e., natural-science or biology teachers in Galápagos who might be expected to teach evolution) completed the survey. Teachers took an average of 40 min to complete the survey. No teacher expressed any reservations about the survey before, after, or while taking the survey.

What do biology teachers in Galápagos know about the core tenets of evolutionary thought?

Table 8.1 summarizes the teachers' responses to the KEE. The teachers' average score on the KEE was 36%. Most (i.e., 71% of) teachers could identify the definition of natural selection, but only 42% could identify the definition of evolution. Only about one-third (i.e., 32%) of the teachers could identify the most-fit individual from a group exhibiting a range of reproductive success (Table 8.1).

Table 8.1 A summary of teachers' responses to the Knowledge of Evolution Exam (KEE)

KEE Item #	Knowledge of evolution revealed	Percent of respondents
1	Can identify that several lines of evidence support the theory of evolution	39
2	Can identify the occurrence of evolution by natural selection in an altered environment	45
3	Understand that fitness is measured by reproductive success	32
4	Can isolate the steps leading to adaptation	5
5	Can select the correct definition of natural selection	71
6	Realize that genetic evidence suggests common ancestry for all organisms	50
7	Understand that natural selection is not a random process	37
8	Can identify the definition of evolution	42
9	Understand that mutation is the ultimate source of genetic variation	11
10	Realize that natural selection is simply one mechanism that results in evolutionary change	24

Note Numbers in the table are the percentages of respondents who chose the correct answer for each of the 10 questions

Teachers' incorrect answers also revealed important information. For example, on item five of the KEE, in which respondents are asked to select from several scenarios leading to an adaptation, over half (21/38) selected the Lamarckian explanation—that is, the teleological based on adaptations arising to meet an explicit need (in Darwin's words, "adaptations from the slow willing of animals"). Similarly, only 11% of the teachers identified mutation as the ultimate source of genetic novelty; far larger percentages attributed variation to recombination, natural selection, or hybridization. For comparison, the undergraduate students' average score on the KEE was 51%.

To what extent do biology teachers in Galápagos accept the principles of the theory of evolution?

Teachers' average responses to the MATE items are illustrated in Table 8.2. Some data are difficult to reconcile; for example, 14% (5 of 37) of the respondents agreed that "evolution is not a scientifically valid theory," yet 92% (34 of 37) agree that "evolution *is* a scientifically valid theory." These inconsistencies aside, certain themes emerge: the numbers of young-Earth advocates—that is, those agreeing that Earth is less than 20,000 years—is relatively low (20%), albeit significantly higher ($p < 0.05$) than the same numbers in a comparison population of United States undergraduates (7%). Also, the 30% of Galápagos teachers who agree that "the theory of evolution cannot be correct since it disagrees with the Biblical account of creation" is significantly greater ($p < 0.05$) than the 12% agreement in the U.S. comparison group. Nevertheless, significantly more Galápagos teachers (86%)

Table 8.2 The average amounts of agreement with items on the Measure of Acceptance of the Theory of Evolution (MATE)

MATE Item	Biology teachers in Galápagos		College students (non-biologists) in the United States	
	Percent who agree/ strongly agree	Average on 5-point scale	Percent who agree/ strongly agree	Average on 5-point scale
Modern humans are the product of evolutionary processes that have occurred over millions of years	76	2.3	76	2.05
The theory of evolution cannot be tested scientifically	32	3.42	12	3.65
Organisms existing today are the result of evolutionary processes that have occurred over millions of years	84	1.81	75	2.13
The theory of evolution is based on speculation and not valid scientific observation and testing	19	3.7	13	3.69
Most scientists accept evolutionary theory to be a scientifically valid theory	84	1.87	72	2.13
The available data are unclear as to whether evolution actually occurs	26	3.39	16	3.52
*The age of the earth is less than 20,000 years	**20**	**3.6**	**7**	**4.11**
There is a significant body of data that supports evolutionary theory	72.9	2	72	2.12
Organisms exist today in essentially the same form in which they always have	17	3.97	12	3.81
Evolution is not a scientifically valid theory	14	3.89	9	3.82
**The age of the earth is at least 4 billion years	**50**	**2.88**	**66**	**2.21**
*Current evolutionary theory is the result of sound scientific research and methodology	**86**	**1.95**	**60**	**2.35**
Evolutionary theory generates testable predictions with respect to the characteristics of life	78	2.27	58	2.40
*The theory of evolution cannot be correct since it disagrees with the Biblical account of creation	**30**	**3.32**	**12**	**3.92**
Humans exist today in essentially the same form in which they always have	19	3.78	14	3.76
Evolutionary theory is supported by factual historical and laboratory data	76	2.14	62	2.35

(continued)

Table 8.2 (continued)

MATE Item	Biology teachers in Galápagos		College students (non-biologists) in the United States	
	Percent who agree/ strongly agree	Average on 5-point scale	Percent who agree/ strongly agree	Average on 5-point scale
*Much of the scientific community doubts if evolution occurs	33	3.25	7	3.78
**The theory of evolution brings meaning to the diverse characteristics and behaviors observed in living forms	92	1.7	68	2.21
With few exceptions, organisms on earth came into existence at about the same time	42	3.12	18	3.40
**Evolution is a scientifically valid theory	92	1.73	66	2.23

Note Numbers in the table are presented as percentages of teachers or students who agree or strongly agree with the statements and as averages on a 5-point Likert scale. The averages were calculated by assigning numeric values to each survey option, with 1 = strongly agree, 2 = agree, 3 = undecided, 4 = disagree, and 5 = strongly disagree. Bold denotes items in which biology teachers ($n = 34$–38 per item) in Galápagos agreed, on average, significantly (*$p < 0.05$; **$p < 0.01$) more or less than college students in the United States ($n = 529$–535 per item)

agreed that "current evolutionary theory is the result of sound scientific research and methodology" than did U.S. undergraduates (60%; $p < 0.05$).

What do biology teachers in Galápagos know about Charles Darwin's connection to the archipelago?

Several of teachers' responses to questions about Charles Darwin's link to Galápagos are presented in Table 8.3. All of the teachers identified *On the Origin of Species* as Darwin's masterwork on natural selection. However, none knew that mockingbirds (i.e., not finches) were the birds that most impressed Darwin during his time in Galápagos. Indeed, Darwin's arguments about adaptive radiation are based on mockingbirds; finches, meanwhile, are not mentioned in *Origin*.

Again, teachers' incorrect answers are telling. Many teachers chose 1535, not 1835, as the year Darwin visited the islands, and "his ship was blown off course from Peru," not "as part of a voyage to survey the coast of South America," as the reason for Darwin's visit. In fact, Panamanian Bishop Tomás de Berlanga, in 1535, visited the islands when his ship was blown off course from Peru.

How do biology teachers perceive their role as evolution educators, and do they value the role that Galápagos has played in the history of evolutionary thought?

Table 8.4 summarizes teachers' views of their roles in evolution education and the potential conflict of religion and science. Almost all (i.e., 97%) of biology teachers

Table 8.3 Summary of teachers' responses to questions regarding "Darwin in Galápagos"

Darwin Item #	Knowledge of Darwin revealed	Percent of respondents
1	Can identify when (roughly) Charles Darwin visited the islands	50
2	Can identify why Darwin visited the islands	53
3	Are aware that mockingbirds—not finches—were the birds that most impressed Charles Darwin	0
4	Know that Darwin visited four islands during the *Beagle's* stopover in the archipelago	32
5	Can identify *The Origin of Species* as the book Darwin wrote, describing his theory of evolution and using examples from Galápagos	100

Note Numbers in the table indicate the percent of respondents (out of 38) who chose the correct answer for each of the questions

in Galápagos claimed to be "confident in my understanding of evolution." A large majority (i.e., 87–95%) of biology teachers in Galápagos enjoys teaching about Galápagos and the history of evolutionary thought. However, a comparable majority (79–82%) of the teachers are uncomfortable with teaching about evolution

Table 8.4 Teachers' views of their roles as evolution educators and the potential conflicts between science and religion

"Value of Evolutionary Thought" Item	% Agree or strongly agree	Average on 5-point scale
There is a conflict between religion and science when it comes to teaching about evolution	82	2.11
I am uncomfortable teaching about evolution	79	2.16
My students tell me that they cannot agree with the science of evolution because of their religious beliefs	47	3
My students' parents tell me that they cannot agree with the science of evolution because of their religious beliefs	29	3.34
I am confident in my understanding of evolution	97	1.66
Galapágos is closely connected to the history of evolutionary thought	95	1.53
I am proud of the connection between Galapágos and evolutionary thought	89	1.74
I enjoy teaching about evolution	87	1.93
I enjoy teaching about Galapágos and the history of evolutionary thought	95	1.62
I would like to know more about Galapágos and the history of evolutionary thought	95	1.63

Note Numbers in the table are percentages of teachers (n = 38) who agreed, or strongly agreed, with each statement. The averages were calculated by assigning numeric values to each survey option, with 1 = strongly agree, 2 = agree, 3 = undecided, 4 = disagree, and 5 = strongly disagree and perceive a conflict between religion and science. According to these teachers, smaller percentages of students (47%) and students' parents (29%) have expressed a similar conflict.

8.4 Discussion

Galápagos is part of the Republic of Ecuador, a representative democracy in northwest South America. Ecuador's Gross Domestic Product (GDP) is approximately $20 billion, and Spanish (Ecuador's official language) is spoken by most of Ecuador's 16 million residents. Ecuador, a secular country, is predominantly Catholic. In 2008, Ecuador's constitution became the first in the world to formally recognize "Rights of Nature," thereby acknowledging that people have the legal authority to enforce the rights of nature and all of life to exist, persist, maintain and regenerate its vital cycles. Ecosystems themselves can be named as defendants in legal challenges.

Prior to interpreting any findings, we must first caution against reckless comparisons or excessive extrapolation. The University of Minnesota population is one that has been discussed, in light of the MATE and the KEE, in other work (e.g., Walker et al., (2017); Cotner, Brooks, & Moore, 2010; Moore & Cotner, 2009), and these metrics have been validated for this population. However, it is not possible to know, after the fact, exactly how all the MATE and KEE items were interpreted by the Galápagos teachers; the validity of the MATE, for example, has been questioned for cross-cultural comparison (Wagler & Wagler, 2013). In fact, some of the inconsistencies we found may reflect problems with the instruments themselves (e.g., Ashgar, Wiles, & Alters, 2014). We can, however, make note of trends observed in these 38 biology teachers, and attempt to make sense of any confusion by considering the interesting example posed by Galápagos itself.

Although biology teachers in Galápagos are exceedingly confident that they understand evolution, their average score on the KEE (i.e., 36%) is the lowest yet reported for any group (e.g., see Moore, Brooks, & Cotner, 2011). This "disconnect" may be due to the interaction of religiosity with public welfare in Galápagos. Indeed, Galápagos is heavily religious and predominantly (79%) Catholic (PEW, 2014). Catholics have, in recent history, made allowances for evolution (e.g., Pope Francis, 2014); for example, in 1996, Pope John Paul II told the Pontifical Academy of Sciences that "truth cannot contradict truth" (Pew Advent, 2017). However, a literal interpretation of the Bible usually predicts young-Earth creationists' claim that Earth is only about 6,000 years old (Abelson, 1982). One of the earliest modern proponents of this idea was George McCready Price, a Seventh-Day Adventist (Moore & Decker, 2008; Whitcomb & Morris, 1961). Until recently in Galápagos, tourists and every student at the Seventh-Day Adventist school in Puerto Ayora were greeted by a large billboard proclaiming Genesis 1:1. How can it be true that "in the beginning"—six thousand years ago—"God created the heavens and the earth," while it is also true that we are all the product of over 3 billion years of slow, sometimes gruesome and often capricious natural laws? This prominent billboard, which stood for decades on Charles Darwin Avenue, supports our earlier claim that Earth's age may be the issue on which the evolution-creationism controversy balances (Cotner, Brooks, & Moore, 2010).

Although creationists' claims are often interwoven with ignorance and rejection of evolution (Moore et al., 2011), the status of evolution education in Galápagos is more complicated. The Seventh-Day Adventist church is the largest anti-evolution organization in Galápagos. However, creation museums and creationism-based tourist sites (e.g., Ark Encounter), which reject (and usually vilify) evolution by presenting claims contradicting those of modern science, are not present in Galápagos (or South America, for that matter; Visit Creation, 2017). Nor are anti-evolution organizations such as Answers in Genesis and the Institute for Creation Research, which produce a vast number of television shows, radio shows, books, magazines, home-school curricula, conferences, DVDs, and other materials in which Biblical literalism replaces evolution as the explanation for life's diversity. Rather, in Galápagos, the economy is based on tourism, and this tourism is closely linked with evolution; teachers and other residents of the islands are proud of their archipelago's link with evolution, and evolution is taught in virtually all of the public and private schools. Indeed, several travel agencies bring their tour groups to K-12 schools in Galápagos, where the tourists are entertained by student presentations showcasing Darwin, evolution, and the islands' biodiversity. Thus, while the failure to teach and learn the basics of evolution are often linked to religiosity (*a la* Rissler et al., 2014), religiosity (except for Seventh-Day Adventism) in Galápagos may not be connected to the outright rejection of evolution itself. Most teachers in Galápagos accept that evolution is supported by scientists and is closely connected to the islands' economy that sustains themselves and their families.

Interestingly, the love of *the idea of* evolution by Galápagos' biology teachers is not accompanied by clear knowledge of evolution or an acceptance of certain evolutionary principles. This acceptance of evolution, *alongside* a literal interpretation of Genesis, in Galápagos is fascinating, and suggests that ideas about evolution have been decoupled from social and economic priorities in the islands. In attempting to shed some light on this sort of cognitive dissonance, we may find clues in the Identity Protective Cognition (IPC) work of Walker et al., (2017), Kahan (2010), Kahan, Braman, Gastil, Slovic, & Mertz (2007), and McCright & Dunlap (2011). According to the IPC hypothesis, individuals may exhibit a form of *motivated cognition*, in which they are motivated to [mis]interpret scientific findings in a way that protects their in-group identities (Kahan et al., 2007). In the Galápagos context, biology teachers may have competing identities at work—that of their religious affiliation, and that of a biologist teaching evolution at a site known for evolutionary thought.

8.5 Conclusion and Suggestions for Improving Evolution Education in Galápagos

In June 2013, scientists from around the world convened on San Cristóbal for the third World Evolution Summit (Paz-y-Miño-C & Espinosa, 2013). Given the omnipresence of evolution in Galápagos, few attendees probably suspected that the archipelago's teachers—virtually all of whom are enthusiastic about evolution— know so little about evolution. Although this is concerning, we are encouraged by the fact that these teachers enjoy teaching about Galápagos and evolution, and 95% of the teachers want to learn more about the islands and the history of evolutionary thought. For a discussion of how teachers' views of evolution align with those of guides in Galápagos National Park, see Cotner et al., (2017).

Biology teachers in Galápagos would benefit greatly from pre-service and in-service workshops and other training programs focused on identifying and correcting the teachers' misconceptions about evolution. These workshops, which virtually all of the teachers said they would welcome, would be an important way for other teachers, researchers, and scientific organizations—presumably in coordination with the Charles Darwin Research Center and/or Galápagos Conservancy —to help the archipelago's biology teachers and, in the process, improve evolution education in Galápagos.

References

Abelson, P. H. (1982). Creationism and the age of the earth. *Science, 215*(4529), 119.
Ashgar, A., Wiles, J. R., & Alters, B. (2014). Discovering international perspectives on biological evolution across religions and cultures. *International Journal of the Diversity, 6*(4), 81–88.
Barnes, M. E., & Brownell, S. E. (2016). Practices and perspectives of college instructors on addressing religious beliefs when teaching evolution. *Cell Biology Education, 1*(2), ar18–ar18. https://doi.org/10.1187/cbe.15-11-0243.
Blancke, S., Hiermitslev, H. H., & Kjaergaard, P. C. (Eds.). (2014). *Creationism in Europe*. Baltimore, MD: Johns Hopkins University Press.
Bowler, P. J. (2003). *Evolution: The history of an idea* (3rd ed.). Oakland, CA: University of California Press.
Clément, P. (2015). Creationism, science and religion: A survey of teachers' conceptions in 30 countries. *Procedia: Social and Behavioral Science, 167*, 279–287.
Cotner, S., Brooks, D. C., & Moore, R. (2010). Is the age of the earth one of our "sorest troubles"? Students' perceptions about deep time affect their acceptance of evolutionary theory. *Evolution, 64*(3), 858–864.
Cotner, S., Brooks, D. C., & Moore, R. (2014). Science and society: Evolution and student voting patterns. *NCSE Reports, 34*(6), 1–11.
Cotner, S., Mazur, C., Galush, T., & Moore, R. (2017). Teaching the tourists in Galápagos: What do Galápagos National Park guides know, think, and teach tourists about evolution? *Evolution: Education and Outreach, 10*, 9. https://doi.org/10.1186/s12052-017-0072-4.
Cotner, S., & Moore, R. (2011). *Arguing for evolution: An encyclopedia for understanding science*. Westport, CT: Greenwood Press.

Council of Europe Parliamentary Assembly. (2007). Resolution 1580. Retrieved March 3, 2015, from http://assembly.coe.int/main.asp?link=/documents/adoptedtext/ta07/eres1580.htm.
Creation Museums and Learning Centers. (2017). Retrieved January 20, 2017, from www.nwcreation.net/museums.html.
Darwin, C. (1859). *On the origin of species*. London: J. Murray.
Gervais, W. M. (2015). Override the controversy: Analytic thinking predicts endorsement of evolution. *Cognition, 142*, 312–321.
Hameed, S. (2008). Bracing for Islamic creationism. *Science, 322*, 1637–1638.
Harmon, K. (2011). Evolution abroad: Creationism evolves in science classrooms around the globe. *Scientific American*. Retrieved February 28, 2015, from http://www.scientificamerican.com/article/evolution-education-abroad/.
Heddy, B. C., & Nadelson, L. S. (2012). A global perspective of the variables associated with acceptance of evolution. *Evolution: Education and Outreach, 5*(3), 412–418. https://doi.org/10.1007/s12052-012-0423-0.
Honey, M. (2008). *Ecotourism and sustainable development: Who Owns Paradise?* (2nd ed.). Washington, DC: Island Press.
IAP Statement on Teaching Evolution. (2006). Retrieved March 10, 2015, from http://www.interacademies.net/File.aspx?id=6150.
Kahan, D. M. (2010). Fixing the communications failure. *Nature, 463*, 296–297.
Kahan, D. M., Braman, D., Gastil, J., Slovic, P., & Mertz, C. K. (2007). Culture and identity-protective cognition: Explaining the white-male effect in risk perception. *Journal of Empirical Legal Studies, 4*, 465–505.
Laddaran, K. C. (2015). Poll says Charles Darwin's On the Origin of Species is most influential book. Retrieved January 19, 2017, from http://www.cnn.com/2015/11/11/world/charles-darwin-irpt/.
Larson, E. J. (2004). *Evolution: The remarkable history of a scientific theory*. New York: Modern Library.
McCright, A. M., & Dunlap, R. E. (2011). Cool dudes: The denial of climate change among conservative white males in the United States. *Global Environmental Change, 21*, 1163–1172.
Miller, J. D., Scott, E. C., & Okamoto, S. (2006). Public acceptance of evolution. *Science, 313*, 765–766.
Moore, R., Brooks, D. C., & Cotner, S. (2011). The relation of high school biology courses and students' religious beliefs to college students' knowledge of evolution. *The American Biology Teacher, 73*(4), 222–226.
Moore, R., & Cotner, S. (2009). The creationist down the hall: Does it matter when teachers teach creationism? *BioScience, 59*(5), 429–435.
Moore, R., & Cotner, S. (2013). *Understanding Galápagos—What you'll see and what it means*. New York, NY: McGraw-Hill.
Moore, R., & Decker, M. D. (2008). *More than Darwin: An encyclopedia of the people and places of the evolution-creationism controversy*. Westport, CT: Greenwood Press.
Moore, R., Decker, M. D., & Cotner, S. (2009). *No prospect of an end: A chronology of the evolution-creationism controversy*. Westport, CT: Greenwood Press.
Paz-y-Miño-C, G., & Espinosa, A. (2013). Galapagos III World Evolution Summit: Why evolution matters. *Evolution: Education and Outreach, 6*, 28.
Pew Advent. (2017). Truth cannot contradict truth. Retrieved April 24, 2017, from http://www.newadvent.org/library/docs_jp02tc.htm.
Pew Research Center. (2014, November 13). Religion in Latin America: Widespread change in a historically Catholic region. Retrieved January 20, 2017, from http://www.pewforum.org/2014/11/13/religion-in-latin-america/.
Pope Francis. (2014, October 27). Pope Francis' Address at Inauguration of Bronze Bust of Benedict XVI. Retrieved March 30, 2015, from http://www.zenit.org/en/articles/pope-francis-address-at-inauguration-of-bronze-bust-of-benedict-xvi.

Rissler, L. J., Duncan, S. I., & Caruso, N. M. (2014). The relative importance of religion and education on university students' views of evolution in the Deep South and state science standards across the United States. *Evolution: Education and Outreach, 7*, 24.

Rutledge, M., & Warden, M. (1999). The development and validation of the measure of acceptance of the theory of evolution instrument. *School Science and Mathematics, 99*, 13–18.

van Wyhe, J. (2008). *Darwin: The story of the man and his theories of evolution*. London: Andre Deutsch.

Visit Creation. (2017). Retrieved January 20, 2017, from www.visitcreation.org/type/museums.

Wagler, A., & Wagler, R. (2013). Addressing the lack of measurement invariance for the measure of acceptance of the theory of evolution. *International Journal of Science Education, 35*(13), 2278–2298.

Walker, J.D., Wassenberg, D., Franta, G., & Cotner, S. (2017). What Determines Student Acceptance of Politically Controversial Scientific Conclusions?. *Journal of College Science Teaching, 47*, 46–56.

Whitcomb, J. C., & Morris, H. M. (1961). *The genesis flood—The biblical record and its scientific implications*. Ada, MI: Baker Books.

Sehoya Cotner is an Associate Professor in the Department of Biology Teaching and Learning at the University of Minnesota (USA). Her current research interests include strategies for effective teaching of evolution and ecology, science (especially biology) for the non-scientist, and inclusivity in the sciences. She has co-authored several articles and books on evolution, including "Arguing for Evolution" and "Understanding Galápagos," both with Randy Moore.

Randy Moore is a Full Professor in the Department of Biology Teaching and Learning at the University of Minnesota (USA). He has authored numerous articles and books on teaching evolution and the nature of science, including "Arguing for Evolution" and "Understanding Galápagos," both with Sehoya Cotner, and the upcoming "The Grand Canyon: An Encyclopedia of Geography, History, and Culture," with Kara Felicia Witt.

Part III
Europe

Chapter 9
Evolution Education in England

Michael J. Reiss

Abstract Until about the year 2000, evolution was a relatively uncontested area of the school curriculum in England. It occupied a core but fairly modest place within secondary school biology (for 11–18 year-olds) and was also often considered within religious education lessons in the context of the relationship between science and religion. However, the rise of creationism in England has contributed to change this. Evolution is now increasingly seen in England as a site of contestation within the curriculum. Successive national governments have been consistent in their support for evolution as occupying a key and mandatory place in the school science curriculum. Indeed, the current version of the science curriculum now includes evolution at primary level (5–11 year-olds) for the first time. It is too soon to say what the consequences of these various socio-political and cultural forces will be. England remains a country with fairly high levels of acceptance of evolution. However, a not inconsiderable proportion of school science teachers favour presenting creationism as a valid alternative to evolution.

9.1 Public Acceptance of Evolutionary Theory

England is, of course, the country where Charles Darwin lived all of his life, other than his five years on *HMS Beagle*, and his *On the Origin of Species* was first published in England in 1859. It is perhaps unsurprising that while we don't have any hard social science data, there has been a long history in England, one that pre-dates Darwin himself, of quite widespread public acceptance of evolution. Indeed, while not all elements of Victorian society were pleased to hear of Darwin's ideas, these ideas rapidly met with broad acceptance in the Church of England and elsewhere, in part because of the care Darwin himself took to try to minimise any conflict with religion or the mores of the time (cf. Browne, 2002).

M. J. Reiss (✉)
UCL Institute of Education, London, UK
e-mail: m.reiss@ucl.ac.uk

The evolutionary view of life in England led to two main theological responses. The minority approach was the one that eventually gave rise to today's creationism. Perhaps the most ingenious and infamous of these was that of Philip Henry Gosse. In addition to being an outstanding naturalist (he was elected a Fellow of the Royal Society in 1856), Gosse had a deep religious faith and coined what has become known as the 'Omphalos hypothesis' when he published *Omphalos: An attempt to untie the geological knot* in 1857, just 2 years before Darwin's *Origin*.

The Omphalos hypothesis is an attempt to combine a serious reading of the fossil record—which suggests ages before the Garden of Eden—and a literal reading of the Bible. *Omphalos* is Greek for navel and Gosse began by wondering whether Adam had a navel (despite having not been attached by an umbilical cord to a womb). Gosse supposed that he did—just as the trees in the Garden of Eden were presumably created with tree rings. Extrapolating somewhat, the whole of the fossil record could have been created during the biblical days of creation. The critics reacted badly to Gosse, and the book bombed. Most of the first edition was eventually sold as waste paper. Charles Kingsley wrote that he could not believe that God had "written on the rocks one enormous and superfluous lie for all mankind" (as cited in Rendle-Short, 1998).

The majority theological approach to evolutionary thinking arose after the publication of Darwin's *On the Origin of Species*. The mass of evidence, the rigour of its argument, and the care Darwin took to avoid theological confrontation and the issue of human evolution were crucial in the quite rapid Victorian acceptance of evolutionary thinking. The same Charles Kingsley read a pre-publication copy of *The Origin* and wrote to Darwin: "I have gradually learnt to see that it is just as noble a conception of Deity, to believe that he created primal forms capable of self development into all forms needful *pro tempore & pro loco*, as to believe that He required a fresh act of intervention to supply the *lacunas wh.* he himself had made".

More recently, there have been a number of surveys to quantify the public acceptance of evolutionary theory in England or the United Kingdom (according to the 2011 Census, 84% of people in the UK live in England, 8% in Scotland, 5% in Wales and 3% in Northern Ireland). By and large, England/UK has high levels of acceptance of evolution. In the international study of Miller et al. (2006), the United Kingdom was ranked 6th out of 34 countries for public acceptance of evolution (Iceland was 1st, Turkey 34th).

However, in a 2008 survey commissioned by Theos and conducted by the polling company ComRes, only 37% of people in the UK said they believed that Darwin's theory of evolution was "beyond reasonable doubt", 32% said that Young Earth Creationism ("the idea that God created the world sometime in the last 10,000 years") was either definitely or probably true, and 51% said that Intelligent Design ("the idea that evolution alone is not enough to explain the complex structures of some living things, so the intervention of a designer is needed at key stages") was either definitely or probably true (Lawes, 2009; Spencer & Alexander, 2009).

It has become clear that the precise wording of such surveys is important. Indeed, Baker (2012) has argued that the Theos survey suffered from some fundamental flaws including a failure to understand creationism or to appreciate that the definition of the term 'evolution' (e.g. micro-/macro-evolution) is a critical aspect of the debate. More generally, McCain & Kampourakis (2016) have pointed out that data from studies conducted to determine acceptance rates for evolution are often misleading, so that the questions that are asked and compared to one another do not always give an authentic picture of respondents' views.

Overall, the safest conclusion seems to be that while there is broad acceptance in England of the occurrence of evolution, only a minority (albeit quite a large minority) of people accept all aspects of the standard scientific account in which all of life is held to have evolved from inorganic precursors through entirely natural processes, i.e. without any divine intervention, over several billions of years, meaning that all species on Earth share a common ancestor and are thus related.

9.2 Anti-evolution Movements

The two important anti-evolution movements are creationism and the theory of Intelligent design. Creationism exists in a number of different versions but something like 15% of adults in England believe that the Earth came into existence as described by a literal (fundamentalist) reading of the early parts of the Bible or the Qu'ran and that the most that evolution has done is to change species into closely related species (Miller et al., 2006; Lawes, 2009). For a creationist it is possible, for example, that the various species of deer had a common ancestor but this is not the case for deer, bears and squirrels—still less for monkeys and humans, for birds and reptiles or for fish and fir trees (Reiss, 2011a).

Allied to creationism is the theory of intelligent design. While many of those who advocate intelligent design have been involved in the creationism movement, to the extent that the US courts have argued that the country's First Amendment separation of religion and the State precludes its teaching in public schools (Moore, 2007), intelligent design can claim to be a theory that simply critiques evolutionary biology rather than advocating or requiring religious faith. Those who promote intelligent design typically come from a conservative faith-based position. However, in many of their arguments, they make no reference to the scriptures or a deity but argue that the intricacy of what we see in the natural world, including at a sub-cellular level, provides strong evidence for the existence of an intelligence behind this (e.g. Behe, 1996; Dembski, 1998; Johnson, 1999). An undirected process, such as natural selection, is held to be inadequate.

Most of the literature on creationism (and/or intelligent design) and evolutionary theory puts them in stark opposition. Evolution is consistently presented in creationist books and articles as illogical (e.g. natural selection cannot, on account of the second law of thermodynamics, create order out of disorder; mutations are always deleterious and so cannot lead to improvements), contradicted by the scientific evidence (e.g. the fossil record shows human footprints alongside animals supposed by evolutionists to be long extinct; the fossil record does not provide evidence for transitional forms), the product of non-scientific reasoning (e.g. the early history of life would require life to arise from inorganic matter—a form of spontaneous generation rejected by science in the 19th Century; radioactive dating makes assumptions about the constancy of natural processes over aeons of time whereas we increasingly know of natural processes that affect the rate of radioactive decay), the product of those who ridicule the word of God, and a cause of a whole range of social evils (from eugenics, Marxism, Nazism and racism to juvenile delinquency)—e.g. Whitcomb and Morris (1961), Watson (1975), Hayward (1985), Burgess (2008), Carter (2014) and articles too many to mention in the journals and other publications of such organisations as Answers in Genesis (based in the USA), the Biblical Creation Society (based in England), the Creation Science Movement (Based in England) and the Institute for Creation Research (based in the USA).

Of these organisations, which all now have an international reach, the England-based Creation Science Movement deserves a special mention in that it seems to be the oldest creationist movement in the world. It was founded in 1932 as The Evolution Protest Movement. A quotation from its website gives a flavour of its message:

> Today society witnesses to the effect of atheistic humanism which belief in the theory of evolution has brought–fragmented family units, abortion, child abuse etc. In fact in all these intervening years the evidence has mounted up arguing that of course a Creator must have made this planet Earth and the heavens. There is a wealth of further scientific evidence supporting Creation which these eminent men in the early 1930s did not then know. Advances in our knowledge of genetics, biochemistry and information theory are just some areas where progress in the last eighty years has made belief in evolution even less logical. (Creation Science Movement, 2017)

9.3 Evolutionary Theory in the School Curriculum

Until about the year 2000, evolution was a relatively uncontested area of the school curriculum in England. It occupied a core but fairly modest place within secondary school biology (for 11–18 year-olds) and was also often considered within religious education lessons in the context of the relationship between science and religion. However, the rise of creationism in England—due partly to immigration, including from Muslim families, and partly to an increasing polarisation within mainstream Christianity with a growth in fundamentalism—has contributed to change this.

For example, in September 2006 the organisation Truth in Science www.truthinscience.org.uk sent a free resource pack to the Head of Science in each UK secondary school and sixth form college. As stated on the organisation website:

> We consider that it is time for students to be permitted to adopt a more critical approach to Darwinism in science lessons. They should be exposed to the fact that there is a modern controversy over Darwin's theory of evolution and the neo-Darwinian synthesis, and that this has considerable social, spiritual, moral and ethical implications. Truth in Science

promotes the critical examination of Darwinism in schools, as an important component of science education. (Truth in Science, 2017)

At about the same time, many scientists and educators in the UK and in other countries received copies of the first volume of what intended to be a massive seven volume series titled *The Atlas of Creation*. Authored by Harun Yahya (the pen name of Adnan Oktar) these lavish books (volume 1 is 800 pages in length and weighs 5.4 kg) present a creationist critique of the fossil evidence for evolution.

Evolution is now seen in England as a site of contestation within the curriculum. On the one hand, organisation such as the British Humanist Association have successfully campaigned for more teaching about evolution in schools and for prohibitions to be placed on the teaching of creationism. On the other hand, a number (albeit a minority) of religious organisations have campaigned, with somewhat less success, for evolution to be considered as a controversial issue or, at any rate, for individual teachers and schools to have considerable autonomy as to what they teach in this area.

In the summer of 2007, after months of behind-the-scenes meetings and discussions, the DCSF (Department of Children, Schools and Families) Guidance on Creationism and Intelligent design received Ministerial approval and was published (DCSF, 2007). The Guidance points out that the use of the word 'theory' in science (as in 'the theory of evolution') can mislead those not familiar with science as a subject discipline because it is different from the everyday meaning (i.e. of being little more than an idea). In science the word indicates that there is a substantial amount of supporting evidence, underpinned by principles and explanations accepted by the international scientific community. The Guidance goes on to state:

> Creationism and intelligent design are sometimes claimed to be scientific theories. This is not the case as they have no underpinning scientific principles, or explanations, and are not accepted by the science community as a whole. Creationism and intelligent design therefore do not form part of the science National Curriculum programmes of study. (DCSF, 2007)

The Guidance points out that the nature of, and evidence for, evolution must be taught at key stage 4 (14–16 year-olds) as these topics are part of the programme of study for science, while key stages 1 (5–7 year-olds), 2 (7–11 year-olds) and 3 (11–14 year-olds) include topics such as variation, classification and inheritance that lay the foundations for developing an understanding of evolution at key stage 4 and post-16. It then goes on to say:

> Creationism and intelligent design are not part of the science National Curriculum programmes of study and should not be taught as science. However, there is a real difference between teaching 'x' and teaching *about* 'x'. Any questions about creationism and intelligent design which arise in science lessons, for example as a result of media coverage, could provide the opportunity to explain or explore why they are not considered to be scientific theories and, in the right context, why evolution is considered to be a scientific theory. (DCSF, 2007)

This seems to me a key point. Many scientists, and some science educators, fear that consideration of creationism or intelligent design in a science classroom legitimises them. For example, the excellent book *Science, Evolution, and*

Creationism published by the US National Academy of Sciences and Institute of Medicine asserts "The ideas offered by intelligent design creationists are not the products of scientific reasoning. Discussing these ideas in science classes would not be appropriate given their lack of scientific support" (National Academy of Sciences and Institute of Medicine, 2008, p. 52).

As I have argued (Reiss, 2008), I agree with the first sentence of this quote but disagree with the second. Just because something lacks scientific support doesn't seem to me a sufficient reason to omit it from a science lesson. When I was taught physics at school, and taught it extremely well in my view, what I remember finding so exciting was that we could discuss almost anything providing we were prepared to defend our thinking in a way that admitted objective evidence and logical argument. Nancy Brickhouse and Will Letts (1998) have argued that one of the central problems in science education is that science is often taught 'dogmatically'. With particular reference to creationism they write:

> Should student beliefs about creationism be addressed in the science curriculum? Is the dictum stated in the California's *Science Frameworks* (California Department of Education, 1990) that any student who brings up the matter of creationism is to be referred to a family member of member of the clergy a reasonable policy? We think not. Although we do not believe that what people call "creationist science" is good science (nor do scientists), to place a gag order on teachers about the subject entirely seems counterproductive. Particularly in parts of the country where there are significant numbers of conservative religious people, ignoring students' views about creationism because they do not qualify as good science is insensitive at best. (Brickhouse & Letts, 1998, p. 227)

More recently, Thomas Nagel (2008) has argued that so-called scientific reasons for excluding intelligent design (ID) from science lessons do not stand up to critical scrutiny (cf. Koperski, 2008). With reference to the USA he concludes:

> I understand the attitude that ID is just the latest manifestation of the fundamentalist threat, and that you have to stand and fight them here or you will end up having to fight for the right to teach evolution at all. However, I believe that both intellectually and constitutionally the line does not have to be drawn at this point, and that a noncommittal discussion of some of the issues would be preferable. (Nagel, 2008, p. 205)

9.4 Evolutionary Theory in Biology Teacher Education Programs

There appear to be no systematic data on the place of evolution in teacher education programs in England nor am I aware of any research in this area. As a long-standing (indeed, founder) member of the informal network 'Biology Education Research Group' (BERG) and a long, regular attendee of the Annual (UK-based) Conference of the Association for Science Education it is absolutely clear to me that there is overwhelming support among biology teacher educators for high quality teaching of evolution in school biology lessons.

In England, there is no detailed curriculum for biology teacher education (whether initial teacher education or continuing professional development) but the tradition is for those responsible to help beginning teachers teach school biology as well as they can. When I trained to be a biology teacher in 1982–83, I was introduced to a range of ways of teaching evolution and when I, in turned, trained university students to become secondary (1988–94) or primary (1994–2000) biology teachers, evolution was a core component of my teaching, along with other major areas of biology such as biochemistry, physiology, genetics and ecology. Similarly, the standard text in England that is intended to help biology teachers teach their subject has a chapter on evolution in both its first (Reiss, 1999) and second (Reiss, 2011b) editions. A volume currently being written for an international audience similarly has one of its chapters on evolution (Kampourakis & Reiss, forthcoming).

The Labour (1997–2010), Coalition (2010–15) and Conservative (2015-present) governments in the UK have been consistent in their support for evolution as occupying a key and mandatory place in the school science curriculum. (For historical reasons, the curriculum in Scotland is distinct from the curricula of the other three UK nations and is dealt with separately in this volume). As a result, the current version of the science curriculum now includes evolution at primary level (5–11 year-olds) for the first time and this is reflected in the greater emphasis currently being given to evolution education in the education of those training to become primary teachers (Billingsley et al., forthcoming; Russell and McGuigan, 2015). However, successful teaching of evolution in primary schools is hampered by the small proportion of primary teachers who have learnt biology since they were 16 years-old and by political moves since 2010 to reduce the involvement of universities in initial teacher education and continuing professional development of teachers.

9.5 Biology Teachers' Attitudes Toward Teaching Evolutionary Theory

The most valuable work to date in England on biology teachers' attitudes toward teaching evolutionary theory was undertaken by Hanley (2012) who undertook a small national survey and detailed work in four secondary case study schools. After considerable efforts, a national sample of 55 science teachers returned questionnaires and the data from these were supplemented by interviews with science teachers in the four case schools. A key finding was the wide range of views about evolution among science teachers. While the large majority were positive about teaching evolution (the non-representative nature of the sample means that there is little point in providing much quantitative data), some teachers, including one in one of the case study schools did not accept evolution above the level of the species. When asked their opinion about how human beings came into being, 3% of the surveyed science teachers responded 'divinely created in present form', 29% 'developed: some divine involvement', 55% 'developed: no divine involvement' and 12% 'other'.

Whether one finds these results surprising, encouraging or worrying depends on one's point of view. It is worth mentioning that among the general public in the UK there is greater support for the teaching of creationism and Intelligent design as well as evolution in school science than most scientists or science educators realise (BBC, 2006). Nevertheless, the teaching of creationism in UK science classes is deeply controversial (Allgaier, 2014; Williams, 2008) and probably rare, other than in a relatively small number of Christian, Muslim and Jewish Orthodox schools (cf. Baker, 2013; Scaramanga, 2017).

9.6 Improving Evolution Education

Evolution is widely agreed to be the central, key, unifying framework of biology. Yet many school-aged students and adults understand relatively little of the theory of evolution. Recent years have produced valuable work examining the reasons why such understanding is limited, whether for cognitive or socio-cultural reasons (Jones & Reiss, 2007; Kampourakis, 2014; Rosengren et al., 2012).

There is a large literature on the cognitive difficulties that student have in learning about evolution, difficulties that often manifest themselves in student misconceptions (Harms & Reiss, forthcoming). Evolution takes place over long periods of time and the geological notion of 'deep time' is one that is difficult for students. Then there is the problem that understanding the principal driver of evolution, natural selection, requires considerable powers of abstract reasoning. Added to this are difficulties in understanding the origins of phenotypic variation (including the non-directed nature of genetic mutations, independent assortment and the relationship between genotype and phenotype) and the ways in which change at the individual and population level interact.

Good teaching, that combines appropriate practical work with tasks to overcome misconceptions and learn valid concepts, can do much to address these cognitive challenges at both primary (Russell & McGuigan, 2015) and secondary (Ingram, 2011; Mead et al., 2017) levels.

Overcoming standard cognitive misconceptions about evolution is one thing but what about students who come to their biology classes as creationists. Can biology education help here? One way of interpreting the move from creationism to an acceptance of evolutionary theory is to see it as an instance of conceptual change. There is a large psychological literature on conceptual change with an *International Handbook of Research on Conceptual Change* edited by Vosniadou (2008a). As Vosniadou herself points out "The roots of the conceptual change approach to learning can be found in Thomas Kuhn's work on theory change in the philosophy and history of science" (Vosniadou, 2008b, p. xiii). Famously, Kuhn likens the switch from one paradigm to another to a gestalt switch (when we suddenly see something in a new way) or even a religious conversion.

Changing from a position where one sees creationism as valid to one where one sees the evolutionary understanding of life as valid can be very difficult (Long, 2011; Winslow et al., 2011). The science educator Lee Meadows is one who has made this journey. He writes about his collaboration with David Jackson, a science educator at the University of Georgia:

> Our first work together, "Hearts and Minds in the Science Classroom: The Education of a Confirmed Evolutionist" (Jackson et al., 1995), chronicles David's growth as he learned how a different set of life experiences can deeply impact science teachers' approaches to evolution in the classroom. David, an agnostic, had never worked with science teachers who also held to a deep faith until he moved to Georgia in the USA. David was surprised to find some science teachers who were staunchly opposed to teaching evolution in their classes. At first, David tried to correct their beliefs about evolution, but then he began to realize that he had skipped the essential first step of listening to them before trying to influence them. He began to find that, rather than being uninformed, many of these teachers were thinking through their religious beliefs, their scientific beliefs, and the interplay between the two. He began to see that science teachers had to consider the hearts, as well as the minds, of their students. Many of the teachers in the study, and by extension religious students like them in science classes, are actively choosing not to learn about evolution … Evolutionary science pales in importance to the eternal issues of God, Heaven, and salvation.
>
> I know well this tension between the heart and the mind because I've lived it. I was raised in a Christian fundamentalist home and church, and I'm now a science teacher and educator. Working through this tension was a perspective I brought to the Hearts and Minds study. My own faith journey has led me away from fundamentalism, but I do still hold to the view that the Christian scriptures are the inspired words of God. I find truth in both worldviews. Science provides truth from the basis of evidence, but my faith also provides an intellectual, durable system of knowing the world.
>
> (Meadows, 2007, p. 149)

So, when teaching evolution there is much to be said for allowing students to raise any doubts they have (hardly a revolutionary idea in science teaching) and doing one's best to have a genuine discussion. The word 'genuine' doesn't mean that creationism or intelligent design deserve equal time. However, in certain classes, depending on the comfort of the teacher in dealing with such issues and the make up of the student body, it can be appropriate to deal with these matters. If questions or issues about creationism and intelligent design arise during science lessons they can be used to illustrate a number of aspects of how science works such as 'how interpretation of data, using creative thought, provides evidence to test ideas and develop theories'; 'that there are some questions that science cannot currently answer, and some that science cannot address'; 'how uncertainties in scientific knowledge and scientific ideas change over time and about the role of the scientific community in validating these changes' (Reiss, 2008).

Having said that, such teaching is sometimes not easy. Some students get very heated; others remain silent even if they disagree profoundly with what is said. The DCSF Guidance suggests: "Some students do hold creationist beliefs or believe in the arguments of the intelligent design movement and/or have parents/carers who accept such views. If either is brought up in a science lesson it should be handled in a way that is respectful of students' views, religious and otherwise, whilst clearly giving the message that the theory of evolution and the notion of an old Earth/universe are supported by a mass of evidence and fully accepted by the scientific community".

Teachers should take seriously and respectfully the concerns of students who do not accept the theory of evolution while still introducing them to it. While it is unlikely that this will help students who have a conflict between science and their religious beliefs to resolve the conflict, good science teaching can help students to manage it—and to learn more science. Creationism can profitably be seen not as a simple misconception that careful science teaching can correct, as careful science teaching might hope to persuade a student that an object continues at uniform velocity unless acted on by a net force, or that most of the mass of a plant comes from air. Rather, a student who believes in creationism can be seen as inhabiting a non-scientific worldview, that is a very different way of seeing the world. One very rarely changes one's worldview as a result of a 50 min lesson, however well taught.

My hope, rather, is simply to enable students to understand the scientific worldview with respect to origins, not necessarily to accept it (cf. Williams, 2015). Students can be helped to find their science lessons interesting and intellectually challenging without their being threatening. Effective teaching in this area can not only help students learn about the theory of evolution but better to appreciate the way science is done, the procedures by which scientific knowledge accumulates, the limitations of science and the ways in which scientific knowledge differs from other forms of knowledge.

Finally, there is the issue of biology education not in schools but in informal settings, such as museums. One advantage of informal education is that learners in informal settings are often more motivated than they are in school. In addition, they are typically in much smaller groups and are often accompanied by family members. This means that there can be greater opportunity for learning to be personalised and for the subject matter to be an immediate cause of animated conversation. Furthermore, informal settings often provide rare material (e.g. fossils) of a sort rarely available in schools. And those in informal settings responsible for the provision of teaching and information more generally often make a commitment of time to the preparation of these that is way beyond what a school teacher can manage (Reiss, 2017).

Science museums have long had exhibits about evolution. Tony Bennett (2004) examines the history of museum displays about evolution. He looks at nineteenth century studies in geology, palaeontology, natural history, archaeology and anthropology and "trace[s] the development, across each of these disciplines, of an 'archaeological gaze' in which the relations between past and present are envisaged as so many sequential accumulations, carried over from one period to another so that each layer of development can be read to identify the pasts that have been deposited within it" (Bennett, 2004, pp. 6–7). Bennett concludes that evolutionary museums "are just as much institutions of culture as art museums" (p. 187).

In one sense this is obvious—museums and galleries have to make selections about what to display and how to narrate such displays and these are clearly cultural decisions whether one is referring to art, evolution, mathematics or any technology. However, whereas a visitor to an art gallery is unlikely to presume that what is being viewed is the only reading possible, a visitor to a science museum might presume that they are being presented with objective fact (Reiss, 2013).

Monique Scott too has produced a book about evolution in museums (Scott, 2007). Scott's work, unlike Bennett (2004), is more to do with the now than with history. Using questionnaires and interviews, she gathered the views of nearly 500 visitors at the Natural History Museum in London, the Horniman Museum in London, the National Museum of Kenya in Nairobi and the American Museum of Natural History in New York. Perhaps her key finding is that many of the visitors interpreted the human evolution exhibitions as providing a linear narrative of progress from African prehistory to a European present. As she puts it:

> Progress narratives persist as an interpretive strategy because they still function as a conceptual crutch. They are nearly ubiquitous in popular culture (can you imagine human evolution without imagining the cartoonish images of humans evolving single-file toward their destiny?) and they stand largely unchallenged in museum exhibitions which conventionally move case-by-linear-case from Africa to Europe. Many museum visitors, particularly Western museum visitors, rely upon cultural progress narratives—particularly the Victorian anthropological notion that human evolution has proceeded linearly from a primitive African prehistory to a civilized Europe—to facilitate their own comprehension and acceptance of African origins. Overwhelmingly, museum visitors relate to origins stories intimately, and in ways that satisfy or redeem the images they already have of themselves. (Scott, 2007, p. 2)

Scott's work is an important reminder of the fact that it can be difficult to teach well about evolution, for reasons that have nothing to do with religion.

References

Allgaier, J. (2014). United Kingdom. In S. Blancke, H. H. Hjermitsleve, & P. C. Kjærgaard (Eds.), *Creationism in Europe* (pp. 50–64). Baltimore: Johns Hopkins University Press.

Baker, S. (2012). The Theos/ComRes survey into public perception of Darwinism in the UK: a recipe for confusion. *Public Understanding of Science, 21,* 286–293.

Baker, S. (2013). *Swimming against the tide: The new independent christian schools and their teenage pupils*. Oxford: Peter Lang.

BBC. (2006). *Britons unconvinced on evolution*. Retrieved August 13 2017 from http://news.bbc.co.uk/1/hi/sci/tech/4648598.stm..

Behe, M. J. (1996). *Darwin's black box: The biochemical challenge to evolution*. New York: Free Press.

Bennett, T. (2004). *Pasts beyond memory: Evolution*. Colonialism, Routledge, London: Museums.

Billingsley, B., Abedin, M., Chappell K. & Hatcher, C. (forthcoming). Learning about evolution in a cross-curricular session: Findings from an intervention study with primary school teacher pre-services. In Harms, U. & Reiss, M. J. (Eds.) *Implementing and Researching Evolution Education: Understanding what Works*, Springer, Dordrecht.

Brickhouse, N. W., & Letts, W. J., IV. (1998). The problem of dogmatism in science education. In J. T. Sears & J. C. Carper (Eds.), *Curriculum, Religion, and Public Education: Conversations for an enlarging public square* (pp. 221–230). New York: Teachers College, Columbia University.

Browne, E. J. (2002). *Charles Darwin: Vol. 2. The Power of Place*, Jonathan Cape, London.

Burgess, S. (2008). *Creationpoints: In God's image—The divine origin of humans*, Day One Publications, Leominster.

Carter, R. (Ed.). (2014). *Evolution's Achilles' Heels*. Powder Springs, Georgia: Creation Book Publishers.

Creation Science Movement. (2017). *Who We Are*. Retrieved August 13 2017 from https://www.csm.org.uk/whoweare.html.

DCSF. (2007). *Guidance on creationism and intelligent design*. Retrieved August 13 2017 from http://webarchive.nationalarchives.gov.uk/20071204131026/http://www.teachernet.gov.uk/docbank/index.cfm?id=11890.

Dembski, W. A. (1998). *The design inference: Eliminating chance through small probabilities*. Cambridge: Cambridge University Press.

Hanley, P. (2012). *The Inter-relationship of science and religious education in a cultural context: Teaching the Origin of Life (Unpublished doctoral dissertation)*. UK: University of York.

Harms, U. & Reiss, M. J. (forthcoming). The present status of evolution education. In Harms, U. & Reiss, M. J. (Eds.) *Implementing and Researching Evolution Education: Understanding what Works*, Springer, Dordrecht.

Hayward, A. (1985). *Creation and evolution: The facts and fallacies*. London: Triangle.

Ingram, N. (2011). Classification, variation, adaptation and evolution. In M. Reiss (Ed.), *Teaching Secondary Biology* (2nd ed., pp. 215–242). London: Hodder Education.

Jackson, D. F., Doster, E. C., Meadows, L., & Wood, T. (1995). Hearts and minds in the science classroom: the education of a confirmed evolutionist. *Journal of Research in Science Teaching, 32*, 585–611.

Johnson, P. E. (1999). The wedge: breaking the modernist monopoly on science. *Touchstone, 12* (4), 18–24.

Jones, L., & Reiss, M. J. (Eds.). (2007). *Teaching about scientific origins: Taking account of creationism*. New York: Peter Lang.

Kampourakis, K. (2014). *Understanding evolution*. Cambridge: Cambridge University Press.

Kampourakis, K. & Reiss, M. J. (Eds.). (forthcoming). *Teaching Biology in schools: Global research, issues and trends*, Routledge, London.

Lawes, C. (2009). *Faith and darwin: Harmony, conflict, or confusion?*. London: Theos.

Long, D. E. (2011). *Evolution and religion in American education: an ethnography*. Dordrecht: Springer.

McCain, K., & Kampourakis, K. (2016). Which question do polls about evolution and belief really ask, and why does it matter? *Public Understanding of Science*. https://doi.org/10.1177/0963662516642726.

Mead, R., Hejmadi, M., & Hurst, L. D. (2017). Teaching genetics prior to teaching evolution improves evolution understanding but not acceptance. *PLoS Biology, 15*(5), e2002255. https://doi.org/10.1371/journal.pbio.2002255.

Meadows, L. (2007). Approaching the conflict between religion and evolution. In L. Jones & M. J. Reiss (Eds.), *Teaching about scientific origins: Taking account of creationism* (pp. 145–157). New York: Peter Lang.

Miller, J. D., Scott, E. C., & Okamoto, S. (2006). Public acceptance of evolution. *Science, 313*, 765–766.

Moore, R. (2007). The history of the creationism/evolution controversy and likely future developments. In L. Jones & M. J. Reiss (Eds.), *Teaching about scientific origins: Taking account of creationism* (pp. 11–29). New York: Peter Lang.

Nagel, T. (2008). Public education and intelligent design. *Philosophy & Public Affairs, 36*, 187–205.

National Academy of Sciences and Institute of Medicine. (2008). *Science, evolution, and creationism*. Washington, DC: National Academies Press.
Reiss, M. (Ed.). (1999). *Teaching secondary biology*. London: John Murray.
Reiss, M. J. (2008). Teaching evolution in a creationist environment: an approach based on worldviews, not misconceptions. *School Science Review, 90*(331), 49–56.
Reiss, M. J. (2011a). How should creationism and intelligent design be dealt with in the classroom? *Journal of Philosophy of Education, 45*, 399–415.
Reiss, M. (Ed.). (2011b). *Teaching secondary biology* (2nd ed.). London: Hodder Education.
Reiss, M. J. (2013). Beliefs and the value of evidence. In J. K. Gilbert & S. M. Stocklmayer (Eds.), *Communication and engagement with science and technology: Issues and dilemmas* (pp. 148–161). New York: Routledge.
Reiss, M. J. (2017). Teaching the theory of evolution in informal settings to those who are uncomfortable with it. In P. G. Patrick (Ed.), *Preparing informal science educators: Perspectives from science communication and education* (pp. 495–507). Cham: Springer.
Rendle-Short, J. (1998). *Green eye of the storm*. Carlisle, PA: Banner of Truth Trust.
Rosengren, K. L., Brem, S. K., Evans, E. M., & Sinatra, G. M. (Eds.). (2012). *Evolution challenges: Integrating research and practice in teaching and learning about evolution*. Oxford: Oxford University Press.
Russell, T. & McGuigan, L. (2015). *Understanding Evolution and Inheritance at KS1 and KS2: Final Report*. Retrieved August 13 2017 from http://www.nuffieldfoundation.org/sites/default/files/files/Final%20report%20-%20Understanding%20evoluation%20and%20inheritance%20-%20July%202015.pdf.
Scaramanga, J. (2017). *Systems of indoctrination: Accelerated christian education in England (Unpublished doctoral thesis)*. London: University College London.
Scott, M. (2007). *Rethinking evolution in the Museum: envisioning African origins*. London: Routledge.
Spencer, N., & Alexander, D. (2009). *Rescuing darwin: God and evolution in Britain today*. London: Theos.
Truth in Science. (2017). Retrieved August 13 2017 from http://www.truthinscience.org.uk.
Vosniadou, S. (Ed.). (2008a). *International handbook of research on conceptual change*. New York: Routledge.
Vosniadou, S. (2008b). Conceptual change research: an introduction. In Vosniadou, S. (Ed.), *International handbook of research on conceptual change*, Routledge, New York, pp. xiii–xxviii.
Watson, D. C. C. (1975). *The Great Brain Robbery*. Walter, Worthing: Henry E.
Whitcomb, J. C., & Morris, H. M. (1961). *Genesis flood: The biblical record and its scientific implications*. Philadelphia: Presbyterian & Reformed Publishing.
Williams, J. D. (2008). Creationist teaching in school science: A UK perspective. *Evolution Education & Outreach, 1*, 87–95.
Williams, J. D. (2015). Evolution versus creationism: A matter of acceptance versus belief. *Journal of Biological Education, 49*, 322–333.
Winslow, M. W., Staver, J. R., & Scharmann, L. C. (2011). Evolution and personal religious belief: Christian university biology-related majors' search for reconciliation. *Journal of Research in Science Teaching, 48*, 1026–1049.

Michael Reiss is Professor of Science Education at UCL Institute of Education, University College London, a Fellow of the Academy of Social Sciences and Visiting Professor at the Universities of Kiel and York and the Royal Veterinary College. The former Director of Education at the Royal Society, he has written extensively about curricula, pedagogy and assessment in science education and has directed a very large number of research, evaluation and consultancy projects over the past twenty years funded by UK Research Councils, Government Departments, charities and international agencies.

Chapter 10
Evolution Education and Evolution Denial in Scotland

J. Roger Downie, Ronan Southcott, Paul S. Braterman and N. J. Barron

Abstract This chapter begins by tracing Scotland's early encounters with and reaction to evolution, starting with Darwin's time as a medical student in Edinburgh. The multicultural nature of modern Scotland, its education system and the role of religion in that system are discussed. Scotland's education system has long been independent of the rest of the UK. Biology, including evolution, has relatively recently become prominent in Scottish schools. Currently, evolution is introduced in compulsory *General Science* at secondary year three, but a deeper treatment is given in optional *Biology* in years 4-6. Unfortunately, evolution is absent from the optional *Human Biology* curriculum, a course much used by prospective medical students. Our survey of Scottish biology teachers showed that most were confident in their ability to teach evolution, but that a small minority did not accept the theory themselves. Most Scottish universities include evolution as part of their biology foundation courses, and some provide advanced courses in modern evolutionary biology. Our surveys of Glasgow University biology and medical students show low, but still worrying evolution rejection rates. We provide an analysis of students' reasons for acceptance or rejection of evolution and comment on what education can do to improve the acceptance rate. We trace the influence of creationism in Scottish schools, especially links with the USA, and analyse the coverage of Earth history and species origins in the Scottish schools religious education curriculum.

J. R. Downie (✉) · N. J. Barron
School of Life Sciences, University of Glasgow, Glasgow, Scotland G12 8QQ, UK
e-mail: Roger.Downie@glasgow.ac.uk

R. Southcott
School of Education, University of Glasgow, Glasgow, Scotland G12 8QQ, UK

P. S. Braterman
School of Chemistry, University of Glasgow, Glasgow, Scotland G12 8QQ, UK

© Springer International Publishing AG, part of Springer Nature 2018
H. Deniz and L. A. Borgerding (eds.), *Evolution Education Around the Globe*, https://doi.org/10.1007/978-3-319-90939-4_10

10.1 Introduction

In this chapter, we provide some historical context for Scotland's encounter with evolutionary biology. We describe Scotland's education system, both at school level and in higher education, and how it treats the biological sciences, particularly evolution. We also review studies on the level of acceptance of the theory of evolution amongst students and teachers in Scotland (including some new results), and discuss the context and the impact of creationist activity in Scotland, especially in relation to religious education in Scottish schools.

10.2 Vestiges and Origins: Evolution in 19th and Early 20th Century Scotland

Charles Darwin's time as a medical student in Edinburgh (1825–1827) had at least two important influences on his development as the scientific natural historian who later published *On the Origin of Species* (1859). First, by chance, he met and became friends with Robert Grant who was researching the nature of sponges: whether they are plants or animals was still an unsettled issue at that time. Grant favoured the ideas expounded in *Zoonomia*, written by Darwin's grandfather Erasmus, who questioned the fixity of species. Grant's encouragement led Charles to become active in the local natural history society (he needed little persuasion to neglect his medical studies), where he gave his first scientific paper, aged only 18 (Stott, 2012). Second, Darwin learned about the biologically diverse tropical forests of Guyana, and how to preserve biological specimens, from a freed black slave who worked and taught as a taxidermist at the University's museum. This experience, along with his family's role in the campaign to abolish slavery, was important in Darwin's thinking on the unity of the human species and the absence of evidence for the superiority of Europeans (Desmond & Moore, 2009).

In 1844, the London publisher Churchill brought out *Vestiges of the Natural History of Creation*, a book by an anonymous author. This was a controversial bestseller, running into many editions, seriously promoting the ideas of Earth as very old and species being able to change. *Vestiges* was written by the radical Edinburgh writer/publisher Robert Chambers and published anonymously because he was worried about the reactions to his ideas, especially from the Church. He was right to be anxious: despite the book's popularity amongst the general public, it was condemned in England by such as the eminent geologist/cleric Adam Sedgwick, and in Scotland by the popular self-taught geologist Hugh Miller, who was prominent in the Free Church of Scotland. Even Thomas Henry Huxley, who later became Darwin's champion, attacked *Vestiges*. Darwin concluded from these reactions that he needed more evidence before revealing to the public his already written sketch on evolution by natural selection (Stott, 2012).

A useful piece of evidence for Darwin came in correspondence with the Scottish climatologist/geologist James Croll, who calculated that the Earth's age was around 500 million years, a figure much closer to what Darwin needed to account for the slow process of species transformation that he envisaged. This estimate was in strong contrast to the calculation of 20–100 million years by Glasgow physicist William Thomson, Lord Kelvin (Burchfield, 1990; Farrow, 2001). The age of the Earth was a critical issue for all those who wished to use the Bible as their guide to life. Hugh Miller discovered many fossils of animals no longer on Earth, and appreciated the vast amount of time required to produce the thick layers of sedimentary rocks he knew of. As a devout and active member of the Free Church, he came to regard the 'days' of the *Genesis* creation story as geological ages of considerable duration. Sadly, Miller committed suicide in 1856, just two years before Darwin and Wallace's theories became public.

Scowen (1998) has provided a detailed account of the reception to *The Origin* in Edinburgh, the main location of the Scottish enlightenment. Darwin's views, or at least an interpretation of them, were widely discussed in scientific circles, in churches (mainly in the dominant Free Church) and in the many regular periodicals such as the *Edinburgh Review* and *Blackwood's Magazine*. After an initial, mainly hostile reaction, a version of Darwinism became broadly accepted, even by the Church. This, however, was not Darwinism as he himself saw it. There was general acceptance of Earth being very ancient and of new species evolving from the old, but not of Darwin's central propositions that the process was essentially random, with no divine direction. In the University, evolutionary topics were soon incorporated into the curriculum, with Huxley's *Elements of Comparative Anatomy* adopted as a text by 1865. The British Association for the Advancement of Science (BAAS) meeting in Edinburgh (1871) hosted several talks on evolutionary themes.

Livingstone (1994) compares the reaction to Darwin amongst Calvinist religious communities in different cities. In Belfast, there was little initial reaction, but in 1874, the city's theologians mounted a ferocious attack on Darwinism, likely stimulated by that year's BAAS meeting in Belfast which had taken a strongly secular view of evolution. Livingstone contrasts this to the USA where James McCosh encouraged the philosophical section of the General Conference of the Evangelical Alliance (1873) to 'support the scientific enterprise by showing how younger naturalists could retain their old faith in God and the Bible with their new faith in science'. McCosh was an Ayrshire-born, Glasgow- educated theologian and natural historian. He became Professor of Philosophy at Queen's College, Belfast, then President of Princeton College, laying the foundations for its later university status. McCosh co-authored a book on biodiversity and later came to accept the antiquity of the Earth and that species were not fixed. We return briefly to this theme when discussing creationist influences in Scottish education.

In Glasgow, a key figure was John Scouler, who had collected specimens on the Galapagos Islands (in 1825, 10 years before Darwin) as part of a botanical expedition to the western coast of North America (Nelson, 2014). Scouler held academic posts in Anderson's University in Glasgow (later Strathclyde University), and in the Royal Dublin Society. He returned to Glasgow from the mid 1850s and remained there until his death in 1871. He acted as honorary president of the Geological Society of Glasgow, the Natural History Society of Glasgow (NHSG) and the Renfrewshire Natural History Society. Following publication of *The Origin*, Scouler addressed the NHSG twice (1862, 1863) on the transmutation and permanence of species. Scouler did not regard Darwin as having said anything fundamentally new, and was critical of the notion of natural selection. He chose to illustrate the essential permanence of species with extant and fossil molluscs, a fair choice given their later use by Stephen J. Gould to illustrate long periods of evolutionary stasis. From 1866–1902, the Regius Chair of Natural History at Glasgow University was occupied by John Young, principally a geologist. He was progressive in campaigning for women to be admitted to higher education but sceptical of natural selection, giving a public lecture as late as 1892 on the inability of selection to explain the evolution of complex parasitic life cycles. During this time, the NHSG (whose membership included eminent scientists like Lord Kelvin, John Young and Frederick Bower, the Professor of Botany, along with students and enthusiastic amateurs) occasionally heard talks on aspects of evolution, such as whether mimicry could be explained by natural selection and the possible role of Lamarckian inheritance. However, the Society's main business of recording the occurrence and distribution of local species was little affected by the new thinking. The arrival of John Graham Kerr to the Regius Chair of Natural History in 1902 (Natural History was divided into Geology and Zoology in 1903, with Kerr becoming Professor of Zoology) brought a committed Darwinian to Glasgow, with particular interests in the relationship between embryos and phylogeny.

The changes made by Kerr can be followed from the University of Glasgow calendar entries for Natural History, then Zoology, where the syllabus is summarised and previous examination papers published. Up to John Young's retirement in 1902, the word 'evolution' was not used either in the syllabus (which simply listed the groups of animals to be studied that year) or in the examination questions which focused on animal groups and their anatomy. From 1903, however, we find questions such as: *'Discuss the evolution of the breathing organs of vertebrates'; 'Enunciate the Galtonian and Mendelian laws of inheritance and discuss their relative importance in evolution'; 'What are mutations? Discuss the part which may have been played by mutations in evolution'; 'Discuss the question of the extent to which palaeontological evidence is confirmatory of the theory of evolution'*. With minor tweaking, most of these questions could still be set to-day.

This section has not attempted a comprehensive history of evolution's impact in Scotland, but has assembled a number of relevant strands, mostly from the major cities of Edinburgh and Glasgow. We can conclude that the ancient age of the Earth and the appearance of new species over that vast timescale were broadly accepted in Scotland by the late 19th century, even by some churchmen. However, acceptance of Darwin and Wallace's essential process, natural selection based on random variants, with no guiding hand or progressive direction, was less accepted. This, however, was a common situation among biologists until the mechanisms of heredity and variation became clearer in the early decades of the 20th century (Endersby, 2007).

10.3 Modern Scotland: Some Basic Facts

Scotland has a population of 5.3 million (2011 census). The official language is English with Scots and Scottish Gaelic as minority languages, and a wide variety of languages spoken by immigrant communities, notably Urdu and Polish. A little over half of the population adhere to a religion, 53.3% being Christian, principally Church of Scotland, followed by Roman Catholic, with other religions making up around 3%, of which 1.3% are Muslims. Education of children has been compulsory since 1872. Responsibility for education has been devolved from the UK Parliament to the Scottish Parliament since 1998. Primary school begins at age 4.5–5.5 years and lasts for 7 years. Secondary school starts at age 11–12 and lasts compulsorily for 4 years (until a child's 16th birthday), followed by an optional two further years. The transition from primary to secondary school is not dependent on an examination. Education at state schools is free, but 4.5% of children are educated at independent (fee-paying) schools, and a few are educated out of school, usually by parents. State schools use the curricula and qualifications provided by the Scottish Qualifications Authority, but some independent schools follow English curricula or International Baccalaureate. In terms of religion, most schools are non-denominational, not adhering to a particular faith, but 14% of state schools are Roman Catholic and one is Jewish. The subjects studied and grades obtained in secondary schools act as qualifications for entry to Further or Higher Education. Scottish students currently pay no fees for Further or Higher Education.

10.4 Evolution Education in Scotland: Schools and Universities

The education system in Scotland has long been distinct from that of the rest of the UK (Scotland, 1969; Anderson, 2013). Schools were more egalitarian, with no segregation of rich and poor, and girls were taught as well as boys, at least at elementary level. After the 16th century church reformation, the Scottish Parliament was very active in legislating to improve schools, a process which slowed greatly after the Union of Parliaments (1707) abolished the separate Scottish Parliament. Scottish education retained its distinctiveness but was managed from London until the office of the Secretary of State for Scotland took control in 1939. In higher education too, Scotland was distinctive, a small country with four ancient Universities (St. Andrews from 1410, Glasgow from 1451; Edinburgh and Aberdeen soon following) when England had only Oxford and Cambridge until well into the 19th century.

10.4.1 Biology in Schools

In 19th century Scotland, there is evidence that some reformers saw the need for natural history education in schools. James Nichol (1853) used his inaugural lecture at Marischal College (Aberdeen) to extol the benefits of natural history as a subject in 'general education'. These included observational skills, memory training and developing the intellect, imagination and moral feelings. John Young's Natural History Society of Glasgow presidential address (1871) also promoted the benefits of natural history education, but argued that the lack of science training for school teachers was holding this back. A major advance, across the UK, was the publication of Huxley and Martin's (1875) course in elementary biology.

In 20th century Scottish schools, biology as a subject was long over-shadowed by chemistry and physics. According to Souter (2003), prior to the publication of a new syllabus for biology in 1968, fewer than 1000 candidates per year took Scottish examinations in a combination of bioscience subjects; botany, zoology, agriculture and horticulture. However, Day (2013) wrote that 'from humble beginnings in the mid- 1960s, biology rose to become, by 2010, the third most popular subject taken by pupils in Scotland'. In 2011, examination candidate numbers in the biosciences were: *Standard Grade Biology*, 20,315; *Higher Biology*, 9767; *Higher Human Biology*, 4226; *Advanced Higher Biology*, 2288.

As in any country, the science curriculum has developed over time. A major reform in 2012 introduced the Curriculum for Excellence (CfE) which aims to emphasise links across different areas of learning, rather than to keep subjects separate (Brown, 2014). The curriculum is delivered at the following levels: early = nursery; 1, 2 = primary; 3 = secondary years 1, 2; 4 = secondary year 3; 5 = secondary year 4; higher and advanced higher in secondary years 5 and 6. Scrutiny of the general science curriculum shows that in the theme *Planet Earth*, biodiversity and inter-dependence of living organisms are first covered at level 4: prior to this, it is unlikely that evolution is a topic that would be introduced to pupils.

It is at level 5 that Scottish pupils choose subjects for national qualifications. The pre-2012 '5–18' syllabus at this level (Standard Grade) made no mention of evolution, natural selection or Darwin, but the new *National 5 Biology* is a distinct improvement, with a syllabus chapter dedicated to evolution, including natural selection. Although the syllabus does not mention Darwin, many teachers cover the *Voyage of the Beagle* in lessons. For pupils studying biology further, there is a choice, *Higher Biology* or *Higher Human Biology*. The pre-2012 *Higher Biology* course included good coverage of evolution, with a full chapter on natural selection and references in other chapters to aspects of evolution. This level of coverage has been retained in *CfE Biology*, but with modernisation of the treatment of DNA. A seriously disappointing feature is *Higher Human Biology*. The pre-2012 curriculum made no mention of evolution, even of the origins of the human species. This meant that pupils opting for this course (it is particularly favoured by prospective medical students) following the pre-2012 *Standard Grade Biology*, could leave school with no consideration of evolution. The *CfE Human Biology* is no improvement, but at least these students will have been introduced to evolution in *National 5 Biology*. *Advanced Higher Biology*, studied in secondary year 6, is intended to develop investigative skills and includes a set of course options and a research project. Pupils could opt to study evolution as part of this course, but *Advanced Higher Biology* is taken by relatively few.

10.4.2 Biology Textbooks

To support learning, it is normal for school teachers to use classroom textbooks. Although Scottish school teachers are free to use whatever textbooks they consider suitable, in the biosciences, a set of commercially-published texts written by a team of Scottish teachers led by James Torrance has for several decades been the main source, supplemented by other more detailed books held in school libraries. Torrance's team closely aligns their treatment of the different levels and aspects of biology to the arrangements laid out by the curriculum designers. We have analysed the evolution content of this set of books.

Pre-2012 '5–18' curriculum. *Standard Grade Biology* contained 242 pages of course material, none of them on evolution (Torrance, 2001). Similarly, *Higher Human Biology*, 327 pages, had nothing on evolution (Torrance, 2002). In contrast, *New Higher Biology*, 324 pages, devoted four chapters totalling 32 pages to evolution topics: natural selection, speciation, adaptive radiation, extinction and conservation (Torrance, 1999). There was no book designed for *Advanced Higher Biology* because of the flexible content of the curriculum, and this continues to be the case.

CfE curriculum. *National 5 Biology* has 198 pages, 11 covering evolution (Torrance et al., 2013). The new *Higher Human Biology* text continues to omit evolution in its 348 pages (Simms et al., 2013), but *Higher Biology for CfE*, 352 pages, has been substantially revised and now has a 32 page chapter on evolution and a 26 page chapter on biodiversity, including evolutionary aspects (Marsh et al., 2012).

Although the books by the Torrance team are sound on evolution and cover the subject to an appropriate depth, it is worth noting that as recently as the late 1980s, one book widely used in Scottish schools (Mackean, 1988) stated that 'scientific evidence can be found to support the theory of evolution but the evidence can be interpreted in other ways. Evolution is a theory, not a fact'.

10.4.3 School Teachers and Their Attitudes

In Scotland, teacher certification and training are rigorous. To teach science at secondary level, prospective teachers must have literacy and numeracy qualifications and a degree which includes relevant subjects. Training is by means of a one-year postgraduate certificate in education. This focuses on education theory and practice, rather than the particular curricula that the trainee will deliver, and provides substantial mentored classroom experience. After training, teachers must register with the GTCS (General Teaching Council of Scotland). Although the GTCS sets the standards and registers teachers in Scotland, teachers have a fair degree of autonomy in what they deliver in the classroom (Hargreaves, 2000). There is no explicit requirement to teach the whole curriculum for a subject, so a teacher could in theory decide to ignore the evolution sections of the biology courses: we have no evidence of teachers actually doing this and consider it unlikely to be happening.

Although the views on evolution of schoolteachers have been reported from some countries (e.g. USA: Berkman et al., 2008; Turkey; Deniz et al., 2008), we know of no such survey from Scotland. To remedy this, we constructed a ten-question survey (Table 10.1) based largely on the work of Rutledge and Mitchell (2002) and Rutledge and Sadler (2007). This was sent electronically using *Survey Monkey* to Scottish biology teachers using an online sharing platform in December, 2016. We received 149 responses (from a population of 2828 teachers registered to teach biology). Teachers varied greatly in their level of experience: 35% were in their first five years; 39% had taught for 6–15 years and 26% for more than 15 years. The teachers had studied a wide variety of University biological science subjects: *Zoology, Human Biology, Biomedical Sciences* and *Biochemistry* each accounted for 12–15% of respondents. Only 43% reported that their degree course had included a specific module on evolution, and that this had been mostly at first or second year level. However, despite this, 92% were confident in their understanding of the theory. Table 10.2 shows respondents' views on the level of evolution education provided by the four current Scottish biology curricula, with *Higher Human Biology* (not surprisingly) generating the highest level of dissatisfaction. Table 10.3 demonstrates the level of evolution rejection in our sample of Scottish biology teachers. In addition, we asked specifically about human evolution: 78% agreed with the statement: 'humans have evolved over millions of years and God played no part in the process'; 16% agreed on the evolution time-scale, but felt that God had a role; 6% agreed with the statement: 'humans appeared on Earth 10,000 years ago'.

10.4.4 Universities

Scotland has 14 Universities that offer undergraduate degree courses in the biosciences. The Scottish honours degree in science (Bachelor of Science, B.Sc.) normally takes four years of study. Universities design their own curricula, although courses aimed at professional qualifications have externally-set requirements. All courses are expected to follow the broad guidelines established for the biosciences by the UK Quality Assurance Agency (QAA, 2007). These refer to

Table 10.1 Questionnaire sent to Scottish biology schoolteachers

Question	Answer choices/Follow-up questions
Q1. What specialism would you consider your biology degree to be based in? Indicate all that apply	Biomedical, Sports science, Environmental, Human Biology, Botany, Microbiology, Genetics, Zoology, Biochemistry, Psychology, Other (please specify)
Q2. Did your degree course have any specific evolution modules that you took?	Yes, No, If yes what level?
Q3. Do you accept the theory of evolution to be a valid scientific explanation for the occurrence and diversity of organisms (past and present) on Earth?	Show your level of acceptance from 1 = strongly disagree, to 5 = strongly agree
Q4. How confident is your understanding of the theory of evolution?	Indicate your confidence level from 1 = not at all confident, to 5 = very confident
Q5. Select which of the following statements you associate with in terms of human evolution	• Humans have evolved over millions of years and God played no part in the process • Humans have evolved over millions of years and God played a role in this process • Humans appeared on Earth around about 10,000 years ago
Q6. Show your level of agreement with the following statement: Current evolutionary theory is the result of sound scientific research and methodology	Respond from 1 = strongly disagree, to 5 = strongly agree
Q7. Show your level of agreement with the following statement: The theory of evolution cannot be correct since it disagrees with a religious text account of creation which I accept	Respond from 1 = strongly disagree to 5 = strongly agree
Q8. How long have you been a teacher of Biology?	0–5 years, 6–10 years, 11–15 years, 15 years+
Q9. Which Biology courses do you teach? Please select all that apply	National 5, Higher, Higher Human, Advanced Higher, A level, IB, Other (please specify)
Q10. The level of evolution education, in the courses of the Scottish biology curriculum shown below (the course list is given in Table 10.2), is in my opinion	Respond as follows: 1–2 = too little; 3 = adequate; 4–5 = too much

evolution at several points, for example: *'Darwin's theory of natural selection is a major philosophical and scientific step forward for mankind. This theory, bolstered by the study of modern genetics and molecular biology, has brought us to a clearer understanding of life's basic processes'* (paragraph 2.1); *'Qualities of mind appropriate to bioscience learners include an appreciation of the complexity and diversity of life processes through the study of organisms, their molecular, cellular, and physiological processes, their genetics and evolution, and the inter-relationships between them and their environment'* (paragraph 3.3).

Table 10.2 Scottish biology teachers' assessments of the evolution content of different biology curricula

Course	Assessment		
	Too little	Adequate	Too much
National 5 (n = 100)	26	73	1
Higher (n = 75)	20	76	4
Higher human (n = 63)	59	41	0
Advanced higher (n = 65)	34	64.5	1.5

Note Data are presented as % of respondents. Sample sizes (n) differ for the different courses

Table 10.3 Proportions of Scottish biology teachers (n = 141) agreeing with statements on evolution

Statement	Agree	N	Disagree
The theory of evolution is a valid scientific explanation for the occurrence and diversity of organisms (past and present) on Earth	83.0	2.1	14.9
Current evolutionary theory is the result of sound scientific research and methodology	89.4	5.0	5.6
The theory of evolution cannot be correct since it disagrees with a religious text account of creation which I accept	5.7	1.4	92.8

Note Data are shown as % of those giving rankings on a five point scale from 1 = strongly agree to 5 = strongly disagree, consolidated to 1, 2 = agree; 3 = N; 4, 5 = disagree

To assess the coverage of evolution in Scottish University bioscience courses, we consulted the publically available information on University web-sites. Bioscience degree courses can be classified broadly as: 1- biomedical, with an emphasis on human health and disease; 2-organismal or environmental, such as Microbiology, Marine Biology or Conservation Biology; 3- molecular; or 4- applied. In most, but not all cases, the first year includes a general biology module that covers biodiversity and, although the word 'evolution' does not always appear in the brief course descriptors, it is likely to receive some coverage and is explicitly mentioned in several cases. For example: Dundee University states that '*Early modules in the biological science programme have a strong evolutionary theme. They emphasise the fact that it is impossible to understand any aspect of biology unless you have a clear understanding of its most important general theory, evolution by natural selection*'. Such explicit reference to evolution in the first- year programme is most obvious in Dundee and the four ancient Universities (Aberdeen, Edinburgh, Glasgow and St. Andrews). These Universities also have higher level coverage of aspects of evolution such as modules in *Molecular Ecology and Evolution* (Aberdeen), *Evolutionary Ecology of Plants* (Edinburgh), *Evolution: Pattern and Process* (Glasgow) and degree programmes in *Evolutionary Biology* and *Evolutionary and Comparative Psychology* (St. Andrews). From the publically available information, coverage of evolution is least in the more applied biomedical programmes, a pity given the relevance of evolution to human health and disease (Nesse & Williams, 1995). Overall, however, evolution is an active, established and uncontested part of bioscience higher education in Scotland.

10.5 Evolution Acceptance and Rejection in Scotland

10.5.1 The General Population

We know of no survey specifically addressing the Scottish public's attitude to evolution. The nearest results are from the report by Miller et al. (2006) which includes data from the USA, Japan, Turkey and 31 European countries, amongst them the UK as a whole. Subjects were asked to respond to the statement '*Human beings, as we know them, developed from earlier species of animals*' either by rating the statement true or false or by assessing it on a scale from definitely true to definitely false. In 2002, UK responses were classed as 32% definitely true, 37% probably true, 8% probably false, 7% definitely false and 16% not sure. In 2005, using slightly different categorisation, 1308 responses were classed as 76% true, 17% false and 7% unsure. In the 2005 European results, Iceland showed the highest level of acceptance, Cyprus the lowest, and the UK fifth highest, with similar acceptance levels to Japan. We would not expect Scotland as a whole to differ from the rest of the UK, but there is evidence that creationism remains strong in some areas (see later).

10.5.2 Scottish Biology Students

Downie and Barron (2000) and Southcott and Downie (2012) surveyed the attitudes to evolution of University of Glasgow biology students over the period 1987–2010. Here we summarise their results and augment them with the results of a 2016 survey. Downie and Barron (2000) surveyed students over 9 years in the Level 1 biology class which acts as a foundation for all students progressing to degree programmes in a biological science discipline, but is also taken by students in other degree programmes as a complete course in elementary biology. The course includes lectures, tutorials and laboratory sessions related to evolution. Southcott and Downie (2012) surveyed students over two years and compared Level 1 biology students with final year (Level 4) bioscience students in two categories, those who had studied aspects of evolution beyond first year and those who had not. In all cases, the surveys were carried out at the end of classes with students informed that completion of the survey was anonymous and entirely voluntary, though we mentioned how useful the data would be. Overall, this procedure led to very high completion rates: for example, in 2016, from 710 students, we obtained completion by 89.6%.

In the past, the vast majority of students in Glasgow's Level 1 biology course were of Scottish origin, and many had not studied biology prior to university. However, by 2016, the intake was much more international and better prepared by their previous studies. From students' responses about their pre-university qualifications, 61% were educated in Scotland, 18% elsewhere in the UK and 21% outside the UK. When asked about how much they had learned about evolution prior to university, only 11% regarded it as a little.

The starting question students responded to was '*Do you accept that some kind of biological evolution, lasting millions of years, has occurred on Earth?*' The proportions of Level 1 biology students rejecting evolution over the years are shown in Table 10.4. The two highest rejection rates were in the first three years and the data show a downwards trend. The rejection rates are low (2.4% in 2016), and lower than the UK population rate of 17% in 2005 found by Miller et al. (2006), but we need to remember that these are students who have chosen biology, and who would therefore be expected to know more than the general public. We analysed all the 2016 responses to questions from evolution rejectors and a random sample of 200 acceptors.

Religious affiliation. Students were asked to state their religion, if they had one, and to write 'none' rather than leaving a blank (Table 10.5). Effectively all rejectors

Table 10.4 Proportions of Biology-1 and Medicine-1 students in each year who rejected the proposition that a long period of biological evolution has occurred

Year and Class	Total number sampled	Rejectors (%)
Biology-1		
1987–88	221	11.3
88–89	160	7.5
89–90	230	10.0
90–91	210	8.1
91–92	269	5.6
92–93	312	7.1
93–94	417	5.3
94–95	517	6.8
98–99	518	3.9
2008–09	388	7.6
09–10	532	5.6
16–17	623	2.4
Medicine-1		
1999–2000	225	10.2
2002–2003	223	10.8

Note Data for 1987–2000 from Downie & Barron (2000); 2008–10 from Southcott & Downie (2012); Medicine-1 from Downie (2004). The 2016–17 data are previously unpublished

Table 10.5 Religious affiliation and its relationship to evolution acceptance/rejection in Biology-1 and Medicine-1 students

(a) Proportion (%) of the class stating a religious belief

Year and Class Biology-1	Acceptors	Rejectors
1987–99 (mean)	57	86
2008–9	47	100
2009–10	41	96
2016–17	38.5	100
Medicine-1		
1999	60	96
2002	61	100

(b) Proportions (%) of the different religions stated by evolution acceptors and rejectors

	Biology-1				Medicine-1	
Acceptors	91–95	98–99	2008–10	2016	1999	2002
Judaism	0.5	0	2	1	0	2
Islam	3	8	7	11	8	6
Christianity	94	90	84	88	85	92
Buddhist, Sikh & Hindu	2.5	0	7	0	7	0
Rejectors						
Judaism	1	0	1	0	0	0
Islam	18	22	18	27	32	4
Christianity	80	78	80	67	80	96
Buddhist, Sikh & Hindu	1	0	1	7	0	0

Note Data sources as for Table 10.3

stated a religion; for the acceptors, the proportion of students stating a religion has declined steadily and in 2016, fell below 40%. The religions stated were mainly forms of Christianity or Islam. Throughout the years, the proportion of Islamic students in the rejector group (27% of rejectors in 2016) has been higher than in the acceptors (11% in 2016).

Reasons for accepting or rejecting evolution. We offered three reasons for accepting evolution, plus 'other, please state'. In Downie and Barron (2000) and 2016, the statements and instructions were slightly different from Southcott & Downie (2012) (Table 10.6). Few students opted for the 'lecturers know best' option and very few wrote in 'other' reasons. The two most chosen options can be characterised as (a) the evidence is good and (b) I do not know of any good alternatives. The proportions choosing these differed somewhat between the 2000 and 2012 studies, possibly relating to the former study asking students to indicate all the reasons that applied, while the latter asked for only one reason to be chosen.

Rejectors were given a similar range of reasons (Table 10.7): acceptance of a religious creation account was the most chosen reason for evolution rejection, with some differences between the 2000 and 2012 studies, again relating to instructional

Table 10.6 Proportions of students (%) who chose particular reasons for accepting evolution

(a) Reasons for accepting evolution	Year and course		
	Biology-1		Medicine-1
	1987-99 (mean)	2016-17	1999
The evidence for evolution is clear and unambiguous	36	74	26
I tend to accept what my teachers say: they know the evidence much better than I do	11	9	10
I do not think there are any good alternatives to evolution that explain well the origins and distribution of species	78	58	71
Other reasons	8	9	5

(b) Reasons for accepting evolution	Year and course			
	Biology-1		Biology-4	
	2008-9	2009-10	High	Low
The evidence is convincing and well supported	75	72	93	76
No better explanation has been presented to me at this present time	16	21	6	17
I accept that my lecturers have a greater knowledge of the subject than me so I accept what has been taught to me	8	5	1	4
Other reasons	1	2	0	3

Note Data sources as for Table 10.3. Percentages in (a) do not total 100 because students were asked to tick all the reasons that applied to them, but the percentages are based on the number of students in the sample. In (b), students were asked to choose one reason only

Table 10.7 Proportions of students (%) who chose particular reasons for rejecting evolution

(c) Reasons for rejecting evolution	Year and course		
	Biology-1 1987-99 (mean)	2016-17	Medicine-1 1999
The evidence for evolution is full of conflicts and contradictions	33	33	30
I accept the literal truth of a religious creation account that excludes evolution	71	80	96
I think there are good alternatives to evolution that explain well the origins and distribution of species	19	23	17
Other reasons	19	0	9

(d) Reasons for rejecting evolution	Year and course	
	Biology-1 2008-9	2009-10
There is insufficient evidence to prove conclusively to my satisfaction that evolution has occurred	28	21
I have insufficient knowledge about evolution to show me that it has occurred	13	32
I believe there are alternative explanations for the diversity of life seen today (e.g. divine creator, intelligent design)	57	41
Other reasons	2	6

Note Data sources as for Table 10.3. Analysis as for Table 10.5

differences. To probe more deeply into what evolution rejection means, we pointed out that some people reject the idea that species can change from one kind to another, but accept that natural selection can operate within species to adapt them to the environment. Downie and Barron (2000) and the 2016 survey found 83–93% of rejectors accepted this role for natural selection. Southcott and Downie (2012) investigated this point somewhat differently; acceptors and rejectors were both asked to indicate their level of acceptance of three statements. Evolution acceptors gave high levels of acceptance to statements on human origins, new species formation and within species selection. Level 1 rejectors gave very variable responses; 51% rejected human evolution, 47% rejected new species but only 34% rejected within species selection. In 2016, we asked evolution rejectors what evidence would need to be obtained to convince them that evolution has occurred; several of the Muslim students wrote that it is the evolution of human beings that is the sticking point for them.

Scepticism of science. Since our surveys were of science students, we went on to assess their views on how well established several scientific propositions are, including evolution. One aim of this comparison was to test whether evolution rejectors were more sceptical of science in general than acceptors. The preamble to

Table 10.8 Evolution acceptors' and rejectors' ratings of how well several scientific propositions are established

Proposition (full statements below)	Acceptors (n = 200)			Rejectors (n = 15)		
	Poor	N	Well	Poor	N	Well
Climate change	5.3	16.9	77.8	6.7	26.7	66.7
Cigarettes	5.0	10.6	84.4	0	6.7	93.3
Evolution	1.0	8.6	90.4	73.3	6.7	20.0
Tectonic plates	2.1	8.9	89.0	13.3	26.7	60.0
Neonicotinoids[a]	15.6	43.3	41.1	6.7	60.0	33.3

Note Data are shown as % of students giving a ranking on a five-point scale from 1-poorly to 5 = well established, with the five-point scale consolidated to Poor = 1, 2; N = 3; Well = 4, 5
The full statement of each proposition, as given to the students, was
Climate change: current climate change is mostly the result of the emission of greenhouse gases, generated by human activities, into the atmosphere
Cigarettes: cigarette smoke causes lung cancer
Evolution: biological evolution lasting many millions of years has occurred on Earth
Tectonic places: the continents are not fixed in position, but move relative to one another
Neonicotinoids: neonicotinoid pesticides are a major cause of declines in bee populations. [a]This was the only statement where significant numbers did not respond, or wrote in 'don't know'

the question pointed out that 'few ideas in science are based on certainty; most major scientific generalisations are theories based on well-established but not certain evidence'. Results from the 2016 survey are shown in Table 10.8. The biggest difference between acceptors and rejectors was in their judgment of the evidence for evolution and continental drift, the two propositions at variance with Christian and Muslim creation accounts. On the other propositions, there was no evidence that evolution rejectors are sceptical of science in general, although the 2000 and 2012 studies did find such an effect. We can conclude, therefore, that evolution rejectors are judging the evidence on evolution to be poor, not on the basis of a scientific judgment, but on the basis of their religious convictions.

Understanding how evolution happens. Students may claim to accept or reject evolution, but do they really understand what it involves? Southcott and Downie (2012) asked Level 1 students to choose which of six definitions most accurately described Darwinian evolution. The definitions were: (A) living systems are so complex that they must have been designed by some kind of intelligent agency; (B) during their lives, organisms adapt to their environments and these useful adaptations are passed on to the next generation; (C) all living and extinct species were created at one time less than 10,000 years ago; (D) all living and extinct species were created over a long period of time, with species made extinct by catastrophic events replaced by new sets of created species; (E) all species are the result of a long period of gradual change, with favourable variations becoming more common in populations as a result of conferring reproductive advantages; (F) an organism mutates and then changes to be fitter for its environment. Of these, only E correctly describes Darwinian evolution; A corresponds to intelligent design; B, Lamarckian evolution; C, young Earth creationism; D, a version of old Earth creationism; F, a mistaken version of Darwinian evolution. We found that 82% of Level 1 evolution acceptors correctly chose E, with 11% choosing B. However, 48% of rejectors chose B, 30% E and 18% F, so it is clear that evolution rejectors had an insecure grasp of what Darwinian evolution actually is.

How effective is teaching in changing views on evolution? Southcott and Downie (2012) compared evolution acceptance/rejection in Level 1 and final year (Level 4) students, with Level 4 from two groups: 'high', those whose courses beyond Level 1 included more advanced treatments of evolution (mainly students in Genetics and Zoology degree programmes); 'low', students in degree programmes with minimal advanced coverage of evolution (human biology, sports science, molecular biology etc.). All students continuing to reject evolution in Level 4 were in the 'low' group. A small number (n = 7) of Level 4 students had altered from rejecting to accepting evolution, but not so much based on their learning (none gave this as a reason) as on realising that their religious beliefs did not really conflict with evolution (n = 6). Learning, however, did make a difference to the Level 4 acceptors; 94% of them now felt that the evidence for evolution is convincing and well supported, compared to 72–75% at Level 1. It should be noted that this Level 1/Level 4 comparison was made in the same calendar year, so it did not examine change in a cohort of students over four years.

10.5.3 Scottish Medical Students

Downie and Barron (2000) and Downie (2004) surveyed the attitudes to evolution of Level 1 medical students in 1999 and 2002. In addition, Downie (2004) reported on the reactions of Level 3 students to an optional course on *Evolution in Health and Disease*. Although the training of medical students at Glasgow University formerly included courses on general biology with some evolution content, curricular changes have concentrated the programme on medicine and medical practice with evolution playing no part. Medical students' knowledge of evolution therefore derives only from their pre-university education except for those for whom a medical degree follows a degree in a biological science (much less common in the UK than in some other countries).

As shown in Table 10.3, about 10% of medical students were evolution rejectors, noticeably higher than in science students around the same years. In most respects, medical student rejectors were similar to those in science; all were religious and nearly all rejected evolution because of belief in a religious creation account. Probing into a matter of medical importance, Downie (2004) asked acceptors and rejectors whether they agreed that natural selection could explain the increase of drug resistance in microbial populations; nearly all students in both groups accepted this. Downie (2004) then noted that the evolution of drug resistance posed a possible equity problem between present and future generations i.e. heavy use of a drug to-day could render it ineffective in future. Students were asked whether their duty should be mostly to present or to future generations.

Interestingly, a higher proportion of acceptors (8 from 59) than rejectors (1 from 23) were prepared to consider a duty to the future, so evolution rejection seemed to have a potential impact on medical practice. This was a small sample, but worth further investigation.

Nesse and Williams (1995) made a strong case for the importance of evolutionary biology to medicine and therefore to its inclusion in medical curricula. As a response to this, a five- week optional course on evolution in health and diseases was offered to Level 3 Glasgow medical students from 1999 to 2011 (the course then sadly ended following major timetabling changes). Downie (2004) reported on the content of the course and student reactions to it over its first four years: 90% of students regarded the course as very suitable and interesting; 60% regarded evolution as important or very important to medicine, compared to only 35% of Level 1 students, indicating that a course which demonstrated the relevance of evolutionary ideas to medicine could make a difference.

10.5.4 Glasgow Physics, Astronomy and Geology Students

Most focus on evolution rejection has been on biology students. However, religious creation stories clearly also conflict with a modern understanding of the origins and age of the Universe and the geological processes that have shaped the Earth. We asked two colleagues, both heavily involved in undergraduate teaching in the University of Glasgow's Physics, Astronomy and Geology departments whether they knew of any evidence of evolution and ancient Earth denial amongst Scottish students in their subjects. The astronomer had experienced such problems when teaching in the USA, but not in Scotland; the geologist reported that no student had ever raised the issue with him, but also that geology teachers do not go looking for such attitudes. Neither knew of any survey of students in their subjects of the kind we report for biology and medical students.

10.6 Creationist Influence in Scottish Education, from Primary Schools to University

In Scotland, as in the rest of the United Kingdom, the Churches are closely involved with the public educational system. *Religious Observance* is a statutory aspect of Scottish education. However, recognising the diversity of the modern population, Government guidance suggests that schools may prefer the term 'time for reflection' to religious observance, at least in non-denominational schools. 'Time for reflection' is intended to involve the whole school community in a communal activity such as an assembly. This may, or may not, have a religious theme. In addition, *Religious and Moral Education* is a core curriculum theme, delivered by teachers trained in the subject, and taught at both primary and secondary schools (compulsory up to level 3). In secondary schools, there is an optional national examination subject, *Religious, Moral and Philosophical Studies* at National 5, Higher and Advanced Higher levels.

Creationists have learnt to avoid critical scrutiny of their activities, making it difficult to obtain more than anecdotal information, but there are clear hints of widespread creationist influence. The Catholic Church has accepted the material fact of evolution and common descent for decades, and this acceptance has been clearly reaffirmed by Pope Francis (Francis, 2014). The other churches are divided, especially in the Highlands and the Western Isles, mirroring the broader conflict between fundamentalists and modernisers worldwide. The Church of Scotland accepts the science, but many of its theologically conservative ministers do not, since it conflicts with the literalist-infallibilist approach to the Bible. This approach is widely accepted in some areas of the Highlands and Western Islands, both by local Church of Scotland congregations and by what is now known as the Free Church of Scotland, as well as other Calvinist churches, making the topic politically sensitive.

The nineteenth century Free Church of Scotland was more theologically liberal than the Church of Scotland itself, and led the way in acceptance of evolution. Hugh Miller, already mentioned, accepted an old Earth, and even a prolonged sequence of emergence of plants of increasing complexity (Miller, 1857), although he rejected the evolutionary theory put forward in *Vestiges of Creation*. Sadly, we cannot know how he would have reacted to the more powerful arguments put forward by Darwin and Wallace shortly after his death. Most noteworthy is Henry Drummond, naturalist, theologian, and lecturer at the Free Church College in Glasgow, who argued for the unity of religion and natural law, and in 1894 published *The Ascent of Man*, in which he argued that altruism increased fitness. In this, eight years before Kropotkin's *Mutual Aid*, he warned against those invoking a 'God of the gaps', or as he originally put it, 'gaps which they will fill up with God', and expressed the view that 'an immanent God, which is the God of Evolution, is infinitely grander than the occasional wonder-worker, who is the God of an old theology' (Drummond, 1894). Most congregations of the nineteenth century Free Church took part in a series of mergers that eventually led, in 1929, to the formation of the present-day Church of Scotland, but a theologically conservative faction survives as a separate entity, the present day Free Church, which is strongly opposed to evolution and predisposed to young Earth creationism, as are many Baptist and other evangelical churches.

A 21st century phenomenon is the development of strong links between such churches and the creationist and Intelligent Design movements in the United States. Two examples are particularly relevant, Glasgow's Centre for Intelligent Design (C4ID), and Highland Theological College. C4ID, founded in 2010, was inspired by, and has extremely strong links with, the Seattle-based Discovery Institute, whose Center for Science and Culture (established 1996) is devoted to the Intelligent Design concept, as part of a programme aimed at replacing naturalistic with God-acknowledging science (Rosenau, 2008). Intelligent Design maintains that biological complexity is too great to have arisen through natural processes alone (C4ID, 2010). Its advocates, almost without exception, also deny the material facts of evolution, common descent, and specifically human descent from non-human animals. There is, however, one remarkable difference here between the US and the UK. Intelligent Design advocates in the US are predominantly old Earth creationists, believing in separate creation of kinds, but accepting the antiquity of the Earth. In the UK, however, they are almost uniformly young Earth creationists, basing their chronology on *Genesis*, or else completely evasive when challenged on this question (Braterman, 2014).

The Centre for Intelligent Design has only had occasional success in gaining entry to schools, but has more indirect influence. Its director has worked (Noble, 2017) for School Leaders Scotland and for CARE (Christian Action Research and Education) Scotland, self-styled (CARE, 2017) 'servant of the Scottish Church', which is influential in the development of *Religious Observance* and *Religious, Moral and Philosophical Studies* in schools. He has also (Braterman, 2014) lectured on Intelligent Design at one of Glasgow's mosques, an ominous development given the overrepresentation of evolution denial among Muslim students at Glasgow University (see earlier).

Highland Theological College (HTC), Dingwall, formerly Highland Theological Institute, was founded in 1994. It has been a component of the UHI Higher Education Institute, now the University of the Highlands and Islands, since 2001, and has been awarding its own degrees since 2008. It describes itself (HTC, 2017) as 'both an independent college run by its own Board and one of thirteen colleges and research institutions which together make up the University of the Highlands and Islands … giving HTC a unique opportunity to impact on the training of ministers from a number of denominations.'

The College had, when one of us (PSB) enquired in 2013, only two academic theologians among its governors and trustees. These were Rev Dr J. Ligon Duncan III, Professor of Systematic and Historical Theology at Reformed Theological Seminary, Jackson, Mississippi, and Rev Dr Douglas F. Kelly, Professor of Theology at Reformed Theological Seminary, Charlotte, North Carolina. The Reformed Theological Seminary movement was started in 1966 by theologically conservative Southern Presbyterians, and both Duncan and Kelly are committed young Earth creationists. Kelly in *Creation and Change* (2015) maintains that the literal *Genesis* account is scientifically viable, while in the book *The Genesis Debate,* Ligon Duncan & Hall (2001) also argue in favour of creation within six days of 24 h.

The Church of Scotland recognized Highland Theological College as a training seminary for its own ministers in 2006, possibly as a result of the overall negotiations between liberal and conservative wings of the Church, which at the time was wrestling with the issue of blessings for same-sex unions. Thus College graduates will already be serving as ministers and as school chaplains, and eligible for those positions on Local Authority Education Committees that are reserved for the nominees of the Church of Scotland. Given their training, it is likely that at least some of these graduates promote creationism.

Many readers will be astounded to learn that there are such nominees; that every Local Authority Education Committee in Scotland must, by law, include three unelected Church appointees as full voting members. Of these, one comes from the Catholic Church, one from the Church of Scotland, and one from a third denomination chosen by rather haphazard procedures, leading fringe Evangelical or creationist Baptist churches to be over-represented (Braterman, 2013a). The law in its present form dates back to 1994, but has its roots in the 1872 re-organisation of the Scottish education system, which incorporated many previously Church-run schools. The Education Committees administer local policy regarding *Religious Observance* and *Religious, Moral and Philosophical Studies*, which can involve discussion of creationism, and the Church appointees sometimes form part of the committees interviewing teachers for senior posts. At the time of writing, the position of these Church appointees is under active discussion, with a petition (Scottish Parliament, 2016) under consideration in committee by the Scottish Parliament.

At both primary and secondary school level, creationism creeps in through extracurricular activities, and the involvement of outside bodies with a conservative religious agenda. For example, People With A Mission Ministry (PWAMM), funded by the Scottish transport tycoon Brian Souter, offers lessons and experiences within the *Religious, Moral and Philosophical Studies* curriculum (PWAMM, 2017a) and has a fleet of buses (PWAMM, 2017b) that visit schools throughout Scotland. It is doctrinally close to *Answers in Genesis*, whose literature it formerly showed on its bookshop website. There are also Scripture Unions, not devoted, alas, to critical analysis of the glorious complexity and tangled history of the biblical text, but to something much closer to a blinkered literalism, with all that that implies (Scripture Union Scotland, 2017).

Senior teachers themselves are enormously influential, both in deciding which school visitors to allow, and more directly. They play a leading role in choosing school chaplains, shaping the curriculum, and determining the nature of *Religious Observance* and *Religious, Moral and Philosophical Studies*. This places a substantial responsibility on the teaching staff, not all of whom are well versed in either biology or in the complexity of religious thought. Occasionally, this leads to scandal. At one primary school near Glasgow, the curriculum fell into the hands of a US-based creationist sect, and the situation only attracted attention when children were sent home with books saying that evolution is an unscientific lie used to promote immorality (Gardner, 2013; Braterman, 2013b). More often, especially at Primary school level, creationism is taught by default, because pupils are taught Bible stories with no clear distinction being made between history and myth; children might hear about Noah's Ark in the same way that they hear about the Roman invasion of Britain. In one recent example, the school set up a 'Creation corner', with each class being invited to contribute artwork representing one of the Six Days. Interestingly, after a parent asked why there was no 'evolution corner', this display disappeared and the school denies that the incident ever happened, despite photographic evidence. Children will frequently encounter teachers who invoke God as explanation ('why?' 'Because God made it that way'), with no awareness that they are shutting down the kind of enquiry that education is meant to encourage. The primary science curriculum itself mentions the difference between inherited and non-inherited characteristics, but makes no reference to change in species over time.

In England, pressure on the Government led it to issue clear guidance in 2014 stating that creationism and Intelligent Design should not be taught as part of the science curriculum in publicly funded schools. This followed concern that English 'Academy' and 'Free' schools, which are often sponsored by religious organisations, might teach creationism as a valid scientific proposition (BHA, 2014). However, a petition seeking to get similar language from the Scottish Government led to only partial success, in the form of a statement that these are not part of the science curriculum, and should not be taught in the science classroom (Scottish Parliament, 2015). This, of course, leaves room for them to be taught (as opposed to taught *about*, which will clearly be appropriate; see below) in other classrooms, including *Religious, Moral and Philosophical Studies*. The problem, as usual, is political; the then Minister for Schools had his constituency in the Highlands and Islands, the most theologically conservative party of the country.

The Secondary School *Religious, Moral and Philosophical Studies* (RMPS) curriculum includes, as it should, discussion of evolution and of religious responses to its inconsistency with the literal Biblical account (similar remarks apply to the Big Bang theory, which is also included in the syllabus). These topics present a challenge to the RMPS teacher and textbook writer. It is in the nature of the subject that pupils be invited to make up their own minds, and yet the science is completely unambiguous. This has presented problems in the past, but the most recent approved textbook (Walker, 2016) does a very good job here. There are familiar faults in the presentation, such as excessive emphasis on Darwin, the running together of the origins of life, a largely unsolved problem, with the evolution of life, which has been mapped out in considerable detail, and confusion about the meanings of the word 'theory'. The text also conveys the unfair impression that the nineteenth century Churches were committed to a young Earth. However, and much more importantly, the textbook makes it clear that the scientific consensus in favour of evolution is overwhelming, with explicit reference to the fossil record, biogeography, the analogy with artificial selection, cross-species comparisons, and vestigial features. Moreover, there is a clear distinction between describing past evolution, and providing a mechanism, and there is a brief but good discussion of mutation and adaptation. There is a short description of the argument from irreducible complexity, but it is immediately answered both scientifically and philosophically; complexity can arise incrementally, and the argument from allegedly irreducible complexity to a 'designer' is circular reasoning.

There has been steady progress in Scottish school science teaching, but there remains a need for basic science grounding for Primary school teachers generally and for Secondary school *Religious, Moral and Philosophical Studies* teachers on what current evolution science actually says, and how its claims are justified.

We would also recommend drawing their attention to resources such as the 'talkorigins' catalogue of creationist arguments (Isaak, 2016), to prepare them for dealing with creationist objections, which may be a matter of considerable importance in some parts of the country.

10.7 Conclusion

Scotland's encounter with evolutionary ideas began early; Charles Darwin's career as a natural historian started in Edinburgh, and *Vestiges,* a popular pre-Darwinian evolution proposal, was a product of the Scottish enlightenment. Scotland's education system has long been independent of the rest of the UK. A basic treatment of the theory of evolution is provided in the secondary school biology curriculum but is damagingly absent from the human biology course studied by many intending medical students. The religious studies curriculum also deals with the origins of life and its diversity, and this provides scope for some mixed messages to be taught. All Scottish biology schoolteachers have studied some branch of biology at University prior to teacher training. However, the coverage of evolution they experience depends on their course and can be very slight. Despite this, most teachers are confident in their ability to teach the theory well. Our survey found 6% of biology teachers do not accept the theory of evolution, but we do not know how much this affects their teaching practice. It would be worth considering the use of evolution acceptance as a pre-requisite for admission to a biology teacher training programme. However, a better strategy would be to ensure that teacher training programmes, both for biology and for religious studies, include strategies for how to cope with pupils who come to school with strong anti-evolution beliefs. One of us (RS) received no specific guidance on this during his training, but notes that teachers are encouraged to be sensitive to pupils' viewpoints in general.

Although most of Scotland's religious faiths have come to terms with the theory of evolution, young Earth creationism is active in some Protestant Christian groups and amongst some Muslims. It is an anomaly deriving from the early 20th century political settlement between churches and government that faith representatives have statutory places on local authority education committees. In our view, it is time for the abolition of this undemocratic role of faith groups.

The level of education rejection in the Scottish population has not been assessed, but among students of biology at one Scottish university, the level of rejection has declined from about 10% to around 3% over the last 20 years. However, among medical students, the rejection level is higher and this may have consequences for medical practice. It is regrettable that curriculum change in medical schools has removed the coverage of evolution and its relevance to medicine.

10.8 Acknowledgements and Disclosure

We thank: all the students and teachers who took the time to complete our questionnaires, and Biology Level 1 coordinator Chris Finlay who gave time and permission for the class to be surveyed; Martin Hendry and Iain Neill who commented on physics and geology students respectively; staff of the University of Glasgow Archives and Special Collections departments for access to books and records; Clare Marsh and Colin Gambles for comments on the manuscript; Lorna Kennedy for help with manuscript preparation. PSB discloses his personal involvement with petitions to the Scottish Parliament on religious influences on school management.

References

Anderson, R. (2013). The history of Scottish education, pre-1980. In T. G. K. Bryce, W. M. Humes, W. M. Gillies, & A. Kennedy (Eds.), *Scottish education* (4th ed., pp. 241–250). Edinburgh: Edinburgh University Press.

Berkman, M. B., Pacheco, J. S., & Plutzer, E. (2008). Evolution and creationism in America's classrooms: a national portrait. *PLoS Biology, 6*, e124.

BHA (2014). Government bans all existing and future Academies and Free schools from teaching creationism as science. British Humanist Association 18 June, 2014. https://humanism.org.uk/2014/06/18/

Braterman, P. S. (2013a). Retrieved Mar 1, 2015 from https://paulbraterman.wordpress.com/2013/10/21/petition-to-abolish-church-seats-on-scottish-education-committees-9-good-reasons-to-sign/.

Braterman, P.S. (2013b). Retrieved Sept, 15 2013 from https://paulbraterman.wordpress.com/2013/09/16/reviewed-young-earth-creationist-books-handed-out-in-scottish-primary-school/.

Braterman, P. S. (2014). Retrieved 13 Dec 2014 from https://paulbraterman.wordpress.com/2014/12/04/glasgows-intelligent-design-director-has-open-mind-on-age-of-earth/.

Brown, S. (2014). The 'curriculum for excellence': A major change for Scottish science education. *School Science Review, 352*, 30–36.

Burchfield, J. D. (1990). *Lord Kelvin and the age of the Earth*. Chicago: University of Chicago Press.

C4ID (2010). Retrieved 10 January 2017 from http://www.c4id.org.uk/, updated sporadically.

CARE (2017). Retrieved 10 January 2017https://www.care.org.uk/about-us/where-we-work/Scotland.

Day, S. P. (2013). Biology education. In T.G.K. Bryce, W. M. Humes, W. M. Gillies, & A. Kennedy (Eds.), *Scottish education* (4th ed.) (pp. 518–523). Edinburgh: Edinburgh University Press.

Deniz, H., Donnelly, L. A., & Yilmaz, I. (2008). Exploring the factors related to acceptance of evolutionary theory amongst Turkish pre-service biology teachers; toward a more informative conceptual ecology for biological evolution. *Journal of Research in Science Teaching, 45*, 420–443.

Desmond, A., & Moore, J. (2009). *Darwin's sacred cause: race, slavery and the quest for human origins*. London: Allen Lane.

Downie, J. R. (2004). Evolution in health and disease; the role of evolutionary biology in the medical curriculum. *Bioscience Education eJournal, 4*, 3.

Downie, J. R., & Barron, N. J. (2000). Evolution and religion; attitudes of Scottish first year biology and medical students to the teaching of evolutionary biology. *Journal of Biological Education, 34*, 139–146.

Drummond, H. (1894). *The Lowell lectures on the ascent of man*. New York: James Pott & Co.

Endersby, J. (2007). *A guinea pig's history of biology*. London: William Heinemann.

Farrow, G. E. (2001). James Croll; a 19th century pioneer of climate change. *The Glasgow Naturalist, 23*(6), 9–18.

Francis (2014). Address of his holiness Pope Francis on the occasion of the inauguration of the bust in honour of Pope Benedict XVI, 27 October 2014. https://w2vatican.va/content/francesco/en/speeches/2014/October

Gardner, C. (2013). Retrieved 10 January 2017http://www.scotsman.com/news/politics/sect-s-preacher-spent-8-years-at-primary-school-1-3081113.

Hargreaves, A. (2000). Four ages of professionalism and professional learning. *Teachers and Teaching: Theory and Practice, 6*, 151–182.

HTC (2017). Retrieved 10 January 2017 from https://www.htc.uhi.ac.uk/about-us/faqs,.

Huxley, T. H., & Martin, H. N. (1875). *A course of practical instruction in elementary biology*. London: Macmillan & Company.

Isaak, M. (2016). Retrieved 10 January 2017 from http://www.talkorigins.org/indexcc/list.html, last updated 5 November 2006.

Kelly, D. F. (2015). *Creation and change: Genesis 1:1–2.4 in the light of changing scientific paradigms: discovering God in creation*. Fearn, Scotland: Mentor.

Ligon Duncan III, J., & Hall, D. W. (2001). A short study of evangelical views on creation in Genesis. In D.C. Hagopian (Ed.), *The Genesis debate* (pp. 47–52).Mission Viejo, California: Cruxpress.

Livingstone, D. (1994). Science and religion; foreword to the historical geography of an encounter. *Journal of Historical Geography, 20*, 367–383.

Mackean, D. G. (1988). *Life study, a textbook of biology*. London: John Murray.

Marsh, C., Simms, J., Stevenson, C., Torrance, J., & Fullarton, J. (2012). *Higher biology for CfE with answers*. Paisley, Scotland: Hodder Gibson.

Miller, H. (1857). *The testimony of the rocks, or, Geology in its bearing on the two theologies, natural and revealed*. Edinburgh: Thomas Constable & Co.

Miller, J. D., Scott, E. C., & Okomoto, S. (2006). Public acceptance of evolution. *Science, 313*, 765–766.

Nelson, E. C. (2014). *John Scouler (c1804 1871): Scottish naturalist, a life with two voyages. The Glasgow Naturalist* (Supplement). Glasgow: Glasgow Natural History Society.

Nesse, R. M., & Williams, G. C. (1995). *Evolution and healing; the new science of Darwinian medicine*. London: Weidenfeld and Nicolson.

Nichol, J. (1853). *On the study of natural history as a branch of general education*. Inaugural lecture, Aberdeen: Marischal College.

Noble, A. (2017). Retrieved 10 January 2017 from https://www.linkedin.com/in/alastair-noble-47a7127.

PWAMM (2017a). Retrieved 10 January 2017 from http://www.challengereducation.com/

PWAMM (2017b). http://www.pwamm.com/challenger-outreach/.

QAA (2007). *Benchmark for the biosciences*. Quality Assurance Agency for Higher Education. www.qaa.ac.uk.

Rosenau, J. (2008). Retrieved 10 January 2017 from https://ncse.com/creationism/general/wedge-document.

Rutledge, M. L., & Mitchell, M. A. (2002). High school biology teachers' knowledge structure, acceptance and teaching of evolution. *American Biology Teacher, 64*, 21–28.

Rutledge, M. L., & Sadler, K. C. (2007). Reliability of the measure of acceptance of the theory of evolution (MATE) instrument with university students. *American Biology Teacher, 69*, 332–335.

Scottish Parliament. (2015). PE01530: *Guidance on how creationism is presented in schools*, Education and Culture Committee, Official report, 12 May 2015 http://www.parliament.scot/parliamentarybusiness/report.aspx?r=9951&i=91368#ScotParlOR.

Scottish Parliament. (2016). *PE01623: Unelected church appointees on Local Authority Education Committees*, https://www.parliament.scot/GettingInvolved/Petitions/ChurchAppointees; regularly being updated.

Scotland, J. (1969). *The history of Scottish education (two volumes)*. London: University of London Press.

Scowen, J. (1998). A study in the historical geography of an idea: Darwinism in Edinburgh 1859–75. *Scottish Geographical Magazine, 114*, 148–157.

Scripture Union Scotland. (2017). https://www.suscotland.org.uk/about-us/our-beliefs/, Retrieved 10 January 2017.

Simms, J., Fullarton, J., Marsh, C., Stevenson, C., & Torrance, J. (2013). *Higher human biology for CfE with answers*. Paisley, Scotland: Hodder Gibson.

Souter, N. (2003). Biology education. In T. G. K. Bryce & W. M. Humes (Eds.), *Scottish education* (2nd ed., pp. 477–482). Edinburgh: Edinburgh University Press.

Southcott, R., & Downie, J.R. (2012). Evolution and religion: attitudes of Scottish bioscience students to the teaching of evolutionary biology. *Evolution Education and Outreach, 5,* 301–311

Stott, R. (2012). *Darwin's ghosts: In search of the first evolutionists*. London: Bloomsbury Publishing.

Torrance, J. (co-ordinator). (1999). *New higher biology* (2nd ed.). London: Hodder & Stoughton Educational.

Torrance, J. (co-ordinator). (2001). *Standard grade biology* (3rd ed.). London: Hodder & Stoughton Educational.

Torrance, J. (co-ordinator). (2002). *Higher human biology* (2nd ed.). Paisley, Scotland: Hodder Gibson.

Torrance, J., Fullarton, J., Marsh, C., Simms, J., & Stevenson, C. (2013). *National 5 biology: applying knowledge and skills*. Paisley, Scotland: Hodder Gibson.

Walker, J. (2016). *Higher RMPS; religious and philosophical questions*. Paisley, Scotland: Hodder Gibson.

Roger Downie is a semi-retired Professor of Zoological Education at the University of Glasgow, Scotland. His research interests include animal development, marine turtle conservation, amphibian reproductive ecology and conservation, and ethics, fieldwork and evolution education. He has a long term association with Trinidad and Tobago where much of his field-work has been done, mainly on staff-student expeditions.

Ronan Southcott is a secondary school biology teacher working in Scotland. He has previously collaborated with Roger Downie on research into the attitudes of University of Glasgow students to evolution. He is currently studying for a doctorate in education.

Paul Braterman is Honorary Senior Research Fellow at the University of Glasgow, and Professor Emeritus at the University of North Texas. His research on topics related to conditions on the early Earth was supported by NASA's Astrobiology Program and by the [US] National Science Foundation. He is on the committees of the British Centre for Science Education and the Scottish Secular Society, and in these capacities has been instrumental in persuading both Westminster and Scottish governments to oppose the teaching of creationism as scientifically valid.

Naomi Barron graduated in zoology and taught biology and mathematics in secondary schools for 20 years. Since marrying Roger Downie, she has participated in research on frogs and turtles in Trinidad and Tobago, and in evolution education in Glasgow.

Chapter 11
Teaching Evolution in Greece

Panagiotis K. Stasinakis and Kostas Kampourakis

Abstract In this chapter, we provide an overview of the teaching of evolution in the Greek educational system. We discuss issues relating to the education, training and professional development of teachers; the educational policies that determine the content of teaching, of textbooks, and of the exams that provide enrolment to university studies; the interests, the choices, the social environment, and the priorities of students. Whereas there is research in teaching and learning evolution in Greece, as well as on Greek teachers' pedagogical content knowledge, it has not really been taken into account in policy decisions relevant to the teaching of evolution. It should be noted that no particular religious influences against the teaching of evolution exist in Greece, and this is important to note if one considers that it is a country with no separation between church and state. However, it has been claimed or implied that it is due to such influences that content about evolution has only been included in the last chapters of textbooks. This is the case but not the major problem in our view. Even in the textbook-driven and exam-focused Greek educational system, teachers who are appropriately trained and who feel confident to teach evolution should be able to do so in any biology chapter. Therefore, the major problem in our view is that neither systematic pre-service and in-service training, nor a robust undergraduate education on the teaching of biology are offered to Greek biology teachers. Given the current difficult fiscal environment that impacts educational policy choices and decisions, we suggest actions that could enhance the teaching of evolution and contribute to the efforts of improving biological literacy in our country.

P. K. Stasinakis
4th High School of Zografou, Athens, Greece

K. Kampourakis (✉)
Section of Biology and IUFE, University of Geneva, Geneva, Switzerland
e-mail: Kostas.Kampourakis@unige.ch

11.1 Introduction

During the last 15 years, there has been a growing interest in research on the teaching of evolution in Greece. The outcome has been a series of empirical studies with significant findings, on which conclusions that could improve the teaching and learning of evolution could be drawn. This might facilitate Greek students' understanding of one of the greatest intellectual achievements of humanity: the theory of evolution. Dobzhansky (1973) famously described it as the central unifying theory of biology in his article "Nothing in biology makes sense except in the light of evolution"; he correctly pointed out that without evolution, biology is a pile of sundry facts that make no meaningful picture as a whole.

The Greek educational system is characterized by the lack of systematic pre-service and in-service teacher training programmes. Therefore, teachers' professional development depends on their willingness to attend seminars or graduate programs in science education, if they feel that didactics can provide them with effective tools for their teaching practice. Scientific associations, such as the PanHellenic Union of Bioscientists, produce educational materials, promote partnerships among teachers, as well as among teachers and researchers in science education, and advocate the teaching of biology—especially of evolutionary theory. Because of such efforts, during the last six years Greek students have excelled in the International Biology Olympiad.[1]

At the same time, the Ministry of Education via the Institute for Educational Policy, an institution responsible for education policy and an advisory body for each Minister of Education, is trying to bring about changes in the educational practice, either by making it less teacher-centered (through reducing the content to be taught, proposing teaching strategies that allow teachers more liberty in their selection of materials for teaching) or by contributing to the production of educational material (for instance, the Computer Technology Institute and Press 'Diophantus' has produced digital material and numerous learning objects).

However, despite efforts like these to rectify the educational culture in Greece, there are weaknesses that arise from the centralized structure, the exam-focused and single-textbook character of the Greek educational system, the gap between the findings of empirical research in the teaching of biology and its application in curriculum design, the resistance of students and teachers towards any radical changes, and the pace by which any educational changes are implemented that are either too fast or too slow to be effective.

This chapter aims at describing various aspects of evolution education in Greece, and consists of three sections. In Sect. 11.2, we provide a detailed description of the current Greek educational system and how it does not support any effective teaching of evolution. In Sect. 11.3, we present findings from the empirical research

---

[1]IBO—International Biology Olympiad, http://www.ibo-info.org/.

on evolution education in Greece, outline its weaknesses, and discuss the opportunities that exist. Finally, in Sect. 11.4 we propose immediate solutions and long-term strategies that could improve both the teaching of evolution and of biology.

11.2 The Educational System and the Teaching of Evolution in Greece

This section is about secondary education (grades 7–12), since biology is not taught as a distinct subject in primary school or preschool. Instead, all natural sciences are taught together. Additionally, we present the undergraduate studies that universities offer to pre-service biology teachers.

Secondary education in Greece consists of two levels, the lower secondary (gymnasium, ages 13–15 years old) and the upper secondary one (lyceum, ages 16–18 years old), each consisting of three grades (A, B, C). Biology is taught in all grades. Grade C of gymnasium is the last one of compulsory education, and grade C of lyceum is the preparatory one for students taking the national exams in order to enroll in university studies. It should be noted that although there is an official national curriculum, teachers rather follow the single textbook provided by the Ministry of Education for each course. The reason for this is that the content that teachers ought to teach and assess is prescribed in terms of textbook pages, decided by the Institute for Educational Policy. Therefore, the main textbook practically replaces the curriculum.

According to a Eurobarometer study (2005), when Greek citizens were asked to comment on the statement "Human beings, as we know them today, developed from earlier species of animals", 55% of them replied that it was 'True', 32% replied that it was 'False' and 14% that they did know (the average in EU25 was 70, 20, 10%, respectively). Only about half of the 1000 participants had a good knowledge about human evolution.

The book 'On the Origin of Species', was translated into Greek language for the first time in 1915, by the great Greek writer Nikos Kazantzakis (1915). The first recorded reaction from the Church was in 1934 (Xirafas, 2009), in the official magazine 'Ecclesia', noting that "…there is a mistake, the contradictory contrast between science and philosophy of religion, which often lead to exaggerations" (p. 4, Ecclesia). In the same volume we can also find have the first negative reaction of the Church towards Darwin's theory: a report criticizing a Professor of Zoology in his speech about *the evolution of living life*, mentioned that "this theory is out of date and that is unscientific in this era such lectures to be given" (Xirafas, 2009, p. 432). Xirafas (2009), concluded: "… reactions were milder than those in Western Europe and Western Church. As in every scientific field, in theology too there are always opposing views that have been kept on a mild level with few exceptions.

According to a statement by the representative of the Church of Greece, the latter never officially condemned Darwin's theory" (p. 434).

We also tend to belief that overall the Greek Church and its representatives have not had any significant influence on hindering Greek students' and citizens' misunderstanding of evolutionary theory. Even though it could be the case that biology teachers may have faced religious resistance from individual students and their families, there has never been a religious-founded organized anti-evolution movement in Greece that has had any significant effect upon the teaching of evolution.

11.2.1 The Greek Secondary Biology Curriculum

Greece does not have a robust, stable educational system. Strangely, changes rather occur in response to exogenous, not education-related, reasons. For instance, various changes have taken place recently because of the fiscal problems in Greece. As a result, during the current school year (2016–2017) Biology is taught at all gymnasium grades only for a single period (45 min) per week. What is worse, changes take place so abruptly that the time allocated to the teaching of Biology in lower-secondary school has changed three times within the last three years (Table 11.1).

Apparently, the time allowed for the teaching of biology is very limited and so it is impossible to teach all topics appropriately. Therefore, although a chapter about evolution should be taught in grade C of gymnasium, which is the last grade of compulsory school, this is often not done due to lack of time.

During the last three years, the Ministry of Education has provided teachers with specific guidelines (Table 11.2).[2]

As we can see in guidelines one and two, teachers are prompted to discuss aspects of evolutionary theory (adaptations/diversity) with their students, even though the respective chapters are not about evolution (*Organization of Life* and *Nutrition* respectively). In guidelines 3 and 4, teachers are advised to consider students' previous knowledge and clarify the concept of adaptation. There is also supplementary material about adaptations and more guidelines about discussing figures of the main textbook that refer to diversity and evolution. These guidelines seem to be in the right direction: teachers should teach evolutionary theory as the unifying theory of biology, and could refer to it in any chapter of the textbook. However, it is uncertain whether teachers will be able to fully cover the content to be taught in the limited teaching time allowed.

For the school year 2016–2017, teachers in gymnasium ought to teach the following textbook chapters:

---

[2]15/09/2016, 150022/Δ2.

Table 11.1 The time devoted to Biology in periods (45 min) per week in lower secondary school (biology/natural science in total—natural science comprises Physics, Chemistry, Biology, Geography)

	2014–2015[a]	2015–2016[b]	2016–2017[c]
A	2/5	2/5	1/3
B	1/6	1/6	1/6
C	2/5	1/4	1/4
Total	5/16	4/15	3/13

[a]Government Gazette (GG) 2121/28-08-2013 (A, B), Government Gazette (GG) 1890/04/09/2009 (C)
[b]Government Gazette (GG) 2121/28-08-2013
[c]Government Gazette (GG) 1640/09-06-2016

Table 11.2 Guidelines about teaching evolution in gymnasium

#	Class	References	Chapters
1	A	… to be discussed … that in any environment … the best adapted organisms survive … best adapted in an environment are not always the most "powerful" organisms	1. Organization of life
2	A	… focus on similarities/differences of digestive systems several organisms … highlight the evolutionary aspect	2. Nutrition
3	C	… at the beginning discuss further concepts such as species and population	7. Evolution
4	C	… discuss the adaptations … highlight that the adaptations are properties, structures, attributes, behaviors that have been acquired or preserved by natural selection because they provided to individuals the best chance of survival and/or reproduction success in competition with others in a particular environment	7. Evolution

- Grade A: 1. Organization of Life; 2. Digestion and nutrition; 3. Transport and excretion; 4. Respiration.
- Grade B: 2. Organisms in their environment; 3. Metabolism; 4. Diseases and the factors associated with their occurrence.
- Grade C: 1. Organization of life—Biological Systems; 5. Genetics; 7. Evolution (this year, there is supplementary material about human evolution, with an extra activity).

In all these chapters (the contents of which are described in detail in Sect. 11.2.2), apart from the one that is explicitly about evolution, teachers can teach about evolutionary theory and its various concepts such as common ancestry, biodiversity, adaptation, shared characteristics, and natural selection while teaching about the topics included in the other chapters.

Regarding biology courses at the lyceum level, there have been fewer changes in the three last years (Table 11.3).

Table 11.3 The time devoted to Biology in periods (45 min) per week in upper secondary school (biology/natural science in total—natural science comprises Physics, Chemistry, Biology, Geography)

	2014–2015[a]	2015–2016[b]	2016–2017[c]
A	2/6	2/6	2/6
B	2/9	2/9	2/9
C	1 + 2/9	2 + 2/10	2 + 2/10
Total	5 + 2/24	6 + 2/25	6 + 2/25

[a]Government Gazette (GG) 193/17-09-2013 (A, B), Government Gazette (GG) 921/05/07/2005 (C)
[b]4186/2013—Government Gazette (GG) 193/A/17/9/2013 (A, B), 4327/2015- Government Gazette (GG) 50/A/14-5-2015 (C)
[c]4186/2013—Government Gazette (GG) 193/A/17/9/2013 (A, B), 4327/2015- Government Gazette (GG) 50/A/14-5-2015 (C)

During the past three years, the Ministry of Education has provided teachers with specific guidelines.[3] There are no guidelines about teaching evolution in grades A and B (5–6 years ago, there were recommendations by the Ministry of Education about teaching evolution as a unifying theory in B grade). Textbooks in upper secondary school have not changed since 2000 (no changes have been proposed by the Institute for Educational Policy), whereas those in lower secondary school have not changed since 2007. Since 2009, the chapter on evolution taught in grade C of lyceum has been included in the subject matter required for the national exams. Between 2009 and 2015, evolution as a fact, natural selection and the differences between the Darwinian and Lamarckian theories were taught the introductory section only. Since 2015, the New Synthesis, Speciation, Human Evolution, Human Phylogenetic Trees, Genes and Evolution have been taught as well. This seems to be an evolution-rich curriculum, but there is a problem: students are taught all these topics just a couple of months before the end of the school year, whilst they are preparing for the national exams. Only a minority of students (less than 5% on average) will be examined in this subject during the national exams because this is an elective course, and so most students do not pay attention to these topics.

The last grade of compulsory education is C in gymnasium. According to the latest data, during the school year 2013–2014, 99,000 students (51,316 males and 47,684 females) attended this grade. Grade A of Lyceum[4] of the next school year 2014–2015 was attended by 82,009 students (38,710 males and 43,299 females), which means that approximately 17,000 students, or 1 out of 5, did not enroll in upper secondary education. These are future citizens who have most likely completed the compulsory education without having any basic content knowledge about evolutionary theory.

---

[3]15/09/2016, 150658/Δ2.

[4]Hellenic Statistical Authority, Retrieved 27 December 2016 from https://goo.gl/mVZViw.

11.2.2 The Greek Secondary Biology Textbooks

Six different textbooks are used in the six grades of secondary education, two at the lower and four at the upper secondary school. Their contents for the current academic year, 2016–2017, are presented in Table 11.4.

11.2.3 Biology Education at the University Level

In Greece, there are six departments the graduates of which are referred to as "Biologists". During the academic year 2016–2017, 765 students were enrolled in these departments (115 in Aristotle University of Thessaloniki, 110 in the National and Kapodistrian University of Athens, 135 in the University of Crete, 150 in the University of Patras, 125 in the University of Ioannina, 130 in the Democritus University of Thrace). All departments focus on research-centered studies in the life sciences.

According to the law,[5] all students who graduated from any of the above departments before 2014, are considered to be pedagogically competent and could be appointed as teachers in public schools, even though they may have not have attended any course on didactics or pedagogy during their undergraduate studies. If one looks at the courses offered by the various departments (Table 11.5), one department has no such course (Patras), one has a core course, and three have only one elective course. In three departments, there are relevant postgraduate studies, but only the departments of Athens and Thrace provide a Master's program focused on Didactics of Biology. In contrast, those who graduated from the various biology departments after 2014 ought to have a certificate of pedagogical competence, but to the best of our knowledge no department provides such a certificate on the basis of an appropriate teacher training program.

One might think that even if biology teachers have had insufficient studies on biology didactics and thus do not possess the necessary pedagogical content knowledge, they should at least have the necessary factual knowledge about general biology and specific subjects such as evolution. However, there is another problem. According to the law,[6] all science teachers (biologists, physicists, chemists, geologists, naturalists, agronomists and foresters) and health care professionals (physicians, dentists, pharmacists, and nurses), are allowed to teach biology at any grade. This would be no problem if the non-biologists had a relevant, secondary specialization. However, many of them, especially physicists and chemists, were never enrolled in any biology course during their undergraduate studies (this out-of-discipline teaching occurs very often, especially in lower level and at schools

---

[5]1894/1990 (Government Gazette (GG) 110 A'), 3194/2003 (Government Gazette (GG) 267/20-11-2003), 1566/1985, 3699/2008.
[6]29/09/2015, 151893/Δ2.

Table 11.4 The content of Greek secondary Biology textbooks (*C-g (general): an elective course in the last grade of the lyceum, offered to all students, some of whom (less than 5%) will be examined during the national exams in Biology-g; C-sc (science): a course offered to all students intending to enroll in science studies, less than half will be examined in Biology-sc during national exams the other half will be examined in Mathematics)

	Class	Chapters title	Content
Gymnasium	A	1. Organization of life	Characteristics of organisms, Cell, Organization of multicellular organisms, Interactions and adaptations
		2. Digestion and nutrition	Photosynthesis, Unicellular organisms, Animals, Humans
		3. Transport and excretion	Plants, Unicellular organisms, Humans
		4. Respiration	Plants, Unicellular organisms, Humans (cellular respiration, respiratory system, transpiration)
	B	2. Organisms in their environment	Equilibrium in biological systems, Ecosystems, Energy, Human's interference
		3. Metabolism	Humans and energy, Enzymes
		4. Diseases and the factors associated with their occurrence	Homeostasis, Bacteria, Antibiotics, Protozoa, Infection, Vaccines, Inflammation
	C	1. Organization of life—Biological Systems	Molecules, Cells, Organisms, Ecosystems
		5. Genetics	Chromosomes, Genes, Replication-Transcription-Translation, Alleles, Mitosis-Meiosis, Mendelian inheritance, Mutations
		7. Evolution	Biochemical evidence, Human Evolution
Lyceum	A	1. From cell to organism	Human tissues, Organs, Systems
		3. Circulatory system of human	Structure and function of heart, Blood, Vessels, Circulation
		9. Nervous system of human	Neural cell, Peripheral, central autonomous nervous system
		12. Reproduction-Development of human	Male, female reproduction organs, Menstrual cycle, Spermatogenesis, Oogenesis, fertilization, Embryo-Birth, Sexually transmitted diseases, contraception
	B	1. Molecules	Macromolecules (protein, nucleic acid, lipids, carbohydrates),
		2. Cell	Cellular membrane: structure-function, nucleus, endomembrane system, chloroplasts-mitochondria,
		3. Metabolism-Energy	ATP, Enzymes, Cellular respiration-anaerobic & aerobic, Photosynthesis (light-dark reaction), Leaf structure-stomata, Lipids and protein break down

(continued)

Table 11.4 (continued)

Class	Chapters title	Content
	4. Molecular Biology	Central dogma (replication-transcription-translation), chromatin-chromosome, Mitosis-Meiosis
*C-g	1. Human & health	Homeostasis, Microorganisms, Infection, Antibiotics, Sexual transmitted diseases, Immunity, Vaccines, drip of antibodies, B and T-cells, allergy, autoimmune diseases, AIDS, addiction (drugs, smoking, alcohol)
	2. Human & environment	Ecosystem, Trophic structure, food webs, Trophic levels, energy flow, biochemical cycles, pollution
	3. Evolution	Classification, Species-populations, phylogenetic, Darwinian-Lamarckian theories, Natural Selection, industrial revolution, evolutionary synthesis, human evolution-primates, hominids, diversity in human populations-genetic variation
*C-sc	1. Genetic material	DNA structure, Double-helix, chromatin-chromosome, Karyotype, Mitochondrial—Chloroplastic DNA, Prokaryotic and Viral genetic material,
	2. Genetic information	Central dogma (replication-transcription-translation), Regulation of genes, Lac-Operon
	4. Recombinant DNA	EcoRI, c-DNA library, genomic library
	5. Mendelian inheritance	Monohybrid cross, Dihybrid cross, Multiple alleles, sex linkage, pedigrees
	6. Mutations	Gene mutations, chromosomal abnormalities
	7. Biotechnology	Microbial cultures, aerobic and anaerobic culture methods
	8. Biotechnology in medicine	Insulin, monoclonal antibodies, gene therapy, pharmaceutical proteins, human genome
	9. Biotechnology in plant and animals	Ti-plasmid, gene pharming, animal cloning-Dolly

in rural areas). Therefore, they are totally unprepared to teach evolutionary theory, because they do not even possess the fundamental biology content knowledge. This is a big problem, especially for teaching at the lyceum level.

Since 1998, a national contest for the recruitment of teachers in secondary education has been taking place. The relevant content includes basic pedagogy and didactics, but there is no requirement for previous teaching experience, a training certificate or university courses on teaching and pedagogy. In addition, no training programs for pre-service secondary school teachers exist. In the past, in-service teachers were obliged to attend an initial training program about didactics and

Table 11.5 Courses about didactics and pedagogy for future biology teachers

	Undergraduate	Graduate
University of Patras, Biology Department[a]	No course about 'Didactics of Biology'	No course about 'Didactics of Biology'
Aristotle University, School of Biology[b]	Environmental Education and Public Awareness (Elective course), Didactics of Biology (Elective course)	Specialization in Environmental Education[c]
University of Crete, Department of Biology[d]	Practice in Didactics of Biology (Elective course)	No course about 'Didactics of Biology'
National and Kapodistrian University of Athens, Faculty of Biology[e]	Introduction to Pedagogy, Learning theories—Didactic methodology (Elective course)	Didactics of Biology[f]
University of Ioannina, Department of Biological Applications & Technology[g]	Didactics of Natural Science (Elective course), Pedagogy I and II (Elective courses), Educational Sociology I and II (Elective courses)	No course about 'Didactics of Biology'
Democritus University of Thrace, Department of Molecular Biology & Genetics[h]	Pedagogy, Didactics (Micro-Teaching) I and II (Elective courses), Teaching Methodology (Elective course), Adult Education (Elective course)	Didactics of Biosciences[i]

[a]Program of Study for 2016-2017, Retrieved 9 May 2017 from https://goo.gl/S7910r
[b]Website of department, Retrieved 26 December 2016 from https://goo.gl/uCbmTa
[c]Website of graduate program, Retrieved 26 December 2016 from https://goo.gl/XNEUZs
[d]Program of Study for 2016-2017, Retrieved 26 December 2016 from https://goo.gl/2bo5xd
[e]Program of Study for 2016-2017, Retrieved 26 December 2016 from https://goo.gl/Dv3MU9
[f]Website of graduate program, Retrieved 26 December 2016 from https://goo.gl/fvV5Ae
[g]Program of Study for 2016-2017, Retrieved 16 January 2017 from https://goo.gl/6NMJfP
[h]Program of Study for 2016-2017, Retrieved 16 January 2017 from https://goo.gl/urxOEt
[i]Website of graduate program, Retrieved 16 January 2017 from https://goo.gl/J9kzaf

pedagogy, in three training cycles that lasted about a month in total. After this, there are no training programs for teachers, except from training meetings organized by school consultants.

All in all, biology teachers in Greece lack the necessary knowledge and experience relevant to teaching and learning, because universities usually do not train them for this purpose and because there are no pre-service and in-service training programs to compensate for this. This means that not only teachers are not as competent as they could be in the beginning of their careers, but also that they rarely have any opportunities to acquire the necessary knowledge and experience. We think that this is a major issue and a cause for the low public understanding of evolution in Greece.

11.3 Research on Teaching and Learning Evolution in Greece

During the last 10 years, research on students' conceptions about evolution or relevant topics has taken place at all levels of education, from the kindergarten to the university level. In this section, we briefly review this research and highlight some important conclusions for teaching evolution.

Kampourakis, Pavlidi, Papadopoulou, & Palaiokrassa (2012) investigated whether second-grade Greek students (n = 149, seven to eight years old) provided teleological explanations for particular organisms, artifacts and natural objects, as well as if there was any relation between their explanations and their familiarity with these objects. Overall, students seemed to provide teleological explanations for organisms and artifacts, but not for natural objects. In a second study, which extended the previous one, Kampourakis, Palaiokrassa, Papadopoulou, Pavlidi, & Argyropoulou (2012) also included pre-school (74 participants, five to six years old) and first- grade (n = 153, six to seven years old) children. They identified a shift from a non-discriminative (teleological explanations for all objects) to a discriminative (teleological explanations for organisms and artifacts, but not for natural objects) teleology during the ages from five to seven years old. Whereas pre-school students provided teleological explanations in a non-discriminative manner, first-grade children and second-grade children were progressively more discriminative. Therefore, a developmental shift might exist from non-discriminative to discriminative teleological explanations. This might have significant implications for teaching biology, since teleological explanations play an important role for the understanding of evolution and natural selection (Kelemen, 2012; Kampourakis, 2014).

Prinou, et al. (2008) analyzed the responses of 411 10th grade students from 12 different schools. They used a questionnaire that consisted of open-ended and multiple-choice questions. Most of the students appeared to accept the idea of evolution, since six out of ten accepted that humans have evolved from simpler life forms. About half of them accepted the common origin of species, although three out of ten did not accept it and two out of ten did not know/reply. In addition, 34.5% of students agreed with the statement "At one time, people co-existed with dinosaurs" and 18.7% responded with "Don't know/No reply". Finally, only three out of ten clearly understood the term 'theory', and almost none of them considered natural selection as an explanation for the changes in organisms.

In another study, 98 students (9th Grade, 14–15 years old) completed an open-ended questionnaire aimed at identifying their preconceptions about evolution. It was found that the greater was the amount of information students were provided with, the less were the teleological explanations that they provided (Kampourakis & Zogza, 2008). Moreover, it seemed that students could overcome their teleological preconceptions when the teaching of biology was organized in the following sequence: levels of biological organization (cells, organisms, ecosystems), mechanisms of heredity and origin of genetic variation, i.e. when the whole course was structured so as to support the teaching of evolution. This was tested with the same group of students, who were taught biology through a specific teaching sequence, which was based on a conceptual conflict strategy and highlighted the idea of contingency in evolution. The analysis of their explanations showed that this teaching sequence was effective, in terms of conceptual change and evolutionary explanations (Kampourakis & Zogza, 2009).

Stasinakis & Athanasiou (2012) conducted interviews with secondary teachers, in order to document problems and issues relevant to the teaching of evolution. They found that even though teachers had the intention to teach evolutionary theory, their lack of teaching skills and their inadequate factual knowledge precluded them from teaching effectively. Interestingly, teachers often had the same alternative conceptions about evolution as their students (for example, teleological explanations, the meaning of the term "theory"). This should be no surprise as several of the teachers had never had any evolution course during their undergraduate studies, and as none of them had ever attended any course on the Didactics of Biology or Science Teaching during their university studies. As a result, most of them were not aware of fundamental teaching approaches, such as considering their students' preconceptions and prerequisite knowledge, teaching for conceptual change, etc. Finally, the researchers concluded that as far as religion is concerned, it is not an issue that affects the way teachers teach biology in general or evolution in particular.

In another study with Greek secondary teachers (Stasinakis & Athanasiou, 2016), researchers found a lack of Pedagogical Content Knowledge (PCK) for teaching evolution. The analysis of the responses of 181 in-service teachers indicated low scores in several components of PCK, and therefore insufficient competence for teaching evolution. Because, as mentioned above, teachers of different specialties are allowed to teach biology in Greece, one would expect that non-biologists would not be able to teach evolution appropriately. However, it seems that biologists would not necessarily perform better than non- biologists in teaching basic aspects of evolutionary theory. In conclusion, teachers' inability to transform scientific knowledge to school level knowledge (using appropriate teaching strategies, examples, models, explanations, etc.), because of lack of the necessary PCK and the weaknesses of the educational system, can result in insufficient teaching of evolution. The researchers suggested a training program based on enrichment of PCK components that could improve biology and evolution teaching.

An e-learning training program about evolution teaching was also implemented (Stasinakis & Kalogiannakis, 2017). It was based on the model of PCK, and aimed at improving the individual PCK of participants regarding the teaching of evolutionary theory. Teachers managed to improve in various aspects of their teaching, as well as in all the individual PCK components (such as using examples, the concepts of evolution, teaching strategies, etc.). At the same time, they also collaborated with one another and exchanged ideas about lesson plans, addressing misconceptions in evolution teaching, teaching evolution as a unifying theory, etc.

Athanasiou & Papadopoulou (2012) studied the acceptance of evolutionary theory among Greek university students, intending to become teachers in early childhood education. Overall, 350 participants attended a course of general biology that employed evolution as its central unifying theme. One hundred and twelve students completed both the pre- and the post-test. It was found that after the course these participants had increased their level of acceptance of evolution, even though they did not change their views about human evolution and nature of science. It seems that open—minded students were more likely to accept evolutionary theory, and perceived no conflict with their religiosity. The main conclusion was that a biology course organized around the unifying concept of evolution could increase the acceptance of evolution.

Athanasiou, Katakos and Papadopoulou (2012) surveyed students (prospective primary teachers) who attended a course that employed evolution as its central unifying theme. Among them, 113 students completed both the pre-course and the post-course survey. The researchers found a significant improvement of understanding and acceptance of evolution after teaching, with a correlation between the two parameters. It was also found that the variety of teaching and inquiry methods seemed to play an important role for this change. Finally, no conflict with participants' religious beliefs seemed to exist.

In another study, the biology 'self-efficacy beliefs' of 202 teachers working in public primary schools were measured (Mavrikaki & Athanasiou, 2011). Greek primary school teachers' 'self-efficacy' in biology teaching was found to be moderate to high, even though performance was found to be better as the years of teachers' experience increased. No impact of having a PhD or Master's degree was found. Those teachers who had taken more science courses in school (science-oriented curriculum) achieved higher scores than the others (humanities-oriented curriculum). The number of biology courses teachers had attended during their undergraduate studies seemed to positively affect their 'self-efficacy' beliefs in biology teaching. Similar was the case for the number of science teaching courses that they had attended. Greek primary teachers felt less efficient in teaching concepts related to evolution, molecular biology and microbiology, but they felt more 'self-efficacious' in teaching concepts related to plants, ecology and human biology (i.e., human anatomy).

Using a well-known questionnaire, the Conceptual Inventory of Natural Selection (CINS), Athanasiou & Mavrikaki (2014) analyzed the level of knowledge of evolution by means of natural selection (ENS) of 352 biology and non-biology students. Greek undergraduate biology majors and non-majors' CINS scores increased along with the number of evolution-related courses; their understanding of ENS also increased with the number of biology courses they had attended. Comparisons between least and most evolutionary educated university students showed that those who had taken more evolution-related courses also gave more evolutionary answers, even though advanced biology students showed an improvement only in 14 out of the 20 CINS items compared to novice biology students. Education students who had just entered the university, and had thus taken no biology course, scored very low. As all of them had just finished secondary education, one could conclude that the Greek educational system had not provided them with the necessary knowledge about evolution that would help them to better understand the living world.

11.4 Discussion

In our chapter, we have presented research on teaching and learning of evolution or of evolution-related concepts in all educational levels in Greece. We have described the difficulties in teaching evolution, which derive from structural, organizational, and from academic weaknesses of the Greek educational system. We must note that there is no evidence for any kind of exogenous interventions, e.g., from the Greek Church or religious groups, which might impede the educational process. This is important to note because there have been voices implying that the problems with teaching evolution in Greece may have been due to anti-religious sentiments (e.g. Nicolaidis, 2014). In fact, during the last 40 years, there has been no explicit discussion about the possibility of teaching creationism in biology courses in Greece, and there have been no explicit movements against the teaching of evolution in schools.

The main problem in our view is that the Greek higher education system does not effectively prepare biology graduates to teach in secondary schools, as well as that there are no pre-service and in-service training programs. Those who graduate from the departments of primary education are not expected to have received a solid scientific training and often have difficulties in the teaching of evolution-related concepts. Furthermore, the courses they are expected to teach at school are about science in general. Even when primary teachers are supposed to teach biological concepts, such as classification, this is not done from an evolutionary perspective. Worse than that, the biology university departments, whose graduates are supposed to teach biology at secondary schools, do not prepare them for this purpose as they either do not offer any courses on biology didactics or when they do so, these are elective ones and only a few students take them.

Therefore, a first, crucial step would be a reorganization of university curricula so that they offer didactics courses that are necessary for future teachers. This is especially important for biology departments, as there are no departments of secondary education in Greece. What is needed is that biology undergraduates attend a combination of compulsory and elective courses in which they will learn about biology didactics and develop the skills that are necessary for teaching biology at school.

We also need curricula that are less content-centered and that focus on core concepts, such as biodiversity, evolution, historicity, nature of science, etc. There are several necessary changes. For instance, Greek students are not taught anything about Nature of Science (NOS) in secondary education (Kampourakis, 2017). In contrast, NOS ideas should be discussed throughout all biology courses. In addition, only one chapter on evolution is included in the last grades of the gymnasium and the lyceum. This is not adequate in order for students to understand evolution. What is needed, in contrast, are biology curricula where evolution is the central unifying theory. Now, and this is the important point we want to make, all these require teachers that are sufficiently trained in order to be able to teach effectively about evolution and NOS.

There are other problems, too. The teaching of biology at the gymnasium and the lyceum levels is based on curricula that were written years ago (2003 for Gymnasium[7] and 1999 for Lyceum[8]). These are far from up-to-date. In 2014, the Ministry of Education published a new curriculum for C Gymnasium.[9] This suggested some innovations that could benefit the teaching of evolution. In particular, a chapter about NOS was included that could facilitate students' understanding of concepts such as theory, hypotheses, as well as the historical nature of biology, etc. Moreover, in the last chapter about evolution, new topics were introduced such as the distinction between micro- and macro-evolution, the Hardy-Weinberg equilibrium, etc., as well as several activities that could be used for learning purposes. Unfortunately, this curriculum was never widely used, because of changes in the government.

Furthermore, during the past 15 years, several studies on the teaching of biology have been conducted in Greece that should have been considered in curriculum design; but this has not been the case. Modern curricula are less analytic and are based on core ideas in contrary to extant ones that are content-centered and do not allow teachers to organize their own way of teaching core concepts of biology. This effort requires the cooperation of all stakeholders (scientific community, educators, researchers in didactics) and should also take into account the current literature about science teaching, both at the Greek and at the international level. Pilot studies should be conducted to evaluate the new curricula, in order to revise them appropriately, before they are implemented in classrooms. The experience so far is that curricula had been designed within three to four months, in closed groups in which external participants were not admitted, without solid scientific or didactic documentation. Therefore, curricula designed in this manner could not bring about any desired effect on improving biological knowledge of students.

Teachers should teach biology in an evolutionary framework, and should present evolution as its central unifying theory. Evolution should be presented as a core, concept, underlying all biological phenomena, structures and functions. Teachers should also talk about certain aspects of evolutionary theory (such as historicity, diversity, common descent, natural selection, etc.) whenever possible in their ordinary lessons. For instance, the current utility of adaptations should be presented but reference to their evolutionary history is also important in order for students to better understand their origin (Kampourakis, 2013). Stasinakis & Athanasiou (2012) noted that Greek teachers do not teach evolution as a unifying theory

---

[7]Government Gazette (GG) 304B/13-03-2003.
[8]Government Gazette (GG) 366/B, 13-04-1999, Γ2/1096.
[9]Government Gazette (GG) 97/22-01-2014.

because nobody told them to teach in such way, or because they did not think of this option. Kampourakis & Zogza (2009) found that when lower secondary students were taught biology through a specific teaching sequence based on evolution as a unifying theory, students explained structures and functions in more evolutionary terms.

Greek biology teachers face many obstacles in teaching evolution effectively: insufficient time to teach effectively, presence of only one chapter on evolution at the end of school textbooks, and the decreased interest of students. Therefore, in order to deal with these obstacles, they need the support of the State. This could be achieved by regular teaching seminars. In this spirit, scientific associations, such as the PanHellenic Union of Bioscientists could play a crucial role in supporting the pedagogical training of teachers and in helping them enrich their teaching. Finally, as biology is taught by teachers who do not necessarily have undergraduate studies in biology, the Greek Ministry of Education should provide specific and detailed guidelines for teaching. These should not be limited to proposals about how to teach the prescribed content, but also about how to deal with particular difficulties. The Ministry of Education should also produce educational materials for teaching, in the form of lesson plans that promote the active participation of students. As we see that the mandatory teaching of evolution in grade C of lyceum obliged teachers and students to talk about this, we believe that this could be a solution for all grades: make the teaching of evolution mandatory could help overcome common weaknesses and excuses such as, 'it is the last chapter of textbook', 'there is not enough time', 'there is only one chapter', etc.

Biology is in constant interaction with society, and understanding the issues at stake when it comes to environment, food, health, bioethics, etc. requires citizens who are able to keep up with current developments in biological research and who are able to understand the respective socio-scientific issues. However, the current limited teaching of biology in Greek lower secondary-compulsory education (one period per week per grade) is not consistent with these aims and cannot contribute to the development of scientific literacy of Greek students. Therefore, to improve biology education in Greece, we need a clear long-term policy for education, which will be solely determined by pedagogical criteria, and no short-term solutions. Evolutionary theory has a central place in such a change, and designing biology curricula from an evolutionary perspective might be a first crucial step in improving biology education in Greece and in educating biologically literate citizens.

References

Athanasiou, K., Katakos, E., & Papadopoulou, P. (2012). Conceptual ecology of evolution acceptance among Greek education students: The contribution of knowledge increase. *Journal of Biological Education*, 46(4), 234–241. http://doi.org/10.1080/00219266.2012.716780.

Athanasiou, K., & Mavrikaki, E. (2014). Conceptual Inventory of natural selection as a tool for measuring Greek University Students' Evolution Knowledge: Differences between novice and

advanced students. *International Journal of Science Education, 36*(8), 1262–1285. http://doi.org/10.1080/09500693.2013.856529.
Athanasiou, K., & Papadopoulou, P. (2012). Conceptual ecology of the evolution acceptance among Greek Education Students: Knowledge, religious practices and social influences. *International Journal of Science Education, 34*(6), 903–924. http://doi.org/10.1080/09500693.2011.586072.
Eurobarometer. (2005). Europeans, society and technology. Special Eurobarometer 224/ Wave 63.1—TNS Opinion & Social.
Dobzhansky, D. (1973). Nothing in Biology Makes Sense except in the Light of Evolution. *The American Biology Teacher, 35*(3), 125–129.
Kampourakis, K. (2013). Teaching about adaptation: why evolutionary history matters. *Science & Education, 22*(2), 173–188.
Kampourakis, K. (2014). *Understanding Evolution.* Cambridge: Cambridge University Press.
Kampourakis, K. (2017) Nature of science representations in Greek secondary school biology textbooks. In C. McDonald, F. Abd-El-Khalick (Eds.) *Representations of Nature of Science in School Science Textbooks: A Global Perspective* (pp. 118–134). London: Routledge.
Kampourakis, K., Pavlidi, V., Papadopoulou, M., & Palaiokrassa, E. (2012). Children's teleological intuitions: What kind of explanations do 7–8 year olds give for the features of organisms, artifacts and natural objects? *Research in Science Education, 42*(4), 651–671. http://doi.org/10.1007/s11165-011-9219-4.
Kampourakis, K., Palaiokrassa, E., Papadopoulou, M., Pavlidi, V., & Argyropoulou, M. (2012). Children's intuitive teleology: Shifting the focus of evolution education research. *Evolution: Education and Outreach, 5*(2), 279–291. http://doi.org/10.1007/s12052-012-0393-2.
Kampourakis, K., & Zogza, V. (2008). Students' intuitive explanations of the causes of homologies and adaptations. *Science & Education, 17*(1), 27–47. http://doi.org/10.1007/s11191-007-9075-9 .
Kampourakis, K., & Zogza, V. (2009). Preliminary evolutionary explanations: A basic framework for conceptual change and explanatory coherence in evolution. *Science & Education, 18*(10), 1313–1340. http://doi.org/10.1007/s11191-008-9171-5 .
Kazantzakis, N. (1915). *Darwin, Charles. On the origin of species (title in Greek: « Περί της γενέσεως των ειδών »).* Publishing House G.D. Fexis, Athens.
Kelemen, D. (2012). Teleological minds: how natural intuitions about agency and purpose influence learning about evolution. In K. Rosengren, S. Brem, E. M. Evans, and G. M. Sinatra (Eds.) *Evolution Challenges. Integrating Research and Practice in Teaching and Learning About Evolution* (pp. 66–92). Oxford: Oxford University Press.
Mavrikaki, E., & Athanasiou, K. (2011). Development and application of an instrument to measure Greek Primary Education Teachers' Biology Teaching Self-efficacy Beliefs. *Eurasia Journal of Mathematics, Science & Technology Education, 7*(3), 203–213.
Nicolaidis, E. (2014). Greece. In S. Blancke, H. Henrik Hjermitslev, & P. C. Kjærgaard (Eds.), *Creationism in Europe* (pp. 144–161). Baltimore: Johns Hopkins University Press.
Prinou, L., Halkia, L., & Skordoulis, C. (2008). What conceptions do Greek School Students Form about biological evolution? *Evolution: Education and Outreach, 1*(3), 312–317. http://doi.org/10.1007/s12052-008-0051-x.
Stasinakis, P., & Athanasiou, K. (2012). Greek teachers' attitudes, beliefs, knowledge and context, concerning evolution teaching. In C. Bruguiere, A. Tiberghien, & P. Clement (Eds.), *Science learning and citizenship* (pp. 179–185). Lyon, France: European Science Education Research Association.
Stasinakis, P. K., & Athanasiou, K. (2016). Investigating Greek Biology Teachers' attitudes towards evolution teaching with respect to their pedagogical content knowledge: Suggestions for their professional development. *Eurasia Journal of Mathematics, Science and Technology Education, 12*(6), 1605–1617. http://doi.org/10.12973/eurasia.2016.1249a.

Stasinakis, P. K., & Kalogiannakis, M. (2017). Analysis of a Moodle-based training program about the pedagogical content knowledge of evolution theory and natural selection. *World Journal of Education, 7*(1), 14–32.

Xirafas, V. (2009). The reactions of the church and the academic theology of Greece, in Darwinian Theology. In V. Zogza, K. Kampourakis, & D. Notaras (Eds.), *Evolution theory teaching: theoretical and pedagogical issues*. Geitonas Schools, Athens (book in Greek language).

Panagiotis K. Stasinakis is a high school biology teacher with a PhD in Biology Didactics. Currently he is supervisor in one of the Laboratory Centers for Natural Sciences (EKFE) in Greece. His research interests focus on science education and particularly on the understanding of biology (special topics of interest are, evolution, phylogeny, human body, etc), as well on the use of technology as a means of effective science teaching. He has participated in several training seminars for secondary education teachers and in many national working groups for the production of educational material related to biology curriculum in low-secondary education, learning objectives for biology teaching, public promotion of biology through communication.

Kostas Kampourakis is a researcher at the University of Geneva, where he teaches courses at the University Teacher Training Institute and the Section of Biology. He is the Editor-in-Chief of the journal Science & Education and the book series Science: Philosophy, History and Education (Springer). He is also co-editor of the books Teaching Biology in Schools (Routledge, 2018) and Newton's Apple and Other Myths about Science (Harvard University Press, 2015), as well as editor of The Philosophy of Biology: A Companion for Educators (Springer, 2013); he is also the author of Turning Points: How Critical Events Have Driven Evolution, Life and Development (Prometheus Books, 2018), Making Sense of Genes (Cambridge University Press, 2017), and Understanding Evolution (Cambridge University Press, 2014). His main research interests include the teaching, learning and the public understanding of nature of science,evolution, and genetics.

Chapter 12
Evolution Education in France: Evolution Is Widely Taught and Accepted

Marie-Pierre Quessada and Pierre Clément

Abstract In 1898, the topic "Evolution" emerged for the first time in the French curriculum. The teaching of evolution remained installed in the science syllabus during the 20th Century until today. That probably explains the very good acceptance of evolution by French people compared to other countries: 80% of French adults accepted evolution. In the beginning of this 21st Century, some anti-evolution movements were attempted in France. In the name of secularism, the French Ministry of Education actively reacted to counteract their possible influence at school. Today, in France, biological evolution is clearly present in the science syllabus, starting by an initiation at the Primary School, a development in Lower Secondary School and a large deepening in the scientific section of High Schools. At University the biology teacher education programs are focused on Life and Earth Sciences and are in accordance with Dobzhansky's claim (1973): "Nothing in biology makes sense except in light of evolution". In this chapter, we present original analyses of our data related to conceptions of evolution in two French Regions (Rhône-Alpes and Languedoc-Roussillon). The results of our research show that more than 98% of French teachers and 94% of French students accept evolution. We also present suggestions to improve evolution education.

12.1 Introduction

In June 2006, the 68 member countries of the InterAcademy Panel (IAP), published a joint statement: "*We, the undersigned Academies of Sciences, have learned that in various parts of the world, within science courses taught in certain public systems of education, scientific evidence, data, and testable theories about the origins and*

M.-P. Quessada (✉)
LIRDEF, Montpellier University, Montpellier, France
e-mail: quessada.mp@gmail.com

P. Clément
ADEF, Aix-Marseille University, Marseille, France
e-mail: clement.grave@free.fr

evolution of life on Earth are being concealed, denied, or confused with theories not testable by science. We urge decision makers, teachers, and parents to educate all children about the methods and discoveries of science, and to foster an understanding of the science of nature."(p. 1).

What is the situation of France? France is a country of 67 million inhabitants with a single national language, French. France is a secular state which respects the freedom of conscience and religious practice of each citizen (law of separation between the Churches and the State, December 9, 1905). In 2012, 34% of French people said they are non-religious, 29% said they are atheists and 37% identified themselves with a religion, making France the fourth most atheistic country in the fifty countries of the survey, behind China, Japan and the Czech Republic (WIN/Gallup International Survey, 2012 quoted by Marchand, 2015). France does not have official statistics of religious communities. In 2011, at the question "You personally, are you…?", 61% of French people answered Catholic, 4% Protestant, 25% without religious affiliation, 7% Muslin and 1% Jewish (IFOP, 2011).

International surveys about general public beliefs show a very good acceptance of evolution by French people, compared to other countries. For example, Miller, Scott, & Okamoto (2006) mentioned that 80% of surveyed French adults agreed with the assertion that "*Humans as we know them were developed from earlier species of animals*". France ranks fourth among 34 countries in the survey, just after Iceland, Denmark and Sweden. How to explain today these good results in France? What lessons can be drawn for educational policies relating to Evolution Education? Our research tried to answer these questions, using different approaches: an epistemological and historical approach, as well as a didactic approach about curricula, textbooks and conceptions of teachers and students.

12.2 A Historical Approach

The historical approach analyzed in parallel the advancement of knowledge related to the sciences of evolution, the changes in the curricula concerning biological evolution and the incorporation of these changes into school textbooks. We measured the delay for the knowledge to enter into syllabi and textbooks: the DTD (Didactic Transposition Delay) is the interval of time between the emergence of a scientific concept and its appearance in syllabi or textbooks. Our research covers more than two centuries. We compared the changes in scientific knowledge on human origins (40 scientific articles, treatises or syntheses written by scientists) with the changes in French secondary school syllabi (a comprehensive study of fifty official texts) contents concerning this same topic during the 19th and 20th centuries (Quessada & Clément, 2007).

From the beginning of the 19th century, French scientists were at the center of debates about biological evolution. Lamarck and Cuvier, in their zoological studies, were led, like Linnaeus, in 1735, to classify living species according to their differences and similarities. At the same time, paleontological studies lead to the elaboration of a geological calendar with a succession of periods. Some scientists like Lamarck (1802) saw biological evolution occurring over geological times spans while others like Cuvier (1817) did not accept this idea. The controversy between Lamarck and Cuvier raged at the Natural History Museum of Paris. During this period, the Ministry of Education was influenced by natural theology: the mission of *"Natural History"* is to demonstrate *"Divine Providence"*. Cuvier had a great influence on the school system, as a member at the Royal Council of Public Education. In particular, he was directly involved in preparing syllabi. His zoological conception of man (isolated from monkeys in a separate Bimanes order) was introduced into French programs in 1833. Before 1833, the syllabi showed a creationist conception of humans: mankind originated from a particular creation, which separates it from other living species. In 1833 there is movement from a creationist conception to a zoological conception with a DTD of a hundred years after Linnaeus's publication, and a DTD of less than twenty years after Cuvier's publication.

The continuity of geological time, the succession of fossils during the different geological periods and the existence of prehistoric men were admitted gradually in the syllabi of the last part of the 19th century. A highly romanced account of human origins was still given, in which mankind had first to strive to survive and then progressively developed higher and higher levels of civilisation. The difference between prehistoric man and contemporary man was still seen to be purely cultural. Thus the biological evolution of humans was not yet clearly taken into account in the French syllabi in 1885. This secular mythical account of human origins replaced the biblical creationist version. Prehistoric man was introduced in the French school syllabus in 1885, only 25 years after the French Academy of Sciences and the British Royal Society finally recognised its existence (Cohen, 1999). This DTD is much shorter than the delay observed during the first part of the 19th Century.

The introduction of the idea of prehistoric humans was probably linked to the political decision to secularise republican schools. Jules Ferry, Minister of Public schooling declared (1880): *"Gentlemen, the Government believes that the religious neutrality of schools is a necessary principle whose time has arisen, and the application of which can no longer wait"*. Meanwhile, Edmond Perrier, zoologist and evolutionist (1882) wrote: *"Natural history ... has fought in close combat with ancient philosophies, doing away with old legends one by one, and is now preparing for its toughest battle yet, the most profound revolution ever achieved in the philosophical, political and religious orders"*. These two citations are good examples of the context and the role of Natural History in the secularisation of schools in France during the end of the 19th Century. During this period, it is mostly for political reasons that the DTD of prehistoric man in French syllabi was relatively short, assuming a radical break with the idea of a biblical creation of man.

The introduction of an evolutionist conception of species in the new syllabi also attested to the opposition of scientists like Edmond Perrier and Louis Mangin against the creationist ideas of Cuvier. In 1898, Louis Mangin, member of the Higher Council of Public Education, declared in a report: *"It is possible to give pupils clear ideas on evolution and to show that transformism regulates the succession of beings"*.

The evolution of mammals was present in the 1902 syllabi. The general idea of evolution of plants and animals was introduced in the 1912 syllabi. The teaching of evolution remained firmly set into the French science syllabi during the 20th Century until today. The high level of acceptance of evolution by French people today is probably rooted in this continuity of teaching evolution in the French School, linked to the School secularization with a clear separation between Church and State, and the promotion of *"laicity"* (secularism) since 1905. This permanence of biological evolution in French education since the beginning of the 20th century can also be related to the active French research in zoology, palaeontology and prehistory since the 18th century in the framework of the Philosophy of Enlightenment. For example, French evolutionary scientists were involved in the ambitious reform of science education in 1902.

12.3 The Recent and Limited Influence of Anti-evolution Movements in France

During the 20th Century, most French biologists gradually accepted the Darwinian theory of evolution, after some initial points of resistance. One instance of this was the persistent influence of Lamarck. Loison (2008, 2010) has analyzed how Lamarckian transformism, where the unit of evolution is the individual organism, represents an obstacle to the Darwinian theory based on a population approach. Famous French biologists, such as Grassé (1973), have defended this non-Darwinian theory of evolution, which is based on essentialist thinking and on a difficulty in accepting the role of chance.

An important obstacle is finalism, the doctrine that natural processes are directed towards some goal. Teilhard de Chardin (1956) provides a good illustration of this double movement in the 20th century: he defended evolutionary processes, to which his own work in paleontology contributed, all the while considering (in agreement with his Christian faith, he was a priest) that evolution is directed towards the emergence of man. Similarly, in 1996, Pope Jean-Paul II recognized that "new knowledge leads to recognizing that the theory of evolution is more than a mere hypothesis", but still maintained that this biological evolution had the goal of leading to the emergence of the human species. At the end of the 20th century, this finalism was effectively combatted at the international level (for example Gould, 1984, 1997, 2000) and in particular by French biologists and philosophers (among others Gayon, 1993; Kupiec, 2008; Lecointre, 2009, 2014), some of whom are themselves Catholics (Arnould, 2007; Perru, 2010). In France, only a small minority of Catholics continue to criticize evolutionary processes. Among the latter, we may note the attempt of those who defend the idea of intelligent design, as in the UIP (the Interdisciplinary Union) of Paris (Baudoin & Brosseau, 2013, pp. 143–160).

French Protestants and other Calvinists are traditionally evolutionists, but the evangelical Protestants, who are increasingly numerous and active in this beginning of the 21st century, are more divided in their opinions. However, as one of them, Sébastien Fath (quoted by Baudoin & Brosseau, 2013, p. 181), recognizes: "*simple-minded anti-Darwinism, which is widespread among American evangelists, is not frequent in France*". The same authors (still on p. 181) cite an IFOP opinion poll which indicates that "*only one French evangelist in five considers the story of the Creation as a historical truth*". Jehovah's witnesses—who are actively creationist, freely spreading pamphlets such as "Life—How Did It Get Here? By Evolution or by Creation?" (Témoins de Jéhovah, 2002)—are not very numerous in France. One of us has published a critical analysis of their arguments (Clément, 2002).

There are a larger number of fundamentalist Muslims who openly criticize Darwinian evolution, including young people in high school. They are inspired by documents such as the work of Keskas (1994). But it was the widespread distribution of the Atlas of Creation (Yahya, 2008) which provoked an immediate reaction of the French Ministry of National Education: schools (which had all received a free copy of the very luxurious creationist work) were forbidden to put it in their libraries; the Ministry organized national symposia, such as "Enseigner l'évolution" (Teaching evolution) (2008) where teachers of biology and philosophy from the whole of France were invited; the Ministry aids for the publication of evolutionist works and documents. The French media were unanimous in criticizing this creationist propaganda, which thus had a limited impact, as confirmed by the analysis of the conceptions of teachers and pupils presented in the last part of this chapter. But before that we will take a closer look at the place of evolution in school curricula and teacher training.

12.4 Place of Evolutionary Theory in the Curriculum

The French education system is centralized with a national curriculum and a final national examination at the end of the secondary school curriculum, "le baccalauréat" (students 18–19 years old). 83% of establishments are public, 17% are private, mostly Catholic. The schooling starts in Primary School: 3 years in nursery schools (students 3–6 years old) and 5 years in elementary school (students 6–11 years old). Then there are four years in Lower Secondary School (students 11–15 years old) and finally 3 years in High School for students 15–18 years old. School attendance is compulsory for students from 6 to 16 years old.

Today, in France, biological evolution is clearly present in the science syllabi, starting with an initiation in the Primary School, further development in Lower Secondary School and deeper presentation in the scientific sections of High School (Quessada & Clément, 2013; Quessada, 2016). The place of evolutionary theory in the curriculum is central. The school curricula are published by the Ministry of National Education which regulates the teaching of all French public schools, and private schools under contract with the state, i.e. the vast majority of schools in France.

The primary school and lower secondary school syllabi (Official Bulletin n2 of 26 March 2015, Special Official Bulletin n11 of 26 November 2015) start with a discovery of living animals and plants in order to distinguish the living and the non-living (pupils 3–6 years old), and then their interactions and biodiversity (pupils 6–8 years old). Pupils aged 9–11 (the last two years of primary school and first year of secondary school) learn how to classify living organisms using an evolutionary perspective. Students are introduced to the notion of the long time (at the geological time scale), identify changes in species on Earth over time and study the present and past diversity of species and the evolution of living species. During history courses of the first year of the Secondary School, students study prehistory together with the long history of humanity and the human migrations. In the three following years, students from 12 to 14 years old learn to distinguish scientific facts from beliefs in order to enter into a scientific relationship with the living world. They study the great groups of living beings, including *Homo sapiens*, their relationship and their evolution. They address the mechanisms of evolution, genetic mixing, chance and natural selection.

After Lower Secondary School, students' paths diverge between Professional High School and Classical High School. In French Classical High School, for 15-years-old pupils, one-third of the Life and Earth Sciences syllabus is about *"The Earth in the Universe, Life and Evolution of Life"* (Special Official Bulletin n4 of 29 April 2010). This coverage deepens the knowledge acquired previously, specifically by studying the characteristics of the Earth that make it possible to understand how life is developing there, characteristics of life, biodiversity and its evolutionary origins. Biodiversity in this case includes ecosystem diversity, species diversity and genetic diversity within species. According to the syllabi, students must learn that: the current state of biodiversity corresponds to a stage in the history of the living world: today's species represent a tiny fraction of the total number of species that have existed since the beginning of life; biodiversity is changing over time as a result of many factors, including human activity; within biodiversity, kinship relationships are the basis for the classification of living beings; a common organization of species in a group suggests that they all share a common ancestor; the diversity of alleles is one aspect of biodiversity; natural selection and genetic drift can lead to the emergence of new species.

In French High School, students' paths diverge at 16: They must choose for the last two years between a literary section, scientific section, economic and social section or technical section. For French pupils, 16-year-olds in a non-scientific section, science education does not focus on Evolution. However during a discussion of vision, the comparative study of retinal pigments makes it possible to place human species among the Primates. For French 16-year-old pupils in a scientific section, half of the Life and Earth Sciences syllabus is about *"Earth in the Universe, Life and Evolution of Life"*. It examines the fundamental aspects of genetic heritage (replication, transcription, translation, mutation and genetic variability), with a molecular approach that makes it possible to progress in the understanding of the living world (Special Official Bulletin n9 of 30 September 2010).

In the final year of secondary school (pupils 17–18 years old), only science section students have a life and Earth sciences course. Half of the syllabus focuses on the topic *"The Earth in the Universe, Life and Evolution of Life"*: the genetic mixing associated with sexual reproduction and some aspects of the mechanisms of evolution are studied. An important part of this study is devoted to the human species: "*A look at the evolution of Humankind*". Students learn that *Homo sapiens* can be regarded, in an evolutionary approach, as any other species: humans have an evolutionary history and are continuously evolving; the human story is part of the more general history of primates (Special Official Bulletin n8 of 13 October 2011).

Evolution education has long been established in French curricula. However, in 2008, a reform in primary school resulted in the disappearance of the study of fossils, the major stages in the history of the Earth and the term "Evolution". The announced objective of these modifications was a simplification of the official texts. In the draft program, the idea of diversity of the living was preferred to the idea of Evolution. After the reactions of the scientific community, the idea of the unity of life was added, the studies of fossils and of major stages in the history of the Earth were not re-introduced (Quessada, 2008, p. 378). The presence of the word Evolution in programs was pushed back to the fourth year of lower secondary school, with the risk that this introduction for 14-year-old pupils comes too late after cultural and religious influence have cemented opposition to this evolutionary conception of the living world. In 2015, these various elements were reintroduced with the new reform of primary school curricula (Special Official Bulletin n11 of 26 November 2015).

A study of French science textbooks from 1994 to the present, focused on human evolution in the last class of secondary education (Quessada and Clément, 2013), reveals clearly the very important development of the phylogenetic approach. This approach can be linked to the many publications of researchers in this field at the beginning of the 21st century. Several French publications by researchers at the Museum of Natural History in Paris popularize this new knowledge on the classification of living species for teachers of primary and secondary education (Lecointre & Le Guyader, 2001; Lecointre, 2004, 2008). We show in this study a trend toward textbooks that focus on phylogenetic data, patterns and methods that supplants the more historical approach and mitigates some associated problems (e.g., the problems of dating, fossil study, study of the phenomena of the past). On the other hand, we can note that science textbooks do not include a discussion of current and past controversies in the sciences of evolution.

12.5 Emphasis Given to Evolutionary Theory in Biology Teacher Education Programs

In France, teachers are recruited by a competitive entrance examination. The winners of the competitive examination have one year of paid training, at the end of which they are tenured if they validate their training and if they hold a master's degree.

In primary education, the laureates have a university education in a wide variety of disciplines, mastery of scientific knowledge is not compulsory to pass the competitive examination for primary school teachers. However, during the year of training following the competitive entry examination, the future teacher will be required to upgrade his/her knowledge in science in order to be able to teach the science curriculum in primary school. Selected individuals have, in particular, to master the characteristics of living organisms and their classification according to evolutionary relationships.

In secondary education, competitive entry examinations are held at the national level. Candidates in the competitive entry examination of Earth and Life Sciences teacher must have passed a first year of master's degree in sciences. Programs for the entry examination include all the science syllabi of the secondary school classes that the future teacher of life and Earth sciences will have to master and eight more specialized subjects. One of the eight subjects is the phylogenetic classification of living species, an inclusion that indicates its importance.

12.6 French Teachers' Conceptions of Evolution

The BIOHEAD-Citizen project (Biology, Health and Environmental Education for better Citizenship) investigated, in 18 countries, teachers' conceptions related to six topics including evolution (Carvalho et al., 2008). We had the responsibility of the team for the topic Evolution (Clément, 2008; Clément & Quessada, 2008, 2009; Quessada & Clément, 2011; Quessada, Munoz, & Clément, 2007). This research was then expanded to include more than thirty countries (Clément & Quessada, 2013; Clément, Quessada, & Castéra, 2013; Clément, 2014, 2015). We present here some of results from France.

The validated questionnaire (Clément, 2008) contains 144 questions, of which 15 are related to the topic Evolution:

- 6 questions tested the possible creationist conceptions of teachers
- 2 questions detected finalism (goal-ended evolution).
- 7 questions were related to the teachers' knowledge of some processes of evolution.

Several other questions were related to some characteristics or opinions of each teacher: gender, age, topic and level of training, religion, degree of belief in God, and practice of religion, political and social opinions.

A total of 732 teachers filled out the questionnaire in two French Regions: 424 in Rhône-Alpes Region, mainly in the towns of Lyon and Saint-Etienne; 308 in Languedoc-Roussillon Region, mainly in the towns Montpellier and Nîmes. For the sake of simplicity we will call the two Regions "Lyon" and "Montpellier". There are 2.6 million inhabitants in the Languedoc-Roussillon Region, which is a little more agricultural and touristic (sea beach for number or tourists and retired persons) than the Rhône-Alpes Region. This one is bigger (6 million inhabitants), more industrial and with the Alps mountains for tourism. The regional GDP is about

60.5 billion Euros (3.3ù of the national GDP) in Languedoc-Roussillon, while 197 billion Euros in 2012 for the Rhône-Alpes Region (9.7% of the national GDP).

There is no significant difference between the two Regions related to the declared religion of their inhabitants: about 65% declared to be Catholic; this percentage decreases with time: they were more than 80% in 1970 and 90% in 1905. 25% of French people declared to be Agnostic in 2006, 6% Muslim and 4% Protestant (IFOP/La Croix 2006 quoted in Machelon, 2006; p. 8). Our sample of teachers (Table 12.2) confirms the absence of difference between the two Regions but also the specificity of teachers: more Agnostic/Atheist, and consequently less for each religion.

In each Region, we applied the questionnaire to a balanced sample of in-service teachers (In) and pre-service teachers (Pre), practicing at Primary schools (P), or teaching Biology (B) or French Language (L) at secondary school, which yielded six sampling groups (InP, PreP, InB, PreB, InL, PreL): Table 12.1. Pre-service teachers are students who are at the end of their studies to become a teacher (first and second years of master's degree).

The characteristics of the samples of the two Regions are very similar. For example, the percentage of women, their ages, and their level of instruction, do not differ between the regions. Their religions are presented in Table 12.2.

In the questionnaire, a question prompted each teacher to indicate his/her religious or philosophical positioning in a list. Some have chosen the proposal "I don't want to answer".

Table 12.1 Number of teachers in each group sampled

Nb (%)		Total	Lyon	Montpellier
InB	In service Biology teachers	100 (13.7%)	51	49
InL	In service French Language teachers	110 (15.0%)	62	48
InP	In service Primary Schools teachers	114 (15.6%)	64	50
PreB	Pre-service Biology teachers	149 (20.4%)	99	50
PreL	Pre-service French Language teachers	101 (13.8%)	50	51
PreP	Pre-service Primary Schools teachers	158 (21.6%)	98	60

Table 12.2 Number of teachers for each religion (Total = 732)

Religion	Total	Lyon	Montpellier
Agnostic/Atheist	370 (50.6%)	217	153
Catholic	279 (38.1%)	157	122
Protestant	14 (1.9%)	9	5
Orthodox	2 (0.3%)	1	1
Muslim	11 (1.5%)	8	3
Other Religion	20 (2.7%)	12	8
No answer	36 (4.9%)	20	16

Teachers completeded the questionnaire anonymously in a dedicated room, in the presence of project research fellows. The answers were then analyzed using the statistical software R with package ade4 (Dray & Dufour, 2007) for multivariate analyses to compare the two French Regions and the possible effects of the controlled parameters (Munoz et al., 2009). Multivariate analysis from the 15 questions related to Evolution, shows no significant difference between the two Regions. Thus, the data from the two Regions are grouped in the following analyses. Nevertheless, Figs. 12.1, 12.2 and 12.3 show the answers for each of the two Regions, to illustrate the absence of significant differences between them.

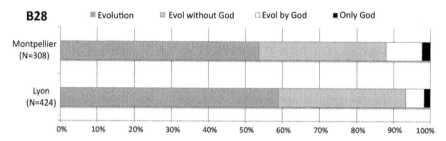

Fig. 12.1 French teachers' answers to the question B28 (human origin), grouped by French region: Montpellier = Languedoc-Roussillon; Lyon = Rhône-Alpes

B28. Which of the following four statements do you agree with most? Select **ONLY** one sentence:
- It is certain that the origin of the humankind results from evolutionary processes.
- Human origin can be explained by evolutionary processes without considering the hypothesis that God created humankind.
- Human origin can be explained by evolutionary processes that are governed by God.
- It is certain that God created humankind

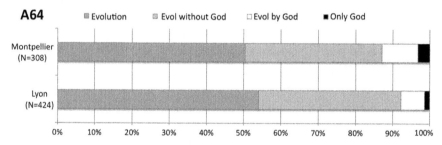

Fig. 12.2 French teachers' answers to the question A64 (origin of Life), grouped by French Region: Montpellier = Languedoc-Roussillon; Lyon = Rhône-Alpes

A64. Which of the following four statements do you agree with the most? (tick only **ONE** answer)
- It is certain that the origin of life resulted from natural phenomena.
- The origin of life may be explained by natural phenomena without considering the hypothesis that God created life.
- The origin of life may be explained by natural phenomena that are governed by God.
- It is certain that God created life

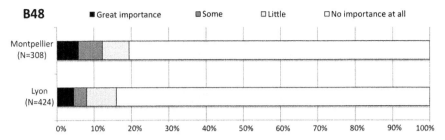

Fig. 12.3 French teachers' answers to the question B48 (importance of God in species evolution), grouped by French Region: Montpellier = Languedoc-Roussillon; Lyon = Rhône-Alpes
Indicate your evaluation of the importance of the following factors in species evolution (tick only **ONE** box for each line):

		Great importance	Some importance	Little importance	No importance at all
B48	God				

This absence of difference is not a surprise. As mentioned earlier, the French educational system has a teacher training and recruitment program which ensures great homogeneity at the national level.

Most French Teachers Accept Evolution

Questions B28 and A64 (Figs. 12.1 and 12.2) propose four propositions:

- the first one corresponds to an assertive evolutionist proposition;
- the second proposition is also evolutionist but formulated in a less dogmatic way by pointing out the religious register;
- the third one is evolutionist and creationist, accepting evolution while thinking it is controlled by God, a possibility proposed by the famous evolutionist Dobzhansky who wrote (1973, p. 127): "*I am a creationist and an evolutionist. Evolution is God's, or Nature's, method of Creation;*"
- the fourth proposition is creationist without evolution acceptance.

When answering these questions, nearly all French teachers (> 98%) accepted evolution (propositions 1, 2 and 3). Very few French teachers (< 2%) ticked the creationist item without evolution acceptance for the origin of humankind (question B28: Fig. 12.1) and for the origin of life (question A64: Fig. 12.2). About 8% accepted evolution while thinking it is controlled by God (item 3: Figs. 12.1 and 12.2).

There is no significant difference between the two French Regions, and the teachers' answers are totally similar for the origin of life (Fig. 12.2) and the origin of humankind (Fig. 12.1).

The comparison with the answers to question B48 (importance of God in species evolution: Fig. 12.3) is interesting. While 90% of teachers ticked a clearly evolutionist item for the origin of humankind (Fig. 12.1) or for the origin of life (Fig. 12.2), 82% ticked "no importance at all of God, 8% "little importance" and about 10% "some" or "great importance": that confirms the teachers' conceptions shown in the Figs. 12.1 and 12.2. The positive correlation between the answers to the different questions about creationism asserts that the creationist conceptions are not conceptions induced by a biased formulation of the question because they emerge from different kinds of questions (Clément, 2010).

Who are the Most Creationist French Teachers?

A Co-Inertia Analysis shows interesting and significant correlation between the PCA (Principal Components Analysis) from the 15 questions related to evolution, and the PCA from 17 questions related to religious, social or political opinions of the teachers. Only the questions related to creationism are at the origin of the correlation: the most creationist conceptions are clearly correlated with a high degree of belief in God and practice of religion, a preference for private schools (rather than public ones), and an opinion against a separation between science and religion, or between politics and religion.

Between-class analyses show no significant differences related to several of the controlled parameters: gender, age, level of instruction. This last absence of effect differs from that found in other countries, where the more a teacher learned at university (whatever the type of training), the less he/she is creationist (Clément & Quessada, 2008, 2013; Quessada et al., 2007). Presumably, this effect is absent in France because most of the teachers have the same degree of instruction.

As we will show below, the conceptions of Biology teachers significantly differ from those of their colleagues in other disciplines, but mainly with respect to knowledge, rather than acceptance of evolution.

The only differences observed for the creationist variables come from the teachers' religion. A between-class analysis, followed by a randomization test, shows significant differences of conceptions ($p < 0.001$) depending on the teachers' religions, mainly linked to the creationist variables. Concerning the question B28 origin of humankind, without surprise, all the atheist or agnostic teachers (50.6% of the sample) are clearly evolutionist, 2/3 ticking the item 1, and 1/3 the item 2 (less dogmatic). A large proportion of Catholic teachers (85.3%) are also only evolutionist, 11.8% being evolutionist and creationist (item 3) and 2.9% only creationist (item 4). The numbers are very similar for Protestant teachers: with a total of fourteen, six are clearly evolutionist, three evolutionist and creationist and only one is creationist. There were only two Orthodox teachers, one being evolutionist and the second one creationist. Concerning the eleven Muslim teachers, five are clearly evolutionist, three are evolutionist and creationist, and three are only creationist.

Finally, fourteen teachers (2%) were only creationist: eight Catholic, one Protestant, one Orthodox, three Muslim and one who ticked "Other religion". That means French religions are not fundamentalist. For instance, French Protestants differ from US Protestants: French are traditionally Calvinist, sometimes Lutheran, and even the growing Evangelical sector is less radical than in the US, and generally agrees with evolution, science and Darwinism.

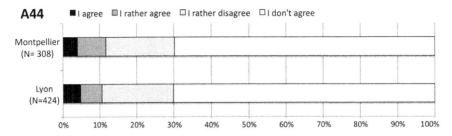

Fig. 12.4 French teachers' answers to the question A44, grouped by French Region: Montpellier = Languedoc-Roussillon; Lyon = Rhône-Alpes

| A44. | The emergence of the human species (*Homo sapiens*) was the aim of the evolution of living species | I agree | | | | I don't agree |

Finalist Conceptions (Goal-ended Evolution)
Two questions addressed this aspect of evolution, which was recognized as a major challenge when teaching evolution (Fortin, 2009; Tidon & Lewontin, 2004). A majority of teachers (70%) are in complete disagreement with the finalist proposals, and 20% rather disagree (Fig. 12.4 for the question A44). In France, there is a great amount of work by scientists explaining and popularizing that man is not the summit of evolution (among others, translation of Gould 1984, 1997, 2000; Lecointre, 2009; Picq, 2007) and they are well relayed by the media and by teachers at schools.

Nevertheless, 32 of the 732 teachers (4.4%) agreed with the finalist proposition in A44, including 15 Catholic and 14 Agnostic or Atheist. That finding may be explained by the position of the Catholic Church for Christian teachers and more broadly by a finalist conception linked with the Judeo-Christian roots of French society. It is interesting to notice that not a single Muslim teacher agreed with the finalist proposition: the three Muslim teachers who ticked the creationist item 4 are more opposed to evolution and Darwin than to a non-finalist view of the origin of species.

Conceptions Related to the Teachers' Knowledge
Six questions tested the teachers' knowledge of the importance of different factors to species evolution. 98% of teachers ticked "great or some importance" for natural selection and for the environment. This high score (98%) corresponds to the teachers who ticked an evolutionist item for the origin of humankind and for the origin of life, as well as those who ticked the third item, evolutionist and creationist (Figs. 12.1 and 12.2). Only the 2% of teachers who ticked the creationist item did not also accept the importance of natural selection and environment for the evolution of species. These results illustrate an interesting interaction between the knowledge and values of teachers.

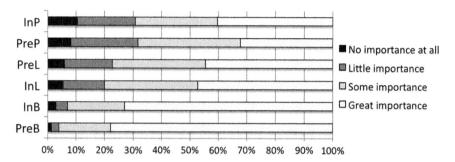

Fig. 12.5 French teachers' answers (N = 732) to the question related to the importance of CHANCE to species evolution, grouped by sample: In = in service; Pre = pre-service; P = Primary Schools; L = French Language (Secondary Schools); B = Biology (Secondary Schools)

		Great importance	Some importance	Little importance	No importance at all
B42	Chance				

20% of teachers, several of them being evolutionist, assigned "no or little importance" to chance, while chance is actually considered a very important part of evolution. This percentage is lower for biology teacher (Fig. 12.5): only 4% of future biology teachers, and 7% of in-service biology teachers, are not convinced of the importance of chance. This percentage is 32% for Primary Schools teachers (in service and pre-service) and 20 and 22% for the French Language teachers (in service and pre-service). This result may indicate a positive effect of the training in biology.

Finally, the most important result is the low impact of creationism on French teachers' conceptions. When we compare the answers to the question B28 (origin of mankind) obtained in 28 countries, French teachers are primarily evolutionist, as are Danish, Swedish, Spanish and Estonian teachers (Fig. 12.6). These results are consistent with the results of Miller et al. (2006) survey showing the good acceptance of evolution in French public opinion.

12.7 French Students' Conceptions of Evolution

We analyze here French students' conceptions of evolution, using the same BIOHEAD-Citizen questionnaire as for teachers.

A first sampling was done in Lyon: two urban high schools (Center: N = 102 and 8th Arrondissement: N = 34), and one in a economically-disadvantaged suburb (N = 57). These data were recently analyzed (Clément, 2017). A second sampling was done partly in agricultural schools in the rural area north of Lyon (data collected by J. Castéra: Roanne, N = 84, Villefranche/Saône, N = 94), partly in the

B28

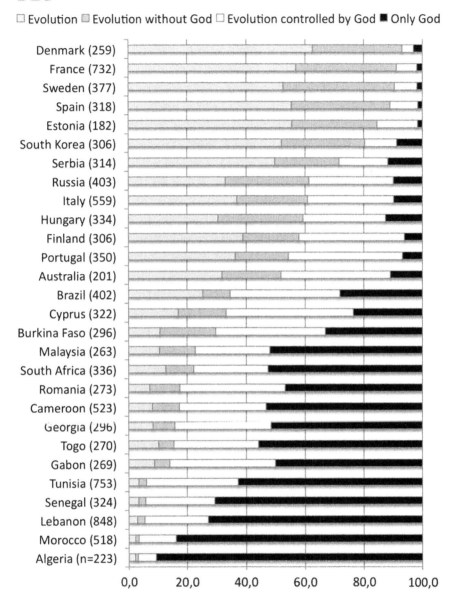

Fig. 12.6 Teachers' answers to the question B28 (Human origin), grouped by country

Table 12.3 Students' Declared Religion (Total = 690)

Religion	Total	Lyon Center	Lyon 8eme	Lyon suburbs	Villefranche (rural)	Roanne (rural)	Lunel (South France)
Agnostic/Atheist	323	47.1%	47.1%	17.5%	61.7%	44.0%	48.3%
Christian	204	35.3%	29.4%	15.8%	22.3%	38.1%	30.1%
Muslim	65	3.9%	11.8%	54.4%	4.3%	1.2%	6.6%
Other Religion	32	5.9%	5.9%	3.5%	4.3%	4.8%	4.4%
No answer	66	7.8%	5.9%	8.8%	7.4%	11.9%	10.7%
Total	**690**	**102**	**34**	**57**	**94**	**84**	**319**

south of France, in Lunel between Nîmes and Montpellier (data collected by P. Clément: N = 319). The analysis of these data has not previously been published.

Students were 16 to 19 years old, from the last level of high school (52.9%) or the year before (45.4%), and predominantly female (57.4%); around half were in a scientific path.

The students' religious affiliation is presented in Table 12.3. In the suburb of Lyon, where most of the students come from a poor immigrant background (mainly from Algeria and Morocco), more than half of them are Muslim. In the other high schools, the percentage of declared religions is similar to the samples of teachers presented above, yet with a few more Muslims in Lunel (6.6%).

Only part of the BIOHEAD-Citizen questionnaire was used in these high schools: most of the questions related to evolution (i.e., creationism, finalism, knowledge), and several characteristics of the student (gender, age, religion, socio-cultural feature of parents, etc.).

Students filled out the questionnaire anonymously, in their classroom in the presence of the researcher who immediately gathered the completed questionnaires. The data were then analyzed in the same way as the teachers' answers to the questionnaire.

We present here the results for question A64 (origin of life). All the students answered this question, while the question B28 (origin of humankind) was not included in the questionnaire for some schools. Nevertheless, when both questions were present (origin of life and origin of humankind), the students' answers were very similar for both, as it was the case for teachers (Figs. 12.1 and 12.2).

The answers of students reveal predominantly evolutionary conceptions. However, significant differences can be noted between high schools. The main difference comes from the suburban high school where 35.1% of students ticked the exclusively creationist item, versus approximately 6% of students in the other high schools (Fig. 12.7). More students of the suburb of Lyon also ticked item 3 (evolutionist and creationist: 33.3%) than in Lunel and in the two rural high schools (6% in each), and in the two urban high schools of Lyon (18%): Fig. 12.7.

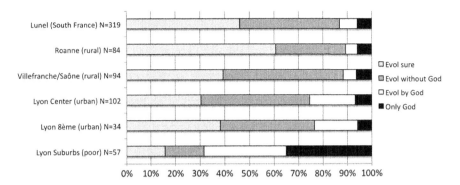

Fig. 12.7 Students' answers to the question A64 (origin of life), grouped by high school
A64. Which of the following four statements do you agree with the most? (tick only **ONE** answer)
- It is certain that the origin of life resulted from natural phenomena.
- The origin of life may be explained by natural phenomena without considering the hypothesis that God created life.
- The origin of life may be explained by natural phenomena that are governed by God.
- It is certain that God created life

If we compare with the teachers' answers (Figs. 12.1 and 12.2), the main result is that the students are generally as evolutionist as the teachers: in most of schools, 94% are evolutionist (if we include the item 3 evolutionist and creationist), but only 65% in the suburban high school. This suburban exception is correlated with the poor socio-economic level and with the religious background of the families in this suburb, where most of them are immigrants from predominantly Muslim countries (mainly Maghreb) (Fig. 12.8).

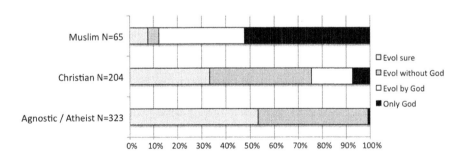

Fig. 12.8 Students' answers to the question A64), grouped by their declared religion
A64. Which of the following four statements do you agree with the most? (tick only **ONE** answer)
- It is certain that the origin of life resulted from natural phenomena.
- The origin of life may be explained by natural phenomena without considering the hypothesis that God created life.
- The origin of life may be explained by natural phenomena that are governed by God.
- It is certain that God created life

Religion and socio-economic level may also explain the difference between the students' acceptance of evolution in this suburban high school and other populations (the general public, teachers and students of other high schools). Nevertheless, Muslim French students are much more evolutionist than teachers and future teachers in Algeria and Morocco, their main countries of origin (Clément, 2017, Fig. 12.6 compared with Fig. 12.8): the socio-cultural influence of their country of origin appears to be counterbalanced by French education and other socio-cultural influences.

The answers to the question on the importance of chance show a real problem for all students: 56.9% of Muslin students, 27.9% of Christian students, and 21.4% of Agnostic or Atheist, consider that chance has no importance at all, or little importance. That is a challenge to improve the French teaching of the importance of stochastic processes in biology.

12.8 Conclusion

In conclusion, we underline the great acceptance of the concept of evolution in France by the general public as well as by teachers and students. These results are related to the secular French school system, which for more than one hundred years included the concept of evolution in its programs. For several decades, curricula include, since primary school, the idea of relationships between species, Deep Time and evolution of species. These concepts are explored several times during the course of schooling.

Furthermore, there is teacher training in mastering the concept of evolution. However, some obstacles to the learning of this concept remain. Epistemological and didactic obstacles relate to the nature of science, the mechanisms of evolution, the importance of chance and stochastic processes in biology, the absence of finalism (no goal-ended evolution). While centered on the phylogenetic approach, the French textbooks often under develop some of these dimensions.

At the end of this chapter, we can present some suggestions to improve evolution education in France, for instance the maintenance of evolution, including human evolution, at several grade levels in the curriculum. We recommend developing an interdisciplinary approach (biology, geology, philosophy, etc.) and particularly an epistemological and historical approach to help students to understand the nature of scientific knowledge related to evolution and to distinguish the scientific and religious registers. Teachers' training needs also to be improved to help them to facilitate debates between students.

Acknowledgements Thanks to Nour Eddine Sellamna and John Stewart who improved the quality of English of the first draft of this text.

References

Arnould, J. (2007). *Dieu versus Darwin. Les créationnistes vont-ils triompher de la science?*. Paris: Albin Michel.
Baudouin, C., & Brosseau, O. (2013). *Enquête sur les créationnismes. Réseaux, stratégies et objectifs politiques,* Paris, Belin.
Carvalho, G., Clément, P., Bogner, F., & Caravita, S. (2008). *BIOHEAD-citizen: Biology, health and environmental education for better Citizenship, Final Report.* Brussels: FP6, Priority 7, Project N CITC-CT-2004–506015.
Clément, P. (2002). Methods to analyse argumentation in (more or less) scientific texts. An example: analysis of a text promoting Creationism. In D. Krnel, *Proceedings of the 6th ESERA Summerschool, 25-31 August* (CD-ROM), Publications of Faculty of Education, Univ. of Ljubljana (7 pp.).
Clément, P. (2008). Human Evolution: Objectives, methodologies, main achievements and implications. In G. Carvalho et al., *BIOHEAD-citizen: Biology, health and environmental education for better Citizenship, Final Report.* (FP6, Priority 7, Project N CITC-CT-2004-506015), Brussels: European Community, pp. 54–67.
Clément, P. (2010). Conceptions, représentations sociales et modèle KVP. *Skholê (Univ. de Provence, IUFM), 16,* 55–70.
Clément, P. (2014). Les conceptions créationnistes d'enseignants dans 30 pays. Varient-elles en fonction de leur religion? *Education et Sociétés, 33,* 2014/1, 113–136.
Clément, P. (2015). Muslim teachers' conceptions of evolution in several countries. *Public Understanding of Science,* 24 (4): 400–421. Published online before print August 13, 2013. https://doi.org/10.1177/0963662513494549. http://hal.archives-ouvertes.fr/hal-01024550
Clément, P. (2017). Conceptions créationnistes de lycéens musulmans français. *Carrefours de l'Education* (in press).
Clément, P., & Quessada, M. P. (2008). Les convictions créationnistes et/ou évolutionnistes d'enseignants de biologie: une étude comparative dans 19 pays. *Natures Sciences Sociétés, 16,* 154–158.
Clément, P., & Quessada, M. P. (2009). Creationist beliefs in Europe. *Science, 324,* 1644.
Clément, P., & Quessada M.P. (2013). Les conceptions sur l'évolution biologique d'enseignants du primaire et du secondaire dans 28 pays varient selon leur pays et selon leur niveau d'étude. *Actes AREF 2013* (en ligne: symposium 188/3, 19 pp.). http://hal.archives-ouvertes.fr/hal-01026095.
Clément P., Quessada M.-P., & Castéra J. (2013). Creationism and innatism of teachers in 26 countries. In Abrougui M. et al., *Science & Technology Education for Development, Citizenship and Social Justice (IOSTE-14), Journal INEDP,* Vol. 1, No. 1. http://hal.archives-ouvertes.fr/hal-01026102.
Cohen, C. (1999). *L'Homme des origines.* Paris: Seuil.
Cuvier, G. (1817). *Le règne animal distribué d'après son organisation, pour servir de base à l'histoire naturelle des animaux et d'introduction à l'anatomie comparée.* Paris: Deterville.
Dobzhansky, T. (1973). Nothing in biology makes sense except in light of evolution. *American Biology Teacher, 35,* 125–129.
Dray, S., & Dufour, A.-B. (2007). The ade4 package: implementing the duality diagram for ecologists. *Journal of Statistical Software,* 22(4), 1–20.
Enseigner l'évolution (2008). Colloque des 13 et 14 novembre, Paris, CSI et Collège de France
Ferry, J. (1880). In A. Monchaton (Ed.), *1789 Recueil de textes et documents du XVIIIe siècle à nos jours* (pp. 222–223). Paris: Centre National de Documentation Pédagogique.
Fortin, C. (2009). La théorie de l'évolution: réception et enjeux d'éducation, in *Guide critique de l'évolution,* G. Lecointre G. (Dir.), Paris, Belin, 162–177.
Gayon, J. (1993). La biologie entre loi et histoire. *Philosophie, 38,* 30–57.
Gould, S. J. (1984). *Quand les poules auront des dents.* Paris: Fayard.
Gould, S. J. (1997). Non-overlapping Magisteria. *Natural History, 106,* 16–22.

Gould, S. J. (2000). *Et Dieu dit: que Darwin soit*. Paris: Seuil.
Grassé, P. P. (1973). *L'évolution du vivant, matériaux pour une nouvelle théorie transformiste*. Paris: Albin Michel.
IAP, InterAcademy Panel. (2006). *IAP Statement on the Teaching of Evolution*. [Page Web]. Accès: http://www.interacademies.net/File.aspx?id=6150
IFOP, Institut Français d'Opinion Public. (2011). *Les Français et la croyance religieuse*. Accès: http://www.ifop.com/media/poll/1479-1-study_file.pdf
Jean-Paul II. (1996). L'Eglise devant les recherches sur les origines de la vie et son évolution. *Message aux membres de l'assemblée générale plénière de l'Académie pontificale des sciences*, 22 octobre 1996.
Keskas, M. (1994). *La théorie de Darwin: Le hasard impossible*. Paris: Le Figuier.
Kupiec, J.-J. (2008). *L'origine des individus*. Paris: Fayard, Le temps des sciences.
Lamarck, J.B. (1802). Recherches sur l'organisation des corps vivans et particulièrement sur son origine, sur la cause de ses développements et des progrès de sa composition, et sur celle qui, tendant continuellement à la détruire dans chaque individu, amène nécessairement sa mort; précédé du discours d'ouverture du cours de zoologie, donné dans le Muséum national d'Histoire Naturelle. Paris: Maillard. Site Lamarck - www.lamarck.net.
Lecointre, G. (2004, 2008). *Comprendre et enseigner la classification du Vivant*. Paris: Belin, Guides Pédagogiques
Lecointre, G. (2009). *Guide critique de l'évolution*. Paris: Belin.
Lecointre, G. (2014). *L'évolution, question d'actualité ?*. Versailles: Editions Quæ.
Lecointre, G., & Le Guyader, G. (2001). *Classification Phylogénétique du Vivant*. Paris: Belin.
Loison, L. (2008). Lamarck fait de la résistance. *Les Dossiers de La Recherche, L'héritage Darwin, n, 33*, 40–45.
Loison, L. (2010). *Qu'est-ce que le néolamarckisme ? Les biologistes français et la question de l'évolution des espèces, 1870-1940*. Paris: Vuibert.
Machelon, J.P. (2006). *Rapport sur les relations des cultes avec les pouvoirs publics*. La Documentation française. Accès: http://www.ladocumentationfrancaise.fr/var/storage/rapports-publics/064000727.pdf
Mangin, L. (1898). In B. Belhoste (Ed.), *1995, Les sciences dans l'enseignement français: Textes officiels, t I: 1789-1914* (p. 574). Paris: INRP/Economica.
Marchand, L. (2015). Plus de la moitié des Français ne se réclament d'aucune religion. http://www.lemonde.fr/les-decodeurs/article/2015/05/07/une-grande-majorite-de-francais-ne-se-reclament-d-aucune-religion_4629612_4355770.html
Miller, D., Scott, E., & Okamoto, S. (2006). Public acceptance of evolution. *Science, 313*(5788), 765–766.
Munoz, F., Bogner, F., Clément, P., & Carvalho, G.S. (2009). Teachers' conceptions of nature and environment in 16 countries. *Journal of Environmental Psychology*, 29: 407–413. http://hal.archives-ouvertes.fr/hal-01024981
Perrier, E. (1882). Histoire Naturelle, in F. Buisson (éd.), *Dictionnaire de pédagogie*, 1°partie, tome 1, 1274.
Perru, O. (2010). *La création dans le créationnisme?*. Paris: Kimé.
Picq, P. (2007). Créationnisme et dessein intelligent. *Pour la Science, 357*, 50–54.
Quessada, M.P. (2008). L'enseignement des origines d'Homo sapiens, hier et aujourd'hui, en France et ailleurs: programmes, manuels scolaires, conceptions des enseignants. PhD thesis, University Montpellier 2, http://tel.archives-ouvertes.fr/tel-00353971/fr/.
Quessada, M.P. (2016). Les migrations à l'origine de l'espèce humaine actuelle dans les manuels de sciences français de 2001 à aujourd'hui. In B.Maurer, M.Verdelhan, A.Denimal, *Migrants et Migrations dans les manuels scolaires en Méditerranée*. Paris: L'Harmattan, p. 27–44.
Quessada, M.P., & Clément, P. (2007). An epistemological approach to French curricula on human origin during the 19th & 20th centuries. *Science & Education*, 16, 9–10, 991-1006. [Page Web]. Accès: http://dx.doi.org/10.1007/s11191-006-9051-9
Quessada, M.P., & Clément, P. (2011). The origins of humankind: A survey of school textbooks and teachers' conceptions in 14 countries. In: A. Yarden, & G.S. Carvalho (Eds.). *Authenticity*

in *Biology Education: Benefits and Challenges. A selection of papers presented at the 8th Conference of European Researchers in Didactics of Biology (ERIDOB)*, Braga, Portugal, p. 295–307.

Quessada, M.P., & Clément, P. (2013). L'évolution humaine dans les programmes et les manuels scolaires de science français de 1994 à aujourd'hui: interactions entre connaissances, valeurs et contexte socioculturel, *Actes du Congrès international AREF 2013 Actualité de la Recherche en Education et en Formation*, Montpellier, 29 août-1er septembre: 10 pages dans les Actes CD-Rom.

Quessada, M.P., Munoz, F., & Clément, P. (2007). Les conceptions sur l'évolution biologique d'enseignants du primaire et du secondaire de douze pays (Afrique, Europe et Moyen Orient) varient selon leur niveau d'étude. *Actes Colloque AREF (Actualité de la Recherche en Education et en Formation)*, Strasbourg, 407 (12 pp.)

Teilhard de Chardin, P. (1956). *Le groupe zoologique humain*. Paris: Albin Michel.

Témoins de Jéhovah. (2002). La vie : comment est-elle apparue? Evolution ou création? (Life – How Did It Get Here? By Evolution or by Creation? French Sce-F). Published by the Watchtower Bible and Tract Society of New York, Inc., International Bible Students Association, Brooklyn, New York, U.S.A.

Tidon, R., & Lewontin, C. (2004). Teaching evolutionary biology. *Genetics and molecular biology, 27,* 1–8.

Yahya, H. (2008). *Atlas de la Création, vol. 1,* Istanbul, éd. Global.

Marie-Pierre Quessada an Associate Professor of Science at the University of Montpellier (France), taught master and doctoral level courses in science and in epistemology, history and didactics of science for the training of primary schools teachers and biology teachers. She directed the Nîmes site of the Faculty of Education at the University of Montpellier from 2008 to 2016. Currently retired from teaching, she pursues her research on science-society relations in the teaching of evolution in France and also at the international level. She studies the didactic transposition in curricula and textbooks and also the teachers and students' conceptions of biological evolution.

Pierre Clément initially taught and did researches in Animal Biology, then in Didactics of Biology, mostly in France (University Lyon 1), but also, more punctually, in Algeria, Tunisia, Lebanon, Senegal, Morocco, Denmark and Sweden. He is member of the board of IOSTE. He coordinated the BIOHEAD-Citizen research project (Biology, Health and Environmental Education for better Citizenship, 19 countries). He is now retired but still active in research, extending to new countries (a total of 34 today) his research on teachers' conceptions of evolution, human genetics, sex, health and environmental education.

Chapter 13
Evolution Education in the German-Speaking Countries

Erich Eder, Victoria Seidl, Joshua Lange and Dittmar Graf

Abstract This chapter discusses evolution education and the acceptance of evolution in Germany, Austria, Switzerland, South Tyrol, and Luxembourg. Ernst Haeckel first introduced Darwin's concepts to a broad public in Germany. His interpretation of evolution was more teleological, hierarchical, and definitely more anti-religious than intended by Darwin. Later, pseudo-evolutionary arguments were adopted by the Nazis to pursue racism and mass murder. These historical developments still influence the perception of evolution in the German-speaking countries. Acceptance of evolution has been generally increasing since 1970, and correlates negatively with religious faith and positively with attitudes on science and the understanding of evolution, respectively. Students' preconceptions are frequently faulty, anthropomorphic and teleological, e.g. they regard adaptation as an intentional process. Knowing these preconceptions helps to deal with them in order to help students properly understand evolution. Anti-evolution movements exist, but not in a comparable intensity to the United States. Occasional support of "Intelligent Design" (ID) by church dignitaries is usually mocked by the media, and ignored by a vast majority. Parts of the scientific community are alert to creationist initiatives like fake textbooks with ID contents. Creationism and ID are not intended to be taught at school or universities. In the curricula for primary school, evolution is not mentioned. Later, it usually appears once at lower and higher secondary school each. Students with general qualification for university entrance should at least know the basic evolutionary concepts and relationships.

E. Eder (✉) · J. Lange
Sigmund Freud University Vienna, Medical School (SFU MED),
Freudplatz 3, 1020 Vienna, Austria
e-mail: erich.eder@med.sfu.ac.at

V. Seidl
Core Facility Botanical Garden, University of Vienna, Rennweg 14,
1030 Vienna, Austria

D. Graf
Institut für Biologiedidaktik, Justus-Liebig-University Giessen,
Karl-Gloeckner-Str. 21c, 35394 Giessen, Germany
e-mail: Dittmar.Graf@didaktik.bio.uni-giessen.de

© Springer International Publishing AG, part of Springer Nature 2018
H. Deniz and L. A. Borgerding (eds.), *Evolution Education Around the Globe*, https://doi.org/10.1007/978-3-319-90939-4_13

Teacher education has been heterogenous for the different countries, federal states, and school types, but is being standardized at university courses according to the Bologna process recently. Finally, we suggest improvements for evolution education. We agree with the German Academy of Science that a framework curriculum should be developed to offer evolutionary biology as a "red thread" through all biological phenomena, for all types of schools and for all grades.

13.1 Introduction

Public understanding of science and the acceptance of paradigmatic changes largely depend on the individuals promoting these changes in the public. How evolution has been perceived in the German-speaking countries since its foundations is therefore relevant if we want to deal with long time trends, such as creationism or misunderstood evolutionary concepts. In Germany, the first write-up of Darwin's "Origin of Species" was published within two months (Peschel, 1860); the first German translation by Heinrich Bronn (1800–1862) was published in the same year, followed by several further editions. Ernst Haeckel (1834–1919), also known as "the German Darwin", was the first major biologist devoted to spreading the new ideas not only to scientists and textbooks, but also to the common public, as he considered the new theory "not only threatening to thoroughly unsettle a century-old commonly accepted doctrine, but also to deeply interfere in the personal, scientific and social beliefs of each and every reader, […] an insight that will indeed change the whole worldview" (Haeckel, 1863: p. 3, translation by the authors). However, his interpretation of evolution differed significantly from Charles Darwin's elaborate and cautious approach. Unlike Darwin, Haeckel considered evolution to be a teleological process from "lower" to "higher" creatures, with humans on top (cf. Fig. 13.1). He promoted a free religious, rather pantheistic "evolutionary monism" (Haeckel, 1898), and was proclaimed as the "Antipope" at a freethinker convention in Rome in 1904. In 1906, Haeckel was among the founders of the Deutscher Monistenbund (German Monism Society) and became its honorary president. Monism is the view that existing things can be explained in terms of a uniform basic principle of reality. These activities likely contributed to the fact that evolution was frequently considered to be an offence against faith and religion in German speaking countries, up until the present day.

The early 20th century was the era of popular education in the German-speaking world. The "Kosmos-Gesellschaft der Naturfreunde" (founded in 1904) spread generally intelligible evolutionary ideas in the popular scientific journal "Kosmos". Wilhelm Bölsche (1861–1939), who was not a biologist and interpreted evolution much in Haeckel's way, published illustrated descriptive booklets there (Fig. 13.2).

The ideological background that accompanied the introduction of evolution to Germany and other German-speaking countries, willingly or unwillingly, favoured the development of social Darwinism, eugenics and racism (Bayertz, 1998), which culminated in the mass murders of the Nazi regime.

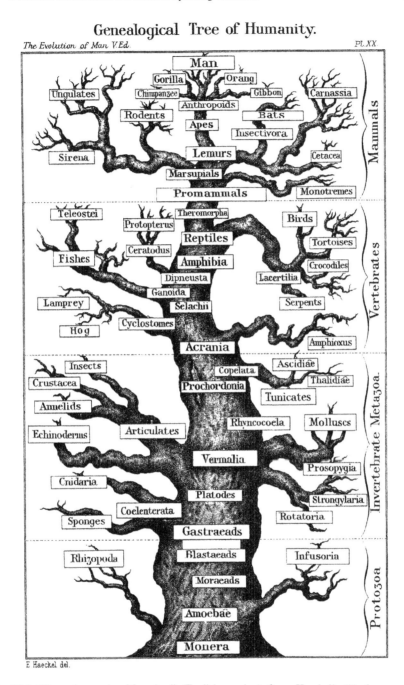

Fig. 13.1 "Stammbaum des Menschen" (English version) from Haeckel's "Anthropogenie" (1874). Compared to Darwin's (1859) exemplary tree (only illustration in "Origin of Species"), this tree insinuates hierarchy and a teleological progress from "lower" towards "higher" evolutionary stages, seemingly culminating in the "crown of creation"

Fig. 13.2 Cover sheet of "Stammbaum der Tiere" (family tree of animals) by Bölsche (1905). The text starts with an ornate description of North Sea dunes, and does not contain any chart of the animal tree at all

In the second half of the 20th century, a few renowned scientists such as Konrad Lorenz (1903–1989), Hoimar von Ditfurth (1921–1989), and Rupert Riedl (1925–2005) wrote popular scientific books popularising novel evolutionary approaches to behaviour and cognition (e.g. Lorenz, 1973; Ditfurth, 1976; Riedl, 1976), which appeared in the mass media and significantly influenced politics and public opinion.[1] Twenty years after Lorenz' death, Föger & Taschwer (2001) revealed that he had joined the NSDAP (Nationalsozialistische Deutsche Arbeiterpartei, National Socialist German Workers' Party) in 1938. This might explain why biologists and other proponents of evolution are frequently confronted with their alleged "Nazi" background, not only by creationists (Bergman, 1986), but also by clergymen (Der Spiegel, 2009) and possibly still by a considerable proportion of the population in the German-speaking countries.

In this chapter, we will give some general information about the acceptance of evolution in the German-speaking education system as well as provide an overview on evolution-relevant curricula, anti-evolution movements, and teacher education in Germany, Austria, Switzerland, Luxembourg, and South Tyrol, an autonomous province of Italy (general data, see Table 13.1). In Belgium, German is also an official language, but is spoken in a very small part of the country only. Liechtenstein, where German is spoken, too, has only 40.000 inhabitants and is oriented toward the education system of Switzerland.

The educational systems of the German speaking countries are quite different and complicated. There are primary schools up to grade 4 or 6 (age 6–10/12 years), and secondary schools up to grade 12 or 13 (age 18–19 years). Degrees can be awarded after grade 9 or 10 (intermediate school-leaving qualification) and 12 or 13 (university-entrance diploma).

13.2 Public Acceptance of Evolution

In this chapter, population-representative surveys are presented first. They are followed by studies with middle and high school students, and finally with teacher students.

For Germany, Austria, and Switzerland, several representative surveys on attitudes towards evolution have previously been carried out. Luxembourg is only mentioned in one of the studies, South Tyrol not at all. Unfortunately, the questionnaires differ between the polls, so that the results are heterogenous, and a comparison is difficult.

Starting in 1970, the Allensbach Institute for Demoscopy has repeatedly asked the question of the relationship of man and "ape" in West Germany. The close kinship of man and other apes seems to be increasingly accepted. However, it is still

---

[1]On the other hand, neither Willi Hennig (1913–1976) nor Ernst Mayr (1904–2005), to mention two important German evolutionary biologists, attained public awareness.

Table 13.1 Basic data on the German-speaking Countries

	Germany	Austria	Switzerland	Luxembourg	South Tyrol (Part of Italy)
Population	82.2 Mio.	8.7 Mio.	8.4 Mio	0.58 Mio.	0.52 Mio.
Languages spoken	German	German	German, French, Italian	Luxembourgish, German, French	German, Italian
Religions	Roman Catholics: 29%; Protestants: 27%; Muslims: 5.5%; Non-Religious: 34%	Roman Catholics: 59%; Protestants: 3.4%; Orthodox: 6%; Muslims: 7%; others or Non-Religious: 24%	Roman Catholics: 38%; Protestants: 27%; other Christians: 4%; Muslims: 5%; Non-Religious: 22%	Roman Catholics: 67%; other Christians: 7%; Muslims: 3%; Non-Religious: 20%	Roman Catholics: 96.1%; Muslims: 2.3%
Constitution	Secular	Secular	Secular	Secular	Secular

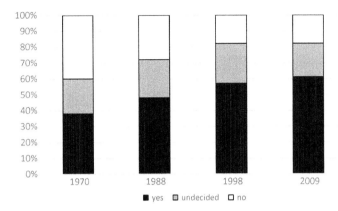

Fig. 13.3 Answers to the question: do humans and apes have a common ancestor or not? Representative surveys (1970, 1988, 1998, and 2009) in West Germany (Allensbach Institute for Demoscopy, 2009)

rejected by almost 20% in the most recent survey (Fig. 13.3). In 2009, 20% of the 1,807 German interviewees agreed with the statement "Man was created by god as it is written in the bible", and 61% agreed with "Man has evolved from other forms of life" (Allensbach Institute for Demoscopy, 2009).

Surveys in 2002 and 2005 examined the acceptance of evolution in Japan, the USA, and 32 European countries, among them 518 respondents from Luxembourg, 999 from Switzerland, 1,034 from Austria, and 1,507 from Germany (Miller et al., 2006).

In both years, the German results were very similar to the Allensbach study: 71% agreed with "Human beings, as we know them, developed from earlier species of animals", and 21% denied this statement. For Austria, in 2002, two thirds of the respondents agreed with the evolution statement, and less than 20% denied the existence of evolution. Three years later, the overall agreement with the above statement was less than 60%, while almost 30% of respondents denied evolution. For Luxembourg and Switzerland, only the 2005 data are available: in Luxembourg, one-third accepted evolution and 25% denied it, and in Switzerland, two thirds agreed with the evolution statement, and 30% denied the existence of evolution (Miller et al., 2006).

In a survey by the opinion research institute "forsa", which was carried out in Germany for the "Forschungsgruppe Weltanschauungen in Deutschland" in 2005, an eighth of the respondents were close to a creationist position ("Man was created by God as it is written in the Bible"), and 25% believed in a theistic evolution ("Life on earth was created by a higher being or by God, but passed through a lengthy developmental process, which was controlled by a higher being or by God") (Fowid, 2007).

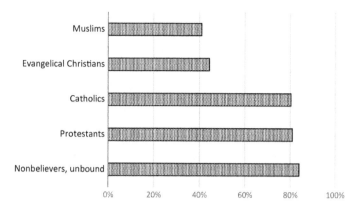

Fig. 13.4 Agreement with the statement: "The theory of evolution is a scientifically acknowledged theory" pre-service teachers in Germany (n = 1,055) (Fowid, 2007). "Evangelical Christians" refers to fundamentalists taking the Bible literally

There are differences between the different religious denominations in Germany: the rejection of evolution is most widespread in Muslims and evangelical Christian fundamentalists (Fig. 13.4).

In 2009, a large oral opinion survey about the Austrians' attitude towards creationism and evolution (n = 1,520) asked participants "What should be taught at Austrian schools?" Only 50% agreed with a naturalistic definition about the "origin of the world", while 21% advocated creationism (GfK Austria, 2009). Austrians generally regarded the topic of evolution as very popular: 80% of the interviewees agreed with the statement that humans and monkeys have a common ancestor. Nevertheless, respondents' answers suggest that evolution is often understood as a directional optimization principle (GfK Austria, 2009).

A representative survey (n = 1,500; 500 each in Switzerland, Austria, and Germany) carried out by a Swiss opinion research institute (IHA-GfK) on behalf of the anti-evolution institute "Progenesis" and the Swiss religious magazine "Factum" showed that in Switzerland more than one in five (21.8%) is convinced that life has been created by God within the last 10,000 years. In Austria (20.4%) and Germany (18.1%) the rates are slightly lower. It is notable that significantly more women than men believe in divine activity (25–14.3%) (Höneisen, 2003). In a further representative survey (n = 1,100) by the same institute in Switzerland, 75% of the respondents advocated in favour of equal rights for both creation and evolution to be taught in biology at school (Höneisen, 2007).

Lammert (2012) conducted a comprehensive study including the acceptance of evolution by middle school students[2] (grades 9 and 10) in North Rhine-Westphalia (a federal state in Germany). The results show that the acceptance of evolution is particularly influenced by their attitude towards science (Fig. 13.5). Subjects who

---

[2]Sekundarstufe I in German.

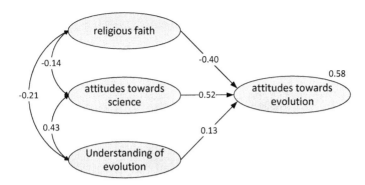

Fig. 13.5 Structural equation model; attitudes towards evolution as a dependent variable; explained variation: 58%, n = 3,969; middle school students (Lammert, 2012)

accept science as meaningful and important also show a higher acceptance of evolution. On the other hand, the religious faith of the participants has a negative impact on their attitude towards evolution: very strong believers show a low acceptance of evolution. In addition, the acceptance is only slightly influenced by the understanding of evolution. In other words: Subjects who understand evolution and their mechanisms better show only a slightly higher acceptance of evolution. Her results indicate that a positive influence cannot be achieved until the learners have passed a certain level, or threshold, of understanding of evolution. Only then can "evolution" provide answers to different questions and promote acceptance through its explanatory potential (Lammert, 2012).

Lammert's assumption is supported by Fenner (2013). She did a design research study with younger students (5th and 6th grade, 10–11 years old; topic evolution, n = 710), and did not find any connection between "understanding of evolution" and "attitude towards evolution" as the dependent variable by using regression analyses, before she performed a didactical intervention. The intervention consisted of an 8-h teaching session regarding evolution and evolutionary theory. After the intervention, the connection was considerable (beta = 0.23).

A survey on more than 2,100 secondary school students in Austria revealed similar data (Eder et al., 2011). Little more than half of the students agreed with naturalistic evolution, and 28% with creationism. Theistic evolution or ID (s. the legend of Fig. 13.6) seemed to be a compromise for many students; more than a third agreed to that statement (Fig. 13.6).

In general, acceptance of evolution correlated negatively with religious belief (r = –0.29, p < 0.00001). Muslim students showed both the highest values in religious belief and the lowest acceptance of evolution, but the agreement with evolution of catholic students was not much higher, although this group exhibits significantly lower religiosity (Fig. 13.7).

What are the predictors of attitudes on evolution? In a study with 729 first-semester teacher students from Germany, Graf and Soran (2011) evaluated factors that predicted "attitudes towards evolution" as a dependent variable.

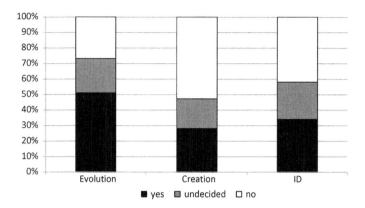

Fig. 13.6 Acceptance of evolution of Austrian secondary school students ("Life on earth has emerged without the influence of any supreme being and has evolved through a natural developmental process"), creationism ("God directly created life on earth, including all species, as described in the Bible"), and a statement equivalent to theistic evolution /ID ("Life on earth was created by a supreme being (God), and has undergone a long developmental process directed by this supreme being (God)"). N = 2,129. From Eder et al. (2011), modified

With the help of a regression analysis it was found that the factors "attitudes towards science" (beta = 0.399), "understanding of science" (beta = 0.169), and "understanding of evolution" (beta = 0.213) have positive effects, but religious faith (beta = −0.212) has a negative effect on the attitudes towards evolution.

Großschedl et al. (2014) identified creationism (beta = −0.39), but not attitudes towards religion (beta = 0.09), by using regression analyses in a study with 180 German biology pre-service teachers, as a predictor for the acceptance of evolutionary theory and the preference for teaching evolution. Attitude towards science was positively correlated with the acceptance of evolutionary theory (beta = 0.29).

13.3 Anti-evolution Movements

Within Darwin's lifetime, there was a conflict that lead to the prohibition of biology teaching in upper secondary school in Prussia for fear of "materialistic" influences at school. The occasion was a school lesson held by flower biologist and biology teacher Hermann Müller at a school in Lippstadt in 1877. He had read from a book by Carus Sterne (real name Ernst L. Krause, 1839–1903) "Werden und Vergehen" which states: "A modern chemist who wanted to translate the history of creation into his formula language would not begin: 'In the beginning was the word', but would have to exclaim: 'In the beginning was carbon with its strange internal forces'" (in Morkramer, 2010, p. 119). Müller was accused of presenting hypotheses as proven facts, and of giving anti-religious lessons. The matter came into the media and was finally addressed in the Prussian parliament. In January 1879, a Freiherr von

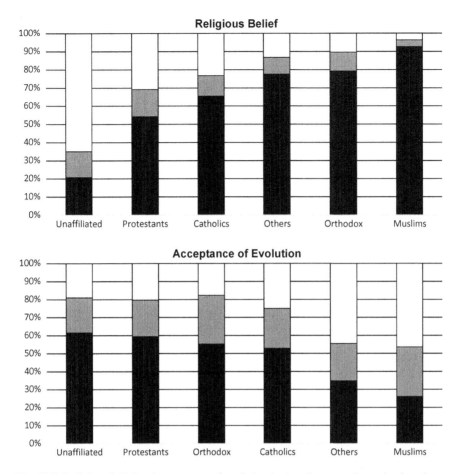

Fig. 13.7 Religious belief and acceptance of evolution in Austrian secondary school students: acceptance (black), neutral (grey) and rejection (white) with the statements "I believe in God" (top graph) and "Life on earth has emerged without the influence of any supreme being and has evolved through a natural developmental process" (bottom graph). Total n = 2,129, Catholics n = 1,191, Unaffiliated n = 342, Muslims n = 269, Protestants (not including evangelical fundamentalists) n = 133, Serbian and Greek Orthodox n = 96, others (including Jehovah's Witnesses, evangelical fundamentalists, and 6 jewish students) n = 98. Religious denominations were ranked according from lowest to highest in religiosity and from highest to lowest in agreement with evolution; note that the order stayed almost the same

Hammerstein reported against Müller: "in fact, the teacher, Müller, read from the book of Carus Sterne… And, indeed, parts in which Christianity was characterized as a fantasy of the mind and the holy trinity as polytheism." (Nummert, 2001, p. 64).

The following month, Müller wrote to Darwin, with whom he was in an occasional exchange of ideas, that Darwin's ideas had come under pressure in Germany. In fact, in the Prussian curriculum adopted in 1882, biology became completely forbidden in the higher classes (Keckstein, 1980). One year later, the elaboration of the curricula explicitly forbade the treatment of Darwin's ideas: "The introduction of the new Darwinian hypotheses, etc., is not one of the tasks of the school, and is therefore to be kept out of the classroom" (Keckstein, 1980, p. 38). Biology teaching in higher education was forbidden in Prussia for more than a quarter of a century, until 1908, when it was re-admitted by decree. The issue of "evolution" was not mentioned in the decree.

In 1925, biology became a compulsory subject in the advanced levels of schooling, although only with a few lessons (Keckstein, 1980). After the Second World War, individuals repeatedly raised their voices against evolution, but there was no organized opposition. The British chemist Arthur Ernest Wilder-Smith, who changed from an atheist to a Christian, was particularly influential. From 1946 onwards, he went on numerous lecture tours in Germany and Switzerland and published books that criticised the theory of evolution (Kotthaus, 2003).

In Germany, creationist ideology gained more influence again in the 1980s, after the foundation of "Studiengemeinschaft Wort und Wissen" (WuW = Word and Knowledge). People involved in WuW are mostly academically educated and are young earth creationists, believing the earth to be less than 10,000 years old. WuW publishes numerous materials for students, including a book designed as a textbook for the upper secondary school. However, it is not accepted for use in schools in any federal state of Germany or other German-speaking countries. The book is elaborately designed and now available in the seventh edition (Junker & Scherer, 2013). It mainly contains evolution criticism. Explicit creationist positions are largely hidden behind ID arguments, but explicit creationism is clearly visible in internal WuW papers.

The book received a "school textbook award" in 2002, although it is not an approved schoolbook at all, by the Christian-oriented "Kuratorium Deutscher Schulbuchpreis". Dieter Althaus, then Prime Minister of the federal state of Thuringia, praised the book as a "very good example of value-oriented education and education". He further hoped, "that your book is not only used by school biologists, but also by a far-reaching readership." Since then, Althaus has distanced himself from these statements. However, in 2005, he invited the co-author of the textbook, Siegfried Scherer, to a discussion on "Evolution and Creation" in the Thuringian State Chancellery, but then disinvited him due to public pressure.

Recently, a second book critical about evolution has appeared in Germany, which in style and design resembles a school textbook (Vom Stein, 2017), but is not approved for use in schools. The book is aimed at middle school students, and, contrary to Junker & Scherer (2013), unvarnishedly argues from a biblical perspective.

An employee of the Max Planck Institute for Plant Breeding Research in Cologne, W.-E. Lönnig, published hundreds of ID-oriented Creationist pages on the web space of the institute in the 1990s, without being criticized by the Max Planck Society. Although the department director H. Saedler distanced himself from the pages, he did not see any reason to remove them from the net. This caused massive disputes between the Association of German biologists (chairman of its working group "Evolutionary Biology": Ulrich Kutschera), and Saedler. It culminated in an article in "Nature" (Kutschera, 2003), with resultant an embarrassment for the Max Planck Society. As a consequence, Lönnig had to remove his pages from the Society's website. The case is documented in detail by Kutschera (2007), along with a number of other clashes between ID and evolutionary biology.

Although it is not permitted to teach creationist ideas as facts in Germany, there are many examples of where this is done in evangelical private schools (Kutschera, 2014), or with parents who illegally[3] homeschool their children (Graf & Lammers, 2011).

In Austria, the Archbishop of Vienna, Christoph Cardinal Schönborn, has considerable influence on public and political opinion, including a regular column in the newspaper with the widest circulation in the country. However, his article "Finding Design in Nature" in the New York Times (Schönborn, 2005), where he emphasized the negative attitude of the Roman Catholic Church towards the naturalistic worldview of evolutionary theory, was picked to pieces by the media, and provided an occasion to discuss international trends of anti-evolutionism. Later, Schönborn backpedalled, and embraced the possibility that evolution theory and faith could coexist (ORF, 2009). Immediately after Schönborn's article, the German pope Benedict XVI explicitly supported it by talking of a cosmic "intelligent plan" (Der Standard, 2005). Two years later, however, Benedict XVI rejected creationism and acknowledged scientific evidence of evolution (see Brasseur 2009).

In Switzerland, teaching material for use from the 7th grade with the name "NatureValue - Plants - Animals – Men" was published for teaching biology, in summer 2007. One of the topics deals with "Creation and Evolution - the Origin of Life". The biblical creation story is told there, and creation myths are mixed together with evolution and presented as equivalent. When studying the accompanying materials for teachers, it becomes clear that the authors believe wrongly that by contrasting beliefs with evolution they are contributing to an "ethical" debate: "young people should get to know different opinions and views in this topic, those of people who believe in a creator or those who regard the origin and development of life as a process of evolution" (Wittwer et al., 2007, p. 19). However, their assumption that the alternative between creation and evolution must be assessed ethically is wrong, as the substance of scientific theories is not to be validated ethically, but empirically. After massive protests, the publisher finally withdrew the worksheet "creation and evolution of life", and replaced it by a general entry text in 2008. Overall, we regard this revision as half-hearted, since the teachers' instructions continue to include the section "Evolution and Creation". This section pretends that evolution should be taught as an "opinion among several" at school, and that students should find their own private theory about the origin and development of life.

About 10 years ago, Islamic creationism has started entering the German-speaking countries (Yahya, 2014). Its direct influence on the Muslim communities has not been investigated yet, but the above data (Figs. 13.4, 13.6)

---

[3]In Germany, attending school is compulsory.

show that this group has religious problems with the acceptance of evolution (see also Ichner, 2017).

Similarly to other European countries and the USA, political right-wing populism is on the rise in the German-speaking countries (Wodak et al., 2013). Although many of these politicians endorse conspiracy theories (e.g. chemtrails, lying press, anti-vaxxers), alternative medicine and new age esotericism (Eder, 2017), we do not know of any politically relevant anti-evolution representatives in right-wing populist parties.

13.4 Preconceptions in Students

Although the acceptance of evolution seems to be relatively well established in the German-speaking countries, the understanding of basic evolutionary concepts is not.

In fact, the success of the school-based teaching effort is quite limited. From Germany, data on the knowledge of evolutionary mechanisms from freshmen students are available (n = 729). It was found that on average, only 8 of 18 simple multiple-choice tasks for the understanding of evolution could be answered correctly (Graf & Soran, 2011). In order to explain evolutionary changes students often use finalistic/teleological (adaptation as a goal oriented process) and occasionally Lamarckian arguments (Fig. 13.8). The fact that some students had undisclosed Lamarckian ideas is shown in Fig. 13.9.

An appropriate concept of "deep time" is also missing in many freshmen. The idea that dinosaurs (which are highly complex eukaryotes) lived at the beginning of evolutionary development is widespread. The period of existence of *Homo sapiens*

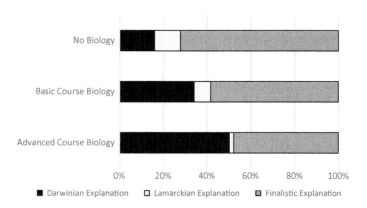

Fig. 13.8 Comparison of frequencies of Darwinian, Lamarckian and finalistic explanations, Students were asked the question: How would a biologist explain how the ability to run fast evolved in cheetahs (Graf & Hamdorf, 2012). The task was first published in Bishop & Anderson (1986)

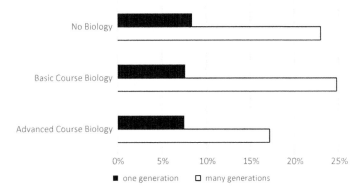

Fig. 13.9 Lamarckian answers to the question: "At the end of the 19th century, the zoologist August Weismann conducted the following experiment: He completely cut off the tails of mice to determine the effects this would have on the direct offspring. (A) How would the children of these mice have looked by considering the evolution theory? (black bars) (B) And let us suppose that Mr. Weismann would have cut off the tails of the offspring, and their offspring, etc., for a total of twenty generations. How do the 21st-generation mice look like by considering the evolution theory? (grey bars)". N = 1,055 (Graf & Hamdorf, 2012)

is regularly overestimated. On occasion, one even finds the idea that dinosaurs and humans would have met each other alive (Graf & Hamdorf, 2012).

The concept of fitness is also often misunderstood by many freshmen. Only 20.9% of those taking part in the study recognize that a crucial measure of biological fitness is the proportion of an individual's offspring that also survive and reproduce (Graf & Soran, 2011).

Fenner (2013) noted that ideas of adaptation are usually wrongly accompanied by ideas of a reduced variation in populations. She examined 710 5th and 6th grade students in North Rhine—Westphalia.

In Austria, recent qualitative investigations with 14-year old secondary school students revealed severe misconceptions of basic evolutionary concepts, although they had lessons about evolution the year before (Seidl, 2017).

The process of adaptation is often seen from an anthropomorphic or teleological point of view. In some cases, it is considered an intentional striving for evolution, which is comparable to an action of everyday life with a goal (Kattmann, 2015). In other cases students only refer to the necessity of adaptation as a requirement for evolution. Concerning this conception, most pupils believe that the individuals have a certain will to live and can choose to change (Hammann & Asshoff, 2015). Sometimes students even claim that the organisms can carry out targeted crossing to change in a certain and beneficial way. This means that the individuals would realize a disadvantageous situation, consciously or unconsciously (Baalmann et al., 2004). A study participant explains this the following way: "It's not like one individual suddenly looked differently or did something else, but they first realize how to do certain things. Animals in which these characteristics were particularly remarkable had a better chance of survival" (Seidl, 2017, p. 76).

Other students claim that an organism's body is able to notice external environmental influences and can react to them with adaptation ("Körperweisheit", which means some kind of inner wisdom of the body). According to this conception, evolutionary changes affect individuals and not populations (Kattmann, 2015): "An animal has problems in a certain situation, for example when it is cold and the animal has problems with this cold, it can slowly develop fur. It adapts itself until it fits" (Seidl, 2017, p. 75).

Some students have Lamarckian views about adaptive changes during evolutionary process where traits can change due to use or disuse (Hammann & Asshoff, 2015; Stover & Mabry, 2007): "I think humans will also evolve physically in the following years. I imagine that some bones will get stronger or recede. I want to give an example with the cell phone. I'm sure that the thumb will get longer and the other fingers will recede because we use our thumb so often while using our smartphones" (Seidl, 2017, p. 74).

In many cases students start to personify nature and claim that it is responsible for the adequate adaptation of species and thus provides a biological balance. It is considered an anthropomorphic phenomenon that many students search for harmony and sense in evolution (Kattmann, 2013).

Furthermore, the competence of freshmen to read phylogenetic trees is only weak (cf. Meir et al., 2007). Information on relatives cannot be taken from the phylogenetic trees in a targeted manner: Many participants assumed that organisms are closely related to each other because they were drawn closely together. A large proportion was unable to draw the timeline correctly into the phylogenetic tree. (Graf, in prep.). Many students interpreted the phylogenetic trees that were shown during the interviews as if they were pedigrees of human families or constellations of connected persons: "If you have pedigrees of human families, you always have father, mother, and the arrows leading to the daughter or something."—"The gorilla is more related to the gibbon than to the human. (…) Like my grandmother is not as close related to me as my father is" (Seidl, 2017, p. 84; cf. Baum et al., 2005).

A very common misconception while observing phylogenetic trees is that similarity shows relatedness of species, or the opposite view that visible differences between species means that they are only distantly related (Gregory, 2008). During the interviews, all of the students repeatedly explained the phylogenetic trees and their perception of relatedness in this way: "It looks like dinosaurs and birds are closely related which is irritating. I would never consider birds to be reptiles because they have feathers and snakes for example have scales. (…) I think reptiles and amphibians are related because they are really similar" (Seidl, 2017, p. 98).

Most students expected an evolutionary development from "lower" to "higher" stages. In particular, humans are considered as the "crown of creation" which is supported by the conceptional depiction of many textbook trees, even though it is possible to randomly rotate every internal node without changing the typology.—"I want to put the human up on the top of the tree because the formation of living things until now, until the presence of humans, the highest stage of life until now" (Seidl, 2017, p. 81).

Some of the studies in the German-speaking areas showed that many students even do not consider humans as animals and draw a clear line between those groups: "I think the gorilla is more closely related to the gibbon, because the gorilla is rather an animal and the human is actually a human. A chimpanzee is also an animal but not as much as the gorilla, because you can educate him to act like a human (Seidl, 2017, p. 81).

13.5 School Curricula and Textbooks

Evolution is not mentioned as a topic in primary school in any of the evaluated German-speaking countries.

Comparing the different biology curricula at secondary school is challenging, because the situation is complex (cf. Skoog, 2005). There are not only several different school types for lower and higher secondary school in these countries, but also regionally different curricula in the German federal states and the cantons of Switzerland.

Generally, the topic evolution is studied twice during secondary school, initially in lower secondary education (grades 5 to 9/10, students' age see Table 13.1) and for a second time in upper secondary education (grades 9/10 to 12/13). This applies to all German-speaking countries. In Luxembourg, there is the specific feature that evolution—like other topics and subjects—is taught in German in the lower secondary level, but in French in the upper level. This language switch could possibly lead to difficulties of understanding evolution.

Generally, the following topics of evolutionary biology appear in a comparable way throughout the different curricula (and/or school textbooks) in the German-speaking countries:

Lower secondary school:

- Development of the universe and the earth
- Deep time concept
- Scientific methods in evolutionary biology
- Fossils and fossil formation
- Evidences for evolution
- Phylogenetic trees
- Phylogeny of organisms
- Hominid evolution

Higher secondary school:

- Evolution history (particularly Lamarck and Darwin)
- Natural selection
- Sexual selection
- Selective breeding
- Speciation mechanisms
- Creation myths, critical analysis

Table 13.2 Evolution as a topic in German-Speaking countries

Grade	5	6	7	8	9	10	11	12
Age	10–11	11–12	12–13	13–14	14–15	15–16	16–17	17–18
Austria			X					X
Germany	5	8	5	4	6	9	9	10
Luxembourg				X				X[b]
South Tyrol		X[a]	X[a]		X[a]	X[a]		
Switzerland			X[a]	X[a]	X[a]		X[a]	X[a]

Note School years where evolution is mentioned in the curriculum are marked with an "X". For Germany, the number of federal states is given. Note that in Germany, some schools have 13 years
[a]Adjacent grades where evolution can be taught alternatively
[b]Grade 13, in French

The time at which evolution is taught differs greatly between and within countries (Table 13.2). In addition, some current curricula no longer assign topics to specific grades. The material to be taught is not identical at all. In lower secondary level, the focus is usually more on delivering the fact of evolution (phylogeny, phylogenetic trees, fossils, evidence of evolution, historical aspects (see above), whereas in upper secondary level, evolutionary theory and evolutionary mechanisms are discussed (natural selection, species development, sexual selection). In some curricula, explicit reference is made to the subject of "creationism" or to creation myths, which are to be discussed critically from the viewpoint of science. Overall, there is a tendency to teach the subject in depth at the end rather than at the beginning of an education course (grades 12 or 13).

If students leave school with a degree after grade 12 (13) in one of the German-speaking countries, they should have at least basic knowledge about evolutionary mechanisms, the deep time concept, and the general outline of animal evolution, including hominids. Students that leave school early, after finishing the compulsory years only, had their last contact with basic evolutionary information at approximately 13 years of age. We doubt that they would be able to outline the basic process of evolution and to explain its causes correctly.

In the 21st century, German Biology textbooks have greatly improved, becoming more illustrated and interactive, and including recent scientific results. However, some factual errors in textbooks can still be found that potentially cause understanding problems in students. Such errors sometimes seem to be a consequence of teleological/finalistic interpretations, anthropomorphism, and the idea of a development from "lower" to "higher" stages (cf. introduction of this chapter). An Austrian Biology textbook made the laudable attempt to provide evolutionary information to grade 5 students by giving a simplified phylogenetic tree of vertebrates, including coloured homologous bones (Fig. 13.10). Regrettably, this tree is flawed in several ways; mammals and birds are presented as sister groups, possibly due to the intention to regard mammals as the "crown of creation". With such pictures, students' faulty preconceptions will not be changed, but possibly reinforced.

13.5.1 Biology Teacher Education

Although the Bologna process has been unifying standards of higher education in Europe since 1999, formal teacher education still varies between the German-speaking countries, and even among the Swiss cantons and the federal states of Germany (Ostinelli, 2009).

In Germany, there are six different types of teaching qualifications, from primary school to higher secondary school; in Switzerland and Austria at least three basic types, primary school, lower secondary school, and higher secondary school level. All German-speaking countries are currently reorganizing and unifying their teacher education curricula according to the Bologna system.

We do not know of any specific evolution courses for primary school teachers.

Education for secondary school biology teachers at university level depends on the specific curriculum at the particular universities. A few universities offer general introductory courses, such as "Evolution" or "Molecular Evolution" (University of Vienna). Much more common are courses with combined subjects, where the theory of evolution supposedly plays a minor role, such as "Anatomy, systematics, and evolution of plants and animals" (Justus-Liebig-University Giessen), "Diversity and evolution of eukaryotic organisms" (Ludwig-Maximilians-University Munich), and "Evolution, structure and function of plants" (Humboldt University Berlin). Such courses will focus on the phylogeny of the specific organisms, which is important, but we consider lectures exclusively on evolution theory as essential, too. Above all, there is a particular lack of specific courses for biology teacher students on how to address and to deal with creationism and anti-evolutionist arguments in the classroom. At the University of Vienna, the seminar "Evolution and Ethics" (held by the Director of the Botanical Garden, Michael Kiehn, and the first author of this chapter) offers the possibility to train these skills.

In all German-speaking countries, teacher students have to choose at least one additional subject. In Austria, it was possible to choose biology as an exclusive topic until 2000. Since then, instead of further deepening the biological knowledge of teacher students, recent university curricula have omitted parts of the original syllabus to give place to courses of the second subject.

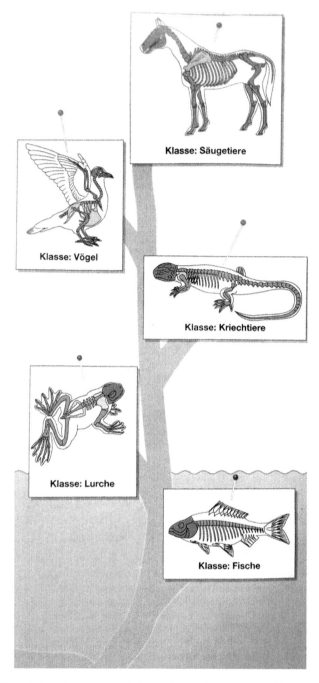

Fig. 13.10 Flawed simplification of a phylogenetic tree in an Austrian biology textbook. Bio Logisch 1, Edition Dorner, 4th Ed. 2007, p. 107

13.6 Suggestions for Improvement and Outlook

More evolutionary biology has to be taught at school. "But there are a lot of other subjects to cover, and if students are going to understand something about history, chemistry, literature, and geometry, then at least some of the details on evolutionary biology may have to wait until college" (Asher, 2012, p. 221). Exponentially growing knowledge in all scientific subjects (Fahrbach, 2011) and the economic need for mathematics, languages and computer skills increase the pressure both on students and on the secondary school curricula. Additional lessons in evolution will be difficult to squeeze in an already tight syllabus.

But evolution does not necessarily need to be treated as a separate chapter[4] at all. It is the "red thread" that permeates all biological processes and life science concepts. We therefore suggest "weaving in" evolutionary concepts into all biological chapters, and from the very beginning of school education. When teaching about domestic animals at primary school, mentioning the wild ancestors, the timeline of domestication, and the method of selective breeding will bring early insight into evolutionary mechanisms. When teaching human blood circulation or blood cells at secondary school, explaining the transformation of the circulatory system from fishes to amphibians to mammals, or discussing the advantages and disadvantages of the loss of the erythrocyte nucleus from amphibians to mammals, will provide deeper insight not only in the evolution of these structures, but also in their function. Good biology teachers already include these aspects in their lessons, but to become all-encompassing didactic knowledge, such concepts need to be incorporated in teacher education, official curricula, and school textbooks.

For the Leopoldina—German Academy of Science, the university education in the topic of evolution is generally unsatisfactory. The Academy highlights a great discrepancy between the importance of evolutionary biology and its communication at German schools and universities. One of the decisive causes for this is the abuse of supposedly evolutionary-biology concepts in National Socialism and the resulting decline in evolutionary biology during the decades after the Second World War at German universities. The academy calls for in-depth knowledge of evolutionary biology processes for those who wish to be active as biologists and also for future teachers of biology and sciences in all courses of education (Leopoldina, 2017). Specifically, the following suggestions are made:

- Development of a framework curriculum covering all grades, "delivering evolutionary biology for all types of schools and for all age groups", with the earliest possible introduction of the topic.
- Development of teaching materials which consider the current didactical and scientific knowledge of evolutionary biology. This requirement is supported by empirical findings, according to which the intention of German teachers to teach

---

[4]This refers to the understanding of evolutionary concepts and mechanisms, not e.g. to the history of Darwin's discoveries or specific topics like human evolution.

evolution, in addition to their attitudes to evolution, was particularly due to the existence of appropriate teaching materials.
- Increased use of extracurricular offerings in the field of life sciences (Science Labs, Natural History Museums) for the improvement of evolutionary biology education (Leopoldina, 2017).

The Evokids project group in Germany tries to ensure that the topic "evolution" should be included in the canon of compulsory topics in primary schools, just as in England or France. In a resolution the demand is justified: "Given the fundamental importance of understanding evolutionary sciences for the development of a modern worldview, it is disconcerting that children at primary school learn so little about this subject—especially where creation myths are treated in the classroom, a topic which can be easily misinterpreted without prior knowledge of evolution. Educationally, this is unjustifiable. Public schools should not unilaterally influence their pupils in terms of a particular religion or belief, but rather allow them access to the central findings of science! From a policy perspective, it is therefore imperative to treat the 'fact of evolution' in class much earlier and more comprehensively than is provided in current curricula".[5]

Specific teaching materials were developed by the Evokids project group (Graf & Schmidt-Salomon, 2016), also a children's book "Big family" (Schmidt-Salomon, 2015) and a children's film[6] of the same content. In the other German-speaking countries such initiatives do not yet exist.

Van Dijk and Kattmann (2008, 2009) suggest a general Biology Curriculum in schools, which is oriented to the perspective of natural history. The main objective of this approach is to improve the understanding of evolutionary processes. Therefore, the different biological topics should be enriched with a natural history perspective. The authors' hypothesis is that the integration of historical narratives in the teaching of biology enables teachers to intertwine the different biological topics and to create a substantial context for a better understanding of evolutionary processes (van Dijk & Kattmann, 2009). The authors have also published some concrete proposals for implementing in German (van Dijk & Kattmann, 2008).

13.7 Conclusion

At present, besides the creationist activities mentioned above, we do not see major political threats to evolution education in the German-speaking countries. However, the quality of evolution education can and should be continuously improved. Both scientific and pedagogical knowledge are crucial to cope with the difficulties of understanding evolutionary concepts. We urge both teachers and curricula to

---

[5]Evokids-Resolution: https://evokids.de/content/resolution-evolution-grundschule#Resolutionstext, translated into English by Dustin Eirdosh.

[6]English version: https://youtu.be/xK1nv9kEhOw.

address evolution in the context of all biological fields, starting with the very first school year. But long before that, the development of a naturalistic worldview starts with an appropriate reaction to the first "why?" of a child.

References

Allensbach Institute for Demoscopy (Ed.). (2009). *Weitläufig Verwandt: Die Meisten glauben inzwischen an einen gemeinsamen Vorfahren von Mensch und Affe*. Allensbach: Allensbacher Berichte.
Asher, R. J. (2012). *Evolution and belief: Confessions of a religious paleontologist*. Cambridge University Press.
Baalmann, W., Frerichs, V., Weitzel, H., Gropengießer, H., & Kattmann, U. (2004). Schülervorstellungen zu Prozessen der Anpassung - Ergebnisse einer Interviewstudie im Rahmen der Didaktischen Rekonstruktion. *Zeitschrift für Didaktik der Naturwissenschaften, 10*, 7–28, 272 K.
Baum, D. A., Smith, S. D., & Donovan, S. S. (2005). The tree-thinking challenge. *Science, 310* (5750), 979–980.
Bayertz, K. (1998). Darwinismus als Politik. Zur Genese des Sozialdarwinismus in Deutschland 1860–1900. In E. Aescht, G. Aubrecht, E. Kraußе, & F. Speta (Eds.), *Welträtsel und Lebenswunder. Ernst Haeckel – Werk, Wirkung und Folgen* (Vol. 56, pp. 229–288). Stapfia.
Bergman, J. (1986). *The influence of evolution on nazi race programs*. Retrieved January 20, 2017, from CSSHS Quarterly Journal: http://www.creationism.org/csshs/v08n3p24.htm.
Bishop, B. A., & Anderson, C. W. (1986). *Evolution by natural selection: A teaching module* (Occational Paper No. 91). East Lansing: Michigan State University.
Bölsche, W. (1905). *Der Stammbaum der Tiere*. Stuttgart: Kosmos, Franckh'sche Verlagshandlung.
Brasseur, A. (2009). Gefahren des Kreationismus für die Bildung. Die Sicht der parlamentarischen Versammlung des Europarats. In O. Kraus (Ed.), *Evolutionstheorie und Kreationismus—ein Gegensatz* (pp. 119–126). Stuttgart: Franz Steiner Verlag.
Darwin, C. (1859). *On the origin of species by means of natural selection*. London: John Murray.
Der Spiegel. (2009). Erzbischof zu Köln: Meisner vergleicht Biologen Dawkins mit Nazis. Retrieved January 20, 2017, from *Spiegel Online*: http://www.spiegel.de/panorama/erzbischof-zu-koeln-meisner-vergleicht-biologen-dawkins-mit-nazis-a-658589.html.
Der Standard. (2005). Papst spricht vom "intelligenten Plan" des Kosmos. Retrieved January 20, 2017, from *Der Standard*: http://derstandard.at/2238621/Papst-spricht-vom-intelligenten-Plan-des-Kosmos.
Ditfurth, H. v. (1976). *Der Geist fiel nicht vom Himmel. Die Evolution unseres Bewußtseins*. Hamburg: Hoffmann & Campe.
Eder, E. (2017). Heil? – Heil! Alternativmedizin, Esoterik und Rechtsextremismus. In: *Gesundheit - Heilung - Heil. Religiös-spirituelle Implikationen medizinischer und alternativmedizinischer Handlungslogiken*. Akademie für Ethik in der Medizin, Universität Wien, 27.-28.1.2017.
Eder, E., Turic, K., Milasowszky, N., van Adzin, K., & Hergovich, A. (2011). The relationships between paranormal belief, creationism, intelligent design and evolution at secondary schools in Vienna (Austria). *Science & Education, 20*(5–6), 517–534.
Fahrbach, L. (2011). How the growth of science ends theory change. *Synthese, 180*(2), 139–155.
Fenner, A. (2013). *Schülervorstellungen zur Evolutionstheorie - Konzeption und Evaluation von Unterricht zur Anpassung durch Selektion*. Dissertation: Gießen.
Föger, B., & Taschwer, K. (2001). *Die andere Seite des Spiegels. Konrad Lorenz und der Nationalsozialismus*. Czernin: Vienna.
Fowid. (2007). *Evolution und Kreationismus*. Retrieved January 08, 2017, from Forschungsgruppe Weltanschauungen in Deutschland: https://fowid.de/meldung/evolution-und-kreationismus.

GfK Austria. (2009). *Einstellungen der ÖsterreicherInnen zur Evolution*. Retrieved January 08, 2015, from http://www.oeaw.ac.at/shared/news/2009/pdf/pk_presseunterlagen_web.pdf.

Graf, D., & Lammers, C. (2011). Evolution und Kreationismus in Europa. In D. Graf (Ed.), *Evolutionstheorie - Akzeptanz und Vermittlung im europäischen Vergleich*. Berlin: Springer.

Graf, D., & Hamdorf, E. (2012). Evolution Verbreitete Fehlvorstellungen zu einem zentralen Thema. In D. Dreesmann, D. Graf, & C. Witte (Eds.), *Evolutionsbiologie. Moderne Themen für den Unterricht*. Heidelberg: Spektrum Akad. Verl.

Graf, D., & Schmidt-Salomon, M. (Eds.). (2016). *Evolution in der Grundschule - Materialien für den Unterricht*. Gießen: Oberwesel.

Graf, D., & Soran, H. (2011). Einstellung und Wissen von Lehramtsstudierenden zur Evolution – ein Vergleich zwischen Deutschland und der Türkei. In D. Graf (Ed.), *Evolutionstheorie - Akzeptanz und Vermittlung im europäischen Vergleich* (pp. 141–161). Berlin: Springer.

Gregory, T. R. (2008). Understanding evolutionary trees. *Evolution: Education and Outreach, 1* (2), 121.

Großschedl, J., Konnemann, C., & Basel, N. (2014). Pre-service biology teachers' acceptance of evolutionary theory and their preference for its teaching. *Evolution: Education and Outreach, 7* (1), 182.

Haeckel, E. (1863). *Über die Entwicklungstheorie Darwins*. Amtlicher Bericht über die 37. Versammlung der deutschen Naturforscher und Ärzte zu Stettin, am 19. September 1862: 17.

Haeckel, E. (1874). *Anthropogenie oder Entwicklungsgeschichte des Menschen. Gemeinverständliche Vorträge über die Grundzüge der menschlichen Keimes-und Stammes-Geschichte*. Lepzig: Wilhelm Engelmann.

Haeckel, E. (1898). *Der Monismus als Band zwischen Religion und Wissenschaft: Glaubensbekenntniss eines Naturforschers*, vorgetragen am 9. October 1892 in Altenburg beim 75jährigen Jubiläum der Naturforschenden Gesellschaft des Osterlandes. Verlag von Emil Strauss.

Hammann, M., & Asshoff, R. (2015). *Schülervorstellungen im Biologieunterricht: Ursachen für Lernschwierigkeiten* (2nd Ed.). Seelze: Klett Kallmeyer.

Höneisen, R. (2003). Gott hat die Hand im Spiel. *Factum*, (3), 24–17. Retrieved January 08, 2017, from http://www.progenesis.ch/diverses/umfrage/Umfrage_factum.pdf.

Höneisen, R. (2007). Schweizer wollen Schöpfungslehre in der Schule. *Factum, 6*, 38–39.

Ichner, B. (2017). Austrotürken: "Die Evolution war Gottes Wille". *Kurier*, January 28, 2017. Online: https://kurier.at/chronik/wien/die-evolution-war-gottes-wille/243.544.826.

Junker, R., & Scherer, S. (2013). *Evolution: Ein kritisches Lehrbuch* (7th Ed.). Gießen: Weyel.

Kattmann, U. (2013). Glaube an die Evolution? Darwins Theorie im Spiegel der Alltagsvorstellungen von Schülern, Lehrern und Wissenschaftlern. *Evolutionstheorie und Schöpfungsglaube: neue Perspektiven der Debatte* (pp. 201–227). Göttingen: V & R Unipress, Vienna UnivPress.

Kattmann, U. (2015). *Schüler besser verstehen: Alltagsvorstellungen im Biologieunterricht*. Hallbergmoos: Aulis Verlag.

Keckstein, R. (1980). Die Geschichte des Biologieunterrichts in Deutschland. *Biologica didactica, 3*(4), 80–99.

Kotthaus, J. (2003). *Propheten des Aberglaubens: Der deutsche Kreationismus zwischen Mystizismus und Pseudowissenschaft*. Forum Religionskritik: Bd. 4. Münster: Lit.

Kutschera, U. (2003). Designer scientific literature. *Nature, 423*(6936), 116.

Kutschera, U. (2007). *Streitpunkt Evolution: Darwinismus und intelligentes Design* (2., aktualisierte und erw. Aufl.). Naturwissenschaft und Glaube: Bd. 2. Berlin, Münster: Lit.

Kutschera, U. (2014). Germany. In S. Blancke, H. H. Hjermitslev, P. Kjærgaard, & R. L. Numbers (Eds.), *Creationism in Europe* (pp. 105–124). Baltimore: Johns Hopkins University Press.

Lammert, N. (2012). *Akzeptanz, Vorstellungen und Wissen von Schülerinnen und Schülern der Sekundarstufe I zu Evolution und Wissenschaft*. Dortmund: Dissertation.

Leopoldina - German National Academy of Sciences (Ed.) (2017). Teaching evolutionary biology at schools and universities. Halle (Saale). Available at http://www.leopoldina.org/uploads/tx_leopublication/2017_Stellungnahme_Evolution_ENG.pdf.

Lorenz, K. (1973). *Die Rückseite des Spiegels. Versuch einer Naturgeschichte menschlichen Erkennens.* München: Piper.

Meir, E., Perry, J., Herron, J. C., & Kingsolver, J. (2007). College students' misconceptions about evolutionary trees. *The American Biology Teacher, 69*(7), e71–e76.

Morkramer, M. (2010). *Der Lippstädter Fall - Hermann Müller und der Kampf um die Lippstädter Schule.* In Ostendörfler e. V. (Eds.), Hermann Müller - Lippstadt (1829–1883) - Naturforscher und Pädagoge. Ransdorf: Basilisken Presse.

Miller, J. D., Scott, E. C., & Okamoto, S. (2006). Public acceptance of evolution. *Science, 313,* 765.

Nummert, D. (2001). Schreiben für die Wahrheit. Ernst Krause alias Carus Sterne (1839-1903). *Berliner Monatsschrift, 2,* 60–65.

ORF. (2009). *Schönborn distanziert sich vom Kreationismus.* Austrian Broadcasting Company, March 5, 2009. Available at http://religionv1.orf.at/projekt03/news/0903/ne090305_schoenborn_fr.htm.

Ostinelli, G. (2009). Teacher education in Italy, Germany, England, Sweden and Finland. *European Journal of Education, 44*(2), 291–308.

Peschel, O. (1860). Eine neue Lehre über die Schöpfungsgeschichte der organischen Welt. *Das Ausland, 5,* 97–101, 6, 135–140.

Riedl, R. (1976). *Die Strategie der Genesis. Naturgeschichte der realen Welt.* München: Piper.

Schmidt-Salomon, M. (2015). *Big Family: Die phantastische Reise in die Vergangenheit* (1. Aufl.). Newel.

Schönborn, C. (2005) *Finding design in nature.* The New York Times. Available at http://www.nytimes.com/2005/07/07/opinion/07schonborn.html?_r=1.

Seidl, V. (2017). *Phylogenetische Stammbäume im Biologieunterricht: Erhebung von Schülervorstellungen als Ausgangspunkt für die Entwicklung von Unterrichtskonzepten.* Master thesis, University of Vienna.

Skoog, G. (2005). The coverage of human evolution in high school biology textbooks in the 20th century and in current state science standards. *Science & Education, 14,* 395–422.

Stover, S. K., & Mabry, M. L. (2007). Influences of teleological and Lamarckian thinking on student understanding of natural selection. *Bioscene: Journal of College Biology Teaching, 33* (1), 11–18.

van Dijk, E. M., & Kattmann, U. (2009). Teaching evolution with historical narratives. *Evolution: Education and Outreach, 2*(3), 479–489.

van Dijk, E. M., & Kattmann, U. (2008). Biologieunterricht in naturgeschichtlicher Perspektive: Zur Reform auf der Sekundarstufe I. *Der Mathematische und Naturwissenschaftliche Unterricht, 61*(2), 107–114.

vom Stein, A. (2017). *Creatio: Lehrbuch zur Schöpfungslehre* (3. Aufl.). Lychen: Daniel-Verl.

Wittwer, S., Bachmann, B., & Kohl, D. (2007). Schöpfung und Evolution - Entstehung des Lebens. In Kommission für Lehrplan- und Lehrmittelfragen der Erziehungsdirektion des Kantons Bern (Ed.), *NaturWert Pflanzen – Tiere – Mensch. Hinweise für Lehrerinnen und Lehrer.* Bern: Schulverlag Plus.

Wodak, R., Khosravinik, M., & Mral, B. (2013). *Right-wing populism in Europe. Politics and discourse.* London: Bloomsbury.

Yahya, H. (2014). *Der Evolutions-Schwindel. Der wissenschaftliche Zusammenbruch der Evolutionstheorie und ihr ideologischer Hintergrund* (The evolution deceit, German translation) (2nd Ed.). Global Publishing: Istanbul.

Erich Eder is an Assistant Professor of Biology at Sigmund Freud University Vienna, Medical School (SFU MED), where he is in charge of the Science Lab and teaches undergraduate medical students. At the University of Vienna, he holds seminars for teacher students (evolution and ethics, didactics). Originally devoted to temporary pools, freshwater crustaceans and Conservation Biology, Erich's recent research interests include students' beliefs in alternative medicine, paranormal claims, and their attitudes towards evolution. publicationslist.org/erich.eder

Victoria Seidl is currently a high school Biology and French teacher in Vienna, Austria. For her Master of Science thesis she studied students' perception of phylogenetic trees and the according evolutionary processes. The students' perspectives were inquired by conducting interviews using partly standardised qualitative guidelines, followed by a qualitative content analysis of those interviews.

Josh Lange is Director of the Advanced Language Institute and lectures in English for Medicine at Sigmund Freud University Vienna, Medical School. He holds a Doctor of Education in International Education from Exeter University and has taught courses at UCL, Reading and Vienna universities in English and Critical Thinking from undergraduate to PhD level. His research interests include team approaches to teaching, social entrepreneurship, and the digitalization of education across disciplines.

Dittmar Graf holds the chair for Biology Didactics at the Justus Liebig University of Giessen, Germany. He instructs teacher students in all areas of Biology didactics. His research interests lie in didactics of evolution, concept learning in biology and the demarcation problem between science and pseudoscience, especially in the field of pseudo medicine. He is a member of the working group for Evolutionary Biology of the German Academy of Science, Leopoldina.

Part IV
Middle East

Chapter 14
An Insight into Evolution Education in Turkey

Ebru Z. Muğaloğlu

Abstract Appreciation of science is one of the major aims of science education. In the context of the theory of evolution, it refers to acceptance of the theory or recognizing its value. The topic of evolution was introduced into school textbooks in Turkey as early as the 1930s. However, anti-evolution movements in the country have historically been widespread and strong. In the 2000s, public acceptance of evolution in Turkey was very low (Miller, Scott, & Okamoto, 2006; Peker, Comert, & Kence, 2010). This chapter elaborates on the reasons for the devaluation of the theory of evolution in the educational context in Turkey. It presents the state of evolution in educational contexts described in Turkish studies on evolution education published in journals from 2000 to 2015. National and international databases were searched using the key word "evolution" in both Turkish and in English. The search yielded 30 publications. Those published in national journals focused mostly on preservice teachers' understanding and acceptance of evolution and their attitudes to and pedagogical content knowledge about teaching evolution. In international journals, the papers generally reflected cultural perspectives, including anti-evolution movements, and factors that affect acceptance and understanding of evolution theory. Drawing on these studies, this chapter presents insights into the teaching of evolution in Turkey's educational contexts.

14.1 Appreciation of the Evolution Theory

Public acceptance of evolution theory in Turkey was shown to be lower than 25% in the early 2000s (Miller et al., 2006). Peker, Comert, and Kence (2010) also showed that the acceptance of evolution as a scientific theory in Turkey was low,

---

E. Z. Muğaloğlu (✉)
Department of Mathematics and Science Education, Boğaziçi University, Istanbul, Turkey
e-mail: akturkeb@boun.edu.tr

E. Z. Muğaloğlu
Visiting Researcher, Copenhagen University, Copenhagen, Denmark

not only by the general public at large but also among university students. Considering the essence of evolution as an explanatory theory in biology, these studies pointed to a problem, if not a crisis, in evolution education in Turkey.

Darwin proposed natural selection as the mechanism for evolution in 1859, since which time a growing body of supporting evidence explaining the origin of species has been accumulated from fossil records and research in comparative anatomy, geographical distribution, artificial selection, and genetics (Mugaloglu, 2014). Gee, Howlett, and Campbell (2009) cite 15 articles in *Nature* that lay out evidence from fossil records, habitats, and molecular processes. Today no reputable biologist denies the theory. Dobzhansky (1973) maintains that without the theory of evolution, many aspects of biology make no sense. Unfortunately, many people in Turkey do not appreciate this central of evolutionary theory in biology.

Taking the sound body of scientific evidence into account, one would expect a high level of public acceptance of evolution in any scientifically literate society, where individuals should be able to distinguish science from non-science and rely on the former when making decisions and taking actions. In such a society, one would also expect individuals to appreciate the value of evolutionary theory in explaining the origin of species. "Appreciation" here refers to recognizing the value of evolution theory (Mugaloglu & Erduran, 2012) in making decisions, even if people hold simultaneous alternative religious beliefs such as creationism or intelligent design. The phrase "appreciation of the theory of evolution" is not equivalent to acceptance of evolution theory. For instance, among those who do accept evolution as a scientific theory, there might be some who also accept intelligent design as a valid scientific theory. Mugaloglu and Erduran (2012) find that Turkish preservice teachers accept evolutionary theory, but they equate it with intelligent design and argue that students have the right to learn all relevant "theories" in a science class. Zemplen (2009, p. 43) points out the dilemma in teaching evolution: "Today one of the gravest problems in the public appreciation of scientific issues is that certain organizations pretend to claim scientific expertise where they are only spreading ideologically motivated messages". Such messages are common, not only in social contexts in Turkey but also in formal educational settings, making evolution education in Turkey an interesting case to examine.

This chapter aims to present a comprehensive picture of appreciation of evolution in formal education contexts in Turkey and analyzes the content of 30 published articles on evolution education in the country. The first section describes the cultural, sociopolitical and educational context, focusing on the issue of evolution theory. The second section describes the framework of the content analysis of the selected studies. In the final section, acceptance, understanding and attitudes towards teaching and learning theory of evolution are examined and discussed in relation to the appreciation of evolution theory.

14.2 The Cultural, Sociopolitical and Educational Context in Turkey

The influence of Turkey's cultural, sociopolitical and educational contexts with respect to public acceptance of evolution are intertwined, and these have naturally influenced evolution education in schools and public acceptance of evolution in Turkey. In their often-cited survey, Miller, Scott, and Okamoto (2006) found that Turkey has the lowest public acceptance of evolutionary theory among 34 countries included in the study and Turkey is the only country debating the issue of secular versus theocratic government. Underpinnings of this ongoing dichotomy lie mainly in the cultural, political and educational contexts of the late Ottoman Empire and the early Republic of Turkey.

The Republic of Turkey was established by Ataturk in 1923 as a secular parliamentary republic after the collapse of the Ottoman Empire. Ataturk and his colleagues were educated in schools that were considered modern at the time. The schools were mostly military and medical schools, whose curriculum included secular topics unlike the traditional educational institutions called madrasa, which taught the traditions of Islam. Many students of the modern schools were influenced by the philosophy of the Enlightenment and positivism. Ataturk himself was secular in his thinking and highly influenced by positivist and Enlightenment philosophies, as reflected in his famous saying: "The truest guide in the world is science. To seek guidance in other things is heedlessness, ignorance, and deviation from the right path".

When the Republic of Turkey was founded, the literacy rate was extremely low, at around 10% (Boran, 2000). Speaking language was Turkish. Most of the population lived in rural areas, and Islamic conservatism was widespread. Secular worldviews were not common, except in the narrow circles of the well-educated soldiers, doctors, journalists and bureaucrats. The first reforms in the earliest years of the republic were aimed at secularizing society and the education system, starting with the abolishment of the madrasa (Berber, 2013). Compulsory religious classes including Qur'an and Arabic classes also came to an end. In July 1924, right after these reforms, John Dewey was invited to Turkey to examine the education system. He stressed the general principles of education in a democratic society, the development of individual enterprise and capability. The Ministry of Education incorporated Dewey's ideas into a new curriculum between 1925 and 1929. In the 1940s, still taking into account the overwhelming rural population and Dewey's recommendations, village institutes were established to train teachers in rural areas. These institutes played a significant role in providing a secular education to the rural population. In 1950, however, when the conservative Democrat party came to power, the village institutes were abolished (Vexliard & Aytac, 1964, p. 45).

Textbooks were also reformed in the early years of the Republic. One important example was history textbooks, which had previously included religious topics such as the story of creation. Ataturk was highly inspired by H. G. Wells' *Esquisse de l'histoire universelle* (Toprak, 2012). In the light of the book, all religious issues in history textbooks, including the creation story were replaced with the theory of evolution (Toprak, 2012). In the early 1930s, all elementary and secondary school textbooks in Turkey featured lessons on Darwin's evolutionary theory (Toprak, 2012). Considering the widespread religiosity in the society, it is quite likely that such reforms were not welcomed by all segments of the society. In the 1940s, just a few years after Ataturk's death in 1938, Darwin's evolution theory was removed from textbooks (Toprak, 2012).

The theory of evolution was embraced by the scientific community in Turkey. The Scientific and Technical Research Council of Turkey (TUBITAK), for instance, established in 1963, published a number of works on evolution and Darwin, including articles in its official magazine for children. After the new conservative party came to power in the early 2000s, anti-evolution positions in the scientific community started to appear. For instance, in 2009, a cover story about Darwin in a popular science magazine published by TUBITAK was immediately censored (Nature, 2009).

In the realm of formal education, it is important to note that education in Turkey is centralized under the surveillance of the Ministry of Education. The Ministry formulates the national curriculum, policies, and the content of the curriculum. An important point to note is that the curriculum has been subject to numerous changes in recent years. Major curriculum reforms were implemented in 1999, 2005, and 2013. In January 2017, the draft of a new curriculum was disclosed to public for comment. The most notable change with regard to evolution in elementary and secondary education is that it includes neither the word "evolution" nor "Darwin", a striking departure from the previous curriculum. To justify the change, the head of the education ministry's curriculum board, Durmus, stated that students were too young to understand controversial subjects (Girit, 2017). Regarding this, Yigit, a board member of the secular education union Egitim-Sen, said "The curriculum change in its entirety is taking the education system away from scientific reasoning and changing it into a dogmatic religious system" (Tuysuz, 2017).

14.3 Content Analysis of the Evolution Education Studies

The main criteria for an article to be included in the content analysis were a focus on biological evolution education in the Turkish educational context and a publication date between 2000 and 2015. The search was conducted using national databases, namely *Ulakbilim* and *Dergipark*, and the databases of ten international journals.[1] Articles both in Turkish and English were searched using the key word "evolution", (*evrim* in Turkish). After eliminating unrelated papers, 30 publications

---

[1] *International Journal of Science Education, Journal of Research in Science Teaching, Journal of Science Education and Technology, Research in Science Education, Research in Science & Technological Education, Science Education, Studies in Science Education, Journal of Biology Education, Science & Education* and *Cultural Studies of Science Education.*

remained, seven of which were in Turkish and the rest were in English. The articles were then categorized into five main groups:

1. Overview of the social and political context of Turkey
2. Science curriculum in Turkey with respect to evolution
3. Acceptance and understanding of evolution
4. Interest, attitudes and views on evolution
5. Teaching evolution.

14.4 Studies Focused on Overview of the Social and Political Context of Turkey

Edis (2009) examines the uneasy relationship between modern science and conservative Islam. He states that evolution theory did not penetrate into Muslim lands. The Qur'an states clearly that the world and all its creatures were created by Allah, for which reason the vast majority of Muslims maintain a creationist perception of life. Lack of divine purpose and intervention in evolution is seen as an obstacle to acceptance of evolution, so the theory of evolution generated controversy in Turkey, where traditional Islamic practices are common. After the secularization of education in the 1920s, the topic of evolution was included in the textbooks. However, underground religious movements rejected Darwinian thinking. In the 1980s, religious instruction in Sunni Islam was made mandatory in schools. The Ministry of Education translated American books on creationism for high school biology textbooks such as *Scientific Creationism* (Edis, 2009). In the 1990s, creationism flourished in Turkey as a result of efforts of Science and Research Foundation (Bilim ve Arastirma Vakfi) led by Harun Yahya. Harun Yahya (2006), in his *Atlas of Creation*, argued that life forms on Earth had never evolved. Harun Yahya's work had global reach as well. His book was translated into many languages and distributed to libraries, universities and researchers all around the world, free of charge.

Focusing on the widespread presence of religious ideologies, both Yalçınoğlu (2009) and Çetinkaya (2006) elaborate on how anti-evolution ideas evolved from creationism and intelligent design in Turkey. They discuss the impact of anti-evolutionist movements on educational policies and the public acceptance of evolutionary theory. Yalçınoğlu (2009) claims that both the USA and Turkey have been influenced by heavy propaganda targeting the theory of evolution. She argues that political propaganda is the main obstacle to the unifying nature of the evolution theory in explaining origins of species. In Turkey, other obstacles for the teaching of evolution in the classroom include curriculum, the understanding of science and the personal beliefs of school administrations, teachers, and students. Overcoming such obstacles is not an easy task and requires challenging efforts by scientists and science educators in Turkey.

In a study comparing the education contexts of Turkey and the United States, Titrek and Cobern (2011) focus on valuing science. They point out that Turkish students embrace both science and religion, and that public opinion and belief are more oriented toward an American-style religious-modernism compatibility path. With regard to the theory of evolution, they ask two important unanswered questions: Will Turkey's budding anti-evolution movements grow, or stall? Will evolution become as broadly accepted as in Europe, or not?

14.5 Studies Focused on the Science Curriculum with Respect to Evolution

Formal education is one of the most common ways people learn about evolution. Of 415 preservice teachers surveyed by Akyol, Tekkaya, Sungur, and Traynor (2012), 81.1% stated that their main source of information about evolution was science class at school. Yalçınoğlu (2009) investigated the place of evolution in Turkey's middle school science curriculum and the secondary biology curriculum after the reform in 2005. One striking result was that, in the middle school curriculum, there was not a single unit or even section of a unit with an "evolution" heading. Moreover, although topics such as fossils, heredity and adaptation were featured, none of the suggested activities or learning outcomes referred to evolution. The term evolution appears only a few times. In the biology curriculum, the term evolution appeared in the title "Biological Variation, Genetics, and Evolution". Yet the content was inadequate. Yalçınoğlu (2009) concludes that coverage of the theory of evolution in the 2005 curriculum was quite unsatisfactory. Comparing the then-current curriculum with its 1997 predecessor, she found that the 1997 curriculum had more learning outcomes related to the theory of evolution.

Bakanay and Durmus (2013), in their examination of the 2007 biology curriculum vis-à-vis the scientific education standards of the US National Academy of Sciences (1998), determined that the learning outcomes in the Turkish curriculum did not meet the US science education standards in evolution education. For one thing, evolution was presented as a view rather than a theory. Creationism was also presented as a view, at the same level of importance as evolution. Learning outcomes for important concepts such as natural selection, adaptation, and variation were inadequate and unclear. Bakanay and Durmus argue that the program is "not complete and consistent by means of evolutionary biology" (p. 92).

Tekkaya and Kılıç (2012) emphasize the curriculum as an important part of teacher training. Preservice biology teachers' knowledge of the curriculum was found to be insufficient. The preservice teachers indicated two main reasons for their lack of knowledge: first, the curriculum had changed frequently, and second, their training did not include information about the details of the most recent curriculum. Köse (2010) investigated biology teachers' (n = 38) and 11th graders' (n = 250) views about the place of evolution and creationism in the biology curriculum. They found that 84.2% of the teachers and 58% of the students agreed that both evolution and creationism should be presented. Over 80% of the students and teachers claimed that creationism should be part of the curriculum. It was clear that, for the biology curriculum, creationism was considered more essential than evolution. Moreover, the way evolution appears in the curriculum is important for biology teachers' acceptance of it as a valid scientific theory. Kılıç (2012) observed that while all German teachers stated that evolution was part of the curriculum and that there was no reason not to teach it. For their counterparts in Turkey, however, frequent changes in the curriculum led to challenges to teaching the topic, and not all the teachers were teaching it.

14.6 Studies Focused on Acceptance and Understanding of Evolution

Among the 30 studies reviewed, 12 dealt with acceptance and/or understanding of evolution as a variable. In general, the samples for these studies were selected from undergraduates in teacher training programs in fields such as science, biology, physics, math, and social science. With regard to acceptance of evolution, in six of nine studies, versions of the Measure of Acceptance of the Theory of Evolution (MATE) were used to assess acceptance level. In the MATE, participants are asked to indicate their views on items such as "Evolution is not a scientifically valid theory". In the Turkish version, some of the items were adapted. The word Bible was changed to Qur'an, for instance. Most studies in this category indicated low acceptance of evolution (Apaydin & Sürmeli, 2009; Peker et al., 2010; Taşkın, 2013) or undecided position about whether evolution occured (Irez & Bakanay, 2011; Kozalak & Ates, 2014; Peker et al., 2010). For instance, Köse (2010) found that only 21.1% of biology teachers and 26.8% of students in his study accepted the theory of evolution.

Deniz et al. (2008) examined factors that affect acceptance of evolution among preservice biology teachers in Turkey. The understanding of evolution was measured by an adapted version of a scale developed by Rutledge and Warden (2000). Thinking dispositions were measured using the actively open-minded thinking (AOT) scale (Sa, West, & Stanovich, 1999). Acceptance of evolution was measured by the MATE. A regression analysis showed that in terms of preservice teachers' understanding of evolution, their thinking dispositions and the education level of the parents altogether explained 10.5% of the variance in acceptance.

The understanding of evolution refers to conceptual comprehension of content related to evolution, such as adaptation and natural selection. Some studies reveal misconceptions about evolution. For instance, Keskin and Köse's (2015) investigation of 117 preservice biology teachers' misconceptions about adaptation and natural selection reveals that 55 had misconceptions such as "Natural selection is seen in living organisms who want to adapt to life" and "Natural selection always chooses the best for organisms." Both statements contradict scientific explanation. Evolution by natural selection is not a result of a purposeful action. It is an outcome if a population of beetles, for instance, has variation, differential reproduction and heredity. Apaydin and Sürmeli (2006) too examine the knowledge levels of preservice science teachers and biology majors about natural selection, adaptation and mutation. Both biology majors and preservice science teachers in their sample had insufficient knowledge about natural selection, adaptation and mutation.

In another study, Erdoğan et al. (2014) investigated 162 preservice biology and science teachers' understanding on genetics. A survey consisting of 27 multiple choice questions was used to measure six factors, namely the nature of genetic material, transmission, expression of gene, gene regulation, evolution and genetics and society. They found that participants did not have a sufficient understanding of genetics either.

Peker et al. (2010) examined the understanding of evolution among Turkish undergraduates; 1,098 undergraduates completed 12 multiple-choice questions. The questions were selected from the existing evolution test (Johnson 1985 in Rutledge & Warden, 2000). For instance, "The evolutionary theory proposed by Charles Darwin was: (A) Change in populations through time as a result of mutations. (B) The spontaneous generation of new organisms. (C) The passing on of genes from one generation to the next. (D) Change in populations through time as a response to environmental change. (E) The development of characteristics by organisms in response to a need". The findings indicated that, in general, the understanding of evolution was quite low. For instance, the mean score of 326 senior biology students was 5.54 out of 12.

Some of the studies reviewed explored the relationship between acceptance and understanding (Annaç & Bahçekapılı, 2012; Deniz et al., 2008; Peker et al., 2010), acceptance and parents' education level (Annaç & Bahçekapılı, 2012; Deniz et al., 2008; Peker et al., 2010), acceptance and ideological position (Peker et al., 2010), and acceptance and religiosity (Annaç & Bahçekapili, 2012). Deniz et al. (2008), for instance, examined factors that affect acceptance of evolution among preservice biology teachers. Similar to Peker et al. (2010), they found a significant positive correlation between acceptance and understanding of evolution and between acceptance and parents' education level. In the Annaç and Bahçekapılı (2012) study, understanding did not correlate with acceptance, attitude, or rationality. Their findings indicate the one major factor that negatively correlates with acceptance of evolution is religiosity, while positive attitudes towards science and rationality correlate positively with acceptance of evolution. To promote acceptance of evolutionary theory, Annaç and Bahçekapılı (2012) underscores the importance of critical thinking. They also state that since religiousness seems to be a hindrance to accepting evolution, discussion of creationist ideas should not be avoided in the classroom.

Akyol et al. (2012) proposed a path model of relationships among understanding and acceptance of evolution, views on nature of science, and self-efficacy beliefs with regard to teaching evolution. The sample consisted of 415 preservice science teachers, who completed a series of self-report. They used the questions adapted by

Deniz et al. (2008) to measure understanding of evolution. The test was generally difficult for the examinees in that out of 21 items, the top score was 18 and the median sum score was only 8. The study also emphasized the significance of nature of science (NOS) views for understanding evolution. Moreover, understanding of evolution was crucial for acceptance and self-efficacy beliefs. The path analysis suggested that sophisticated views of NOS and higher levels of understanding of evolution were associated with higher levels of acceptance of evolutionary theory.

Briefly, the analysis of the 12 articles on acceptance and understanding of evolution showed, first of all, a low acceptance level of evolution—approximately 20–30% among education faculty undergraduates and high school students, a finding that is consistent with that of Miller et al. (2006). Among the factors affecting a low level of acceptance of evolution, religiosity and ideological views are highlighted. Other factors include understanding evolution, NOS views and parents' education level. The results also indicate that one's acceptance level of evolution is unlikely to change at university level. Second, the understanding of evolution is also low among undergraduates, who exhibit misconceptions related to basic concepts of evolution and related topics. Not all studies supported that there was a correlation between understanding and acceptance.

14.7 Studies Focused on Interest, Attitudes and Views of Evolution

This section describes the studies that focused on interest in evolution, attitudes toward evolution, and views about evolution. Akyol et al. (2012) investigated 415 preservice science teachers' interest in evolution. They found that 68.7% had little or no interest in evolution. Similarly, Çakmakcı et al. (2012) investigated self-generated questions of primary school students to reveal their interest in science. The results showed that although biology was the most popular scientific discipline for students (32.8%), only 11 students out of 739 (0.6%) showed interest in the topic of evolution. Both studies confirmed that teachers and students were not very interested in evolution.

In some studies, evolution was used to elicit NOS views of preservice teachers. For instance, Irez (2006) explored 15 preservice teachers' NOS beliefs through two sets of interviews. In the interviews, participants were asked to state their ideas about atomic theory and evolution with regard to the tentativeness of a scientific theory. He found that participants gave different values to different theories. For instance, one participant clearly stated that atomic theory was more reliable and valid than evolutionary theory. Irez pointed out that almost all participants were skeptical about the validity of evolutionary theory due to lack of direct evidence for evolutionary theory and participants' naïve NOS views about inferential nature of scientific knowledge. Irez contemplated that inadequate NOS views might have contributed to the lower acceptance levels of evolutionary theory among their participants.

Religious views are often considered an important variable in developing an attitude towards evolution. Taşkın (2014) focuses on the Islamization of science education and argues against the idea that Islam and science can co-exist. He highlights the polarization between religious and secular scientists in their explanations of the real world. According to many Muslim scientists, the Qur'an is the only source of absolute truth about the real world. In examining how 38 biology teachers and 250 students in Turkey accommodate evolution in their religious beliefs, Köse (2010) found that most of them agreed that evolution "explains the diversity and the similarity of life, not how life first arose," and that evolution is an idea with limited evidence and support. Participants stated that evolution and creationism are mutually incompatible. 69.7% of the students disagreed that they can accept the validity of the theory of evolution and believe in the God at the same time. Köse (2010, p. 197) argued that many students and teachers believed that "if they accept theistic creation, they must reject evolutionary theory".

14.8 Studies Focused on Teaching Evolution

Three studies discussed the attitudes toward teaching evolution among preservice teachers (Akyol, Tekkaya, Sungur, & Traynor, 2012; Kahyaoğlu, 2013; Köse, 2010). Kahyaoğlu (2013) administered a questionnaire to determine 236 preservice primary and science teachers' attitudes toward teaching evolution. He found that that teacher candidates generally had a negative attitude towards teaching evolution. Köse (2010) examined the attitudes of biology teachers (N = 38) and students (N = 250) toward teaching evolution; 58% of the teachers and 60% of the students indicated that teaching evolution was affected by teachers' attitudes and that it was not important to teach it. In contrast, in Akyol et al. (2012)'s study, 68.2% of 415 preservice teachers stated that teaching evolution was as important as other science topics. The difference in the findings might be attributable to the fact that samples were selected from different universities.

Mugaloglu (2014) investigated 48 preservice science teachers' intention to teach evolutionary theory and intelligent design. Two participants preferred to teach only intelligent design, while 19 indicated they would teach both. So, of the 48 preservice science teachers 21 of them had the intention to teach intelligent design in their science classes. Explaining their positions, the preservice teachers who wanted to teach only intelligent design argued that evolutionary theory is similar to intelligent design with respect to having religious background. They argued that evolutionary theory can also be describe as a kind of religious view espoused by atheism and humanism. According to these participants, therefore, having a religious connection was not a good reason to exclude intelligent design from science classes. Mugaloglu (2014) attributes the position of the participants on this matter to the lack of appreciation for the scientific status of evolutionary theory and the intrusion of pseudoscience into science classrooms.

Another issue relevant to evolution education is teachers' pedagogical content knowledge (PCK) about evolution. PCK refers to "... the ways of representing and formulating the subject that make it comprehensible to others" (Shulman, 1986, p. 9). Even if teachers have a positive attitude and intention to teach evolutionary theory, their lack of PCK might be an obstacle to teaching it effectively. In two studies (Bektas, 2015; Tekkaya & Kılıç, 2012), preservice teachers' PCK of evolution was investigated. Bektas (2015) examined seven preservice science teachers' PCK on the topic of reproduction, growth and evolution. Participants were asked to write down possible student misconceptions, the source of the misconceptions, and methods and strategies that could overcome these misconceptions. Regarding the topic of *Reproduction, growth, and evolution*, preservice teachers stated ten misconceptions. The most common misconception stated by the preservice teachers was that growth and evolution are the same. To overcome the misconception, five of them suggested teacher-centered instruction such as lecturing. Bektaş concluded that preservice teachers were inadequate in stating misconceptions.

Tekkaya and Kılıç (2012) interviewed seven senior biology teacher candidates' PCK about evolution. Unlike the participants in the Bektaş (2015) study, participants had a fair knowledge about potential student difficulties and they had a general idea about instructional methods and assessment. The difference between the findings of the two studies may be due to a difference in the samples. The participants in the Tekkaya and Kılıç (2012) study were biology majors, had completed courses on evolution and attended a higher-ranked university. Tekkaya and Kılıç (2012) also found that participants' knowledge about evolution and its place in the curriculum were insufficient. Participants stated that they had had difficulties as a student when they were learning about evolution and that their students might face similar challenges. They had concerns about teaching evolution in terms of dealing with student prejudice and misconceptions, lack of effective teaching, the influence of the media, the prejudice of parents, conflicts with students' religious beliefs, social pressure from school administration, insufficient knowledge (inability to answer student questions), and the approach of the Ministry of Education to the teaching of evolutionary theory. Despite these challenges, these seven participants had a positive attitude toward teaching evolution and believed in its significance in biology education.

In four of the reviewed studies, an activity for teaching evolution developed by the National Academy of Sciences (NAS) in the USA (1998) was used in preservice science teacher training (Ozdem, Ertepinar, Cakiroglu, & Erduran, 2013; Ozgelen et al., 2013a, 2013b; Özgelen & Yilmaz-Tuzun, 2011). In the activity, as part of their teacher training, participants were asked to hypothesize the morphological tree that shows the relationship between gorillas, chimpanzees, and humans. In three of the studies, the activity was part of a program that aimed to improve preservice teachers' understanding of the theory-laden nature of scientific knowledge (Ozgelen et al., 2013a, 2013b; Özgelen & Yilmaz-Tuzun, 2011). Ozdem, Ertepinar,

Cakiroglu, and Erduran (2013) conducted the activity to support preservice teachers' argumentation. The studies found that the aim of the program was achieved.

Self-efficacy beliefs are defined as "beliefs in one's capabilities to organize and execute the courses of action required to produce given attainments" (Bandura, 1997, p. 3). In one of the reviewed studies, Akyol et al. (2012) measured 415 preservice teachers' self-efficacy beliefs for teaching evolutionary to explore the relationship between understanding, acceptance, and NOS views. The participants completed the Personal Science Teaching Efficacy Beliefs (PSTE) inventory with regard to evolution developed by Enochs and Riggs (1990) and adapted and translated into Turkish by Tekkaya et al. (2004). After a path analysis, the model illustrated that, while acceptance and understanding had a positive effect, NOS views had a negative effect on self-efficacy beliefs about teaching evolution. Given the complexity of NOS studies, this issue needs further research.

14.9 Conclusions and Discussion

The review of studies on this chapter aimed to describe the current status of evolution education in Turkey, the challenges facing teaching evolutionary theory, and what to expect in the future in terms of acceptance of evolutionary theory in Turkey. The case of Turkey is interesting because not only has anti-evolution propaganda been widespread throughout the country, but acceptance of evolution has also been associated with opposing constitutional secularism by those who hold a conservative non-secular view of Islam. With this in mind, the present chapter hoped to shed light on challenges of evolution education in Turkey, which may not be present in countries with a well-established secular democracy. It contributes to the international literature on evolution education by revealing the challenges experienced in countries where secularism is a controversial issue.

The analysis of 30 studies revealed the problems surrounding evolution education in Turkey. One of the most problematic issues was the sociopolitical context, which is not conducive to teaching and learning about evolutionary theory. In Turkey, public propaganda against evolutionary theory constitutes the major challenge for its teaching. Such propaganda is often well financed as seen in free dissemination of Atlas of Creation, and easily supported by some religious groups historically opposing constitutional secularism in Turkey. Above all, the education system is highly centralized in Turkey under the Ministry of Education, and education policies are highly influenced by political orientations and preferences. The studies reviewed in this chapter highlight the fact that conservative Islamic practices and beliefs are widespread in the society, and that this is reflected in education policies through the political system such as the intrusion of creationist beliefs into science classes. On the other hand, there are teachers and students who hold a more moderate view of Islam that is neither anti-secularist nor anti-evolutionist. Clément (2015) also indicates the major influence of sociocultural context on evolution in 18 countries. As he states, the challenge is to help those who are "at the same time evolutionist and creationist".

To overcome this challenge, evolution should be included in the curriculum. If evolution is not included in the curriculum, teaching evolution in science classes will be extremely difficult if not impossible, noting that the main source of information about evolution is school education. This implies that if students do not have an opportunity to learn about evolutionary theory at school, their views about the origins of species will be shaped mainly by their surroundings. In the presence of strong anti-evolution propaganda, this would be a significant obstacle to evolution education. The reviewed articles on curriculum further illustrate that the topic of evolution is inadequately covered and that learning outcomes are poor, or in some cases, the topic is not covered at all.

Yet even if the curriculum includes the topic of evolution, there seem to be problems with teachers not teaching it properly. An important issue is teachers' poor appreciation of evolution in explaining the origin of species. Many teachers refer to evolution as a view rather than a scientific theory, equating evolutionary theory with unscientific explanations such as creationism and intelligent design. The reviewed studies indicate that, in addition to poor appreciation, teachers' acceptance of evolution is low, but even those who accept it are reluctant to teach it. Reasons for their reluctance include self-efficacy beliefs, inadequate understanding, and PCK of evolution. These findings are parallel to those of Glaze and Goldston (2015), who state that many teachers lack the knowledge, understanding and confidence that are necessary to teach evolution correctly. They also indicate that teachers allow their religious and other beliefs to influence what they accept and what they teach. Glaze and Goldston (2015) therefore suggest providing science teachers with better training and tools to help them cope with internal and external conflict.

Among the factors that influence acceptance of evolutionary theory, the reviewed studies cite understanding, parents' education, ideological views, and openness to belief change. However, these factors explain a small fraction of variance in acceptance, a point highlighted by Glaze and Goldston (2015) for the US case suggesting that other factors not yet identified explain another 90% of variance. Further studies on factors that can potentially influence acceptance of evolutionary would therefore be useful in helping to change the public's perception of evolutionary theory. Glaze and Goldston (2015) also recommend further research into the complex relationships between religion and acceptance of evolution.

Despite court decisions in favor of teaching evolution in the US, due to reasons such as religiosity and pressure from parents, churches and school boards, it is either omitted or watered down (Glaze & Goldston, 2015). Such social pressure may have more negative consequences for evolution education in countries where religiosity is deeper and where there is no court decision such as the Kitzmiller versus Dover case favoring evolution teaching. The resistance to evolution education in schools for Turkey seems to be growing. The last draft of a new curriculum announced to the public in early 2017 has not a single word about evolution, so students will not have an opportunity to learn about evolution at school. This is a very unfortunate situation, if not a crisis, for evolution education in Turkey. Considering that evolution was introduced in the curriculum as early as the 1930s, the new science curriculum would represent a huge step backward for evolution education after almost a century. However, the religious and sociopolitical climate that is influencing the anti-evolution today is worse than the climate in the 1930s. As of the date of the writing this chapter, approximately 10,000 formal petitions from the public have been submitted to the Ministry of Education demanding that evolutionary theory be included in the curriculum (Kolcu, 2017). The history of evolution education starting from the early 1930s up to the present illustrates interactions between scientific knowledge, values and social practices and the Didactic Transposition Delay framework described by Quessada and Clément (2007), who lay out socio-political reasons and epistemological obstacles such as creationism to explain the delay of ideas about evolution in the education system. Hopefully, the Didactic Transposition Delay in the case of Turkey will be short and evolutionary theory will take its place in the curriculum again so that students in Turkey will have an opportunity to learn about evolutionary theory properly. Scientific literacy is not an easy task to achieve without good evolution education in biology. A better education policy in favor of evolutionary theory would be one that improves philosophy curriculum and introduces an epistemological approach to evolution education (Aroua, Coquide, & Abbes, 2009). NOS and appreciation of science studies (Mugaloglu, 2014) may help teachers and students to recognize the value of evolutionary theory and to rely on scientific theories in making decisions and taking action in science classes.

References

Akyol, G., Tekkaya, C., Sungur, S., & Traynor, A. (2012). Modeling the interrelationships among pre-service science teachers' understanding and acceptance of evolution, their views on nature of science and self-efficacy beliefs regarding teaching evolution. *Journal of Science Teacher Education, 23*(8), 937–957.
Annaç, E., & Bahçekapılı, H. (2012). Understanding and acceptance of evolutionary theory among Turkish university students. *Doğuş Üniversitesi Dergisi, 13*(1), 1–11.
Apaydin, Z., & Sürmeli, H. (2006). Üniversite öğrencilerinin doğal seçilim, adaptasyon, ve mutasyon ile ilgili görüşleri. *Ondokuz Mayıs Üniversitesi Eğitim Fakültesi Dergisi, 22*, 31–46.
Apaydin, Z., & Sürmeli, H. (2009). Undergraduate students' attitudes towards the theory of evolution. *Elementary Education Online, 8*(3), 820–842.
Aroua, S., Coquide, M., & Abbes, S. (2009). Overcoming the effect of the socio-cultural context: Impact of teaching evolution in Tunisia. *Evolution: Education and Outreach, 2*(3), 474–478.
Bakanay, Ç. D., & Durmuş, Z. Ö. (2013). Lise biyoloji öğretim programında evrim eğitiminin kapsamı ve içeriğinin değerlendirilmesi. *Trakya Üniversitesi Eğitim Fakültesi Dergisi, 3*(2), 92–103.
Bandura, A. (1997). *Self-efficacy: The exercise of control*. Macmillan.
Bektas, O. (2015). Pre-service science teachers' pedagogical content knowledge in the physics, chemistry, and biology topics. *European Journal of Physics Education, 6*(2), 41–53.

Berber, S. G. (2013). Kâzim Karabekir'in saltanat ve hilâfetin ilgasina olan tavrının Düşündürdükleri [Thinking over Kazim Karabekir's attitudes about elimination of the caliphate and the sovereignty]. *International Journal of Social Science, 6*(4), 443–460.
Boran, A. (2000). Cumhuriyet Döneminde Eğitimde Meydana Gelen Gelişmeler. *Erciyes Üniversitesi Sosyal Bilimler Enstitüsü Dergisi, 1*(9), 303–312.
Cakmakci, G., Sevindik, H., Pektas, M., Uysal, A., Kole, F., & Kavak, G. (2012). Investigating Turkish primary school students' interest in science by using their self-generated questions. *Research in Science Education, 42*(3), 469–489.
Çetinkaya, H. (2006). Evrim bilim ve eğitim üzerine. *Ege Eğitim Dergisi, 7*(1), 1–21.
Clément, P. (2015). Muslim teachers' conceptions of evolution in several countries. *Public Understanding of Science, 24*(4), 400–421.
Deniz, H., Donnelly, L. A., & Yilmaz, I. (2008). Exploring the factors related to acceptance of evolutionary theory among Turkish preservice biology teachers: Toward a more informative conceptual ecology for biological evolution. *Journal of Research in Science Teaching, 45*(4), 420–443.
Dobzhansky, T. (1973). Nothing in biology makes sense except in the light of evolution. *American Biology Teacher, 35,* 125–129.
Edis, T. (2009). Modern science and conservative Islam: An uneasy relationship. *Science & Education, 18*(6–7), 885–903.
Enochs, L. G., & Riggs, I. M. (1990). Further development of an elementary science teaching efficacy belief instrument: A preservice elementary scale. *School Science and Mathematics, 90*(8), 694–706.
Erdoğan, A., Özsevgec, L. C., & Özsevgec, T. (2014). A study on the genetic literacy levels of prospective teachers. *Necatibey Faculty of Education Electronic Journal of Science & Mathematics Education, 8*(2), 19–37.
Gee, H., Howlett, R., & Campbell, P. (2009). 15 evolutionary gems. *Nature.* https://doi.org/10.1038/nature07740.
Girit, S. (2017). Passions flare as Turkey excludes evolution from textbooks. Retrieved June 24, 2017, from http://www.bbc.com/news/world-europe-40384471.
Glaze, A. L., & Goldston, M. J. (2015). US science teaching and learning of evolution: A critical review of the literature 2000–2014. *Science Education, 99*(3), 500–518.
Irez, S. (2006). Are we prepared? An assessment of preservice science teacher educators' beliefs about nature of science. *Science Education, 90*(6), 1113–1143.
Irez, S., & Bakanay, Ç. D. Ö. (2011). An assessment into pre-service biology teachers' approaches to the theory of evolution and nature of science. *Education and Science, 36*(162), 39–54.
Johnson, R. L. (1985). The acceptance of evolutionary theory by biology majors in colleges of the west north central states. Doctoral dissertation, University of Northern Colorado, Greeley.
Kahyaoğlu, M. (2013). The teacher candidates' attitudes towards teaching of evolution theory. *Necatibey Faculty of Education Electronic Journal of Science & Mathematics Education, 7*(1), 83–96.
Keskin, B., & Köse, E. Ö. (2015). Understanding adaptation and natural selection: Common misconceptions. *International Journal of Academic Research in Education, 1*(2), 53–63. https://doi.org/10.17985/ijare.53146.
Kılıç, D. S. (2012). Biyoloji öğretmen adaylarının evrim öğretimi niyetleri. *Hacettepe Üniversitesi Eğitim Fakültesi Dergisi, 42,* 250–261.
Kolcu G. (2017). Müfredat taslağına 184.042 öneri geldi: 1. Atatürkçülük. Hürriyet. Retrieved February 13, 2017, from http://www.hurriyet.com.tr/mufredat-taslagina-184-042-oneri-geldi-1-ataturkculuk-40363611.
Köse, E. Ö. (2010). Biology students' and teachers' religious beliefs and attitudes towards theory of evolution. *Hacettepe Üniversitesi Eğitim Fakültesi Dergisi, 38,* 189–200.
Kozalak, G., & Ateş, A. (2014). Üniversite fen bilimleri birinci sınıf öğrencilerinin evrim teorisini kabul düzeyleri. *Asian Journal of Instruction, 2*(1), 135–148.
Miller, J. D., Scott, E. C., & Okamoto, S. (2006). Public acceptance of evolution. *Science, 313,* 765–766.

Mugaloglu, E. Z. (2014). The problem of pseudoscience in science education and implications of constructivist pedagogy. *Science & Education, 23*(4), 829–842.
Mugaloglu, E. Z., & Erduran, S. (2012). Prospective science teachers' appreciation of science: the case of evolution vs. intelligent design. In C. Bruguie`re, A. Tiberghien, & P. Clément (Eds.), *E-Book Proceedings of the ESERA 2011 Conference: Science learning and Citizenship. Part 5* (L. Maurines & A. Redfors), (pp. 100–105) Lyon, France: European Science Education Research Association. ISBN: 978-9963-700-44-8.
National Academy of Sciences. (1998). *Teaching about evolution and the nature of science*. Washington, DC: National Academy Press.
Nature (2009), https://doi.org/10.1038/news.2009. Retrieved March 18, 2009, from http://www.nature.com/news/2009/090310/full/news.2009.150.html.
Ozdem, Y., Ertepinar, H., Cakiroglu, J., & Erduran, S. (2013). The nature of pre-service science teachers' argumentation in inquiry-oriented laboratory context. *International Journal of Science Education, 35*(15), 2559–2586.
Ozgelen, S., Hanuscin, D. L., & Yılmaz-Tuzun, O. (2013a). Preservice elementary science teachers' connections among aspects of NOS: Toward a consistent, overarching framework. *Journal of Science Teacher Education, 24*(5), 907–927.
Özgelen, S., & Yılmaz-Tuzun, O. (2011). The structure of subjectivity with theory-laden of scientific knowledge: The "evolution theories" activity and its results. *Mustafa Kemal Üniversitesi Sosyal Bilimler Enstitüsü Dergisi, 8*(16), 535–550.
Ozgelen, S., Yilmaz-Tuzun, O., & Hanuscin, D. L. (2013b). Exploring the development of preservice science teachers' views on the nature of science in inquiry-based laboratory instruction. *Research in Science Education, 43*(4), 1551–1570.
Peker, D., Comert, G. G., & Kence, A. (2010). Three decades of anti-evolution campaign and its results: Turkish undergraduates' acceptance and understanding of the biological evolution theory. *Science & Education, 19*(6–8), 739–755.
Quessada, M. P., & Clément, P. (2007). An epistemological approach to French curricula on human origin during the 19th & 20th centuries. *Science & Education, 16*(9–10), 991–1006.
Rutledge, M. L., & Warden, M. A. (2000). Evolutionary theory, the nature of science & high school biology teachers: Critical relationships. *The American Biology Teacher, 62*(1), 23–31.
Sá, W. C., West, R. F., & Stanovich, K. E. (1999). The domain specificity and generality of belief bias: Searching for a generalizable critical thinking skill. *Journal of Educational Psychology, 91*(3), 497.
Shulman, L. S. (1986). Those who understand: Knowledge growth in teaching. *Educational Researcher, 15*, 4–14.
Taşkın, Ö. (2013). Pre-service science teachers' acceptance of biological evolution in Turkey. *Journal of Biological Education, 47*(4), 200–207.
Taşkın, Ö. (2014). An exploratory examination of Islamic values in science education: Islamization of science teaching and learning via constructivism. *Cultural Studies of Science Education, 9*(4), 855–875.
Tekkaya, C., & Kılıç, D. S. (2012). Biyoloji öğretmen adaylarının evrim öğretimine ilişkin pedagojik alan bilgileri. *Hacettepe Üniversitesi Eğitim Fakültesi Dergisi, 42*, 406–417.
Tekkaya, C., Cakiroglu, J., & Ozkan, O. (2004). Turkish pre-service science teachers' understanding of science, and their confidence in teaching science. *Journal of Education for Teaching, 30*, 57–66.
Titrek, O., & Cobern, W. W. (2011). Valuing acience: A Turkish-American comparison. *International Journal of Science Education, 33*(3), 401–421.
Toprak, Z. (2012). *Darwin'den Dersim'e cumhuriyet ve antropoloji*. Doğan Egmont Yayıncılık: Istanbul.
Tuysuz, G. (2017). Turkey to stop teaching evolution in high school. Retrieved June 23, 2017, from http://edition.cnn.com/2017/06/23/middleeast/turkey-to-stop-teaching-evolution/index.html.
Vexliard, A., & Aytac, K. (1964). The "village institutes" in Turkey. *Comparative Education Review, 8*(1), 41–47.

Yahya, H. (2006). Atlas of Creation. Global Publishing.
Yalçınoğlu, P. (2009). Impacts of anti-evolutionist movements on educational policies and practices in USA and Turkey. *Elementary Education Online, 8*(1), 254–267.
Zemplén, G. Á. (2009). Putting sociology first—reconsidering the role of the social in 'nature of science' education. *Science & Education, 18*(5), 525–559.

Ebru Z. Mugaloglu is an Assistant Professor of Science Education in Mathematics and Science Education Department at Boğazici University, Istanbul. She was a visiting fellow at University of Reading in 2006, at University of Bristol in 2010, at Indiana University in 2012 and at University of Copenhagen in 2016. She participated in science education projects funded by EU, TUBITAK and British Council. She mainly teaches method courses, teaching practice and field experience courses to preservice science teachers. Her research interest includes science teacher training especially focused on nature of science, socio-scientific issues and evidence based teaching.

Chapter 15
Evolution Education in Iran: Shattering Myths About Teaching Evolution in an Islamic State

Mahsa Kazempour and Aidin Amirshokoohi

Abstract This chapter will examine the teaching of evolution in the public education system of the Islamic Republic of Iran. The goal of this chapter is to examine the status of evolution education in the Iranian education system and address inaccurate presumptions that are seeded in the Western views of Iran as a theocratic state with dilapidated ideals and perspectives. Through examination of existing literature and previous reviews and analyses of Iran's science textbooks and nationally mandated curriculum content, this chapter will attempt to shed light on: (a) the views of nature of science projected in the science education standards, (b) the depiction and description of the evolutionary emergence of life and concepts such as natural selection, mutation, and adaptation in the K-12 science content, (c) the history of science and evolution education in Iran, and (d) possible factors that have contributed to Iran's relatively in-depth and accurate attention to evolution education when compared to neighboring countries in the region. There are areas pertaining to evolution education in Iran that remain unexplored and suitable for future research. Further inquiry is necessary into understanding the implementation of the Iranian evolution curriculum and the students, teachers, and general public's beliefs and attitude with respect to evolution.

15.1 Introduction

The opposition to evolution education by the Christian right groups and their continual attempts to omit evolution from school science curricula or include creationism in the science classroom has been an ongoing source of controversy and

---

M. Kazempour (✉)
Penn State University—Berks, Reading, PA, USA
e-mail: muk30@psu.edu

A. Amirshokoohi
DeSales University, Center Valley, PA, USA

debate in the United States for a number of decades (Alters & Alters, 2001; Pennock, 2002; Trani, 2004). In the context of this continuous tension in the U.S., a democracy, one may be led to believe that comparable, or even more serious, challenges to evolutionary education exist in countries where religious authority governs and that dogmatic religious indoctrination of science education in such theocratic states pose a serious threat to evolution education. One such place that may conjure up images of conservatism and theocratic authority is the Islamic Republic of Iran. The predominant perception of the Iranian government as a conservative authority leads to the dominant perception that science education, and, in particular, evolution education, may be undermined and weakened by faith-based texts and doctrines in Iran. Furthermore, until several years ago, studies focusing on evolution in the Muslim world concentrated on (1) countries such as Saudi Arabia where the state-controlled science curriculum is dominated by creationist ideas, and (2) the recent rise of "Muslim creationism" in public opinion and the educational systems of countries such as Turkey and Egypt (e.g. Hameed, 2008).

However, with respect to science and science education, including controversial topics such as evolution, Iran is the land of many surprises. For example, while STEM education advocates in Western countries, particularly the U.S., attempt tirelessly to increase the number of women in such fields, women account for half the scientific workforce in Iran (Ehsan, 2006). Furthermore, Iran has some of the most flexible and liberal laws on stem cell research in the world (Bouhassira, 2015; Raman, 2006; Stone, 2015) and its *Royan Institute* is actively involved in various forms of government funded stem cell research, with the exception of reproductive cloning of humans which is not permissible. In the face of international isolation, sanctions, an eight-year imposed war, and other challenges, Iran has become increasingly determined to flourish in science and technology and succeeded in building a "surprisingly robust scientific enterprise" consisting of numerous government funded initiatives (Stone, 2015, p. 1038). Science education, specifically evolution education, is no exception to the rule. Iran's official stance on science and evolution education has been described as on par with some of its Western counterparts (Burton, 2011).

Our goal in this chapter is to examine the status of evolution education in the Iranian education system and address inaccurate presumptions that are seeded in the Western views of Iran as a theocratic state with dilapidated ideals and perspectives. Through examination of existing literature and previous reviews and analyses of Iran's science textbooks and nationally mandated curriculum content, this chapter will attempt to shed light on: (a) the views of nature of science projected in the science education standards, (b) the depiction and description of the evolutionary emergence of life and concepts such as natural selection, mutation, and adaptation in the K-12 science content, (c) the history of science and evolution education in Iran, and (d) possible factors that have contributed to Iran's relatively in-depth and accurate attention to evolution education when compared to neighboring countries in the region.

15.2 Education in Iran

Iran has a population of more than 79 million making it the 17th most populous county in the world (Central Intelligence Agency, 2017). The median age in Iran is 30.1 years and 74.8% of the population live in urban areas (UNESCO Institute for Statistic, 2015). Iran is a diverse country consisting of numerous ethnic groups including Persians, Azeris, Kurds and Lurs speaking a wide variety of languages. The official language of the country is Persian (Farsi) and the official religion is Shia Islam with adherents comprising 90% of the total population, followed by Sunni Islam (8%) and Christianity, Judaism, and Zoroastrianism whose adherents make up 2% of the population (Central Intelligence Agency, 2017).

The literacy rate for adults 15 years and older is 87% with 91% literacy rate for males and 83% literacy rate for females (UNESCO Institute for Statistics, 2015). Compulsory education in Iran extends to eighth grade which marks the end of the three-year intermediary school (UNESCO Institute of Statistics, 2015). According to UNESCO Institute for Statistics reports (2015), there are 1.3 million pre-primary students (up to age 5), 7 million primary students (ages 6–11), 6.5 million intermediary and secondary students (ages 12–17), and more than 6.6 million post-secondary students (ages 18–22).

15.2.1 K-12 Education in Iran

Iran's public education system is highly centralized with mandatory use of standardized textbooks, curriculum, and annual cumulative exams developed and continually revised by the national Ministry of Education. The education ministry also oversees the tuition-based private schools which abide by the same regulations as public schools including the standardized curriculum and testing. Biological and physical science education as well as scientific-technological education are part of the ten domains that the Fundamental Reform Document of Education (FRDE) has identified as key components of the Iranian education system (2011).

According to the TIMSS report on the participating countries, science is part of the Iranian curriculum starting in first grade. Weekly instruction time for science in fourth and eighth grade make up 12 and 10% of allotted time respectively (Mullis, Martin, Goh, & Cotter, 2016). A number of goals in the FRDE (2011) focus on "developing and deepening of a culture of research and evaluation, creativity and innovation, theorization and documentation of national scientific and educational experiences in the country's general formal education system" (p. 27) Similarly, the science education standards, developed by the Experimental Sciences Division of the Ministry of Education, focus mainly on the significant influence of science and technology on the country's economy and infrastructure. To accomplish these goals, Iran's science education standards emphasize the importance of developing "practical skills, critical inquiry, and the fostering of scientific literacy" (p. 23).

Interestingly, the science education philosophy and general goals outlined by the Experimental Sciences Division includes no explicit mention of Islam and reaffirm the status of science as a separate and valid realm of knowledge that is indispensable for individual and societal interests. The science education standards and standards-based curricula place special emphasis on empirical evidence, such as the study of fossils in understanding evolutionary history, as opposed to reliance on scripture. Furthermore, geologists, evolutionary biologists, and other scientists are portrayed as the authoritative voices of scientific knowledge (Burton, 2011).

15.2.2 Post-secondary Education in Iran

There are two exit examinations, at the end of grades 6 and 9, to determine students' placement in either the academic, technical, or vocational paths during upper secondary level (World Education Services, 2017). To receive their high school diploma and enter all public and most private universities, students must take the national standardized university admission examination, *Konkur,* and achieve the required scores which vary for different universities and academic majors.

Public institutions of higher education are the only ones to offer teacher education training. Elementary and intermediary school teachers complete their two-year associate degrees at teacher education programs that are responsible for teacher training under the supervision of the Ministry of Education. Those interested in teaching natural sciences at the secondary level must pass the higher education entrance examination and complete a four-year bachelor's degree in one of the teacher training colleges and universities (Mullis et al., 2016). Science teacher candidates are required to complete specialized coursework in natural sciences, including those addressing evolution, along with courses in pedagogy and educational psychology.

15.3 Evolution in the K-12 Curriculum

Iran and several other predominantly Muslim countries such as Egypt, Indonesia, and Pakistan are signatory to the Inter Academy Panel (IAP, 2006) statement proclaiming that evolution is an 'evidence based fact' which has remained irrefutable by scientific evidence and advising that all students be provided opportunities to learn the process of scientific inquiry. Evolution is addressed relatively early and in a more formalized and comprehensive fashion in Iran relative to its neighbors including Turkey, Saudi Arabia, and Israel (Burton, 2011). In Iran, evolution is an integral part of the K-12 science curriculum and students are introduced to evolutionary concepts as early as fifth grade. Table 15.1 compares Iran with other Muslim dominated countries in the region and Israel, the only non-Muslim country

Table 15.1 Coverage of evolution in K-12 curricula

Country	Grade(s) evolution addressed
Iran	• 5th grade • 8th grade • High school biology
Pakistan	• 10th (mandatory biology) • 12th grade (medical school prep only)
Egypt	• 10th grade biology (required)
Syria	• 10th grade biology (required)
Turkey	• 8th grade biology (required) (before the curriculum change in 2017) • 9th and 12th grade (before the curriculum change in 2017)
Malaysia	• Medical school preparation courses only (Gr. 12–13)
Israel	• Middle school (since 2015) • 12th grade biology (optional)
Saudi Arabia	• Explicitly rejected

in the Middle East, with respect to when evolutionary concepts are addressed in the K-12 curriculum.

Table 15.2 summarizes the extent of the coverage of evolution in the Iranian K-12 science education curriculum (Asghar, Hameed, & Farahani, 2014; Burton, 2010, 2011) which has been identified as, at least, at par with the education in most countries in Europe, the Americas, and East Asia. Due to the standardized nature of the Iranian curriculum and assessment system, all students across the country are exposed to and are expected to master all listed content. Therefore, it could be argued that all Iranians receive instruction and are expected to possess content knowledge and skills related to the items enumerated in Table 15.2. In fifth grade (Tehrani et al., 2008) as they are completing their final year of primary school, students learn about the history of the Earth and the evolutionary emergence of life over millions of years with specific references to geological and fossil findings as indicated in the excerpts in the table. In eighth grade (Amani et al., 2008), during their final year in intermediate school, students review material related to the history of the Earth and are introduced to evidence for evolution of organisms, and adaptations in the context of mutations and natural selection. High school biology (Karam al-Dini et al., 2008), often only taken by students in the science career tracks, focuses extensively on evolution and populations genetics and dynamics. Specifics of Darwin's journeys, observations, experiences with the scientific community, and his impact on the field of evolutionary biology are highlighted. Different categories of evidence for evolution are discussed in detail. The concepts of evolutionary rates, punctuated equilibrium, and natural selection are introduced with reference and discussion of commonly cited examples such as the peppered moths and Darwin's finches. The near consensus among the scientific community regarding the significance of evolutionary theory in explaining the world's biodiversity is mentioned clearly in the biology textbook.

Table 15.2 Evolution in Iran's K-12 science curriculum and texts

Grade	Key evolutionary concepts, figures, and history	Science textbook excerpts
5th grade [final year of primary school]	• History of the earth (one chapter) – Changing of continents and seas • A short history of life (sub-section of the chapter) – Evolutionary emergence of life over millions of years – Transition to terrestrial life credited – Emergence of plants on dry land – Age of dinosaurs – Extinction of dinosaurs about 65 million years ago – Diversification of mammals	"Geologists, via studies of fossils, have arrived at the conclusion that life began in the sea." (p. 55) "Afterward, the water and air of planet Earth changed such that a suitable environment for the development of reptiles came to exist." (p. 56) "Geologists say in the beginning only one landmass and one giant ocean existed on earth. About 200 million years ago, this large landmass slowly began to divide." (p. 57)
8th grade [final year of compulsory education]	• Chapter on geology and evolution – review of the history of life clearly illustrated geologic time diagram – Evolution of organisms Evidence of evolution – adaptations in the context of mutations and natural selection Addresses the discrediting of lamarckian hypothesis by Weissman and DeVries' ideas of genetic mutation	"This (the *Archaeopteryx*) is the first bird on earth, which also has some reptile traits." (p. 32) "New traits arising by a mutation are mostly harmful and detrimental to life, [but] sometimes in a rare mutation useful traits also appear. An organism possessing one or more useful traits appears, finds greater compatibility with the environment compared to its conspecifics, and gradually the number of [organisms with those traits] increases in the environment." (p. 35)
High school biology and experimental sciences track in high school	• 40-page chapter on evolution and several chapters on population genetics, population dynamics, and biological communities – Darwin, his journey and influences – Evidence of evolution: paleontology, molecular and structural homology, embryology – Evolutionary rates and punctuated equilibrium (as Darwin's original conception of "gradual" evolution – Natural selection examples: peppered moths and work of Peter and Rosemary Grant on Darwin's finches – New evolutionary scientific research (including Peter and Rosemary Grant and Jonathan Losos)	"nearly all biologists today have accepted that Darwin's theory can explain the basis for the diversity of life on earth." (p. 75)

There are two distinct ways in which the Iranian science education standards with respect to evolution differ from some of the other predominantly Muslim countries, namely Saudi Arabia and Turkey. First, unlike the Saudi standards and textbooks, there is no attempt to include any reference to religious text to either support creationism or depict the Quran as a scientific text providing explanations for various phenomena in the natural world (Edis, 2007). All content is presented as evidence-based arguments with scientific examples and explanations. Similarly, the portrayal of the evolutionary theory being a Western idea propagated by Western scientists witnessed in Saudi textbooks is completely absent in the Iranian texts (Burton, 2011; Jafarzadeh, 2009). The Iranian texts, in their coverage and support for all the presented evolutionary ideas, continuously emphasize the significance of empirical evidence and the authority of the scientists, as opposed to religious texts and figures. For example, the eighth grade science textbook states that "one of the most important applications of fossils" (Amani, et al., 2008, p. 33) is as evidence for the occurrence of morphological modifications in evolutionary lineages.

The only topic that is not directly and explicitly addressed in the Iranian science education standards and textbooks is human evolution. Although human evolution within the bigger scheme of evolution is not explicitly addressed, that is not to say that implicit references to human evolution are similarly absent. On the contrary, the lines of evidence provided in the population genetics chapter, including the examples of "stabilizing selection upon newborn weight and the heterozygote advantage in relation to malaria and sickle-cell anemia", clearly support that natural selection does in fact operate on humans (Burton, 2010, p. 27). One explanation for the implicit, rather than the explicit, addressing of human evolution may be that this is a possible effort by the education ministry to reduce conceivable conflicts between content covered in science and religion courses (Burton, 2011).

15.4 Evolution in Informal Science Education

In addition to its coverage in the K-12 science curriculum, evolution is also a theme that is either directly or indirectly addressed as part of informal science education. School associated field trips as well as family planned visits to the various museums of natural history located in the capital, Tehran, and several other large cities are commonplace and advertised and promoted as educational and entertainment centers for students of all ages and adults. These museums possess numerous exhibits pertaining to biodiversity, fossils, and history of the Earth. School children can view and learn about fossils and geology during their visit to Tehran's museum of natural history' evolution exhibit. As Bohannon (2006) noted, the exhibit has numerous artifacts on display, including a trilobite with an accompanying sign that explains the discovery of this specimen near the Alborz mountains in northern Tehran dating back to 400 million years ago in the Devonian period. Bohannon (2006) describes, "Along the opposite wall, a diorama chronicles the evolution of life on earth. Painted scenes of ancient life look as if they've been copied directly from the latest biology textbooks" (p. 292).

In 2014, Jurassic Park, or the Moving Dinosaurs Park, was established as an educational, recreational, and entertainment center in western Tehran. The 30,000 m^2 park contains 28 moving dinosaur models that are designed to realistic dimensions and can blink their eyes, move their legs, and create realistic sounds. The models are accompanied with information about each dinosaur and its habitat and characteristics. The park also consists of educational workshops, Jurassic film studios with 3D film screenings, coffee shops, and the popular Jura Shop which children and families visit to take pictures with the dinosaurs. A quick search online, and using social media tools available in Iran, it is evident that the park is receiving increased attention from the public and becoming highly popular, especially among parents who share information about the park and recommend it to one another based on their own children's experiences.

15.5 Contributing Factors

Science and the Quran are not intrinsically at odds with one another (Bucaille, 1982); hence, it is essential to understand the factors that shape attitudes of Muslims toward evolution at the individual and societal level. The literature on "Islamic creationism" (Hameed, 2008) has only recently been awakened to the reality of the complexity of the Muslim world and the inability to generalize about Muslims and Muslim governments' stances toward evolution education. Only recently, and as consequence of new literature shedding light on the topic of evolution in Iran (Burton, 2010, 2011), have authors become increasingly cognizant of the unique nature of Iran and how it addresses science and evolution in particular (Asghar et al., 2014; Edis & BouJoude, 2014). The case of Iran, in particular, elucidates the critical need for authors focusing on evolutionary attitude and the apparent growth of creationism in the Muslim world to accentuate the role that state educational systems and educational policies play, as opposed to only focusing on the role of popular religion (Burton, 2011). Because of the tendency for evolution to be particularly controversial in educational contexts, examining educational policy regarding evolution will elucidate official positions on evolution and the alignment of such positions with prevalent evolutionary perspectives in society and the political culture of a country (Edis & BouJaoude, 2014).

15.5.1 Religion and Science

Studies focusing on the relationship between science and religion, particularly with respect to evolution, have primarily focused on Christianity and Western countries. However, within the past two decades, there has been a surge of studies exploring evolution within Islamic contexts. Some of the earlier studies suggested strong public opposition to evolution in Muslim dominated countries (Hameed, 2008; Miller, Scott, & Okamoto, 2006) and identified evolution as a recurring source of discord (Edis, 2007). Furthermore, they argued that the immense influence of "Islamic science" could potentially marginalize science education in these countries (Loo, 2001, p. 64), citing Turkey as an example of creationist groups sensationalizing evolution and attacking evolution education (Hameed, 2008). However, although there may be elements of truth to each of these claims, they are generalizations which do not necessarily hold true for all Muslims and Muslim dominated countries. Muslims' perspectives and practices do not necessarily equate with Islam's stance on such issues. Terms including "Islamic science," "Muslim creationism," and other similar terms are impossible to accurately and consensually define. Islam is not monolithic. The Muslim community is a diverse community with a spectrum of beliefs and interpretations (Asghar, Wiles, & Alter, 2007; Mansour, 2011). While for some Muslims, religious text and knowledge supersede scientific evidence, others view Islam and science as two distinct and equally important domains of knowledge which are not opposed to or incompatible with one another (Mansour, 2011). Similarly, some groups, such as the Wahhabi dominated Saudi government, are preoccupied with the literal and strict reading of the Quran and other religious texts, as evangelical Christian groups do. But, other Muslims, specifically the Shia sect of Islam, rely on a flexible and adaptive interpretation of text through logical and independent analysis of the texts (Saniei, 2013).

Shia jurisprudence, science, and evolution. Iran is a predominantly Shia country with Sunni Muslim, Zoroastrian, Christian, and Jewish minority groups. The extensive interest of the Iranian government and the society to advance science and technology is partially rooted, in the emphasis of Islam, particularly Shia Islam, on advancing one's knowledge of the world and incorporating scientific advances at individual and societal levels. In Iran, "science is not described as simply an outgrowth of Islam or subject to preconceived doctrines of any religion—rather it is affirmed as a separate valid field of knowledge, and one crucial to individual and social welfare" (Burton, 2011, p. 27). The Shia system of jurisprudence (*fiqh*), which is based on independent or original interpretation of the Islamic texts (*ijtihad*), by religious jurists, allows for a more dynamic and flexible interpretation of Islamic texts utilizing intellect and reasoning (*aql*). In this way, interpretation and reasoning allow for the adaptation of the religious texts to modern society and accommodating new scientific discoveries and technological advances. Shia scholars in Iran utilize ijtihad and scholarly consensus (ijma) to find solutions and religious-legal justification to tackle new challenges such as stem cell research (Sachedina, 2009; Saniei, 2012). Consequently, numerous influential Shi'ite scholars in Iran, including those who have held powerful positions, directly or indirectly impacting the legislation and execution of laws and cultural and educational guidelines, are not opposed to otherwise controversial topics including general evolutionary ideas (Burton, 2010).

Lack of public debate. Contrary to the U.S. where issues such as stem cell research and teaching of evolution in schools remain controversial and continually debated in one form or another, there is an absence of this type of public debate in Iran (Aramesh & Dabbagh, 2007; Saniei, 2013). There are a number of reasons for the lack of public debate and sociocultural barriers. It's worth noting that not only scientists and other professionals, but also the general public in Iran, are fully cognizant of the flexible and dynamic nature of jurisprudence in Iran with respect to incorporating scientific and technological advances into daily life and, therefore, typically accepting of what is generally approved by the jurists and scholars (Saniei, 2013). Furthermore, Iran is governed by a centralized Shia authority with legal and religious power resting in the hands of the Grand Ayatollah (the Supreme Leader). The state supervises and makes decisions pertaining to all facets of life based on religious texts and interpretations leading to minimal influence from the general public on such decisions and lack of ambiguity or public debate.

The case of human evolution. The lack of explicit mention of human evolution in the K-12 curriculum may be explained by the fact that, even within Shia Islam, there is no acceptance of the idea of human evolution from a common ancestor. Shia scholars do argue that there were a number of different human species but they argue that the existing human race has descended from Adam and Eve. In his famous work, "Islam and the Contemporary Man," Allamah Tabatabaei (2010), one of the highest ranking Shia scholars and prominent thinkers of philosophy in contemporary Shia Islam, explained that "Adam and Eve being the progenitors of the existing human race is an issue stated in the Qur'an in unequivocal terms and as such cannot be construed in any figurative way unless there be definitive proof to the contrary" (2015, p. 29). However, he stated that the fossil records and other evidence of human existence for millions of years, although not serving as proof that these previous humans belong to the same race, are not in conflict with Islamic ideas and principles. He argued that a number of human races have come into existence and then became extinct and replaced with another human race during Earth's different cycles. His argument is corroborated by some *hadith*s (sayings of the Prophet and important religious leaders within Prophet's progeny) which suggest that the existing human race represents the eighth human cycle on Earth.

15.5.2 Evolution in Iran: Historical, Political, and Socio-economic Contexts

When considering instrumental and determining factors that lead to educational policies that support or oppose science and evolution, theological matters account for only part of the equation. One must also consider the historical, social, political, and economic conditions that may be equally significant in influencing these policies (Burton, 2011). In the case of Iran, its historical context with respect to education, science, and evolution, as well as its current social, political, and economic circumstances and needs deserve particular attention.

Edis and BouJaoude (2014) asserted that the Muslim and Western debates over science and religion are in sharp contrast to one another due to the weaker role of the scientific institutions and the liberal religious communities that support them in Muslim countries. However, as described in the previous section, Iran defies this generalization with its flexible and liberal stance toward scientific research and development. Iran continues to view education, particularly in the STEM and health-related fields, as a critical infrastructure impacting the country's success and development. Iran's recent National Master Plan of Supreme Council (2011) reiterates the country's commitment and focus on "bolstering the promotion of science and technology in the Islamic world" as a means of not only achieving national triumph, but also reviving the "great Islamic civilization" (p. 6) of which Iran was a key component.

Iranian government's current goals of promoting scientific development, especially in health-related biological research, date back to not only the early days of the Islamic Revolution in 1979 (Godazgar, 2008), but far back into Iran's history over many centuries. Geographically located between East and West, Iran has been at the forefronts of scientific and medical advancements in that region. Iranian scientists including Avicenna, al-Biruni, and ibn-Hayyan were influential in shaping Western science, mathematics, medicine, and philosophy. Al-Biruni and Ibn Arabi even proposed arguments explaining the evolution of living organisms (Shanavas, 1999). Along with the rest of the Islamic world, Iran experienced a decline in its scientific development as Europe began their Age of Renaissance in the early 1300s.

Traditional models of education persisted in Iran until the mid-19th century when Western secular education, especially the French model of education, began to emerge as the country was experiencing its first wave of urbanization. The country, including its education system, experienced "massive-scale" secularization and Westernization during the Pahlavi dynasty which began in the 1920s and ended with the Islamic Revolution in 1979 (Godazgar, 2008). It was during this wave of educational secularization that the topic of evolution was introduced in the Iranian curriculum and biology textbooks.

Pursuing higher education degrees in the fields of science and engineering in Western countries began to be promoted during the reign of Reza Shah, the first Pahlavi king. Students completing their advanced study in the U.S. or European countries would often find positions as part of the higher administration and the Tehran University while rarely conducting research and creative scholarship. In the 1960s and 1970s, during the second Pahlavi monarch (Mohammad Reza Shah), a new wave of universities were founded including the University of Sharif and Shiraz University which concentrated on science and technology and were based on American style management. This was critical in educating the new scientific and technological elite who were educated mainly in the U.S. and one of the main contributing factors to increased scientific production during this period (Khosrokhavar, Ghaneirad, & Toloo, 2007).

The Islamic Revolution and the change in government led to a three-year shut down of institutions of higher education, and the K-12 textbooks and curriculum underwent scrutiny and alignment with the values of the Islamic Revolution. However, science textbooks were spared from the Islamization and revolutionization that many other subjects underwent. Due to the lack of opposition by Iranian Shia scholars and revolutionary leaders to evolutionary ideas, evolution remained rather untouched, except for the removal of Darwin's name from the textbooks in 1984 as a result of the anti-Western sentiments that grew after the revolution and at the beginning of the imposed war on Iran by its neighbor, Iraq (Godazgar, 2008).

The end of the eight-year war initiated an intensive push by Iran to improve its educational infrastructure, enhance higher education, and establish more private universities. This allowed for staggering student enrollment, launch of physics Ph. D. program at Sharif University, and the inception of research centers such as *Royan Institute for Stem Cell Biology and Technology* and *Royan Institute for Reproductive Biomedicine*, and overall increased focus on science missions with practical societal connections. It was during this period of rebuilding Iran's infrastructure and former President Khatami's call for dialogue and reconciliation that Darwin's name was added back to the textbooks which now include an amicable description of his journey and revolutionary ideas. Overall, Iran's challenges as a result of the war and the imposed sanctions, have further "prevented the politicization of scientific topics such as evolution to the extent seen in other countries, including Turkey, Israel, and the United States" (Burton, 2010).

15.6 Looking Ahead: Need for Research

The discussion thus far makes it clear that the highly centralized science textbooks and standards in Iran, regulated by the government, virtually guarantee that evolutionary content, as specified earlier in the chapter, is taught and students across the country are equally expected to learn and be assessed on such material. Furthermore, the traditional direct instructional approach to science instruction with teachers presenting information without any deviation from the text remains prevalent in Iranian classrooms today. Therefore, it is expected that the content in the texts are taught rather explicitly and in a prescriptive manner.

Yet, there are areas pertaining to evolution education in Iran that remain unexplored and suitable for future research. For example, studies have shown that American students possess minimal understanding of evolutionary concepts and they commonly cite their religious views as a reason for believing that evolution should not be taught in high school (Donnelly, Kazempour, & Amirshokoohi, 2008). Similarly, students' knowledge, beliefs, and attitude with respect to evolution have also been examined in numerous other countries around the world. However, evidence of Iranian students' understanding of evolution and their beliefs and attitude with respect to evolution have not been explored even though students' knowledge of evolutionary concepts is extensively tested in school and national college admission examination. Iranian students may hold misconceptions about evolutionary ideas or beliefs that are contradictory to the scientific consensus on evolution. Hence, there is a significant need for studies exploring students' understanding of evolutionary concepts as well as their beliefs and attitude toward evolution. Furthermore, little is known about Iranian teachers' beliefs about evolution and their pedagogical approach to teaching evolution. Prior studies have suggested that teachers' personal religious beliefs (PRB), defined as their "views, opinions, attitudes, and knowledge," inform their beliefs about both the nature of science (NOS) and the instructional practices with respect to evolution (Donnelly et al., 2008; Mansour, 2008, 2011). Therefore, it is critical to further explore and understand Iranian science teachers' PRB and possible ways in which their beliefs may influence their students' beliefs and attitude toward evolution. It is currently not apparent whether and to what extent religion may enter the scientific conversation with respect to evolution in the classroom. Moreover, there have been no formal studies examining how science, and evolution, in particular, is taught in Iranian classrooms. Hence, it is imperative to investigate evolution instruction in Iranian classrooms and explore the prevalence and nature of teachers' attempts to either overtly or covertly interject their religious beliefs or faith-based text as part of their instruction. Overall, the nature of classroom instruction as well as students' and the general public's beliefs and attitude with respect to the topic of evolution require further inquiry and closer examination.

References

Alters, B., & Alters, S. (2001). *Defending evolution in the classroom: A guide to the creation/evolution controversy*. Boston: Jones and Barlett Publishers.

Amani, M., Mahmudzadeh, G., Ershadi, N., Hosseini, A., Karam al-Dini, M., & Asbaghi, A. (2008). *'Olum-e tajrobi sevvom rāhnemāyi*. Tehran: Vezārat Amuzesh va Parvaresh.

Aramesh, K., & Dabbagh, S. (2007). An Islamic view to stem cell research and cloning: Iran's experience. *American Journal of Bioethics, 7*(2), 62–63.

Asghar, A., Hameed, S., & Farahani, N. K. (2014). Evolution in biology textbooks: A comparative analysis of 5 Muslim countries. *Journal of Religion & Education, 41*(1), 1–15.

Asghar, A., Wiles, J. R., & Alters, B. (2007). Canadian pre-service elementary teachers' conceptions of biological evolution and evolution education. *McGill Journal of Education, 42*, 189–209.

Bohannon, J. (2006). Science in Iran: Picking a path among the fatwas. *Science, 313*(5785), 292–293.

Bouhassira, E. (2015). *The SAGE encyclopedia of stem cell research*. Sage Publications.

Burton, E. (2010). Teaching evolution in Muslim states: Iran and Saudi Arabia compared. *Reports of the National Center for Science Education, 30*, 28–32.

Burton, E. (2011). Evolution and creationism in Middle Eastern education: A new perspective. *Evolution, 65*, 301–304.

Bucaille, M. (1982). *The Bible, the Quran, and science: The holy scriptures examined in the light of modern knowledge*. New York: Tahrike Tarsile Quran Inc.

Central Intelligence Agency. (2017). Iran. In *The world factbook*. Retrieved from https://www.cia.gov/library/publications/the-world-factbook/geos/ir.html.

Donnelly, L. A., Kazempour, M., & Amirshokoohi, A. (2008). High school students' perceptions of evolution instruction: Acceptance and evolution learning experiences. *Research in Science Education, 39,* 643–660.

Edis, T. (2007). *An Illusion of Harmony: Science and Religion in Islam.* Amherst: Prometheus Books.

Edis, T. & BouJaoude, S. (2014). Rejecting Materialism: Muslim responses to modern science In M. Matthews (Ed.), *The international handbook of research in history, philosophy and science teaching* (pp. 1663–1691). New York, NY: Springer.

Ehsan, M. (2006). Islam and science: An Islamist revolution. *Nature, 444*(7115), 22–25.

Godazgar, H. (2008). *The impact of religious factors on educational change in Iran: Islam in policy and Islam in practice.* Lewiston, NY: Edwin Mellen Press.

Hameed, S. (2008). Bracing for Islamic creationism. *Science, 322,* 1637–1638.

Jafarzadeh, S. (2009). Iran at forefront of stem cell research. *Washington Times.*

Karam al-Dini, M., Bahbudi, B., Nikonam, V., Alavi, E., & Al-Mohammad, S. (2008). *Zist shenāsi.* Tehran: Vezārat Amuzesh va Parvaresh.

Khosrokhavar, F., Ghaneirad, M. A & Toloo, G. (2007). Institutional problems of the emerging scientific community in Iran. *Science, Technology, and Society, 12*(2) 171–200.

Loo, S. (2001). Islam, science and science education: Conflict or concord? *Studies in Science Education, 36,* 45–77.

Mansour, N. (2008). Religious beliefs: A hidden variable in the performance of science teachers in the classroom. *European Educational Research Journal, 7*(4), 557–576.

Mansour, N. (2011). Science teachers' views of science and religion vs. the Islamic perspective: conflicting or compatible? *Science Education, 95*(2), 281–309.

Miller, J. D., Scott, E. C., & Okamoto, S. (2006). Science communication. Public acceptance of evolution. *Science, 313,* 765–766.

Mullis, I. V. S., Martin, M. O., Goh, S., & Cotter, K. (Eds.). (2016). *TIMSS 2015 encyclopedia: Education policy and curriculum in mathematics and science.* Retrieved from Boston College.

Pennock, R. (2002). Should Creationism be Taught in the Public Schools? *Science & Education, 11*(2), 111–133.

Raman, A. (2006). *Iran in the forefront when it comes to stem cell research.* Retrieved from http://www.cnn.com/2006/WORLD/meast/11/16/raman.iranstemcell/index.html?eref=rss_latest.

Sachedina, A. (2009). *Islamic biomedical ethics: Principles and application.* New York: Oxford University Press.

Saniei, M. (2012). Human embryonic stem cell research in Iran: The significance of the Islamic context. In M. C. Inhorn, & S. Tremayne (Eds.), *Islam and assisted reproductive technologies: Sunni and Shia perspectives* (pp. 194–219). New York: Berghahn Books.

Saniei, M. (2013). Human embryonic stem cell science and policy: The case of Iran. *Social Science & Medicine, 98,* 345–350.

Shanavas, T. (1999). Islam does not inhibit science. *National Center for Science Education (NCSE) Report, 19,* 36–37.

Stone, R. (2015). Unsanctioned science. *Science, 349*(6252), 1038–1043.

Tabatabaei, M. (2010). *Islam and the contemporary man.* ABWA Publishing. Retrieved from https://www.al-islam.org/printpdf/book/export/html/30561.

Trani, R. (2004). I won't teach evolution: It's against my religion. *The American Biology Teacher, 66*(6), 419–427.

Tehrani, M., Daneshfar, H., Shamim, M., Hosseini A., Pazashpur, M., Rastegar, T., ... Ershadi, N. (2008). *'Olum-e tajrobi panjom dabestān.* Tehran: Vezārat Amuzesh va Parvaresh.

UNESCO Institute for Statistics. (2015). *Education Systems.* UNESCO Institute for Statistics Data Centre. Retrieved from http://uis.unesco.org/en/country/ir?theme=education-and-literacy.

World Education Services. (2017). *Education in Iran.* Retrieved from http://wenr.wes.org/2017/02/education-in-iran.

Mahsa Kazempour is an Associate Professor of Science Education at the Penn State University, Berks campus in Reading, Pennsylvania. She teaches a number of undergraduate courses including science methods for elementary education students and environmental science course for non-science majors. Her research aims to explore prospective elementary teachers' beliefs, attitudes, and self-efficacy with regard to science and science teaching, and changes in these constructs and participants' teaching practices as a result of enrolling a science methods course or participating in professional development sessions. Her other line of research focuses on undergraduate students' level of environmental literacy and sense of social responsibility with respect to the environment and changes in these constructs as a result of enrolling in a service-learning focused environmental science course. She is also at the forefront of the campus sustainability efforts.

Aidin Amirshokoohi is currently an Associate Professor of STEM Education at DeSales University in Center Valley, Pennsylvania. He is mainly responsible for the instruction of science, mathematics, and technology education and instructional design courses for education majors in undergraduate and graduate programs. His primary area of research interest include preparation of pre-service teachers to adopt a Science, Technology, Society, and Environmental (STSE) framework for teaching science, teacher professional development and adoption of the inquiry-based instructional approach by pre-service and in-service teachers.

Chapter 16
Evolution Education in the Arab States: Context, History, Stakeholders' Positions and Future Prospects

Saouma BouJaoude

Abstract While evolution education does not present itself as a public issue in the Arab states, it is seemingly controversial in education circles in general and science education more specifically, because of the perception that it is anti-religion. Within this context, the purpose of the chapter is to analyze the current status of evolution education in the Arab states and discuss possible ways of addressing the controversy within the educational system. The chapter begins by examining the relationship between Islam and science as a background for the potential effect of this relationship on positions regarding evolution and evolution education in the Arab states in which Islam predominates. This is followed by reviewing research that has investigated the status of evolution in science curricula at the school and university levels and in teacher preparation programs in Arab states. Then, we review research that has investigated the positions of high school and college students, biology teachers, and biology university faculty members toward evolution and evolution education and the relationships of these positions with religious affiliation and religiosity in the multi-religious context of a number of Arab states. The chapter concludes by discussing the possibility of including the teaching of evolution in the science curriculum while taking into consideration the contextual factors and the experiences of Islamic countries such as Iran in teaching evolution at the pre-college level.

Teaching the theory of evolution continues to be controversial in the USA with the controversy surfacing in science classes in the European Union and Southeast Asia (Harmon, 2011). In Turkey and in a number of Muslim countries, the growth of anti-evolution has been associated with the efforts of Harun Yahya, an advocate of Muslim creationism (Hameed, 2008). Yahya, whose actual name is Adnan Oktar, is a Turkish old-Earth creationist. Towards the end of the 1970s he started to give religious sermons in Istanbul focused on refuting the theory of evolution (among other topics) and later institutionalized his efforts by establishing the Science

S. BouJaoude (✉)
Department of Education, American University of Beirut, Beirut, Lebanon
e-mail: boujaoud@aub.edu.lb

Research Foundation in Turkey. Yahya and his foundation are very active on the Internet resulting in the design of a website published in a large number of languages including English (http://www.harunyahya.com/) and Arabic (http://ar.harunyahya.com/). Yahya is a prolific writer whose books were translated into many languages and have gained popularity in Southeast Asia, the Arab states, and among the Muslim diaspora in Europe and North America (Riexinger, 2008). It is worth noting that the Turkish Minister of Education has recently announced that evolution will be removed from the Turkish high school curriculum because "Darwin's work is based only on theory" that requires a "separate discussion outside of the school curriculum".[1]

In the Arab states, while evolution education does not present itself as a public issue, it is seemingly controversial in education circles in general and science education more specifically, because of the perception that it is anti-religion. This controversy persists even though the Inter-Academy Panel statement adopted by the academies of science of over 67 countries, including many Muslim and Arab countries (Egypt, Morocco, and Palestine), affirms that biological evolution is accepted by the scientific community as an 'evidence-based fact,' whose scientific evidence has 'never been contradicted' (Inter-Academy Panel [IAP], 2006) and the renowned biologist Dobzhansky (1973, p. 125) stated: "Nothing in biology makes sense except in the light of evolution."

This chapter analyzes the current status of evolution education in the Arab states and discusses possible ways of addressing the controversy within the educational system. First, the relationship between Islam and science is examined to provide the background for the potential effect of this relationship on positions regarding evolution and evolution education in the Arab states; states in which Islam predominates. Then, research that has investigated the status of evolution in science curricula at the school and university levels and in teacher preparation programs in Arab states is reviewed followed by a review of research on the positions of high school and college students, biology teachers, and biology university faculty members toward evolution and evolution education and the relationships of these positions to religious affiliation and religiosity in the multi-religious context of a number of Arab states. It is important to note that, even though Islam is the predominant religion in Arab states, a number of these states have sizeable Christian minorities and different Muslim sects (For example, Sunnis, Shiites, and Druze[2]) with potentially different positions regarding evolution and evolution education. The chapter concludes by discussing the possibility of including the teaching of evolution in the science curriculum by taking into consideration the contextual factors and the experiences of Muslim countries such as Iran in teaching evolution at the pre-college level.

---

[1]Refer to http://yournewswire.com/turkey-darwin-evolution-theory-schools/.

[2]According to Makarem (1974), the Druzes belong to an esoteric Islamic sect based on a philosophical background that appeared at the beginning of the eleventh century. It differs in many respects from traditional Islam and remains inaccessible to many of its adherents.

16.1 Islam, Science, and Evolution

There are varying complex and multidimensional conceptualizations of the relationships between science and religion in Muslim thought; conceptualizations that have resulted in diverse positions regarding the theory of evolution. In what follows, the positions of a number of Muslim scholars are presented to illustrate these views with no intention of being comprehensive. These scholars were selected because they are Muslim scientists with diverse views about the relationship between science and Islam.

Al Jabiri (1991), a Moroccan intellectual, attributes the thinking of Arab scholars about the relationship between science and religion to the special character of Arab thought and reason. Moreover, he contrasts the a priori/apologetic nature of religion to the generally scientifically-oriented and aposteriori nature of science. According to Al Jabiri as quoted by Bahlul (2009), Arab thought and reason are

> essentially teleological, being committed to belief in divine providence. Thus it always seeks to evaluate and understand nature by reference to values and purposes that lie outside nature. Western reason, on the other hand, is heir to the Greek-European mind, which, since its inception, has been characterized by belief in a direct relationship between human reason and nature, a relationship in which the divine is not required to mediate between the mind and what it knows (as it is the case with Arab reason)

This dichotomous view of the relationship between Islam and science is criticized by Guessoum (2011), who maintains that science and religion represent two different worldviews or two ways of knowing and thus are sources of different types of knowledge: metaphysical versus scientific knowledge. Moreover, a careful survey of the positions of Muslim scholars regarding the relationships between science and religion suggests that these positions lie on a continuum. At one end of this continuum are scholars like Maurice Bucaille, a French physician who is a convert to Islam, who champions the theory of the "miraculous scientific content" of the Quran known as *I'jaz,* the basic tenet of which is that the Quran is the source of all scientific knowledge which can be discovered through the interpretation of Quranic verses (Loo, 2001). A prominent supporter of Bucaillism is Zaghloul El Naggar, an Egyptian geologist by training, a prolific writer, and an authority on scientific facts as revealed in the Holy Quran (refer to http://www.elEzr.com/ and http://www.masress.com/en/ahramweekly/15153).

A second view of the relationship between science and Islam is portrayed in the work of Seyyed Hossein Nasr, an Iranian scholar and an MIT and Harvard graduate, who rejects the attributes of modern secular science and suggests that these attributes have led to the collapse of the sacred view of the universe and to environmental and nuclear disasters (Nasr, 2010). Nasr is considered the founding father of Muslim environmentalism and according to him Islam's role is to re-introduce the sense of the sacred in modern Western science and integrate religion and ethics with science rather than relegating these to policy decisions.

A third, but different, view of the relationship between science and religion from a Muslim perspective is attributed to Abdus Salam who rejects the islamization of science (Setia, 2005), suggests that science is universal and international (Segal, 1996), supports the complementarity of science and religion and the total separation between the spiritual and the physical worlds. According to him, science helps in understanding the physical world while religion helps in understanding the spiritual world. Moreover, there is only one universal science and no such things as Muslim science, Hindu science, Jewish science, Confucian science, or Christian science (Hoodbhoy, 1991).

Guessoum (2015) indicated that the positions described above are still prevalent to a certain extent in the Muslim world; with the "miraculous scientific content" of the Quran establishing a stronghold in the Arab states through individuals such as El Naggar in Egypt and Harun Yahya from Turkey. However, according to Guessoum, a new generation of Muslim scientists who "accept modern science's fundamental methodology, theories, and results, and try to find ways to "harmonize" it with Islam" (p. 855) has appeared. Guessoum, cites Bigliardi (2014) who characterizes these "harmonizers" as having (1) interdisciplinary competence and intercultural education, including the ability to address the subject from the solid basis of competence as practicing scientists; (2) constant appeal to philosophical traditions, both Muslim and non-Muslim; and (3) a "culturally pluralistic approach toward other religious and cultural traditions." (pp. 855–856). The idea of harmonization of Islam and the theory of evolution is echoed by Ayoub (2005) who asserts that

> Darwinian evolution did not create a religious crisis in the Muslim world as it did in the West, Muslim intellectuals have continued the debate, but have generally tried to harmonize the Quranic ideas of creation with modern science, including some form of modified Darwinism (p. 173).

The above positions have been criticized by a number of scholars. Bucaille has been criticized because his predictions are "retrospective" in that they go backwards from the discovery to identifying Quranic verses to support the actual discovery rather than use the Quranic verses as guides to discover new scientific ideas (Abdus Salam as cited in Setia, 2005; Hoodbhoy, n.d.,[3] Qadhi, 1999). Meanwhile, Nasr's ideas are criticized by Muslim scholars such as Ziauddin Sardar because they are Muslim but cannot claim to be scientific (Stenberg, 1995). Finally, Abdus Salam is criticized for his idealistic view that scientists are moral beings who have the interest of society at heart and therefore neither they nor their work should be scrutinized by society; views contradict modern conceptions of the nature of science which is considered socially embedded and value laden (Abd-El-Khalick & Lederman, 2000).

The diversity of positions of Muslim scholars regarding the relationship between science and Islam is reflected in their positions regarding evolution. Similar to the positions of members of other religions, these positions seem to lie on a continuum

---

[3]Refer to http://eacpe.org/content/uploads/2014/02/When-teaching-science-becomes-a-subversive-activity.pdf.

which spans views that are completely theistic to others that are completely materialistic.

Hameed (2008) suggests that Bucaille and Nasr reject the theory of evolution but on different grounds. While Bucaille accepts the evolution of animals up to the early hominid species but not for more modern species of animals and certainly not for humans, Nasr, rejects evolution and considers it an ideology rather than a scientific theory. He encourages Muslims to look at evolution from a Muslim spiritual and intellectual perspective as evident in the Quran and Hadith (record of the traditions or sayings of the Prophet Muhammad). Alternatively, Abdus Salam believes in the separation of the spiritual and the material worlds but does not seem to have a specific position regarding the theory of evolution, and suggests that there are questions that are beyond the comprehension of present or even future scientists. Finally, Guessoum (2015) indicates that only few of the "Harmonizers" have expressed detailed views about the theory of evolution. In his book entitled "Islam's Quantum Question" (2011), he states that the theory of evolution does not contradict the core beliefs of Islam. Guessoum also indicates that "Altaie[4] refers to the Quran both in accepting the facts of evolution (including for humans) and in rejecting the randomness of mutations that the standard theory is built upon" (p. 860). Finally, Dajani (2015), a Muslim Jordanian biology professor teaches evolution to her university students and endorses it as a mechanism to explain diversity and the development of species.

16.2 Place of Evolutionary Theory in Curricula at the School and University Levels

Pre-college education systems in almost all Arab states are centrally controlled by ministries of education.[5] According to Faour and Muasher (2011), "Given the nature of political systems in Arab states, ministries of education assume a highly centralized role and continue to be dominated by authoritarian management systems" (p. 13). Consequently, the contents of curricula and textbooks are sanctioned and controlled by these ministries. Public universities are also controlled by either ministries of education or ministries of higher education. While the ministries have some control over the curricula of public universities, this control is less evident than at the pre-college level. The situation in Arab private universities is more complex as some of these have almost total freedom to determine the content of their programs and courses while others experience the same type of control by ministries as public ones. Below, the status of evolution education in schools is presented first followed by the status in universities.

---

[4]Mohammad Altaie is a Jordanian physic professor who writes on Islam and science.
[5]Refer to https://www.files.ethz.ch/isn/157088/CMEC_27_citizenship_education.pdf.

The theory of evolution in science curricula at the school level. Very few studies have investigated the extent to which the theory of evolution is taught at the precollege level in Arab states. These states can be divided into three groups: Those that ban the teaching of evolution completely, those that teach it but within a religious context, and those that teach it as a scientific theory. Examples of the countries that ban the teaching of the theory of evolution include Saudi Arabia (Burton, 2010, 2011), Oman (Al-Balushi & Ambusaidi, 2015), Algeria and Morocco (Clément, Quessada, Laurent, & Carvalho, 2008). In Lebanon, the curriculum and textbooks originally included the theory of evolution in the 1997 curriculum (Center for Educational Research and Development [CERD], 1997); however, it was removed from the curriculum before it was implemented because of pressures from religious groups. Nevertheless, many private schools in Lebanon have adopted international curricula such as the French Baccalaureate, the International Baccalaureate, and American style curricula in which evolution is taught. Consequently, many Lebanese students are exposed to the theory evolution if they enroll in such schools (BouJaoude, Wiles, Asghar, & Alters, 2011).

Saudi Arabia presents the extreme case of opposition to evolution in Arab states. According to Nielsen (2016), the teaching of evolution is totally banned at the pre-college level in Saudi Arabia. The only time it is mentioned is in the Grade 12 level biology textbook where it is presented as an erroneous and blasphemous theory which contradicts the teachings of Islam. Nielsen also mentions that even animal adaptation is presented within a religious context in Saudi Arabia when it is introduced in grade 10 biology. Nielsen quotes the following from the grade 10 biology textbook when it discusses adaptation:

> There exist structural, functional and behavioral characteristics in organisms that help them to survive in their environment. Allah, glory to him, created for organisms those characteristics and structures that enable them to live in their different environments.[6]

Evolution is taught within a religious context in Jordan. The grade 12 biology textbook discusses evolution in general terms within a religious context and with reference to Quranic verses.

Evolution is included in the curricula of four Arab countries. It is incorporated in the Egyptian secondary level required biology curriculum as one complete unit (Grade 10) entitled "Change in living organisms (evolution)" that is taught at the Grade 10 level (Asghar, Hameed, & Farahani, 2014; BouJaoude et al., 2011) even though creationism is presented in the textbook as a viable theory, a fact that changed in the 2013 version of the textbook (Shohdy & Beshir, 2015). The required textbook published by the Egyptian Ministry of Education for this grade (Duwaider, Harass, & Farag, 2005–2006) states that by the end of the unit students should be able to define evolution, differentiate evolution from other theories of human development of life, explain and critique Lamarck's theory, explain and critique Darwin's theory, explain the integrative theory of evolution, define the concept of

---

[6]Fahd bin Nasir al-Aqiyyal et al. al-Ahyā' lil-ṣaff al-awwal al-thānawī [Biology for Secondary Grade One] Riyadh. Translation by Elise Burton.

hereditary balance, explain variability, define and give examples of natural selection, list different types of evidence in support of the theory of evolution, and understand that science is tentative. Surprisingly, even though evolution is a required topic in Egypt, many teachers tell their students to study it because it is required, but not to believe in it, as it conflicts with religion (Zohny, 2012). Similarly, biology is a required subject in grades 10, 11, and 12 in Syria. The theory of evolution is covered in detail at the grade 10 level while other more advanced evolution related topics are covered in Grades 11 and 12 (Asghar et al., 2014).

In Tunisia, evolution is included as a two-week unit in the secondary level biology curriculum of the natural science stream. The aims of the unit are to provide evidence in support of evolution of living organisms and develop phylogenetic trees (evolution trees) by using basic content on evolution (Hrairi & Coquide, 2002). According to Hrairi and Coquide the unit is taught in a cultural context that is not supportive of evolution and in which teaching is very traditional and positivistic in nature. Finally, according to Alshammari, Mansour, and Skinner (2015), the Kuwaiti curriculum implemented in 2010–2011 included teaching the theory of evolution at the intermediate school level. However, the researchers report that this situation created serious resistance by the teachers who thought that inclusion of the theory of evolution in the science curriculum is not aligned with the cultural and religious contexts of Kuwait.

The theory of evolution in programs and courses at the university level. Systematic research on the extent to which the theory of evolution is taught in universities in the Arab states is almost non-existent. Accordingly, the information presented in this section was collected from a variety of sources, some of which are empirically based and others represent opinions by experts in the field.

The status of teaching evolution at the university level in Arab states is more complex and intriguing than the situation at the school level. To start with, while teaching evolution is banned in schools in Saudi Arabia, it is offered as an elective course in the Master's program in King Abdalla University for Science and Technology (KAUST) in Saudi Arabia (Mustafa, 2015). Moreover, while teaching evolution happens within a religious context in schools in Jordan, as indicated above, Dajani (2015) teaches it to her university students and endorses it as a mechanism to explain diversity and the development of species. Another interesting, but counterintuitive idea, is that even though teaching evolution is banned in most schools and universities in the Arab Gulf states, Determann (2015) has shown in a book entitled "Researching biology and evolution in the Gulf states: Networks of science in the Middle East" that research on evolution has flourished in the Gulf states primarily due to the development of academic and professional scientific networks among biologists in the Gulf and between these biologists and researchers in the Western world. These collaborations have succeeded because of the availability of research funds and have not been controversial because most of the research that has resulted from them is written in foreign languages and published in international research journals to which most of the population has no access.

In terms of teaching evolution in universities, Mustafa (2015) claims that it is taught at the Moroccan Mohammad V University, one of the oldest public universities in Morocco and the only one in which evolution is taught in Morocco. Mustafa continues that Sudanese universities were also places where evolution was taught as a full course a few decades ago but is now taught in an integrated manner in biology courses possibly as a result of the implementation of the Shari'a law in the country. Many other universities in the Arab states teach evolution in a fragmented manner in a variety of biology courses thus depriving the students from a coherent and evidence-based discussion on the theory. Since the theory of evolution is not taught in biology courses at the university level in many Arab universities, it goes without saying that those studying to be teachers do not get exposed to the theory during their preparation. Even in countries in which evolution is taught, such as Egypt, Zohny (2012) indicates that teachers might cover evolution in class but supplement their explanations of the theory with their own opinions; which are typically anti-evolution. Moreover, Zohny reports that faculty members who teach biology at the university level admit that evolution is taught in their classes but rarely do they teach human evolution.

Vlaardingerbroek and El-Masri (2006) conducted one of the few empirical studies in the Arab states on the status and prominence of the theory of evolution in university biology departments. Participants in the study included nine biology department heads in each of Lebanon and Australia who were interviewed regarding evolution education in their departments. Results showed that that 8 out of the 10 Lebanese biology departments did not offer any courses on evolution while one offered an elective course[7] on the topic and one was planning to offer an elective course. Results also showed that all 10 participants agreed that evolution should be a required course in university curricula.

16.3 Positions of Students Regarding Evolution

There has been very limited research about the position of the Arab general population about the theory of evolution. One of the few studies was conducted by Hassan (2007) who surveyed individuals from six Muslim countries, including Egypt. Results of the Egyptian sample showed that 8% of the 786 Egyptians who participated in the study said that Darwin's theory of evolution is probably or certainly true, 67% that the theory was probably false or could not possibly be true, while 25% said that they never thought about the theory before. A more recent survey conducted by the Pew Research Center in 2013[8] showed that the majority of individuals in most Arab countries that participated in the study believe that humans

---

[7]It is now a required course.
[8]Refer to http://www.pewforum.org/2013/04/30/the-worlds-muslims-religion-politics-society-science-and-popular-culture//.

and other things evolved over time (78% of the Lebanese, 67% of the Palestinians, 63% of the Moroccans, 52% of the Jordanians, 45% of the Tunisians). The only country in which participants said that humans and other things always existed in the present form was Iraq (67%). The results of the above two studies seem to present two contrasting images of the Arab population: One accepting and one rejecting evolution possibly due to the different research methodologies used in the studies and the different meanings of the term "evolved" used by participants in the Pew survey.

Recent studies in the Arab states that investigated positions regarding evolution and its relationship to religious affiliation were conducted with Egyptian and Lebanese students at the high school and university levels. Egypt and Lebanon were selected because they have a sizable Christian minority and a Muslim majority, thus providing the opportunity to compare the positions of these two communities regarding evolution.

Dagher and BouJaoude (1997) investigated how Lebanese university biology majors accommodated the theory of evolution with their existing religious beliefs. Sixty-two students enrolled in a required college senior biology seminar responded to open-ended questions that addressed their understanding of the theory of evolution, perception of conflict between this theory and religion, and whether the theory of evolution clashed with their own beliefs about the world. Data analysis showed that students' answers clustered around one of four main positions: (a) For evolution (accepted the theory based on scientific evidence or from a reconciliatory perspective of science and religion, for example saying that religious accounts are metaphorical), (b) against evolution (rejected evolution either on religious or scientific bases), (c) compromise (reinterpreted evolution by accepting evolution for all living organisms except humans), and (d) neutral (not committed to any position or confused about the nature of the theory of evolution). Results also showed that 82% of Christian students were for evolution while Muslim students were divided between accepting evolution (35%), rejecting evolution (47%), and accepting evolution in all organisms but not in humans as a compromise position (18%).

In another study, Dagher and BouJaoude (2005) explored how college students understand the nature of the theory of evolution and evaluate its scientific status. Semi-structured interviews were conducted with 15 college biology seniors in which they were asked to explain why they thought evolution assumed the status of a scientific theory, how it compared to other scientific theories, and what criteria they use to determine if an explanation was scientific or not. Results showed that the students focused on one or more of the following themes describing the theory of evolution: "the nature of evidence underpinning the theory, the degree of certainty, experimentation, method of theory generation, and the ability of the theory to generate predictions that allow reproducibility" (p. 6). Those themes focused on the theory's empirical dimension, which seemed to be a result of students' belief in a generic and simplistic model of physical science theories that valued direct evidence. Demanding that evolutionary theory conform to this model revealed a misunderstanding of the nature of scientific theories.

Hokayem and BouJaoude (2008) investigated the relationship among eleven college biology students' epistemological beliefs about science, their beliefs about religion, and their perceptions of nature and causality and their positions regarding the theory of evolution after having completed a course on evolutionary theory. Questionnaires and semi-structured interviews were used to collect data. Based on the data analysis, students were classified into three categories: Those who accepted the theory completely (7 students), those who were uncertain (3 students), and one who rejected the theory. Among the 11 students, 5 considered religion and science as separate entities, 4 considered them in conflict, and 2 considered them in harmony and complementary. Only two of the students admitted that the relationship between science and religion affected their opinion regarding the theory. These results suggest that students' personal beliefs should not be dismissed when teaching the theory of evolution.

BouJaoude et al. (2011) investigated Egyptian and Lebanese high school students' positions about evolution, the relationships between these positions and students' religious beliefs, and the differences in conceptions of evolution between students belonging to different Muslim sects (Sunni, Shiite, and Druze), between Egyptian and Lebanese Muslim students, and between Christian and Muslim students in Lebanon and Egypt. Participants in this study were 194 Egyptian students (63% females and 37% males; 85% Sunni Muslim and 16% Christian) and 865 Lebanese students (49% females and 51% males; 73% Muslim [Sunni, Shiite, and Druze] and 27% Christian). The data source was a questionnaire that examined secondary school students' scientific and religious understandings of evolution which was adopted from Asghar, Wiles, and Alters (2007). Results showed that students in Egypt and Lebanon, irrespective of their religious affiliations, had inadequate understandings of the nature of theories and evidence and had similar misunderstandings regarding the scientific bases of evolutionary theory. These misunderstandings were evidenced by the relatively high percentage of students who either disagreed or were undecided about the items that state that "Humans exist today in essentially the same form in which they always have"; "Human beings as we know them today developed from earlier species of animals", and "humans and monkeys share a common ancestor." Furthermore, while most Egyptian students disagreed with the item that stated that "The term "evolution" means that human beings have developed from apes or monkeys"; slightly less than 50% of the Lebanese students disagreed with this statement.

A higher percentage of Muslim than Christian or Druze students thought that their religion teaches that the first life and humans on planet Earth were created by God, not gradually but suddenly in their present human form, accurate science includes religious explanations, evolution is best learned from the holy book of their religion, biology classes should include their religion's explanations of human and animal history on Earth, God created human beings pretty much in their present form at one time within the last 10,000 years or so, and that religion influences how they think about evolution. Finally, results showed that Lebanese Sunni and Shiite students and Egyptian Sunni students exhibited high levels of religiosity and that these students reported that their religious beliefs influence their positions regarding evolution. Furthermore, Sunni and Shiite Lebanese students were found to have religious beliefs, conceptions of evolution, and positions regarding evolution similar to those of Sunni Egyptian students but significantly different from those of Druze and Christian Lebanese students.

Students' attitudes toward evolution were also investigated in Tunisia, where evolution is taught at the secondary level. Hrairi and Coquidé (2002) administered two questionnaires to 78 secondary school students enrolled in the life science section of the Grade 12 class. The first questionnaire focused on students' conceptions of learning while the second targeted their attitudes towards evolution. Results showed that 31% of the students rejected the theory of evolution while 23% accepted it. The rest of the students were distributed as follows: 5% were instrumentalist in that they considered learning about the theory necessary for passing exams; 8% were indifferent; 6% assimilated it by suggesting that it offered nothing new because all information about evolution was already available in religious texts and traditions; 2% had a nuanced position in that they accepted the theory as scientific but were still unsure about its validity; 5% were ambivalent in that they accepted the theory in school but not in everyday contexts; and 4% restricted their acceptance of the theory to living things other than humans.

As demonstrated in the studies summarized above, there is some evidence to suggest that most Muslim Lebanese university students see a conflict between the theory of evolution and their religious beliefs (Dagher & BouJaoude, 1997). Similarly, the majority of Sunni and Shiite Muslim high school students do not seem to accept evolution for similar reasons. The exception, Druze students, who are labeled as Muslim in Lebanon, but have major differences with Sunnis and Shiites in terms of religious rituals, seemed to be the most accepting of the theory of evolution followed by Christian students.

16.4 Positions of Science Educators Regarding Evolution

There were three studies that investigated the positions of science educators regarding teaching the theory of evolution in the Arab States. These studies were conducted by Clément and Quessada (2008) and Clément, Quessada, Laurent, and Carvalho (2008) while one study was conducted by BouJaoude et al. (2011). As described below, results of these studies indicate that religious affiliation plays a significant role in determining people's positions regarding evolution.

Clément and Quessada (2008) and Clément et al. (2008) surveyed 7,050 biology and non-biology teachers throughout 19 countries in Europe, Africa and the Middle East, out of which four were Arab countries (Algeria, Lebanon, Morocco, and Tunisia) regarding their conceptions of evolution. Results showed that more than 80% of the teachers in the four Arab countries indicated that "it is certain that God created life" with very small minorities in these countries who said that "It is certain that the origin of life resulted from natural phenomena," or the "The origin of life may be explained by natural phenomena without considering the hypothesis that

God created life", or "The origin of life may be explained by natural phenomena that are governed by God." It is worth noting that the above studies investigated the relationships between positions of teachers regarding evolution and their religious affiliations. Results showed that there were significant correlations between teachers' creationist positions and their belief in God and degree to which they practiced their religion.

Moreover, these results showed the highest percentages of creationists are observed in Muslim states, three of which are Arab states (Tunisia, Morocco, and Lebanon). Additionally, Lebanese Christian teachers were found to have more creationist beliefs than their Christian European counterparts. Finally, results showed that there were significant correlations between the levels of education and acceptance of evolution in the whole sample and that the biology teachers were, in general, more accepting of the theory of evolution than non-biology teachers.

BouJaoude et al. (2011) investigated high school biology teachers' and university professors' positions regarding evolution. Research participants included 20 high school biology public and private school teachers and 7 university biology professors. All teachers held at least an undergraduate degree in biology and 14 had more than three years of teaching experience. Fourteen teachers and four professors were Muslim. Data for this study came from semi-structured interviews with participants. Codes developed by Dagher and BouJaoude (1997) were used to analyze the interview data. These codes, however, were modified to incorporate categorical descriptions based on Scott's (2009) evolution/creationism continuum, which describes a range of religious and philosophical beliefs and denotes corresponding levels of acceptance of evolution. The codes then were used to categorize participants' positions regarding evolution and the relationship between these positions and religious beliefs. Results showed that nine (Christian or Druze) teachers accepted the theory, five (4 Muslim) rejected it because it contradicted religious beliefs, and three (Muslim) had a compromise perspective and thus accepted most aspects of the theory but were less willing to accept evolution of humans. Three teachers who rejected or reinterpreted the evolutionary theory (compromise perspective) said that it should not be taught, two said that evolution and creationism should be given equal time in class. Two professors indicated that they taught evolution explicitly and five said that they integrated it in other biology content. One Muslim professor said that she stressed 'the role of God in creation during instruction on evolution'.

16.5 Discussion

It is clear from the above that evolution education in the Arab states is in a crisis as evidenced by the fact that a relatively high percentage of students at the high school and university levels are not supportive of the theory of evolution. Additionally, relatively high percentages of biology teachers do not accept the theory of evolution, while many others have compromise positions. Finally, it seems that religion plays an important role in the lives of high school and university students and high school teachers, a fact that seems to influence their acceptance of the theory of evolution.

Nevertheless, a careful look at the results of research on teaching of evolution shows that the inclusion of the theory of evolution in the curriculum and teaching it does not seem to alter students' acceptance or rejection of the theory. This can be established from considering the cases of Lebanon and Egypt. Even though evolution is included in the Egyptian high school curriculum and is not included in the Lebanese curriculum, acceptance of the theory among students is different in the two countries. While Muslim students in both countries are similar in their rejection of the theory, Lebanese Christian and Druze students, who have not studied evolution are more accepting of the theory; a fact that can also been seen in the results of studies of teachers' views. One conclusion that can be drawn from the above is that the cultural/religious milieu in which students live plays a significant role is shaping their views about the theory of evolution. Consequently, changing these views requires more than a focus on cognitive variables and conceptual change through specific teaching methods and educational approaches that attempt to improve people's understanding of the theory of evolution. As stated by Shohdy and Beshir (2015), the conflict between the public's beliefs about evolution and the accepted views of the scientific community "will not be resolved by simply piling facts and standards into the curriculum. Without changes outside of the scientific and educational spheres, acceptance of evolution is not likely to expand.'(p. 4.5). What is needed is an in-depth investigation whose purpose is to understand how context, attitudes and worldviews influence people's positions regarding the theory of evolution and attempting to approach this matter based on the results of these investigation while at the same time addressing misconceptions about evolution, the nature of theories, the relationships between evidence and theories.

Fortunately, there is a glimpse of hope that ways can be found that allow teaching the theory of evolution in Arab schools and universities while at the same time respecting the cultural context within which this is happening—a long and difficult but possible process. This relative optimism is based on two successful experiences in teaching evolution. The first experience is in Jordan where Dajani (2015) teaches evolution to her university students as a "mechanism to explain diversity and the development of species" without jeopardizing their religious beliefs. The second comes from Iran, which identifies itself as an Islamic Republic, and in which the theory of evolution is taught in its entirety (Burton, 2010, 2011). It seems that in both cases the issue of human evolution, which is a very sensitive issue, is either not mentioned or not emphasized, making the teaching less controversial among teachers and students (and possibly parents). Moreover, it seems that Iran has managed to "separate" the teaching of science from religion in its attempt to encourage the creation of an indigenous and active scientific community that is necessary for economic development.[9] While both attempts can be labeled as

---

[9]Refer to https://www.quora.com/In-Iran-do-they-teach-evolution-in-school.

"compromise" positions when it comes to understanding the theory of evolution, they are pragmatic in that they provide students with the opportunity to develop meaningful understandings of the theory of evolution because, as Dobzhansky (1973, P. 125) asserts, "nothing in biology makes sense except in the light of evolution".

References

Abd-El-Khalick, F., & Lederman, N. G. (2000). Improving science teachers' conceptions of the nature of science: A critical review of the literature. *International Journal of Science Education, 22*, 665–701.

Al Balushi, S., & Ambusaidi, A. (2015). Science education research in the Sultanate of Oman: The representation and diversification of socio-cultural factors and contexts. In N. Mansour & S. Al-Shamrani (Eds.), *Science education in the Arab Gulf states: Visions, sociocultural contexts, and challenges* (pp. 23–47). Rotterdam: Sense Publishers.

Al Jabiri, M. (1991). *The constitution of the Arab mind*. Beirut: Markaz Dirasat al-Wihdah. (in Arabic).

Alshammari, A., Mansour, N., & Skinner, N. (2015). The socio-cultural contexts of science curriculum reform in the State of Kuwait. In N. Mansour & S. Al-Shamrani (Eds.), *Science education in the Arab Gulf states: Visions, sociocultural contexts, and challenges* (pp. 205–223). Rotterdam: Sense Publishers.

Asghar, A., Hameed, S., & Farahani, N. K. (2014). Evolution in biology textbooks: A comparative analysis of 5 Muslim countries. *Religion and Education, 41*, 1–15. Retrieved from http://dx.doi.org/10.1080/15507394.2014.855081.

Asghar, A., Wiles, J. R., & Alters, B. (2007). Canadian pre-service elementary teachers' conceptions of biological evolution and evolution education. *McGill Journal of Education, 42*, 189–209.

Ayoub, M. (2005). Creation or evolution? The reception of Darwinism in modern Arab thought. In Z. A. Bagir (Ed.), *Science and religion in a post-colonial world: Interfaith perspectives* (pp. 173–190). Adelaide, Australia: ATF Press.

Bahlul, R. (2009). *Toward an Islamic conception of democracy: Islam and the notion of public*. Retrieved from http://www.juragentium.org/topics/islam/law/en/public.htm.

Bigliardi, S. (2014). *Islam and the quest for modern science: Conversations with Adnan Oktar, Mehdi Golshani, Mohammed Basil Altaie, Zaghloul El Naggar, Bruno Guiderdoni and Nidhal Guessoum*. Istanbul: Swedish Research Institute in Istanbul.

BouJaoude, S., Asghar, A., Wiles, J. R., Jaber, L., Sarieddine, D., & Alters, B. (2011a). Biology professors' and teachers' positions regarding biological evolution and evolution education in a Middle Eastern society. *International Journal of Science Education, 33*, 979–1000.

BouJaoude, S., Wiles, J., Asghar, A., & Alters, B. (2011b). Muslim Egyptian and Lebanese students' conceptions of biological evolution. *Science & Education, 20*, 895–915.

Burton, E. K. (2010). Teaching evolution in Muslim states: Iran and Saudi Arabia compared. *Reports of the National Center for Science Education, 30*, 28–32.

Burton, E. K. (2011). Evolution and creationism in Middle Eastern education: A new perspective. *Evolution, 65*, 301–304.

CERD. (1997). *Curricula and objectives of general education*. Beirut, Lebanon: Centre for Educational Research and Development. (http://www.cnrdp.edu.lb/cnrdp/curr10.html).

Clément, P., & Quessada, M.-P. (2008). Dossier Évolution et créationisme Les convictions créationnistes et/ou évolutionnistes d'enseignants de biologie: une étude comparative dans dix-neuf pays. *Natures Sciences Sociétés, 16*, 154–158. https://doi.org/10.1051/nss:2008039.

Clément, P., Quessada, M. P., Laurent, C., & Carvalho, G. (2008). *Science and religion: evolutionism and creationism in education, a survey of teachers' conceptions in 14 countries*. Paper presented at the XIII IOSTE Symposium, Izmir (Turkey). Retrieved 21–26 Sept 2008.

Dagher, Z., & BouJaoude, S. (1997). Scientific views and religious beliefs of college students: the case of biological evolution. *Journal of Research in Science Teaching, 34*, 429–455.

Dagher, Z., & BouJaoude, S. (2005). Students' perceptions of the nature of evolutionary theory. *Science Education, 89*, 378–391.

Dajani, R. (2015). Why I teach evolution to Muslim students. *Nature, 520*, 409.

Determann, J. (2015). *Researching biology and evolution in the Gulf states in the Middle East journal*. London: I.B.Tauris.

Dobzhansky, T. (1973). Nothing in biology makes sense except in the light of evolution. *The American Biology Teacher, 62*, 102–107.

Duwaider, A., Harass, H., & Farag, A. (2005–2006). *Life science*. Cairo, Egypt: Egyptian Ministry of Education, Textbook Section (in Arabic).

Faour, M., & Muasher, M. (2011). Education for citizenship in the Arab world: Key to the Future. *The Carnegie Papers*. Beirut: Carnegies Middle east Center. Retrieved from http://carnegieendowment.org/files/citizenship_education.pdf.

Guessoum, N. (2011). *Islam's quantum question: Reconciling Muslim tradition and modern science*. London: I. B. Tauris.

Guessoum, N. (2015). Islam and science: The next phase of debates. *Journal of Religion and Science, 50*, 854–876. Retrieved from http://onlinelibrary.wiley.com/doi/10.1111/zygo.12213/full.

Hameed, S. (2008). Bracing for Islamic creationism. *Science, 322*, 1637–1638. Retrieved from http://helios.hampshire.edu/~sahCS/Hameed-Science-Creationism.pdf.

Harmon, K. (2011). Evolution abroad: Creationism evolves in science classrooms around the globe. *Scientific American*. Retrieved from http://www.scientificamerican.com/article.cfm?id=evolution-education-abroad.

Hassan, R. (2007). On being religious: patterns of religious commitment in Muslim societies. *The Muslim World, 97*, 437–478. Retrieved from http://onlinelibrary.wiley.com/doi/10.1111/j.1478-1913.2007.00190.x/pdf.

Hoodbhoy, P. (1991). *Islam and science: Religious orthodoxy and the battle for rationality*. London: Zed Books. Retrieved from https://docs.google.com/file/d/0B3PhfUp3GgmKNzZkODcxY2UtMjdmYi00OTQ3LWE5NmEtM2NiMjg5YTAwMWZl/edit?usp=drive_web&hl=en.

Hokayem, H., & BouJaoude, S. (2008). College students' perception of the theory of evolution. *Journal of Research in Science Teaching, 45*, 395–419.

Hrairi, S., & Coquidé, M. (2002). Attitudes d'élèves tunisiens par rapport à l'évolution biologique. *Aster, 35*, 149–164.

IAP. (2006). *Inter-Academy Panel Statement on the teaching of evolution*. Retrieved from http://www.interacademies.net/10878/13901.aspx.

Loo, S. (2001) Islam, science and science education: Conflict or Concord? *Studies in Science Education, 36*(1), 45–77. https://doi.org/10.1080/03057260108560167.

Makarem, S. (1974). *The Druze faith*. New York: Caravan Books.

Mustafa, E. T. (2015). Science education in universities in the Islamic World. Retrieved from http://muslim-science.com/task-force-essay-science-education-in-universities-in-the-islamic-world/.

Nasr, S.H. (2010). *Islam in the modern world: challenged by the west threatened by fundamentalism, keeping faith with tradition*. New York, NY: HarperCollins.

Nielsen, R. (2016). *Teaching evolution in the Middle East*. Retrieved from http://www.nielsenlab.org/author/rnielsen/.

Qadhi, A. Y. (1999). *An introduction to the sciences of the Quran*. Birmingham, UK: Al-Hidaayah Publishing and Distribution. Retrieved from https://theauthenticbase.files.wordpress.com/2010/11/introduction-sciences-of-the-quran-yasir-qadhi.pdf.

Riexinger, M. (2008). Propagating Islamic creationism on the internet. *Masaryk University Journal of Law and Technology, 2,* 99–112.

Scott, E. C. (2009). *Evolution vs. creationism: An introduction* (2nd ed.). Berkeley, CA: University of California Press.

Segal, A. (1996). Why Does the Muslim World Lag in Science? *Middle East Quarterly,* 61–70. Retrieved June, 1996, from http://www.meforum.org/306/why-does-the-muslim-world-lag-in-science.

Setia, A. (2005). Islamic science as a scientific research program: Conceptual and pragmatic issues. *Islam & Science, 3*(1), 93–101. Retrieved from http://www.cis-ca.org/jol/vol3-no1/adi-endmatter.pdf.

Shohdy, K. S., & Beshir, M. (2015). *Scorn, not just rejection: Attitudes toward evolution in Egypt.* Reports of the National Center for Science Education. Retrieved January–February, 2015, from https://www.researchgate.net/publication/277007920_Scorn_Not_Just_Rejection_Attitudes_toward_Evolution_in_Egypt.

Stenberg, L. (1995). *The Islamization of science or the marginalization of Islam: The positions of Seyyed Hossein Nasr and Ziauddin Sardar.* Paper Presented at Third Nordic Conference on Middle Eastern Studies: Ethnic Encounter And Culture Change, Joensuu, Finland. Retrieved June, 19–22, 1995, from https://org.uib.no/smi/paj/Stenberg.html.

Vlaardingerbroek, B., & El-Masri, Y. H. (2006). The status of evolutionary theory in undergraduate biology programmes at Lebanese universities: A comparative study. *International Journal of Educational Reform, 15,* 150–163.

Zohny, H. (2012). *Perceptions of Darwin's theory: Evolution in the Egyptian classroom and beyond, Egypt Independent.* Retrieved from http://www.egyptindependent.com/news/perceptions-darwin-s-theory-evolution-egyptian-classroom-and-beyond.

Saouma BouJaoude completed a doctorate in curriculum and instruction/science education in 1988 at the University of Cincinnati, USA. He is presently professor of science education and director of the Center for Teaching and Learning at the American University of Beirut. His research interests include evolution education, curriculum, teaching methods, and the nature of science. BouJaoude has published in international journals such as the Journal of Research in Science Teaching, Science Education, International Journal of Science Education, Journal of Science Teacher Education, the Science Teacher, and School Science Review, among others. Additionally, he has presented his research at local, regional and international education conferences. BouJaoude is presently an associate editor of the Journal of Research in Science Teaching.

Part V
Asia

Chapter 17
Evolution Education in Hong Kong (1991–2016): A Content Analysis of the Biology Textbooks for Secondary School Graduates

Ka Lok Cheng and Kam Ho Chan

Abstract This chapter documents the changes of the evolution-related content in the official textbooks during the first 25 years of its introduction in the curriculums for the secondary school graduates' biology examination. Content analysis of 14 sets of biology textbooks published between 1991 and 2016 were performed. Several key trends are observed in the textbooks studied: The depth and breadth of coverage of evolution in the biology textbooks was growing, an increasing variety of learning strategies and activities had been deployed, and more nature of science ideas had been included over the years. Attempts are made to explain the trends above. The material in the textbooks is not solely dependent on the evolution-related specifications in the corresponding official curricula, while the framework-level and subject-level curricular directions are also influential. Second, the public examination questions, local and overseas, are found to have affected the materials found in the textbooks. Also, both the official curriculum guides and textbooks are being shaped by the new understandings of the international science education community.

17.1 Introduction

17.1.1 Public Acceptance of Evolutionary Theory Within the Social, Political and Cultural Context of the Territory

As a secular city with a population of 6.9 million predominantly (92.0%) made up of Chinese and using Chinese (Cantonese dialect) as the key spoken language

---

K. L. Cheng (✉) · K. H. Chan
Faculty of Education, The University of Hong Kong, Pok Fu Lam, Hong Kong
e-mail: chengkla@hku.hk

© Springer International Publishing AG, part of Springer Nature 2018
H. Deniz and L. A. Borgerding (eds.), *Evolution Education Around the Globe*, https://doi.org/10.1007/978-3-319-90939-4_17

(Census and Statistics Department, 2017), various aspects of Hong Kong are significantly influenced by Chinese philosophy, and education is no exception. Traditional Chinese philosophers focused their energy on the worldly affairs and harmonious interpersonal relationships rather than speculating on the cosmic order (Hall & Ames, 2000). Although there are 0.86 million Christians (including Catholics and Protestants) in Hong Kong (12% of the local population), there are more than 2 million followers of Buddhism and Taoism combined (Home Affairs Department, 2016). This may well explain why, although 53% of secondary and primary school (equivalent to Grade 1–12) students are enrolled in schools with a Christian background (School Statistics Section, 2016), no serious effort has so far been successfully made so far to remove the evolution content from the school curriculum.

On the contrary, there are certain movements aligned with creationism and other alternative theories in schools. The "Genetic and Evolution" section of the current biology curriculum includes the following concluding statement: "In addition to Darwin's theory, students are encouraged to explore other scientific explanations for the origin of life and evolution, to help illustrate the dynamic nature of scientific knowledge" (CDC & HKEAA, 2015/2007, p. 23); indeed, this statement sparked a heated debate in 2009. A group of local scientists and science educators accused the curriculum of promoting the teaching of creationism; the Education Bureau responded, stating that creationism and intelligent design were not parts of the curriculum. While the related parties pressed for further clarification, the legislative body was satisfied with the administration's response and stepped away from the debate (Heron, 2009). The debate then subsided and was not followed-up by the mass media.

17.1.2 Local (Territorial) Evolution Education

In Hong Kong, all residents are required by law to be in school before they reach the age of 15, and they usually complete 6 years of primary and 3 years of junior secondary (secondary 1–3) schooling during the compulsory education period. Senior secondary education (secondary 4–6) is free (despite not compulsory) in Hong Kong, and almost all below 18 continue their study (Census and Statistics Department, 2017) and take the secondary school graduates' examination after secondary 6. Being a special administrative region of the People's Republic of China (PRC), the curricula used in local schools are autonomously developed by the local (territorial) curriculum authority, which could be much different from the ones used in other parts of PRC. The local curricula for senior secondary subjects also doubled as the syllabi of the examinations operated by the local examination authority.

The examination-oriented culture in Hong Kong and the inclusion of evolution-related questions in public examinations almost guarantee discussion of evolution-related curricular content in classrooms. Even after more than a decade of curriculum reform, the reality of Hong Kong classrooms is still in line with Pong and Chow (2002)'s observation that the presence of certain topics in the examination syllabus implies the in-class coverage of those topics. The inclusion of evolution topics in the local Biology Curriculum and Assessment Guide (see the annex regarding the major evolution topics included), both jointly prepared by curriculum and examination authorities, thus delivers a strong signal—evolution topics in the curriculum are examinable, and teachers will definitely cover these topics because students and parents will consider the teachers incompetent in providing students adequate preparation for the examinations if they fail to do so. At this point, it is fitting to refer to descriptions of increased emphasis on evolution in senior secondary [SS] biology textbooks (see Sect. 17.3 below), from which it is possible to deduce that the evolution content is much more rigorously discussed in biology classrooms than two decades before.

The decline in the number of students taking biology at SS level poses the biggest threat to the learning and teaching of evolution in the local schooling system; this is because SS biology is the first, and very likely the only way for students to encounter evolution ideas in their formal schooling. In 2010, around 31,000 students took SS biology as one of their examination subjects, which accounted for 42% of all first attempters in day schools (HKEAA, 2010, p. 40). However, there were only approximately 15,000 students taking the same subject (including biology as a component of "Combined Science") in 2016; this number accounted for just 28% of the cohort (HKEAA, 2016a, pp. 63–64). The sudden drop was due to a change in the academic structure. Students taking their secondary school graduation examination in or before 2010 usually took four or five elective subjects in addition to the three core subjects (English language, Chinese language, and mathematics). In contrast, those taking the examination since 2012 usually took only two or three electives in addition to the four core subjects (Liberal Studies added). Being an elective subject throughout, SS biology was thus taken by fewer students.

17.1.3 Evolution in Teacher Education

Evolution is one of the content topics often used for teaching pedagogy in context in the initial biology teacher education programme. Taking the Biology Methods course in authors' institution as an example, the general and subject-specific pedagogy is the course focus, while evolution and other content topics serve to provide the required contexts. The topic of evolution is also used to exemplify the key nature of science (NOS) tenets. The limited short course duration (24 contact hours) in the crowded curriculum implies insufficient discussion on the rationales behind the inclusion/exclusion of particular evolution topics, the inter-connections between evolution and current scientific advances and everyday life, the possible use of evolution as a unifying theme, and the best practice for addressing the evolution-creation controversy in the classrooms. Despite a plethora of studies concerning these relevant issues in other parts of the world (e.g. Abrie, 2010; Großschedl, Konnemann, & Basel, 2014; Hermann, 2013), including a few from Asian countries (e.g. Kim & Nehm, 2011), corresponding local evolution education research is very rarely found. However, in terms of the experiences of the authors, it appears safe to say that the teaching of evolution in Hong Kong classrooms exhibits patterns which are similar to those reported elsewhere (e.g. Berkman, Pacheco, & Plutzer, 2008; Tidon & Lewontin, 2004). Teachers often spend no more than three weeks covering the topic, with the focus mostly on the mechanisms of evolution, while human evolution and speciation are usually less discussed.

17.1.4 Research Questions

The widespread use of textbooks in classrooms means that said textbooks can be used as an efficient method by which to understand classroom learning and teaching in Hong Kong. Despite the claim that the "textbooks are not the only learning materials", the *Recommended Textbook List* (RTL) is updated annually and only those textbooks that have passed the quality vetting are included in the list (Secretary for Education, 2016). Such effort from the curriculum authorities demonstrates the key role played by textbooks. Moreover, about 99% of secondary school students are being taught by science teachers who frequently use textbooks in class (Martin, Mullis, Foy, & Stanco, 2012, p. 404). As such, the study of the evolution of content in local biology textbooks serves as a way in which to study how the evolution topics are learned in local classrooms.

In light of such an understanding, two research questions are formed: (1) In what ways has the evolution content in the local SS biology textbook changed since it was first included in the curriculum? (2) What are the factors that have shaped the evolution content in the corresponding textbooks? Documentation of the development of evolution education in a territorial system during its early stages is expected to offer practical suggestions which can be used to foster deeper and more widespread evolution education in territorial/national systems.

17.2 Methods

The current study examined all SS biology textbooks listed in RTL since 1991, the year when evolution was first included in the local SS biology curriculum. Since the textbooks for the sixth-form curriculum (last cohort graduated in the 2011/12 school year) were not included in RTL, one set of textbooks (published in 2002 by the publisher of *M03* and *M09*) intended for use by students taking the then sixth-form examination were excluded from the sample. These 14 textbooks were published by four publishers. All textbooks were marketed as multi-volume sets, and only those volumes with at least one section on evolution were sampled. The

Table 17.1 List of textbooks sampled

Code	Title	Publisher	References
Phase I: Early inclusion of the evolution concepts (1991–2002)			
L92	Biology Today 2	Longman	Yip (1992)
O93	Certificate Biology: Mastering Basic Concepts 2	OUP	Pang and Cheung (1993)
A94	Biology: A Modern Approach 2 (3rd ed.)	Aristo	Chan, Chu, and Kong (1994)
O97	Certificate Biology: Mastering Basic Concepts 3 (2nd ed.)	OUP	Pang and Cheung (1997)
A98	Biology: A Modern Approach 3 (4th ed.)	Aristo	Chan, Chu, and Kong (1998)
O01	Certificate Biology: Mastering Basic Concepts 3 (3rd ed.)	OUP	Pang (2001)
Phase II: Evolution as a separated chapter (2002–2007)			
M03	Biology for Tomorrow 3	Manhattan	Yip (2003)
A04	New Biology: A Modern Approach 3	Aristo	Chan, Chu, and Kong (2004)
O04	Certificate Biology: New Mastering Basic Concepts 3	OUP	Pang and Cheung (2004)
Phase III: Further enrichment of evolution concepts (since 2007)			
M09	Discovering Biology 2: Genetic and Evolution	Manhattan	Yip and Yip (2009)
A10	HKDSE Biology: A Modern Approach 4 (Genetic and Evolution)	Aristo	Chan, Ng, Sy, Fung, and Ngan (2010)
O10	New SS Mastering Biology 4	OUP	Yung, Ho, Ho, Tam, and Tong (2010)
A14	HKDSE Biology—Concepts and Applications 4	Aristo	Chan, Fung, Li, Ng, and Sy (2014)
O14	New SS Mastering Biology 4 (2nd ed.)	OUP	Yung, Ho, Ho, Tam, and Tong (2014)

Note Code for publisher: Aristo = Aristo Educational Press, Longman = Addison Wesley Longman China, Manhattan = Manhattan (2003)/Manhattan Marshall Cavendish Education (2009); OUP = Oxford University Press

publication details of all the sampled textbooks are listed below in Table 17.1 together with the codes to be used when these books are referred to thereafter.

As can be seen in Table 17.1, the textbooks are categorized into phases. Phases I, II and III refer to the period when the SS biology textbooks were written in accordance with the curriculum issued in 1991 (CDC, 1991, 2002) and 2007 (CDC & HKEAA, 2015/2007) respectively. The 2007 curriculum was prepared for the most recent structural curriculum reform in Hong Kong, and the academic structure changed from the British-style 3-2-2-3 system (3 years of junior secondary, 2 years of senior secondary, 2 years of sixth-form, and 3 years of university) to a 3-3-4 system (3 years of junior secondary, 3 years of senior secondary, and 4 years of university). As such, although all textbooks examined in the current study were designed to prepare students for the secondary school graduation examinations, the scope and depth of the content between the first two phases and Phase III are substantially different (see the discussion regarding "Phase III" below for further details). Table 17.1 also shows that, among the four publishers, only three of them have published textbooks for the most recent curriculum; moreover, only two have revised their textbooks recently.

Content analysis was carried out in general accordance with the steps laid out by Krippendorff (2004, pp. 83–87). The pages relevant to the learning of evolution in the sampled textbooks were first identified and unitized, following which the units of coding were identified through holistic consideration of a number of factors, including page layout features (e.g. boxed section, figures with captions), paragraphing and (dis)continuation of meaning. Coding was then carried out. The coding method can be understood as "directed content analysis" as the "variables of interest" are largely pre-determined (Hsieh & Shannon, 2005), despite the fact that only the codes for the NOS dimension were derived from literature, while the codes for the other two dimensions, "Concepts" and "Learning activities", were derived from the official biology curricula. After the first round of coding, the coding scheme was revised in order to more efficiently capture the essence of the coding units before another round of coding. The data were summarized through tabulation, and the salient features of each textbook were captured through theoretical memos. The meanings behind these reduced data were inferred through comparing the tabular data and the theoretical memos.

17.3 Results

17.3.1 Phase I: Early Inclusion of the Evolution Concepts (1991–2002)

During the first decade, the curriculum specification on evolution was very subtle and readers of the syllabus might have easily omitted it altogether given the wealth of information it contained. In the 1991 curriculum, the only curriculum specification relevant to evolution was stated as a third-level topic "Significance [of genetic variation]", which was under the topic of "variation". Somewhat interestingly, the topic of variation was, in turn, under the topic "Genetics", and the corresponding explanatory notes read: "The idea of competition for food, space and mates, leading to survival and reproduction of those best fitted to the environment" (CDC, 1991, p. 29). The term "evolution" is not even found in the syllabus, and the evolution content could be understood as being marginalized.

Such marginalization was reflected in all textbooks published during this period. The coverage of evolution in each textbook is no more than 1.5 pages. In view of the total length of each textbook set (approximately 450–700 pages) and the tight course schedule (no more than 200 forty-minute-periods for two years), one should expect nothing other than 10–20 min of in-class lecturing. The text was also written for speedy coverage—around 200–300 words of expository text on natural selection and the role of genetic variations, and perhaps one illustration with at most one multiple choice question designed to assess students' factual recall. The only (implicit) statement with reference to NOS could be found in *L92*, and related to theory: "The above theory attempts to explain how the large number of species …". In view of the lack of NOS understandings among local teachers in the 1990s, the statement above is not expected to have been taken seriously in class. In short, if this is all that was taught with respect to evolution, the SS biology textbooks were not effective in promoting evolution literacy during this phase.

17.3.2 Phase II: Evolution as a Separate Chapter (2002–2007)

The implementation of the 2002 curriculum was a great leap forward. The term "evolution" first appeared in the local SS biology curriculum and became a second-level topic alongside other genetics topics under the section "Genetic and Evolution". The increased importance was reflected in the textbooks and a chapter on evolution could be found in each of them. The quantitative details regarding the coverage are visible in Table 17.2. With 192 forty-minute sessions of class time allocated for the whole SS curriculum (CDC, 2002, p. 13), there were opportunities for students to learn about evolution for four sessions. The increased emphasis was also reflected through the learning activities. A greater variety of learning and teaching activities were introduced, including reading tasks, information search and concept mapping. More questions were added, although most of them were end-of-chapter questions asking for students' factual recall. These textbook changes signify the rising status of evolution education in Hong Kong.

Table 17.2 Number of pages of evolution-related content in the sampled textbooks

Phase I (1991–2002)			Phase II (2002–2007)			Phase III (since 2007)		
Sample	Page	(%)	Sample	Page	(%)	Sample	Page	(%)
L92	1	0.2	*M03*	17	2.0	*M09*	36	1.9
O93	0.5	0.1	*A04*	15	1.9	*A10*	31	2.0
A94	0.5	0.1	*O04*	15	2.0	*O10*	38	2.7
O97	1.5	0.2				*A14*	52	3.2
A98	0.5	0.1				*O14*	48	3.1
O01	1.5	0.2						

Note Columns under "%" indicate the ratios of the number of pages with evolution content to total number of pages of the respective textbook sets (for the whole two-year course)

The textbooks merely covered the concepts found in the official curriculum and the inclusion of ideas not found in the official curriculum guide was very rare. As a result, any important element which was missing from the curriculum guide would also be missing from the textbooks. For instance, although the production of an excessive number of offspring is a key idea behind the Darwinian model (Aleixandre, 1994), the absence of related specifications from the curriculum guide resulted in the absence of such ideas from two (*A04* and *O04*) out of three textbooks published during this phase. As such, readers could deduce that the official curriculum is the most important influence on the development of evolution education in Hong Kong.

The reading task was first introduced to the local biology textbook in this phase, and such inclusion exhibited the influence of system-level curriculum changes brought about by the curriculum reform in 2000, which led to changes in the textbooks. One of the recommended practices in the reform of 2000 was the inclusion of a slightly elaborated passage (usually no more than a page) with reading tasks, often in the form of questions to be answered (CDC, 2001, pp. 133–134). *M03* aligned with this recommendation by introducing a boxed section on the evolution of antibiotic-resistant bacteria followed by a question which required students to explain the recommended practice of antibiotic rotation (*M03*, p. 214). This, as well as other similar tasks in *M03*, allowed for better alignment with system-level curriculum specifications, which was expected to be valued during the textbook vetting process. In other words, the vetting mechanism ensured that the textbooks reflected the requirements of the system-level curriculum in addition to subject-level curriculum guides.

Other kinds of learning activities also reflected the dual influence of system and subject-level official curricula. The subject curriculum guide (CDC, 2002, p. 54) "suggested" the "information search" on extinct organisms, misuse of antibiotics, and the difference between Darwin's and others' theories. As such, these suggestions were translated into corresponding learning activities. The presence of these activities was also a response to the call for developing students' "information technology skills", one of the "generic skills" emphasized in the reform of 2000, as they "help students to seek, absorb, analyze, manage and present information critically and intelligently in an information age and a digitized world"; indeed, this was pointed out as a vital learning element (CDC, 2001, p. 24). In a similar vein, the use of concept maps in all three textbooks was also the result of the synergistic influence of recommendations from the subject-level curriculum guide (CDC, 2002, p. 69) and the focus on information skills in the system-level reform document (CDC, 2001, p. 25). Both system-level and subject-level curriculum specifications were reinforcing each other and determined the learning activities to be found in the textbooks.

There were some initial attempts to introduce NOS ideas into the textbooks during this phase. In all sections of the 2002 biology curriculum guide, "STS connections" were one of the major components. Such elements were expected to help the students "make connections with scientific knowledge, society around them, developments in science and technology, and the nature of science itself"

(CDC, 2002, p. 10); it was also thought that these elements would provide the opportunities for NOS ideas to be discussed in biology class. The textbook analysis allowed for the identification of 14 instances of NOS ideas, more than half of which were found in *M03*. The most frequently discussed NOS idea (9 instances out of 14) was the empirical nature of scientific knowledge, yet this NOS tenet was mostly presented implicitly as in the case of *O04*: "The theory of evolution is supported by several lines of evidence. Fossils provide direct evidence that evolution has taken place" (p. 254).

Interestingly, although the corresponding elements were usually presented in boxed sections under the heading "STS connections" in the three textbooks, none of these boxed sections found within the samples provided illustrations of the science-society interactions. Such a lack of reference to NOS ideas is in stark contrast with the enriched discussion of NOS in the next time period we examined.

17.3.3 Phase III: Further Enrichment of Evolution Concepts (Since 2007)

The coverage of evolution content in the textbooks is both broadened and deepened during this phase. Reviewing Table 17.2, one can identify the increase in the number of pages of evolution-related content. Comparing the 2007 curriculum with the previous year, only one additional topic, namely the role of isolation in speciation, was added; moreover, the only other change was the replacement of the specification "organisms evolving from simple to complex life forms" (CDC, 2002, p. 54) by "Origins of life" (CDC & HKEAA, 2015/2007, p. 26), the latter of which is less concrete and perhaps more inclusive. These slight changes alone could not account for the doubling of coverage in the textbooks published between 2009 and 2010. Table 17.2 also indicates that, despite the doubling of the number of pages, the percentage of evolution-related content in the textbook sets was not increased in parallel; this implies that the increased coverage of evolution-related content was not a result of the improved status of the evolution content, at least during the period spanning 2009–2010, but instead a result of certain subject-level changes.

The revision of academic structure is the main cause of the deepening of coverage. First, the 2002 curriculum was designed to be completed in two years, while the 2007 curriculum was designed with a three-year completion period in mind. Without a significant increase in breadth, the extended learning time (250 h in the 2007 curriculum vs. 192 h in the 2002 curriculum) implies a demand for greater depth of learning. Second, the new curriculum was benchmarked against a more demanding curriculum than the previous one. While the new secondary school graduates' examination was to be attempted after six years of secondary education, it was comparable to the GCE Advanced Level Examination (HKEAA, 2016b) attempted by UK candidates after *seven* years of their secondary (incl. sixth-form) education; on the other hand, the previous secondary school graduation examination was only comparable to the International General Certificate of Secondary Education (HKEAA, 2015) attempted after *five* years of secondary education. The new school graduates' examination will thus be more cognitively challenging, meaning that a deeper learning will be of no surprise. This shows that, given the similar extent of emphasis in the curriculum, a more in-depth study of evolution could be attained through imposing higher demands on the students.

An increased extent of coverage did not result in a greater variety of learning activities. Reading tasks, which could only be found in the textbooks published by Manhattan in the previous phase, were introduced to the textbooks published by OUP and Aristo in 2010 and 2014 respectively. Learning activities that required students to conduct online information searches and read literary sources could still be found, yet the number of instances per textbook set had decreased slightly on average (8 instances in 3 textbooks during phase II, and 9 instances in 5 textbooks during this phase, with "dinosaurs" not included in the information search/browsing task during phase III). The removal of the suggested activities of fossil record observation resulted in a corresponding absence from the textbooks. Indeed, the suggested activities that required students to use "computer simulations or other simulations to model natural selection" (CDC & HKEAA, 2015/2007, p. 27) in the official curriculum guide resulted in *A14*'s inclusion of a hands-on "simulation of 'natural selection'" (p. 31-7); with this said, however, nothing was included in the other four textbooks. The learning activities which saw the greatest increase were recall-type and end-of-chapter questions; in phase II, there were only 2 or 3 pages of end-of-chapter questions in each textbook, while the number of pages of end-of-chapter questions per textbook ranged from 5 to 15 in this phase. This indicates that the increased coverage of evolution in the textbooks does not necessarily result in more student-centered activities—more examination-type drill and practice was included instead.

On the other hand, more attention has been paid to the discussion of NOS ideas in the textbooks. Table 17.3 shows the six NOS ideas found in the textbooks and the respective numbers of instances of these ideas. The empirical nature of scientific knowledge was often mentioned in the textbooks, which have been published since 2007. The other four NOS ideas first introduced in the previous phase became more

Table 17.3 NOS ideas in textbooks

NOS ideas	Number of instances		
	Phase I	Phase II	Phase III
Empirical nature of scientific knowledge	1	8	32
Theories and hypotheses as (attempted) explanations	0	3	15
Tentative and developmental nature of science	0	1	10
Interactions among scientists and peer review	0	2	9
Science-society interactions	0	1	5
Limitation of science and its methods	0	0	3

Note 6 textbooks from phase I, 3 from phase II, and 5 from phase III examined frequently stated. In addition, discussion regarding the limitation of science was also added during this period. The increased emphasis on NOS in the subject curriculum guide could account for the increased attention in the textbooks. During the previous phase, the NOS-related learning objectives were just implicitly mentioned in the "Values and Attitudes" part of the sectional learning objectives (CDC, 2002, p. 51); however, in the 2007 curriculum, there were specialized sections on "STSE connections" and "Nature and History of Biology". In addition to this, specific emphasis was placed on the necessity to learn about the influence of society on science, the dynamic nature of scientific knowledge, the work of scientists including Darwin, Wallace and Lamarck, and the functions of various methods of science in scientific endeavors (CDC & HKEAA, 2015/2007, p. 24). The foregrounding of the NOS elements in the official subject curriculum has caught the attention of the textbook authors, and thus more NOS ideas were included in the text.

In addition to the increased frequencies of inclusion, the NOS ideas were mentioned in a more contextualized manner. In a reading task in *A14*, which involved discussion of the effect of various fossils unearthed over the years on the classification of *Archaeopteryx*, students were asked to use the given information to support a given NOS tenet by "suggest[ing] one reason why theory that explains the origin of birds is subjected to review and change over time" (p. 30-24). Students were thus not simply given the NOS idea, but were asked to consider the evidence in support of the given NOS tenet. During another reading task in *O10*, students were asked to point out the conflict between Darwin's theory and the prevailing theological thought using the information provided within the given text; they were also asked to reflect on whether experimental evidence was necessary for a piece of work to be considered scientific (p. 31-20). These points, together with several other examples in the examined textbooks, stated the NOS ideas explicitly; moreover, they were also attempts to scaffold students' construction of personal meaning with the NOS ideas. The above-mentioned contextualized learning of NOS could represent a step towards the use of historically and contextually rich learning materials, as advocated by Jensen and Finley (1996).

17.4 Forces Involved in the Shaping of Evolution Content

This section suggests three factors which are thought to have an influence on the writing of evolution content in local biology textbooks. However, Skoog and Bilica (2002)'s assertions regarding the complications brought about by the interactions among factors and the impossibility of reducing curriculum changes to linear causal effect links serve as reminders to the authors and readers alike. The following should be considered as some, but not all, of the elements involved in the textbook development process; indeed, the current state of local evolution education is more likely the result of the synergistic effects of these factors and the local educational context.

17.4.1 Local (Territorial) Curriculum

The presence of particular pieces of evolution-based knowledge in the textbooks is mostly determined by the presence of corresponding specifications in the official curriculum. With this said, however, the learning activities and the knowledge *about* science were more affected by the emphases of the curriculum guide at the subject (vs. topic) level. Of particular note here are the genres of learning and teaching strategies recommended in the designated section of the subject curriculum, such as the use of information technology, the contextualized approach, the historical approach, and project learning (CDC, 2001, pp. 57–60). All of these have motivated the curators of textbooks to include a greater variety of learning activities, such as reading and the previously-mentioned information search tasks. The increased NOS content in the evolution chapters is at least partially attributable to the subject-level curricular recommendations. Moreover, the introduction of the "STS connections" section in the 2002 subject curriculum guide highlighted the necessity for NOS ideas to be discussed during class time, while the statement "nature and history of biology", as one of the three "curriculum emphases" that "should be applied across the curriculum" (CDC & HKEAA, 2015/2007, pp. 12–13) served as a reminder. All these points demonstrate that the broad principles and general recommendations in the subject curriculum affect textbook writing and thus student learning.

Territory-level curriculum reform documents are also extremely influential. "Reading to learn" and "Information Technology for Interactive Learning" were two of the "Four Key Tasks" of the system-level curriculum reform in 2000. This resulted in the inclusion of reading and information search tasks in textbooks. Moreover, the change in the academic structure and thus the school examination led to a more in-depth coverage of the evolution ideas. These textbook changes illustrate the system-level curriculum that affects *what* is included in the textbooks and also the *depth* of coverage.

17.4.2 Public Examinations

Public examination questions, both local and overseas (mostly United Kingdom), were extensively included in the examined textbooks, and these indicate the importance of the evolution topics. Table 17.4 illustrates that the number of examination questions that were included in the textbooks has been increasing over the years. Faced with this substantial number of past examination questions, together with an equal (if not larger) number of publisher-designed examination-type practice questions in the textbooks, students today would appreciate that they need to master evolution topics for examination success; as such, one would expect students' effort in the study of evolution concepts could be ensured, particularly given the examination-oriented milieu.

Table 17.4 Number of past local and overseas examination questions in the examined textbooks

Phase II (2002–2007)			Phase III (since 2007)		
Sample	Local	Overseas	Sample	Local	Overseas
M03	2	3	M09	5	5
A04	1	1	A10	5	4
O04	0	2	O10	6	10
			A14	3	14
			O14	15	8

Note No relevant past examination questions have been included in Phase I textbooks

The impact of examination on textbooks was clearly exhibited by the textbook changes resulting from a recent examination question about evolutionary tree. Prior to 2012 (i.e. all examined textbooks except *A14* and *O14*), phylogenetic trees were used as diagrammatic (and mostly decorative) expressions of how the variety of organisms on earth could be traced to a common ancestor. Moreover, it was not possible to identify any examples of the use of the trees for representing the evolutionary relationship, as inferred from biochemical evidence. However, a multiple-choice question (question 23) was set in the practice paper issued by the examination authority in 2012 (HKEAA, 2012), whereby students were asked to pick the phylogenetic tree that corresponded to the evolutionary relationship inferable from the given amino acid differences. The textbooks responded, almost immediately, to the inclusion of such a question and also another similar one in the 2013 examination. Indeed, *A14* included a one-page worked example that was very similar to the question in the practice paper and provided step-by-step guidance on how to construct a phylogenetic tree based on the amino acid differences between species (p. 30-14). In addition, *O14* also demonstrated how a phylogenetic tree could be drawn in view of the amino acid differences between humans and five other kinds of organisms (pp. 29-13, 29-14). From the textbooks changes that resulted from the inclusion of two multiple choice questions in the examinations (and one of them was included in the practice paper only), it is possible to visualize the significant influence of public examinations on the textbook content.

17.4.3 International Science Education Literature

Science teacher educators involved in the publication of biology textbooks and their participation creates the possibility for the recommendations stated in the international science education literature to serve as the guidance of local science textbook writing. Local biology teacher educators served either as authors or consultants in all SS biology textbook sets published in phase III, and these educators were all interested in the history and nature of science. One of the authors of *O10* and *O14*, and the consultant of *A10* and *A14*, were active in promoting NOS understanding

(Wong et al., 2011); in contrast, an author of *M09* was interested in the processes of science (Yip, 2006). Their expertise could foster the application of international NOS research findings on the writing of the local biology textbooks.

The NOS content in the examined textbooks permits the demonstration of such influence. In addition to the previously-mentioned reading task regarding Archaeopteryx in *A14*, in another reading task in *O10* (p. 30-16), students were asked what they can "tell about the nature of scientific knowledge from the above story" after reading a story about the fake fossils of Piltdown Man. These explicit discussions of NOS are consistent with the updated understanding that NOS instruction should be carried out in an explicit manner (Abd-El-Khalick & Lederman, 2000). In another reading task in *M09* (pp. 183–184) on the societal controversy aroused by Darwin's theory of evolution, a concise discussion was provided within the text. This was packed with the prevailing doctrinal assertions, the reactions from the Church, the differing interpretations of the Bible, the responses of the public, and the representations in the mass media at that time. Despite its succinctness (and possible superficiality), the task could still be considered as a response to McComas (2007)'s suggestion that the history of science instances should be used to illustrate key NOS tenets. This showed that although the local curriculum mostly determined *which* NOS ideas were to be taught, the international literature provided stimuli on *how* these NOS ideas should be presented.

17.5 Conclusions

17.5.1 Summary and Outlook

Triggered by three curriculum changes, the evolution content in the SS biology textbooks has been enriched, not only in terms of breadth, but also its depth. The varieties of learning activities have been improved, although students were given more examination-type questions to work on. More frequent allusion to NOS ideas was observed. Local subject curricula, including topic-based specifications, more general suggestions, system-level curricula and curricular framework, all played important roles in bringing about the textbook changes. The questions set in the local and international biology examinations and the current understanding of the scientific community were found to influence the local SS biology textbooks.

Even in view of these mostly positive changes, the authors are not overly optimistic about the future of local evolution education. As stated in the introduction, the main challenge to evolution education in Hong Kong is the decreasing number of students taking SS Biology, and such decrease is unlikely the result of deepened discussion of evolution in the curriculum. Given the current curricular framework, the authors do not expect a foreseeable sudden surge given the variety of elective subjects (around 20) and the small number (2 or 3) of electives taken by each student. In view of the absence of evolution topics in the recent revision of the junior secondary science curriculum (CDC, 2016), the authors do not expect a sudden introduction of evolutionary ideas in this compulsory course for all students in the coming future. However, since there is no adverse factor demanding the curtailing of evolution content in the SS biology curriculum, the current situation of evolution education in Hong Kong is expected to be maintained for a while.

17.5.2 Suggestions to Improve Evolution Education in the Territory

Inspired by the understandings gained from science education research over recent decades, four possible directions to improve the evolution education of Hong Kong are recommended. First, a shift of focus to human evolution is desirable. Similar to the earlier textbooks examined by Skoog (2005), little emphasis is placed on human evolution in the local textbooks. The evolution of humans could be highlighted to enhance the relevancy of the content. Evolution can be "desegregated" and used as a unifying concept through which to understand the whole subject matter, as recommended by Moody (1996) and Nehm et al. (2009). This could make students more appreciative of the centrality of evolution in the living world.

Second, more effort should be made to challenge students' long-held prior misconceptions. Despite the presence of the lists of possible misconceptions in *M03, O04, M09* and *O14*, they were placed at the end of the respective chapters as remedial support for the students. The positioning of these lists also rendered them prone to students' and teachers' omissions. Instead, as recommended by Nelson (2008), students should be mobilized to consider the commonly held misconceptions critically. The textbooks' chapters might be designed to elicit students' misconceptions before conceptual change effort was made through the main parts of the chapters in line with the suggestion of Andrews, Kalinowki, and Leonard (2011).

Third, the learning activities and NOS content should be further enhanced. Due to the effectiveness of active learning strategies when it comes to developing students' conceptual understandings (Nehm & Reilly, 2007), a greater variety of learning and teaching activities should be designed by the textbook authors. Considering the benefits brought about by enhanced NOS understandings towards the learning of evolution concepts (Campbell & Otrel-Cass, 2011; Lombrozo, Thanukos, & Weisberg, 2008), students could be provided with more enriched historical accounts of the key events and figures in the development of evolution concepts. This would aptly illustrate the key NOS tenets. These changes are expected to be helpful for improving students' understanding of the evolution concepts and the working of science as a whole.

Fourth, teacher education programs should be redesigned to provide better support to the biology teachers. In view of the crowded curriculum, the undergraduate curriculum should welcome inter-disciplinary courses which not only cover the key evolution ideas, but which also prepare students to communicate about evolution in science classrooms and other community settings. Student teachers may also be encouraged to conduct studies on the topic of evolution (e.g. Bravo & Cofré, 2016) during their practice. Evolution-specific teacher professional development courses can be offered for in-service biology teachers in order to foster their reflections on their content selection and pedagogical strategies used in their instruction of evolution. Further pedagogical strategies could also be introduced, such as the use of historical arguments (Jensen & Finley, 1995), and the modeling approach (Passmore & Stewart, 2002), all of which are relevant for teaching about evolution.

In conclusion, the concerted changes in curriculum organization, curriculum focus, conceptual development and handling of misconceptions, learning and teaching activities, and teacher education above should enable the students to become better biology learners, and teachers to become better biology educators with respect to evolution.

Appendix: Coding Scheme Used in Textbook Content Analysis (Abbreviated)

Content topics	Learning activities	NOS ideas
C1. Descent from common ancestor	L1. Rhetorical questions	N1. Empirical basis of science
C2. Evidences for evolution	L2. Recall-type questions	N2. Tentative nature of scientific knowledge
C3. Role of genetic variations	L3. Questions that develop process skills	N3. Theories and hypotheses as (possible) explanations
C4. Natural selection as evolution mechanism	L4. Simulations/Hands-on activities/Observations	N4. Limitation of science and its methods
C5. Isolation mechanisms for speciation	L5. Information search/browse	N5. Interactions among scientists
		N6. Science and society

References

Abd-El-Khalick, F., & Lederman, N. G. (2000). Improving science teachers' conceptions of nature of science: A critical review of the literature. *International Journal of Science Education, 22*(7), 665–701.

Abrie, A. L. (2010). Student teachers' attitudes towards and willingness to teach evolution in a changing South African environment. *Journal of Biological Education, 44*(3), 102–107.

Aleixandre, M. P. J. (1994). Teaching evolution and natural selection: A look at textbooks and teachers. *Journal of Research in Science Teaching, 31*(5), 519–535.

Andrews, T. M., Kalinowki, S. T., & Leonard, M. J. (2011). "Are humans evolving?" A classroom discussion to change student misconceptions regarding natural selection. *Evolution: Education and Outreach, 4*(3), 456–466.

Berkman, M. B., Pacheco, J. S., & Plutzer, E. (2008). Evolution and creationism in America's classrooms: A national portrait. *PLOS Biology, 6*(5), e124. Retrieved from https://doi.org/10.1371/journal.pbio.0060124.

Bravo, P., & Cofré, H. (2016). Developing biology teachers' pedagogical content knowledge through learning study: The case of teaching human evolution. *International Journal of Science Education, 38*(16), 2500–2527.

Campbell, A., & Otrel-Cass, K. (2011). Teaching evolution in New Zealand's schools—Reviewing changes in the New Zealand science curriculum. *Research in Science Education, 41*(3), 441–451.

Census and Statistics Department, HKSARG. (2017). *2016 Population by-census summary results*. Hong Kong: Census and Statistics Department.

Curriculum and Development Council [CDC]. (2001). *Learing to learn—The way forward in curriculum development*. Hong Kong: CDC.

Curriculum and Development Council [CDC]. (2016). *Science (Secondary 1–3) curriculum framework—Supplement to the science education KLA curriculum guide (P1–S6)*. Hong Kong: CDC.

Curriculum and Development Council [CDC] and Hong Kong Examinations and Assessment Authority [HKEAA]. (2015/2007). *Biology curriculum and assessment guide (Secondary 4–6)*. Hong Kong: CDC & HKEAA.

Curriculum Development Council [CDC]. (1991). *Syllabus for biology (Secondary 4–5)*. Hong Kong: CDC.

Curriculum Development Council [CDC]. (2001). *Exemplars of curriculum development in schools*. Hong Kong: CDC.

Curriculum Development Council [CDC]. (2002). *Biology curriculum guide (Secondary 4–5)*. Hong Kong: CDC.

Großschedl, J., Konnemann, C., & Basel, N. (2014). Pre-service biology teachers' acceptance of evolutionary theory and their preference for its teaching. *Evolution: Education and Outreach, 7*, 18. https://doi.org/10.1186/s12052-014-0018-z.

Hall, D. L., & Ames, R. T. (2000). Chinese philosophy. In E. Craig (Ed.), *Concise Routledge Encyclopedia of philosophy* (pp. 355–356). London: Routledge.

Hermann, R. S. (2013). High school biology teachers' views on teaching evolution: Implications for science teacher educators. *Journal of Science Teacher Education, 24*(4), 597–616.

Heron, L. (2009, June 26). Victory for Darwin. *South China Morning Post*.

Home Affairs Department, HKSARG. (2016). *Hong Kong: The facts—Religion and custom*. Hong Kong: Home Affairs Department.

Hong Kong Examinations and Assessment Authority [HKEAA]. (2010). *H.K.C.E.E. Examination Report*. Hong Kong: Hong Kong Examinations and Assessment Authority.

Hong Kong Examinations and Assessment Authority [HKEAA]. (2012). *Hong Kong diploma of secondary education examination practice paper—Biology paper 1*. Hong Kong: HKEAA.

Hong Kong Examinations and Assessment Authority [HKEAA]. (2015). IGCSE (2007). Retrieved from http://www.hkeaa.edu.hk/en/recognition/benchmarking/ce_al/igcse/.

Hong Kong Examinations and Assessment Authority [HKEAA]. (2016a). *HKDSE examination report*. Hong Kong: Hong Kong Examinations and Assessment Authority.

Hong Kong Examinations and Assessment Authority [HKEAA]. (2016b). *New UCAS Tariff (2017) for HKDSE and other international qualifications*. Hong Kong: HKEAA Retrieved from http://www.hkeaa.edu.hk/DocLibrary/Media/Leaflets/2017_UCAS_Tariff_factsheet_eng.pdf.

Hsieh, H. F., & Shannon, S. E. (2005). Three approaches to qualitative content analysis. *Qualitative Health Research, 15*(9), 1277–1288.

Jensen, M. S., & Finley, F. N. (1995). Teaching evolution using historical arguments in a conceptual change strategy. *Science Education, 79*(2), 147–166.

Jensen, M. S., & Finley, F. N. (1996). Changes in students' understanding of evolution resulting from different curricular and instructional strategies. *Journal of Research in Science Teaching, 33*(8), 879–900.

Kim, S. Y., & Nehm, R. H. (2011). A cross-cultural comparison of korean and american science teachers' views of evolution and the nature of science. *International Journal of Science Education, 33*(2), 197–227.

Krippendorff, K. (2004). *Content analysis: an introduction to its methodology* (2nd ed.). Thousand Oaks, CA: Sage.

Lombrozo, T., Thanukos, A., & Weisberg, M. (2008). The importance of understanding the nature of science for accepting evolution. *Evolution: Education and Outreach, 1*(3), 290–298.

Martin, M. O., Mullis, I. V. S., Foy, P., & Stanco, G. M. (2012). *TIMSS 2011 international results in science*. Chestnut Hill, MA: TIMSS & PIRL International Study Center.

McComas, W. F. (2007). Seeking historical examples to illustrate aspects of the nature of science. *Science and Education, 17*(2–3), 249–263.

Moody, D. E. (1996). Evolution and the textbook structure of biology. *Science Education, 80*(4), 395–418.

Nehm, R. H., Poole, T. M., Lyford, M. E., Hoskins, S. G., Carruth, L., Ewers, B. E., & Colberg, P. J. (2009). Does the segregation of evolution in biology textbooks and introductory courses reinforce students' faulty mental models of biology and evolution? *Evolution: Education and Outreach, 2*(3), 527–532.

Nehm, R. H., & Reilly, L. (2007). Biology majors' knowledge and misconceptions of natural selection. *BioScience, 57*(3), 263–272.

Nelson, C. E. (2008). Teaching evolution (and all of biology) more effectively: strategies for engagement, critical reasoning, and confronting misconceptions. *Integrative and Comparative Biology, 48*(2), 213–225.

Passmore, C., & Stewart, J. (2002). A modeling approach to teaching evolutionary biology in high schools. *Journal of Research in Science Teaching, 39*(3), 185–204.

Pong, W. Y., & Chow, J. C. S. (2002). On the pedagogy of examinations in Hong Kong. *Teaching and Teacher Education, 18*(2), 139–149.

School Statistics Section, Education Bureau, HKSARG. (2016). *Student Enrolment Statistics, 2015/16 (Kindergarten, Primary and Secondary Levels)*. Hong Kong Education Bureau.

Secretary for Education, HKSARG. (2016). *Education Bureau Circular Memorandum No. 55/ 2016: Selection of Quality Textbooks and Curriculum Resource for Use in Schools*. Hong Kong: Education Bureau.

Skoog, G. (2005). The coverage of human evolution in high school biology textbooks in the 20th century and in current state science standards. *Science and Education, 14*(3–5), 395–422.

Skoog, G., & Bilica, K. (2002). The emphasis given to evolution in state science standards: A lever for change in evolution education? *Science Education, 86*(4), 445–462.

Tidon, R., & Lewontin, R. C. (2004). Teaching evolutionary biology. *Genetics and Molecular Biology, 27*(1), 124–131.

Wong, A. S. L., Yung, B. H. W., Day, J. R., Cheng, M. M. W., Yam, E. Y. H., & Mak, S.-Y. (2011). Enhancing students' understanding of the Nature of Science and the interconnection between science, technology and society through innovative teaching and learning activities. In M. M. H. Cheng & W. W. M. So (Eds.), *Science education in international contexts* (pp. 83–99). Rotterdam, The Netherlands: Sense.

Yip, D.-Y. (2006). Integrating history with scientific investigations. *Teaching Science: The Journal of the Australian Science Teachers Association, 52*(3), 26–29.

Textbooks examined

Chan, W. K., Chu, S. F., & Kong, S. W. (1994). *Biology: A modern approach 2* (3rd ed.). Hong Kong: Aristo.

Chan, W. K., Chu, S. F., & Kong, S. W. (1998). *Biology: A modern approach 3* (4th ed.). Hong Kong: Aristo.

Chan, W. K., Chu, S. F., & Kong, S. W. (2004). *New biology: A modern approach 3*. Hong Kong: Aristo.

Chan, W. K., Fung, Y. C., Li, C. S. Y., Ng, K. K., & Sy, D. (2014). *HKDSE biology—Concepts and applications 4*. Hong Kong: Aristo.

Chan, W. K., Ng, K. K., Sy, D., Fung, Y. C., & Ngan, F. K. (2010). *HKDSE biology: A modern approach 4 (Genetic and Evolution)*. Hong Kong: Aristo.

Pang, K. C. (2001). *Certificate biology: Mastering basic concepts 3* (3rd ed.). Hong Kong: Oxford University Press.

Pang, K. C., & Cheung, L. M. (1993). *Certificate biology: Mastering basic concepts 2*. Hong Kong: Oxford University Press.

Pang, K. C., & Cheung, L. M. (1997). *Certificate biology: Mastering basic concepts 3* (2nd ed.). Hong Kong: Oxford University Press.

Pang, K. C., & Cheung, L. M. (2004). *Certificate biology: New mastering basic concepts 3*. Hong Kong: Oxford University Press.

Yip, D.-Y., & Yip, P. (2009). *Discovering biology 2: Genetic and evolution*. Hong Kong: Manhattan Marshall Cavendish Education.

Yip, P. (1992). *Biology today 2*. Hong Kong: Addison Wesley Longman China.

Yip, P. (2003). *Biology for tomorrow 3*. Hong Kong: Manhattan.

Yung, H. W., Ho, K. M., Ho, Y. K., Tam, K. H., & Tong, L. P. (2010). *New senior secondary mastering biology 4*. Hong Kong: Oxford University Press.

Yung, H. W., Ho, K. M., Ho, Y. K., Tam, K. H., & Tong, L. P. (2014). *New senior secondary mastering biology 4* (2nd ed.). Hong Kong: Oxford University Press.

Ka Lok Cheng is a Lecturer in The University of Hong Kong and he obtained his PhD in the field of curriculum studies / science education there. He is currently teaching Liberal Studies (a senior secondary interdisciplinary / integrative subject) method course and the inquiry-based education foundations course there. His research mostly focuses on the study of curricular materials, and recently he has initiated a research project that deploys constructionist and textual approaches for the study of socioscientific elements in textbooks.

Kennedy Kam Ho Chan is an Assistant Professor in The University of Hong Kong. He received his BSc and MPhil in science there. Before pursuing his PhD studies in the same university, he had worked as a secondary school science teacher in several local secondary schools. His research interests include pedagogical content knowledge (PCK), teacher noticing, using video to promote teacher learning, formative assessment and biology teaching. He was an invited participant of PCK Summit II held in the Netherlands.

Chapter 18
Evolution Education in Indonesia: Pre-service Biology Teachers' Knowledge, Reasoning Models, and Acceptance of Evolution

Arif Rachmatullah, Ross H. Nehm, Fenny Roshayanti and Minsu Ha

Abstract Indonesia has received little attention in the evolution education research community despite being the world's largest Muslim-majority nation and the third most populous democracy. As such, Indonesia has the potential to test generalizations and shed new light on the ways in which religion, culture, and formal education contribute to evolutionary understanding, reasoning, and acceptance levels. Here, we report on empirical studies of moderately large samples (n > 300) of Indonesian pre-service biology teachers' understanding, reasoning, and acceptance of Evolution. In the first and second study, we compare American and Indonesian student's evolutionary reasoning patterns across a range of tasks using written prompts and clinical interviews. In the third study we investigate Indonesian pre-service biology teachers' acceptance of evolution. Our first and second studies found that Indonesian participants commonly displayed: lower levels of understanding compared to American samples, mixtures of naive and normative concepts in evolutionary explanations, weak cognitive coherence across tasks, and teleological reasoning in explanations of evolutionary change. In the third study, we found that Indonesian participants, like those in other cultures, have greater acceptance of microevolution, followed by lower acceptance of macroevolution and human evolution. Taken together, our studies suggest that cognitive difficulties inherent to thinking about evolution, to a greater extent than cultural and religious influences, are shaping evolutionary reasoning patterns and acceptance levels in

A. Rachmatullah (✉) · M. Ha
Division of Science Education, College of Education, Kangwon National University, Chuncheon-si, Republic of Korea
e-mail: arifraach@gmail.com

R. H. Nehm
Department of Ecology & Evolution, Stony Brook University (SUNY), Stony Brook, NY, USA

F. Roshayanti
Faculty of Mathematics, Science and Information Technology Education, Department of Biology Education, Universitas PGRI Semarang, Semarang, Indonesia

© Springer International Publishing AG, part of Springer Nature 2018
H. Deniz and L. A. Borgerding (eds.), *Evolution Education Around the Globe*, https://doi.org/10.1007/978-3-319-90939-4_18

Indonesian pre-service biology teachers. This finding is notable given the strong religious nature of Indonesian society and the prominent role of religion in Indonesian formal education.

18.1 Introduction

Indonesia has received remarkably little attention in the evolution education research community despite being the world's third most populous democracy and the world's largest Muslim-majority nation (CIA, 2016a). Although Islam is the most widely practiced religion in Indonesia (87.2%), five others are represented in the country (Christian Protestant (7%), Roman Catholic (2.9%), Hindu (1.7%) and Buddhist and Confucian (0.9%)). Choosing one of these six religions is mandatory for Indonesian citizens. Therefore, although Indonesia is a Muslim-majority nation, Parker (2017) noted that it may be more appropriate to refer to Indonesia as a religious country.

Religion is a pervasive aspect of life in Indonesia, and deeply intertwined in the cultural traditions of the more than 300 ethnic groups: Javanese (40.1%), Sundanese (15.5%), Malay (3.7%), Batak (3.6%), Betawi (2.9%), Minangkabau (2.7%), Buginese (2.7%) and Bantenese (2%) (CIA, 2016b). In addition to ethnic diversity, Indonesia has more than 700 local language dialects, with several characterizing individual ethnic groups. For these reasons—diversity of religions, ethnic groups, and languages—Indonesia is appropriately known as the most linguistically and culturally diverse country in the world (Skutnabb-Kangas, 2000). Finally, religion is a central feature of the most recent national curriculum; connections between science content and religion are expected in the classroom. As such, Indonesia provides a unique context for exploring long-standing generalizations about evolution education, such as the roles that religion and culture plan in understanding, reasoning patterns, and acceptance levels. Our study presents the first empirical investigation of moderately large samples of Indonesian pre-service biology teachers using both cognitive (knowledge and reasoning of Evolution) and affective (acceptance of Evolution) measurement instruments and accompanying interviews.

18.2 Evolutionary Theory in the School Curriculum and Biology Teacher Preparation

A brief overview of how evolution is presented in Indonesian secondary schools and teacher education programs will help to provide context for our empirical work on pre-service biology teachers. The most recent formal school curriculum was released in 2013 and modified in 2016 by the Indonesian Ministry of Education and Culture, which oversees education in the country (Mendikbud, 2016). Indonesia's placement at the bottom of international studies of scientific literacy (PISA and TIMSS) spurred the development of this new curriculum. In 2015, for example, Indonesia ranked 61st out of 69 countries on PISA (OECD, 2016), and Indonesian fourth graders ranked 44th out of 47 countries on TIMSS science (Martin et al., 2016).

The new curriculum has been implemented in most Indonesian schools, from elementary to high school. Most schools, particularly public schools, use this curriculum to organize and plan instruction. There are four core competences in Curriculum 2013, required for all school lessons, roughly translated as "spiritual/religious, social, knowledge, and skills." The goal of including these competencies is to foster the development of Indonesian citizens who are religious, sociable, knowledgeable and skillful. Evolution is a required component of Curriculum 2013.

Importantly, previous Indonesian curricula did not identify spirituality/religiosity as a core competence. Implementation of Curriculum 2013 required teachers to connect learning objectives to students' religions, and not only the dominant Islamic religion but to all religions legally recognized and represented in a class. A more diverse student body would require teachers to have a broader understanding of students' religious practices, and to foster positive connections with knowledge-related content. Thus, religion is a required component of teacher knowledge and an important aspect of the official school curriculum. Many studies of evolution education have found that religiosity and religious affiliation can be perceived to be in conflict with evolutionary theory (e.g. Deniz et al., 2008; Ha et al., 2012; Nehm & Schonfeld, 2007). A unique aspect of the Indonesian curriculum, therefore, is that evolutionary theory is delivered in the biology class while being explicitly connected to religious ideas.

In Indonesia, evolution is typically taught in the last semester of 12th grade, after students have learned other foundational biological concepts (such as genetics, development, and metabolism). In high school, four lesson hours (4 × 45 min), or around 6 h for two weeks, are allocated to the topic of evolutionary theory. Teachers use Curriculum 2013 to determine which concepts and theories related to evolution should be taught. These are described in the core competencies of 'knowledge' and 'skill':

> "*Memahami struktur dan fungsi enzim dan materi genetik dalam bioproses dan pewarisan sifat pada makhluk hidup, serta kelangsungan hidup organisme di bumi melalui proses mutasi dan evolusi dengan melakukan investigasi literatur dan mengkomunikasikannya secara lisan dan tulisan.*" [Understanding structure and function of enzymes and genetic material in bioprocess and inheritance of living beings, as well as the viability of organisms on the Earth through a process of mutation and evolution by investigating literature and communicate it verbally and in writing]

Based on the above description, evolution should be integrated with other biological topics, such as biochemistry, genetics, and heredity.

Many teachers begin evolution instruction with the origin of life. Then the history of the emergence of evolutionary theory is discussed, including Darwin's research, with examples including finch beak evolution, and the evolution of the turtle shell. Preceding discussions of natural selection, Lamarck's ideas relating to the development of giraffe's necks, homology and analogy, comparative embryology and vestigial organs in present organisms are also included. Following lessons on the origin of life and an overview of the history of evolutionary thought, mechanisms of evolution are introduced. Natural selection, gene mutation, and geographic isolation are discussed. Students calculate gene frequencies in a population (Hardy-Weinberg's Law) and learn basic population genetics. The last section of the evolution unit focuses on fossil evidence for human evolution.

The evolution section ends with mention of ideas contrary to natural selection, including anti-evolutionary ideas, such as Michael Denton's critiques of natural selection, and Muslim writer Harun Yahya's phenomenal arguments. These ideas are contained in some Indonesian high school biology textbooks in order to show how different opinions about evolutionary theory connect with religion, which in turn align with the curricular goal of linking learning to religious and spiritual growth.

One noteworthy aspect of the Indonesian evolution curriculum is that teachers are encouraged to have students compare and contrast the relevance of evolutionary theory with Intelligent Design. The goal of this activity is to bring a spiritual dimension to the discussion of evolution. In the classroom, teachers provide examples from religious texts that are relevant to evolutionary theory. Thus, rather than avoiding religious discussions, teachers are expected to engage students to consider how evolutionary ideas relate to religious and spiritual perspectives. In terms of pedagogical practices, Indonesian teachers are given full autonomy.

18.2.1 Evolutionary Theory in Biology Teacher Preparation

Although the 2013 Curriculum changed what is taught in elementary to high schools in Indonesia, corresponding changes to biology teacher curriculum have been less dramatic; indeed, the majority of biology teacher education is the same as before the new curriculum was introduced. Of course, discussion of the new national curriculum, and associated core competencies (including the spiritual aspect), are now a part of the teacher curriculum.

Biology teacher education programs in Indonesia encompass four years (eight semesters). Similar to the high school curriculum, most biology teacher education programs in public and private universities provide an evolution course at the end of biology instruction (typically in the last three semesters). Most often, evolution is taught in the sixth or seventh semester because in the eighth semester most students are completing their teaching practicum in public or private middle or high schools. Evolution is typically offered as a one-semester class, with 16 sessions, each lasting around 100 min (~ 27 h).

In addition to completing an evolution course in the sixth or seventh semester, many pre-service teachers complete courses containing evolutionary content (e.g., biochemistry, taxonomy, embryology, ecology, and genetics). These courses are delivered with an evolutionary approach; for example, in the plant and animal taxonomy classes the discussion starts from lower to higher taxa, with explanations of loss and gain of traits. Overall, evolution content is widespread in biology teacher preparation coursework. Conceptual investigations and problem solving approaches are used in concert with lecturing and student presentations (Sudargo, 2009).

18.3 Indonesian Evolutionary Knowledge, Reasoning Patterns, and Acceptance Levels

Many core claims in the growing field of evolution education—such as the relationships among knowledge, acceptance, and religiosity—rely on a relatively large body of work on American, Turkish, and Korean students and pre-service teachers (e.g., Ha et al., 2012; Nehm & Schonfeld, 2007), no published studies to our knowledge have explored both cognitive and affective variables in moderately large samples of Indonesian pre-service biology teachers. Cross-cultural studies can be valuable approaches for teasing apart the roles that religion, culture, and formal education play in the development of evolutionary knowledge and acceptance. Moreover, cognitive studies of evolutionary reasoning processes (that is, how cognitive resources are mobilized to solve different types and forms of evolutionary scenarios—such as the gain and loss of traits in animals and plants) are almost exclusively from American samples (Nehm & Ha, 2011). Given that most studies of evolution education have been conducted on American students and teachers (e.g. Borgerding et al., 2016; Nehm & Schonfeld, 2007) it is reasonable to ask whether these findings extend across cultures and geographic contexts.

18.3.1 *Empirical Investigations of Knowledge and Acceptance*

Our empirical work seeks to gain insights into Indonesians' evolutionary thinking patterns using cross-sectional sampling and mixed methods approaches. We used carefully translated and empirically validated measurement instruments to compare Indonesians' and Americans' reasoning patterns. In order to do so, we performed three studies. The first study (Study 1) was quantitative and statistically compared the composition and structure of Indonesian and American evolutionary reasoning patterns across different open-ended problem types. The second study (Study 2) was qualitative and utilized clinical oral interviews in order to provide a richer and deeper understanding of evolutionary reasoning patterns and to corroborate findings from the written tasks. The third study (Study 3) utilized three different measurement instruments in order to quantify magnitudes of evolution acceptance.

18.4 Study 1: Comparing Indonesian and American Understanding and Reasoning About Evolutionary Change

18.4.1 Methods

Our first study is a comparative, cross-sectional study of the evolutionary reasoning patterns of 529 participants (208 Indonesians and 321 Americans). In terms of Indonesian samples, we recruited the participants who were majoring in Biology Education in one University. For the American samples who were recruited from one University in the eastern United States, we studied participants of comparable ages and educational experiences to the Indonesian pre-service teachers. This sampling approach was used because in many US states, teacher preparation begins after the completion of an undergraduate degree.

In term of genders and ages, the Indonesian sample consisted of 17% males and 83% females, with an age range of 17–33 (M = 20.06). The American sample consisted of 47% males and 53% females, with an age range of 18–40 (M = 20.39). We sampled American students at three time points in their undergraduate degree programs (years 1, 2, and 4, with 106–108 students per year), and Indonesian pre-service biology teachers at four time points (years 1, 2, 3 and 4, with 52 students per year). Overall, despite similar ages, differences in gender are apparent between the two samples.

For Study 1, we used the written form of the ACORNS instrument (Assessment of COntextual Reasoning about Natural Selection; Nehm et al., 2012). The ACORNS is an open-response instrument used to document evolutionary explanations across scenarios differing in contextual features and to identify the degree of conceptual abstraction and cognitive coherence in participants' evolutionary reasoning (Nehm et al., 2012; Opfer et al., 2012). For our current study, we used four ACORNS items differing in two surface features (specifically, different taxa and polarities of trait change). Each item was otherwise isomorphic: "A species of X [plant or animal] [lacks/has] Y. How would biologists explain how a species of X [with or without] Y evolved from an ancestral X species [with or without] Y?" (For our study, X = snail/rose/penguin/elm and Y = poison/thorns/flight/winged seeds).

Two expert graders, who developed the ACORNS instrument and have experience analyzing the responses, scored participants' essays after translation into English by bilingual speakers with training in science education. Each explanation was scored for seven key concepts (KC) (variability, heritability, differential survival, competition, hyper-fecundity (no students ever used this idea, and so it is absent from our results), limited sources, and population distribution change) and four naive ideas or "misconceptions" (MIS) (e.g. needs/goals as causes of trait change, the impact of use/disuse on trait change; intentionality as a cause of trait change; and single-generation adapt/acclimation). The published rubrics of Nehm et al. (2010) were used for scoring. Since participants were given four ACORNS items, the possible score ranges were 0–28 for KCs, and 0–16 for MIS. Inter-rater scoring agreement was strong (κ > 0.8) and disagreements were resolved via deliberation.

We tabulated and quantitatively compared the concepts that participants used in their written ACORNS responses. Furthermore, statistical analyses were utilized in order to examine differences in (1) concept distributions (e.g., core key concepts, other key concepts, and misconceptions) and (2) reasoning patterns (types of explanations). Independent sample t-tests and ANOVAs were calculated using SPSS Statistics V22.

18.4.2 Findings

In their explanations of evolutionary change, American participants used significantly more core concepts of natural selection compared to Indonesian participants (Fig. 18.1). For the concept of *variation*, we found that the Indonesian sample used it significantly less often compared to the American sample, with a medium effect size ($t = -4.66$, $p < 0.01$, $d = 0.42$). Similar results were found for *heritability* ($t = -5.891$, $p < 0.01$, $d = 0.53$) and *differential survival* ($t = -17.768$, $p < 0.01$, $d = 1.58$). Both Indonesian and American participants used three additional key concepts for explaining natural selection: *competition, limited resources* and *changes in population distribution* (Fig. 18.1). Both American and Indonesian participants rarely used *competition* in their explanations ($M = 0.062$, $SD = 0.267$; $M = 0.048$, $SD = 0.236$, respectively), and the difference between samples was not statistically significant ($t = -0.627$, $p = 0.530$, $d = 0.06$). Indonesian participants used the concept of *limited resources* slightly more often than Americans, but the difference was not statistically significant ($t = 1.925$, $p = 0.055$, $t = 0.17$). In contrast, we found statistically significant differences in the use of *population distribution changes* ($t = -5.032$, $p < 0.01$, $d = 0.45$), with American participants using the concept more often than the Indonesian participants.

In terms of naive ideas or misconceptions, we found no significant differences between Indonesian and American participants use of teleological reasoning (frequencies of the *need/goal* concept) ($t = -0.740$, $p = 0.460$, $d = 0.07$) or in their use of the concept of *intentionality* ($t = 0.873$, $p = 0.383$, $d = 0.08$). However, *use/disuse* and *adaptation as acclimation* concepts were significantly different ($t = 6.483$, $p < 0.01$, $d = 0.58$; $t = 10.118$, $p < 0.01$, $d = 0.90$, respectively) and more common in the Indonesian sample ($M = 0.438$, $SD = 0.685$; $M = 1.178$, $SD = 1.046$) than in the American sample ($M = 0.140$, $SD = 0.365$; $M = 0.885$, $SD = 1.029$). These patterns are displayed visually in Fig. 18.2.

In addition to examining individual concepts and total concepts, it is possible to characterize the overall reasoning models used by participants across the four ACORNS items: scientifically normative (only using core or key concepts), mixed (using combinations of key concepts and misconceptions), naïve (only using misconceptions), or no model (not using any ideas relevant to the posed question).

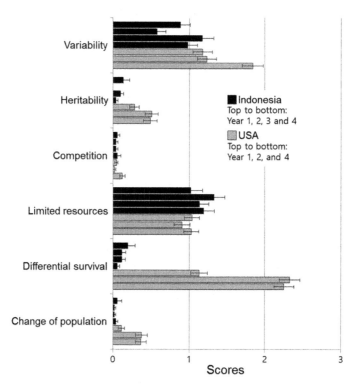

Fig. 18.1 Comparisons of the key concepts used in of Indonesian and American students' ACORNS instrument responses

Looking at Fig. 18.3, it is clear that Indonesian participants (M = 1.091, SD = 0.976; M = 0.952, SD = 0.982; M = 1.005, SD = 1.047, respectively) used significantly fewer scientific models (t = −9.872, p < 0.01, p < 0.01, d = 0.88) and significantly more naïve/no models (t = 4.095, p < 0.01, d = 0.37 and t = 7.452, p < 0.01, d = 0.66, respectively) when compared to their American counterparts (M = 2.128, SD = 1.590; M = 0.623, SD = 0.940; M = 0.399, SD = 0.847, respectively). In contrast, we did not find a significant difference between the Indonesian (M = 0.952, SD = 0.95) and American (M = 0.851, SD = 1.294) samples' uses of mixed models (t = 1.204, p = 0.229, d = 0.11).

Finally, we compared the impact that different surface features (plant vs. animal, trait gain vs. loss) had on participants' evolutionary reasoning (Fig. 18.4). In contrast to novices, evolution experts are not impacted by such context effects, and so context sensitivity provides a measure of expertise (Nehm & Ridgway, 2011). The ANOVA revealed different patterns of context effects in the American and Indonesian samples. For the American sample, there were significant effects of trait polarity (gain/loss) and an interaction effect for key concept use patterns (US: Taxa: $F = 1.1$, $p = 0.29$, $\eta_p^2 = 0.00$, Trait: $F = 75.9$, $p < 0.01$, $\eta_p^2 = 0.19$, Interaction: $F = 40.1$, $p < 0.01$, $\eta_p^2 = 0.11$). In contrast, the Indonesian sample showed a

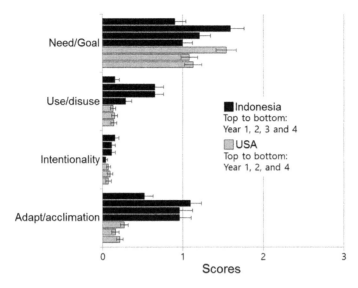

Fig. 18.2 Comparisons of misconception use in the ACORNS responses

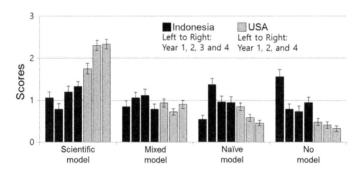

Fig. 18.3 Comparisons of model types by Indonesian and American participants

significant effect of taxon and interaction, but not a significant effect of trait polarity (Taxa: $F = 5.2$, $p = 0.02$, $\eta_p^2 = 0.02$, Trait: $F = 0.0$, $p = 0.95$, $\eta_p^2 = 0.00$, Interaction: $F = 16.3, p < 0.01, \eta_p^2 = 0.07$). It is worth noting that KC use was very low for both gain and loss contexts in the Indonesian sample compared to the American sample.

Examining misconception use patterns by context across the two countries revealed slightly similar patterns for the US and Indonesia. There was a significant effect for both taxa and trait polarity (gain/loss), but no interaction effect (US: Taxa: $F = 49.1$, $p < 0.01$, $\eta_p^2 = 0.13$, Trait: $F = 64.2$, $p < 0.01$, $\eta_p^2 = 0.17$, Interaction: $F = 3.6$, $p = 0.06$, $\eta_p^2 = 0.01$. The Indonesian sample showed significant effects of taxon (animal/plant) and interaction (Taxa: $F = 79.3$, $p < 0.01$, $\eta_p^2 = 0.28$, Trait:

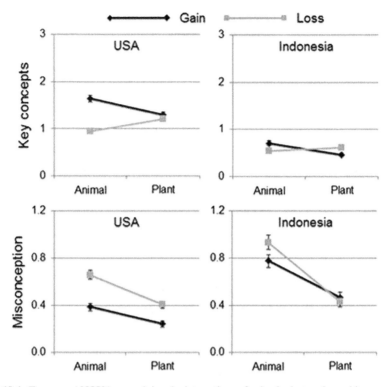

Fig. 18.4 Two-way ANOVAs examining the interactions of animal, plant, gain, and loss contexts on Indonesian participants reasoning

$F = 1.6$, $p = 0.21$, $\eta_p^2 = 0.01$, Interaction: $F = 3.8$, $p = 0.05$, $\eta_p^2 = 0.02$). Overall, it is clear that context is having an impact on some aspects of evolutionary reasoning in both samples (Fig. 18.4).

18.5 Study 2: Clinical Interviews

18.5.1 Methods

In our second study, we performed clinical interviews in order to corroborate findings documented with the written tasks, and to gain deeper insights into Indonesian pre-service biology teachers' understandings and reasoning patterns. Comparable types of studies have been conducted with American samples and have shown that written ACORNS tasks align with oral interviews (Beggrow et al., 2014). However, it is important to confirm this finding with our new sample. 22 Indonesian pre-service biology teachers were randomly chosen from our larger sample. The 22 participants were in their fourth year of the biology teacher preparation program and had completed a course in evolution. Mirroring the demographics of the larger sample, five males and seventeen females participated in the clinical interviews.

Similar to the written tasks, we utilized four ACORNS items in the interview tasks, but varied the surface features (i.e. Snail, Cactus, opossum and lily as taxa, and teeth, thorns, tail, and petals as traits). The interviews lasted between eight and 15 min. Participants' responses were coded similarly to those in Study 1.

18.5.2 Findings

Similar to the written tasks, we found that participants displayed a wide array of reasoning models, ranging from normative to naive, in their clinical interviews. Below we provide two quotes representing a normative scientific model and a naive reasoning model:

> **Normative**: [*translated from Indonesian*] Perhaps, at first there were two variations of opossums, with tail and without tails. Then, unfortunately tailed opossums were more able to survive than opossums without tails. Thus, tailed opossums keep breeding and produced more tailed opossums, while opossum without tail by the time could not survive and finally reached extinction (F17)

> **Naive**: [*translated from Indonesian*] Perhaps, the lily did not need petals to attract insects, so lilies reduced petals because there was no benefit for the life of lily (F12).

As shown in Fig. 18.5, we found that about half of the Indonesian pre-service biology teachers explained evolutionary change using the *need/goal* concept (around 46%) and the *resources* concept (42%). In addition, about 18% of participants used the concept of *use/disuse,* and about 17% used *environmental effects, adaptation/acclimation* and *variation* concepts (Fig. 18.5). Here we also found that participants' thinking about environmental factors was often associated with use-disuse and acclimation misconceptions. Quotes from two participants (F1 and F10) illustrate this point:

> [*translated from Indonesian*] "Perhaps it is influenced by the environment where the opossum lives. It might demand the opossum to not use its tail when they do their activities, thus the tail is gone" (F1)

> [*translated from Indonesian*] "When at first, snail could find the food that easily can be eaten, suddenly the environment where the snail lives was changed and it made the snail's food gone and just remains the food that need more effort for snail to be chewed. Thus, continuously eating that kind of food, over time the teeth, one by one, are grown and it is passed down to their descendants" (F10)

Very few participants discussed *changes in population distribution, differential survival,* and *intentionality* concepts (each about 3–4%). Remarkably, none of the participants used the concepts of *hyperfecundity/over-production of offspring, competition* or *heredity.* Clinical interview findings generally aligned with written tasks in terms of the relative frequencies of concepts (e.g. teleology being the most

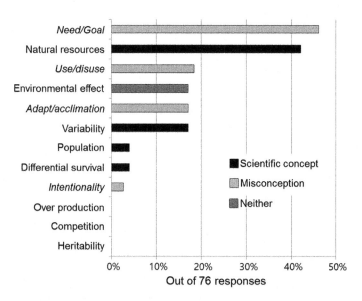

Fig. 18.5 Concept frequencies from the clinical interviews

common naive idea), but the magnitudes of particular ideas differed to some degree (e.g. variation vs. resources). In addition, similar to study 1, in the study 2 we also found that mixtures of normative and non-normative ideas were common.

18.6 Study 3: Acceptance of Evolutionary Theory

18.6.1 Methods

In order to investigate Indonesian pre-service biology teacher's acceptance of evolutionary theory, we employed three different instruments designed to measure the construct of "evolution acceptance." The first instrument, the MATE (Measure of Acceptance of the Theory of Evolution), was developed by Rutledge and Warden (1999) and has been widely used in evolution education research. It was administered to 208 Indonesian pre-service biology teachers early in our study, prior to the availability and translation of two newer instruments, namely the I-SEA (Inventory of Student Evolution Acceptance; developed by Nadelson and Southerland, 2012) and the GAENE 2.1 (Generalized Acceptance of EvolutioN Evaluation version 2.1; developed by Smith et al., 2016). The latter two instruments were translated, checked, and administered to a second sample of 340 Indonesian pre-service biology teachers. Thus, two different participant samples were used in our studies of acceptance.

Demographically, the first sample was comprised of equal numbers of pre-service teacher participants from their first, second, third, and fourth years. In term of genders and ages, the sample was 17% male and 83% female, with an age range of 17–33 years (M = 20.06). The second sample was gathered after the first sample, and also contained equal numbers of pre-service teacher participants from their first, second, third, and fourth years. In terms of genders and ages, the second sample contained 13% males and 87% females, with an age range of 17–23 years (M = 19.40).

We calculated raw means for the MATE so that we could compare our findings to prior MATE work, which has not employed Rasch methods. In contrast, we used ConQuest v.4 to perform raw and Rasch (Partial Credit Model) measures for the GAENE and I-SEA. The Rasch fit-indices for all I-SEA and GAENE items met the suggested cutoffs of 0.6–1.40 (Boone et al., 2014). The internal consistency (Cronbach's alpha and Plausible Value Reliability) were, respectively, 0.839 and 0.795 for the GAENE and 0.831 and 0.727 for the I-SEA.

Unlike the MATE and GAENE, the I-SEA consisted of three different subscales (human evolution, microevolution and macroevolution). Consequently, we tested whether the I-SEA best fit a one-dimensional or three-dimensional model. We used what has been suggested by Bond and Fox (2013) regarding the best model of the data using Rasch analysis. Based on the same data set, we used the value of Deviance and AIC for comparing the I-SEA and GAENE data. We found that the three-dimensional model had lower Final Deviance and AIC (19438.08 and 19638.08) and higher chi-square (χ^2 = 1271.48, df = 21, p < 0.01) compared to one-dimension (19523.55 and 19713.55) and with a lower chi-square value (χ^2 = 279.13, df = 23, p < 0.01). Consequently, we used a three dimensional model to describe the I-SEA results. Recent empirical work on American undergraduates also suggests that the I-SEA is best modeled as a multidimensional instrument (Sbeglia & Nehm, 2017).

18.6.2 Findings: MATE

The average MATE score was 65.06 (SD = 6.76). Based on Rutledge (1996), this score is considered "moderate acceptance" of evolutionary theory, although this average is near the border between moderate and low acceptance. Analyzing the results individually, only 5% of the Indonesian participants had "high acceptance" of evolutionary theory, and more than 40% had "moderate" to "low acceptance" (48% and 45%, respectively). In addition, only 2% of the sample had very low acceptance, and not a single participant had very high acceptance.

Figure 18.6 depicts Indonesian pre-service biology teachers' MATE scores compared to previous studies. The most similar acceptance levels are from Turkish participants studied by Deniz et al. (2008) and Korean pre-service biology teachers in their fourth year (Ha et al., 2012). Indonesian MATE scores are higher than the Turkish scores (N = 132, M = 63.69, SD = 12.2), although the difference is not

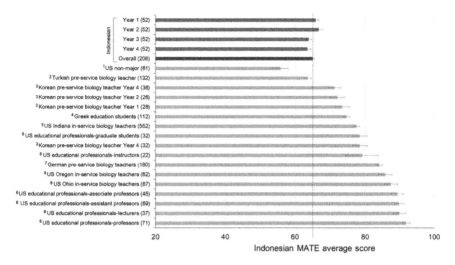

Fig. 18.6 Average scores for MATE ([1]Rutledge and Sadler (2007); [2]Deniz et al. (2008); [3]Ha et al. (2012); [4]Athanasiou and Papadopoulou (2012); [5]Rutledge and Warden (2000); [6]Nadelson & Sinatra (2008); [7]Großschedl et al., (2014); [8]Trani (2004); [9]Korte (2003)). Indonesian = the current study

statistically significant (t = 1.33, p = 0.18, d = 0.15). In contrast, the Indonesian sample has significantly lower scores (t = 4.64, p < 0.01, d = 0.82) than Korean pre-service biology teachers in year 4 (N = 38, M = 71.26, SD = 11.10).

18.6.3 Findings: I-SEA and GAENE

The raw mean of I-SEA scores was 3.27 (N = 340, SD = 0.33) which is significantly lower (t = 7.60, p < 0.001, d = 0.73) than an American sample studied by Nadelson and Hardy (2015; N = 159, M = 3.61, SD = 0.67). Indonesians' acceptance of macroevolution, microevolution and human evolution were, respectively, 3.42 (SD = 0.42), 3.44 (SD = 0.43) and 2.97 (SD = 0.48). Compared to the American sample (N = 159), we found that the Indonesian sample was significantly less likely to accept macroevolution (t = 10.65, p < 0.001, d = 1.04), microevolution (t = 2.51, p = 0.012, d = 0.24) and human evolution (t = 5.78, p < 0.001, d = 0.56), with large, small, and medium effect sizes, respectively (macroevolution, M = 3.92, SD = 0.61; microevolution, M = 3.57, SD = 0.72 and human evolution, M = 3.33, SD = 0.91).

For the GAENE, the raw mean for our sample was 2.96 (N = 340; SD = 0.31). There is only one previously published study using the GAENE (Smith et al., 2016). In this study, 671 American high school students and undergraduates (M = 3.74; SD = 0.35) were given a five point scale (in contrast, we used a four-point scale). In order to make the findings between the two studies comparable, we converted our four-scale data to five scales with the methods suggested by IBM Support (http://www-01.ibm.com/support/docview.wss?uid=swg21482329) and it fell to $M = 3.62$ (SD = 0.40). Indonesians have significantly lower acceptance than documented in the American sample (t = 4.90, p < 0.01, d = 0.33).

In addition to using raw scores, we utilized Rasch person measures from the I-SEA and GAENE to analyze acceptance of evolutionary theory. Based on GAENE scores, we found that the mean person measure was 1.24 (SD = 0.98), indicating a generally positive attitude. Approximately 91% of the sample had positive person values; most individuals were above the zero point. Analysis of the I-SEA scores generally produced similar findings. The average person measure was 0.21 (SD = 0.37), indicating a generally positive acceptance level. Nevertheless, approximately 24% had negative person measures. In addition more than a fifth of the sample ($\sim$23%) had scores near zero.

Looking at the different dimensions of acceptance in Fig. 18.7, it is apparent that the sample is most accepting of microevolution, and less accepting of macroevolution and human evolution. Average person values were positive for macroevolution and microevolution ($M = 0.25$, SD = 0.46 and $M = 0.52$, SD = 0.52 respectively) and negative for human evolution ($M = -0.08$, SD = 0.31). Less than half of the sample (40%) is above the zero point for human evolution. In contrast, only 13% had negative values for microevolution.

The three measurement instruments illustrate somewhat different perspectives on Indonesian pre-service teachers' acceptance of evolution, which is not surprising given that these tools conceptualize acceptance in slightly different ways (see Smith et al., 2016 for a detailed review). The MATE, GAENE and I-SEA results suggest that the Indonesian pre-service teachers we studied have lower levels of acceptance

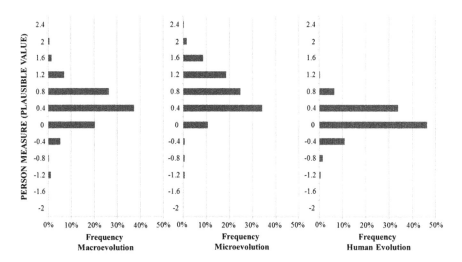

Fig. 18.7 Acceptance of macroevolution, microevolution and human evolution (I-SEA)

compared to American samples. However, the I-SEA results follow expected patterns of acceptance, with human evolution the lowest, followed by macroevolution and microevolution (the highest acceptance). Rasch scores generally corroborate the raw scores, but suggest more positive overall acceptance levels.

18.7 Discussion

Indonesia provides a unique context for exploring long-standing generalizations about evolution education, such as the extent to which religion and culture influence evolutionary understanding, reasoning patterns, and acceptance levels. This Muslim-majority democracy considers religious growth to be a central goal of the national curriculum, and teachers are expected to make explicit connections between science content (including evolution) and religion in the classroom. Given these unique aspects of Indonesian education, our study focused on cognitive and affective measures of large samples of pre-service biology teachers from Indonesia and compared to the findings from other studies.

Qualitatively, our study documented many similarities between Indonesian pre-service biology teachers and American participants' evolutionary knowledge and reasoning patterns despite the different religious and cultural backgrounds of the two samples (e.g., mostly Muslim vs. mostly Christian, Asian vs. Western). For example, Indonesian and American evolutionary reasoning models were similar in their overall form; we found that different explanatory models (e.g., normative, mixed, naive) were employed across the four evolutionary problems presented (i.e. evolutionary gain and loss in plants and animals); purely scientific models were used in some cases, and purely naive models in others (Fig. 18.3). This indicates that consistent or "coherent" mental models of evolutionary causation (naive or normative) do *not* characterize the Indonesian (or American) samples. This corroborates prior work suggesting that evolutionary contexts play an important role in evolutionary thinking and reasoning (Nehm & Ha, 2011). Nevertheless, it is important to note that American participants used scientific models almost twice as often as the Indonesian participants, indicating greater progress towards conceptual abstraction (Nehm & Ridgway, 2011).

When approaching evolutionary change scenarios, Indonesian participants recruited similar types of normative concepts from long-term memory that have been documented in other samples (Nehm & Schonfeld, 2007). For instance, *variability* and *limited resources* were commonly used when explaining change; competition and heritability less so. Like their American counterparts, hyper fecundity did not play a role in explaining change. Many naive ideas or misconceptions discussed in the literature were also found in Indonesian pre-service teachers. Ayala (1970) noted that the concepts of *use/disuse* and *adaptation/acclimation* are misconceptions inappropriate for explaining evolutionary change through time. Nevertheless, *use/disuse* and *adaptation/acclimation* were common in the Indonesian sample compared to the American sample.

Interestingly, the frequencies of the *need/goal* concept were similar in the Indonesian and American samples. Cognitive psychologists have argued that teleological reasoning is a pervasive feature of human cognition that transcends culture and religion, and that teleological reasoning is a major barrier to normative evolutionary understanding (Kelemen, 2012). Our studies of Indonesian pre-service teachers lend further support to both claims; teleological reasoning was commonly documented in the clinical interviews and in the written explanatory tasks, and was found to be associated with the use of fewer normative evolutionary concepts. Thus, like participants from other cultures, religions, geographic regions, and educational levels, teleological reasoning is one of the most problematic features of evolutionary thinking in Indonesian pre-service teachers.

Rather than first focusing on the generation and availability of existing variation (e.g. by mutation, genetic recombination, and heredity), and subsequently discussing environmental sorting of this variation, many participants viewed the environment as the initial event or cause driving change. Participants failed to distinguish between the factors that were the main causes and effects, and those factors having a supporting role (*enablers*). As Sloman (2005) emphasized, in order to obtain a causal model explaining scientific phenomena, one has to successfully distinguish enablers and cause and effect variables.

Based on the findings from our studies of knowledge and reasoning (study 1 and 2), it is clear that many Indonesian pre-service biology teachers did not utilize normative scientific explanations. According to Norris and Phillips (2003) and Bybee (1997), scientific explanation is a fundamental practice emblematic of scientific literacy. Clearly, educational activities that foster broader application of scientific concepts to the task of evolutionary explanation are needed in Indonesian teacher education which was suggested by the large interaction effect found in our first study (Fig. 18.4). As Bybee (1997) emphasized, one cannot be a scientifically literate person without being able to explain evolutionary change.

Although Indonesian pre-service teachers utilized many similar concepts and arranged them in similar explanatory models as American participants, they displayed lower *magnitudes* of evolutionary knowledge and higher magnitudes of some misconceptions. This finding was reflected in both the written tasks and clinical interviews for individual concepts (e.g. differential survival, use-disuse inheritance) and for overall reasoning models (e.g. exclusively normative explanations). Overall, it is possible to view the conceptual development of Indonesian pre-service teachers within the frameworks developed for American samples, and place the Indonesian sample at a lower level of normative understanding. Thus, many of the recommendations for effective evolution education for teachers in the USA and elsewhere (e.g. Nehm & Schonfeld, 2007) could prove valuable for Indonesian teachers (e.g. explicit attention to misconceptions such as teleology, enrichment with inquiry case studies).

Although Indonesian pre-service teachers displayed lower knowledge of evolution than American samples, it is also important to determine if this finding aligns with evolution acceptance levels. In order to address this issue, we used three instruments—the MATE, I-SEA, and GAENE to measure acceptance patterns. As expected, the different instruments provided somewhat different perspectives on evolution acceptance in the Indonesian sample. The I-SEA results were in line with previous work in the USA, confirming anticipated patterns, namely that acceptance of microevolution was highest, human evolution was the lowest, and macroevolution was at a level intermediate between these two extremes. This comparative gradient of acceptance appears robust across religions and cultures.

The MATE results showed moderate to low levels of acceptance in the sample, with values comparable to Turkish pre-service teachers (Deniz et al., 2008), but significantly below values for Korean biology teachers, American biology teachers, and German biology teachers. Borgerding et al. (2016) suggested that people tend to be more accepting of evolution after they have completed more coursework related to evolution. In the Indonesian sample, we did not see any significant increase in acceptance through the four years of the program (Fig. 18.6). In many prior studies (e.g. Nehm & Schonfeld, 2007), self-selection effects could be impacting inferences about the impact of coursework on acceptance (particularly in the USA, where enrollment in biology education programs is often late in one's academic career). The Indonesian sample was not characterized by this possible sampling bias.

The newest evolution acceptance measure, the GAENE, produced scores that do not align with findings from the MATE. Indeed, GAENE scores for the Indonesian sample were found to be comparable with values from American high school and college students. Given that this instrument is new, and comparative results are lacking, it is difficult to interpret this finding.

Our findings on Indonesian pre-service teachers' acceptance patterns align to some degree with the findings from a research project called "Islam and Evolution" presented at a symposium in McGill University, Canada. In a news report by Hoag (2009), it was suggested that Indonesians, especially high school students, had a relatively good understanding on the scientific validity of evolutionary theory. It is an open question as to how teachers' views compare to those of students, particularly in light of the new curriculum. Comparable measures will need to be employed in such studies.

In closing, our studies of Indonesian pre-service teachers represent an important, but very incomplete, first step towards understanding the complex relationships between culture, religion, and evolutionary understanding in this understudied region. Further work in this important sociocultural context is clearly needed, and is likely to enrich our understanding of how best to approach the challenges of evolution education throughout the world. Further work on how teachers and students make sense of the religious connections to science content could be a particularly valuable research direction. Overall, our findings from Indonesia suggest that many cognitive challenges to evolutionary thinking and reasoning transcend religious affiliation, culture, geography, and formal education.

Acknowledgements We thank to the editors for allowing us to contribute to this important volume.

References

Athanasiou, K., & Papadopoulou, P. (2012). Conceptual ecology of the evolution acceptance among Greek education students: Knowledge, religious practices and social influences. *International Journal of Science Education, 34*(6), 903–924.

Ayala, F. J. (1970). Teleological explanations in evolutionary biology. *Philosophy of Science, 37*(1), 1–15.

Beggrow, E. P., Ha, M., Nehm, R. H., Pearl, D., & Boone, W. J. (2014). Assessing scientific practices using machine-learning methods: How closely do they match clinical interview performance? *Journal of Science Education and Technology, 23*(1), 160–182.

Bond, T. G., & Fox, C. M. (2013). *Applying the Rasch model: Fundamental measurement in the human sciences*. New York, NY: Psychology Press.

Boone, W. J., Staver, J. R., & Yale, M. S. (2014). *Rasch analysis in the human sciences*. Dordrecht: Springer.

Borgerding, L. A., Deniz, H., & Anderson, E. S. (2017). Evolution acceptance and epistemological beliefs of college biology students. *Journal of Research in Science Teaching, 54*(4), 493–519.

Bybee, R. W. (1997). *Achieving scientific literacy: From purposes to practices*. Portsmouth, NH: Heinemann.

CIA [Central Intelligence Agency]. (2016a, November 10). *The world factbook: East and Southeast Asia Indonesia*. Retrieved from https://www.cia.gov/library/publications/the-world-factbook/geos/id.html.

CIA [Central Intelligence Agency]. (2016b, November 14). *The World factbook: North America United States*. Retrieved from https://www.cia.gov/library/publications/the-world-factbook/geos/us.html.

Deniz, H., Donnelly, L. A., & Yilmaz, I. (2008). Exploring the factors related to acceptance of evolutionary theory among Turkish preservice biology teachers: Toward a more informative conceptual ecology for biological evolution. *Journal of Research in Science Teaching, 45*(4), 420–443.

Großschedl, J., Konnemann, C., & Basel, N. (2014). Pre-service biology teachers' acceptance of evolutionary theory and their preference for its teaching. Evolution: Education and Outreach, 7(1), 18-33.

Ha, M., Haury, D. L., & Nehm, R. H. (2012). Feeling of certainty: Uncovering a missing link between knowledge and acceptance of evolution. *Journal of Research in Science Teaching, 49*(1), 95–121.

Hoag, H. (2009, April 3). Muslim students weigh in on evolution: In Indonesia and Pakistan, questions about how science and faith can be reconciled. In *Nature News*. Retrieved from: http://www.nature.com/news/2009/090403/full/news.2009.230.html.

Kelemen, D. (2012). Teleological minds: How natural intuitions about agency and purpose influence learning about evolution. In K. Rosengren & E. M. Evans (Eds.), *Evolution challenges: Integrating research and practice in teaching and learning about evolution*. New York, NY: Oxford University Press.

Korte, S. E. (2003). *The acceptance and understanding of evolutionary theory among Ohio secondary life science teachers*. Doctoral dissertation, Ohio University.

Martin, M. O., Mullis, I. V. S., Foy, P., & Hooper, M. (2016). *TIMSS international results in science*. Lynch School of Education, Boston College: IEA TIMSS & PIRLS International Study Center.

Mendikbud [Indonesian Ministry of Education and Culture]. (2016). *Salinan Lampiran Permendikbud No. 21th 2016 tentang Standar Isi Pendidikan Dasar dan Menengah* [A copy of the additional documents of Indonesian ministry of education and culture regulation number 21 year 2016 on Content Standards for Primary and Secondary Education].

Nadelson, L. S., & Hardy, K. K. (2015). Trust in science and scientists and the acceptance of evolution. *Evolution: Education and Outreach, 8*(1), 1–9.

Nadelson, L. S., & Sinatra, G. M. (2008). Educational psychologists' knowledge of evolution. In *Annual meeting of the American psychological association, Boston.*

Nadelson, L. S., & Southerland, S. (2012). A more fine-grained measure of students' acceptance of evolution: development of the inventory of student evolution acceptance—I-SEA. *International Journal of Science Education, 34*(11), 1637–1666.

Nehm, R. H., & Ha, M. (2011). Item feature effects in evolution assessment. *Journal of Research in Science Teaching, 48*(3), 237–256.

Nehm, R. H., & Ridgway, J. (2011). What do experts and novices "see" in evolutionary problems? *Evolution: Education and Outreach, 4*(4), 666–679.

Nehm, R. H., & Schonfeld, I. S. (2007). Does increasing biology teacher knowledge of evolution and the nature of science lead to greater preference for the teaching of evolution in schools? *Journal of Science Teacher Education, 18*(5), 699–723.

Nehm, R. H., Beggrow, E. P., Opfer, J. E., & Ha, M. (2012). Reasoning about natural selection: diagnosing contextual competency using the ACORNS instrument. *The American Biology Teacher, 74*(2), 92–98.

Nehm, R. H., Rector, M. A., & Ha, M. (2010). "Force-Talk" in evolutionary explanation: Metaphors and misconceptions. *Evolution: Education and Outreach, 3*(4), 605–613.

Norris, S. P., & Phillips, L. M. (2003). How literacy in its fundamental sense is central to scientific literacy. *Science Education, 87*(2), 224–240.

OECD. (2016). *PISA 2015 results in focus.* OECD Publishing.

Opfer, J. E., Nehm, R. H., & Ha, M. (2012). Cognitive foundations for science assessment design: Knowing what students know about evolution. *Journal of Research in Science Teaching, 49*(6), 744–777.

Parker, L. (2017). Religious environmental education? The new school curriculum in Indonesia. *Environmental Education Research, 23*(9), 1249–1272.

Rutledge, M. L. (1996). *Indiana high school biology teachers and evolutionary theory: Acceptance and understanding.* Doctoral dissertation, Ball State University.

Rutledge, M. L., & Sadler, K. C. (2007). Reliability of the measure of acceptance of the theory of evolution (MATE) instrument with university students. *The American Biology Teacher, 69*(6), 332–335.

Rutledge, M. L., & Warden, M. A. (1999). The development and validation of the measure of acceptance of the theory of evolution instrument. *School Science and Mathematics, 99*(1), 13–18.

Sbeglia, G., Nehm, R. H. (2017). *Does evolution acceptance differ across biological scales? A Rasch analysis of the I-SEA.* Paper presented at the national association for research in science teaching (NARST) international conference, San Antonio, Texas.

Skutnabb-Kangas, T. (2000). *Linguistic genocide in education—Or worldwide diversity and human rights?* Mahwah, NJ: Lawrence Erlbaum Associates.

Sloman, S. (2005). *Causal models: How people think about the world and its alternatives.* New York, NJ: Oxford University Press.

Smith, M. U., Snyder, S. W., & Devereaux, R. S. (2016). The GAENE—Generalized acceptance of evolution evaluation: Development of a new measure of evolution acceptance. *Journal of Research in Science Teaching, 53*(9), 1289–1315.

Sudargo, F. (2009). *Syllabus of evolution subject for biology education department* [Silabus Mata Kuliah Evolusi untuk Mahasiswa Pendidikan Biologi]. Retrieved from http://file.upi.edu/Direktori/FPMIPA/JUR_PEND_BIOLOGI/195107261978032-FRANSISCA_SUDARGO/SILABUS_MK_06_evolusi.pdf.

Trani, R. (2004). I won't teach evolution; it's against my religion. And now for the rest of the story …. *The American Biology Teacher, 66*(6), 419–427.

Arif Rachmatullah just obtained his M.Ed. in Science Education from Kangwon National University (KNU), Republic of Korea. Prior to enrolling masters degree in Korea, he went to Indonesia University of Education (UPI) for his bachelor degree majoring in Biology Education. His research interests include socio-cultural effects on science learning, especially on the topic of ecology, evolution and genetics, and science learning motivation. He is now a research assistant in the Division of Science Education at KNU.

Ross Nehm is an associate professor of ecology and evolution, and associate director of the Ph.D. Program in Science Education at Stony Brook University (SUNY). His research focuses on student thinking about biological concepts, particularly natural selection and evolution. Dr. Nehm currently serves as Associate Editor of the *Journal of Research in Science Teaching* (JRST), co-Editor-In-Chief of the journal *Evolution Education and Outreach*, Associate Editor for the journal *Science & Education*, Monitoring Editor *for CBE-Life Sciences Education*, and board member of several other journals. Dr. Nehm's major awards include a CAREER award from the National Science Foundation and a mentoring award from CUNY. In 2013-14 Dr. Nehm was named an Education Mentor in the Life Sciences by the U.S. National Academies. He also serves on the research advisory boards of several national science education projects, and has served as Panel Chair for several NSF programs.

Fenny Roshayanti is an Assistant Professor of Biology Education at Universitas PGRI Semarang (UPGRIS), Indonesia. She teaches undergraduates in Biology Education program and masters in Educational Management program. Her research interests include science motivation, nature of science, scientific reasoning and biology and evolution education in the context of Indonesia. She is involved in several research projects, such as investigating the improvement of Indonesian students' science inquiry skills and environmental literacy in coastal areas, funded by Indonesian Ministry of Research and Higher Education.

Minsu Ha is an Assistant Professor of Science Education at Kangwon National University (KNU), Republic of Korea. He teaches undergraduate, master and doctoral level courses in biology and general science education program at KNU. His research interests include evolution education, computer and science learning, science motivation, assessment and learning process in science education. He is engaged in a research projects about the development of computer scoring for science education.

Chapter 19
A Glimpse of Evolution Education in the Malaysian Context

Yoon Fah Lay, Eng Tek Ong, Crispina Gregory K. Han and Sane Hwui Chan

Abstract Darwin's Theory of Evolution is the widely held notion that all life is related and has descended from a common ancestor. Darwin's general theory presumes the development of life from non-life and stresses a purely naturalistic (undirected) "descent with modification". That is, complex creatures evolve from more simplistic ancestors naturally over time. Malaysia is a multi-racial and multi-religion country in the Southeast Asian region. Due to different religious backgrounds, evolutionary theory is always a sensitive and hotly-debated issue in the teaching of biology in Malaysian schools. This book chapter evaluates the place of evolutionary theory biology curriculum in Malaysian secondary schools. Emphases given to evolutionary theory in the biology teacher education programmes at two public universities as well as prospective biology teachers' attitudes towards evolutionary theory are also investigated. It can be concluded that the level of acceptance on evolutionary theory among Malaysian prospective biology teachers is low as evolutionary theory was not fully understood. Hence, the introduction of specific courses on biological evolution that cover its most fundamental principles is crucially needed in the biology teacher education programmes in the Malaysian context.

Y. F. Lay (✉) · C. G. K. Han
Faculty of Psychology and Education, Universiti Malaysia Sabah,
Kota Kinabalu, Sabah, Malaysia
e-mail: layyoonfah@yahoo.com.my

C. G. K. Han
e-mail: crispina@ums.edu.my

E. T. Ong
Faculty of Human Development, Sultan Idris Education University,
Tanjong Malim, Perak Darul Ridzuan, Malaysia
e-mail: ong.engtek@fppm.upsi.edu.my

S. H. Chan
Kian Kok Middle School, Kota Kinabalu, Sabah, Malaysia
e-mail: shanee0212@gmail.com

© Springer International Publishing AG, part of Springer Nature 2018
H. Deniz and L. A. Borgerding (eds.), *Evolution Education Around the Globe*, https://doi.org/10.1007/978-3-319-90939-4_19

Keywords Evolution education · Darwin's theory of evolution
Biology curriculum · Biology teachers · Teacher education programmes

19.1 Introduction

Darwin's Theory of Evolution is the widely held notion that all life is related and has descended from a common ancestor. Darwin's general theory presumes the development of life from non-life and stresses a purely naturalistic (undirected) "descent with modification". That is, complex creatures evolve from more simplistic ancestors naturally over time. In a nutshell, as random genetic mutations occur within an organism's genetic material, the beneficial mutations are preserved because they aid survival—a process known as "natural selection." Natural selection is the preservation of a functional advantage conferred by genetic mutation that enables a species/individuals to compete better in the wild. These beneficial mutations are passed on to the next generation. Over time, beneficial mutations accumulate, and the result is an entirely different organism (not just a variation of the original, but an entirely different creature).

Cavallo and McCall (2008a, 2008b, p. 522) stated that evolution education is multi-purposeful in that 'students may learn the science of what the theory states, the social significance of the theory, and its importance in understanding the very nature of science as tentative and dynamic.' The National Academy of Sciences, NAS (2008) stated that evolutionary theory is supported by empirical, data-driven evidence and explanations. Undoubtedly, new evidence and studies in evolutionary theory have changed scientific understanding through time and will continue as new discoveries and evidence are added to the existing knowledge base. Many people assume that the theory put forth by Darwin in *Origin of Species* is the final say on evolutionary theory, but Darwin's original theory evolved when his ideas were merged with ideas from genetics to become the Modern Synthesis (Rusell, 2011). The Modern Synthesis is a 20th-century union of ideas from different biological specialties which provide a widely accepted account of evolution (Ayala, 2008).

Scientific understanding of biological evolution is complex (Miller, 1999; Nadelson, 2009; Tidon & Lewontin, 2004) as evolution is a continuous process governed by a variety of mechanisms such as mutations and natural selection occurring in no particular direction and multiple levels (micro to macro) (Miller et al., 2006; Rice, Clough, Olson, Adams, & Colbert, 2015). Controversy and misconception exists in the public regarding the validity of evolutionary theory as biological evolution is not directly observable or testable (Miller et al., 2006; Rice et al., 2015). Therefore, it is not surprising that people may hold incomplete knowledge or misconceptions about the processes (Miller, 1999; Miller et al., 2006; Rice et al., 2015). Nonetheless, biological evolution provides an important context as the unifying idea in biology which makes the theory important especially in biological sciences (Borgerding et al., 2015; Rice et al., 2015). First, biological evolution provides a useful context for apprising our current understanding of the natural world especially in various fields such as in conservation, agriculture, environmental change, and forensics (Nadelson, 2009; Rice et al., 2015). For instance, from a medical perspective, biological evolution explains relationships between human health and environment. This is due to the fact that humans are subject to natural selection. Evolution explains the origin of disease, resistance to antibiotics, and viral function, leading to a better understanding on how to deal with current and future pathogens more effectively (Rice et al., 2015). Other than that, biological evolution can be a useful explanation in any biology-based course, whether it is a general biology course for non-majors or a graduate level seminar on molecular biology (Rice et al., 2015).

19.2 Evolutionary Theory in Biology Education

Evolutionary theory plays a central role in science education, particularly, in the biology curriculum because understanding evolution is fundamental for understanding biology. It is the only scientific explanation for the biodiversity of living things as explains the origin of life, the variety of abiotic components, changes that occur within a population, and how new species evolve across time. Furthermore, understanding evolution is an important part of scientific literacy. As stated by Dobzhansky (1973), "seen in the light of evolution, biology is, perhaps, intellectually the most satisfying and inspiring science. Without that light it becomes a pile of sundry facts—some of them interesting or curious but making no meaningful picture as a whole" (p. 129). Without the knowledge about evolution, nothing in biology makes sense. Thus, effective teaching of evolution is essential for students to understand and appreciate Mother Nature.

Although the theory of evolution is the foundation for modern biology (Chinsamy & Plagányi, 2008; Hermann, 2008; Moore & Cotner, 2009a; Rutledge & Sadler, 2011), a review of literature reveals that teaching this vital principle of biology is often considered as controversial (Chinsamy & Plagányi, 2008; Cotner, Brooks, & Moore, 2010; Hildebrand et al., 2008) as are other various constituents of biology, like the cell theory, germ theory, and molecular genetics (Rutledge & Sadler, 2011). The teaching of evolution has become socially controversial in many Western societies. According to Miller et al. (2006), one in three American adults rejects the theory of evolution because of religious background or acceptance of the supernatural as explanations for the diversity of life on earth. This also resonates among many undergraduate students. Moore and Cotner (2009b) has demonstrated in their study that one in ten undergraduate biology majors agree that the account of Evolution as 'not a scientifically valid theory'. Pro-evolution scientists, science educators and their societies are adamant that Biology and its various branches of science cannot be appropriately understood outside the context of evolution (Wiles, 2010). Wiles (2014) inquired whether students' acceptance of evolution should be a goal of science education, and, thus whether widespread rejection of evolution by the society does, in fact, indicate a failure of science education. Studies have shown that there has been some arguments among various educators and researchers regarding the matters (Nehm & Schonfeld, 2007; Smith & Siegel, 2004).

These controversies have also caused a serious debate in the Muslim world (i.e., predominantly Islamic countries, as well as in those countries where their main population consists of Muslim populations). Relatively poor education standards and misconceptions regarding evolutionary theory have caused the majority of Muslims to reject the theory and believe that the theory contradicts Islamic beliefs (Hameed, 2008). Based on a sociological study analyzing religious patterns among the public in Muslim countries (Indonesia, Pakistan, Kazakhstan, Egypt, Malaysia, and Turkey), respondents were asked to complete a questionnaire reflecting their fundamental religious beliefs with respect to evolution. Research findings revealed that only 16% of Indonesians, 14% of Pakistanis, 8% of Egyptians, 11% of Malaysians, and 22% of Turks agreed that Darwin's theory is probably or most certainly true. Conversely, Kazakhstan showed different religious patterns compared to the other five countries. In fact, a majority of Kazakhs believed in the theory, and only 28% of them thought that evolutionary theory is false (Hassan, 2007). It is perhaps not too surprising that an individual's religious and cultural background may impact their acceptance of the evolutionary theory. A person's religious beliefs may influence the learning process and impact one's understanding of evolutionary theory (Asghar, Hameed, & Farahani, 2014).

19.3 Place of Evolutionary Theory in the Malaysian Biology Curriculum

Malaysia is one of the Southeast Asian countries with a total landmass of 330,803 km^2 separated by the South China Sea into two regions, Peninsula Malaysia and East Malaysia. Malaysia is known as a secular and multi-ethnic country with a National Principle of believing in God. With a population comprising three major ethnicities (Malay, Chinese, and Indian), approximately 61.3% of the 30 million person population are Muslims, 19.8% Buddhists, 9.2% Christians, 6.3% Hindu, 1.3% Confucian, Taoist, and other traditional Chinese religions. In this survey, 0.7% declared no religion, and the remaining 1.4% practices other religions (Malaysia Department of Statistics, 2012). With Muslims representing approximately 60% of the population, most governmental policies are influenced by the majority Muslim public opinion. Thus, the religious and cultural backgrounds of the nation influence the national education policy-making.

Since national independence in 1957, Malaysian education has undergone tremendous change and transformation over the years. Consequently, the Malaysian system of education is comprised of two types of primary schools: national primary schools in which Bahasa Malaysia is the language of instruction (this is the Malay language, and also the official language of the country), and the national type Chinese and Tamil schools in which Mandarin and Tamil, respectively, are the languages of instruction. Both the national and the national type schools use the national curriculum.

Malaysia provides 12 years of free pre-primary, primary (Years 1–6) and secondary education (Forms 1–5) (Ministry of Education Malaysia, 2017a). Students begin their pre-primary education at age 6. Primary schooling (Years 1–6) is compulsory for all children. It is divided into two levels: Level 1 (Years 1–3) and Level 2 (Years 4–6). Upon completion of lower secondary education (Forms 1–3), students continue their schooling at the upper secondary level (Forms 4–5) in the arts, sciences, or technical and vocational streams. After completing Form 5, students may opt to enroll in Lower 6 and subsequently Upper 6 (the two final years of schooling typically taken by students who intend to enroll in universities in Malaysia), pre-university foundation/matriculation/diploma programmes (offered at government and private universities in Malaysia), or skill, technical, and vocational programmes.

The education system has been centrally organized and managed by the Ministry of Education in terms of the national philosophy, curriculum, policies and content. According to the *Falsafah Pendidikan Negara* (National Philosophy of Education), the main goal of education in Malaysia is to produce citizens who are intellectually, spiritually, emotionally, and physically balanced and harmonious based on a firm belief in and devotion to God (Ministry of Education Malaysia, 2017b).

High school students are often unwilling to learn about evolution due to a perceived conflict with their religious beliefs. Therefore, one might encounter challenges when teaching evolution to students who have no interest to learn it (Hermann, 2008). Schauer, Cotner, and Moore (2014) stated that poor understanding and low confidence in understanding the tenets of the evolutionary theory do not impair students' abilities to improve over the course of the semester, and these phenomena are also ubiquitous in the Malaysian context.

In regard to the Biology curriculum in the context of Malaysian secondary schools, the curriculum has been designed not only to provide opportunities for students to acquire and apply scientific knowledge, process skills, and critical thinking skills but also to develop moral values and patriotic sentiments. Furthermore, the curriculum enables students to be aware that scientific discoveries are the result of human endeavors for the betterment of mankind.

In Malaysia, evolutionary theory is only covered in the Form Six Biology Curriculum which prepares students to enter higher education institutions (Asghar et al., 2014). While the compulsory science curricula for secondary grades only include a simple introduction to the concept of evolution in the last chapter of the Form Five syllabus (i.e., Chap. 15: Variation) (Table 19.1), it is embedded in the 2.0 Variation learning area, under the learning objectives of 2.2, to understand the causes of variation. The main objective is to explain the importance of variation by introducing the concept of survival. "Survival of the fittest" is a phrase that originated from Charles Darwin's evolutionary theory as a way of describing the mechanism of natural selection in order for an organism to survive. Other evolutionary mechanisms appear to be completely missing from the compulsory science curriculum at secondary level in Malaysia.

Table 19.1 The concept of evolution in form five biology curriculum

No	Topic/Subtopic	Learning outcomes
2.0	Variation	A student should be able to:
2.2	Understanding the causes of variation	(a) Explain the importance of variation in the survival of a species

Source Ministry of Education Malaysia (2006)

For the Malaysian Higher School Certificate Examination (STPM) level taken by 18–19 year-olds, the Form Six Biology curriculum is divided into three terms: (i) First Term: Biological Molecules and Metabolism; (ii) Second Term: Physiology; and (iii) Third Term: Ecology and Genetics. Evolutionary theory is only taught in Chap. 16: Selection and Speciation during the third term. The history of evolutionary theory is presented through Lamarck's theory of inheritance of acquired characteristics together with scientific criticisms and some evidence found in the Malay Archipelago. This includes the illustration of the extensive fossil record for supporting evolutionary theory. Furthermore, evolutionary theory is also presented in the Darwin-Wallace's Theory where examples of natural selection are discussed. In addition, the syllabus also includes three modes of natural selection: stabilizing, directional, and disruptive. Each mode and its consequences are included in the textbook. Evolutionary significance of mimicry or camouflage is also discussed under the subtopic of natural selection. Beyond that, students are taught about sexual selection, polymorphism, and the importance of artificial

Table 19.2 The concept of evolution in form six biology curriculum

No	Topic/Subtopic	Teaching period	Learning outcomes
16.0	Selection and speciation	10	Candidates should be able to: (a) Describe continuous and discontinuous variations in relation to selection and speciation; (b) Explain the modes of natural selection (stabilising, directional, and disruptive) and their consequences; (c) Describe with examples, sexual selection and polymorphism; (d) Explain the importance of artificial selection (gene bank, germplasm bank and sperm bank)
16.1	Natural and artificial selection	6	
16.2	Speciation	4	Candidates should be able to: (a) Explain the processes of isolation, genetic drift, hybridization and adaptive radiation; (b) Explain the importance of speciation in relation to evolution

Source Malaysian Examinations Council (2012)

selection via the gene bank, germplasm bank, and sperm bank. Various evolutionary mechanisms, including isolation, genetic drift, hybridization, and adaptive radiation are also embedded in the curriculum (Table 19.2). Although the Biology curriculum at this level mentions human beings in the context of artificial selection, human evolution is not mentioned in depth.

19.4 Emphasis Given to Evolutionary Theory in the Biology Teacher Education Programmes: A Glimpse at Universiti Malaysia Sabah (UMS)

There is less emphasis on evolution education at UMS as there is no specific or stand-alone course on evolutionary theory offered by the university. Evolution was only taught sporadically in courses offered in the Biology Teacher Education Program at UMS. Table 19.3 shows a summary of biology-related courses offered by biology teacher education program at UMS.

Table 19.3 Biology-related courses taken by pre-service biology teachers at UMS

No	Course code	Course title	Topic/Subtopics	Credit hours
1	TB10103	Cell and structure biology	(a) Introduction to cell organization	3
2	TB10003	Botany	(a) Introduction to botany: natural selection and evolution	3
3	TB20103	Microbiology	(a) Microbial evolution b) Microbial taxonomy and classification	3
4	TB20303	Zoology	(a) Introduction to zoology: natural selection and evolution	3
5	TB20003	Genetic	(a) Genetic population and evolution: allele frequency theory, genetic drift	3
6	TB20203	Ecology	(a) Population ecology: natural selection	3
7	TB30503	Systematics	(a) Evolution and diversity of green and land plants; vascular plants (b) Evolution and diversity of woody and seeded plants (c) Evolution and diversity of flowering plants—monocots and eudicots (d) Principle of systematic zoology	3
8	TB30203	Plant physiology	(a) Introduction to plant kingdom; plant structure and definition terms, physiological evolution	3
9	TB30403	Animal physiology	(a) Introduction to physiological principles, physiological evolution	3

Source Universiti Malaysia Sabah (2017)

Based on Table 19.3, pre-service biology teachers at UMS are exposed to theory of evolution through biology-related courses like Systematics (TB30503), Microbiology (TB20103), Genetic (TB20003) and Ecology (TB20203). These three-credit-hour courses have a few topics related to evolution. Other than that, a brief introduction to evolution is given as a subtopic in courses such as Cell and Structure Biology (TB10103), Zoology (TB20303), Botany (TB10003), Plant Physiology (TB30203) and Animal Physiology (TB30403). Generally, college level evolution is integrated in these courses to equip pre-service biology teachers with fundamental knowledge on the flow of genetic information, chromosome theory of heredity, relationship between genetics and evolutionary theory, evolution of organisms and adaptation of the organisms to their environment. Some evolution history is also taught to contextualize the origins of multi-cellularity, generation and maintenance of species diversity, taxonomic and phylogenetic relationships of the major groups of organisms,, and ecological relationships between organisms and their environments.

19.5 Emphasis Given to Evolutionary Theory in the Biology Teacher Education Programmes: A Glimpse at Sultan Idris Education University (UPSI)

The pre-service biology teachers who pursue the four-year Bachelor of Education (Honours) Program, majoring in Biology Education, have to follow through 17 Biology-related courses as outlined in Table 19.4.

Based on Table 19.4, it is clear that within the four-year Biology Teacher Education Program, only two three-credit-hour courses expose pre-service teachers to the theory of Evolution and its application: (1) Biodiversity and Evolution of Protista and Animalia (Course code: SBB3023), and Biodiversity and Evolution of Archaea, Bacteria, Fungi, and Plantae (Course Code: SBB3043). In the former course (SBB3023), pre-service teachers are familiarized with the brief history of how animal diversity has been organized for systematic study, capitalizing on the current use of Darwin's Theory of Common Descent (Darwin, 1859) as the major principle underlying animal taxonomy. This serves to achieve the two key cognitive skills documented at the outset: pre-service teachers are able to correlate the evolution history so as to locate origins of multi-cellularity and illustrate historical process that generate and maintain species diversity, and to describe perpetual changes on the planet Earth and relate them to the evolution process of animals, from aquatic to terrestrial. In the latter course coded SBB3043, the evolution of organisms is emphasized, particularly with regard to the adaptation of organisms to their environment which results in the ecological dominance of a particular group of plants. Generally, this serves to achieve three cognitive skills documented at the outset: (a) pre-service teachers' ability to describe the flow of genetic information,

Table 19.4 Biology-related courses taken by pre-service biology teachers at UPSI

No	Course code	Course title	Credit hours
1	SBB3023	Biodiversity and evolution of protista and animalia	3
2	SBB3033	Principles in microbiology	3
3	SBB3043	Biodiversity and evolution of archaea, bacteria, fungi and plantae	3
4	SBC3013	Cell biology	3
5	SBC3023	Plant morphology and anatomy	3
6	SBC3033	Animal morphology and histology	3
7	SBC3043	Developmental biology	3
8	SBF3023	Plant physiology	3
9	SBF3033	Animal physiology	3
10	SBI3013	Information and communication technology in biology	3
11	SBK3013	Principles in biochemistry	3
12	SBR3996	Research project	6
13	SBS3013	Biostatistics	3
14	SBT3013	Biotechnology	3
15	SBU3033	Genetics	3
16	SBV3013	Ecology	3
17	SBV3023	Issues in biology and environment	3

Source Adapted from Faculty of Science and Mathematics (2016, p. 17)

the chromosome theory of heredity and the relationship between genetics and evolutionary theory; (b) evaluating the principles of evolutionary biology and justifying the taxonomy and phylogenetic relationships of the major groups of organisms (excluding protists and animals); and (c) recognizing the ecological relationships between organisms (excluding protists and animals) and their environment.

19.6 Prospective Biology Teachers' Understanding of Evolutionary Theory: A Glimpse at Universiti Malaysia Sabah (UMS)

Understanding of biological evolution has been surveyed among biology students (introductory to upper level), high school biology teachers, pre-service secondary instructors, Christian priesthood, and other groups (Barnes et al., 2009; Brem et al., 2003; Colburn & Henriques, 2006; Ingram & Nelson, 2006; Losh & Nzekwe, 2011; Verhey, 2005). Assessment on views and understandings of evolution across different regions and groups is important in terms of providing valuable insights to improve evolution education (Rice et al., 2015). Nonetheless, less attention has been given to the understanding of biological evolution in higher education institutions (Paz-y-Miño & Espinosa, 2011, 2012; Rice et al., 2015) especially for biology teacher education programs. In addition, limited connections have been proposed or investigated between these teachers' views of evolution and those of the scholars who trained them (i.e., educators of future educators) (Paz-y-Miño & Espinosa, 2012).

To obtain a glimpse of Malaysian prospective biology teachers' acceptance of evolutionary theory, a 20-item English version of Measure of Acceptance of the Theory of Evolution (MATE) instrument was administered to 58 pre-service biology teachers at Universiti Malaysia Sabah and 41 pre-service biology teachers at Sultan Idris Education University. Table 19.5 shows the means and standard deviations of MATE scores obtained from 58 prospective biology teachers at Universiti Malaysia Sabah.

Table 19.5 shows the descriptive statistics for evolution acceptance among a group of 58 pre-service biology teachers at Universiti Malaysia Sabah. It was found that there are consistencies in the responses to certain complementary items. For example, 29.3% of the pre-service teachers agreed, 34.5% undecided, and 36.2% disagreed with the statement that.

"Evolution is a scientifically valid theory" (Item 20). However, when the item was reworded negatively as "Evolution is not a scientifically valid theory" (Item 10), only 34.4% of the pre-service biology teachers agreed with a majority of them (36.2%) disagreeing while 29.3% were undecided. A majority of the pre-service biology teachers (48.3%) disagreed that "Current evolutionary theory is the result of sound scientific research and methodology" (Item 12) while more than the majority (67.3%) agreed that "The theory of evolution cannot be correct since it disagrees with the religious account of creation" (Item 14). Similarly, a great majority of the pre-service biology teachers (74.2%) disagreed that "Organisms existing today are the result of evolutionary processes that have occurred over millions of years" (Item 1) while a majority of the pre-service biology teachers (37.9%) were undecided that "With few exceptions, organisms on earth came into existence at about the same time" (Item 19). Equally, a majority of the pre-service biology teachers (60.3%) disagreed that "Most scientists accept evolutionary theory to be a scientifically valid theory" (Item 5) but more than the majority (56.9%) disagreed that "Much of the scientific community doubts if evolution occurs" (Item 17). Similarly, 17.2% of the pre-service biology teachers agreed that "There is a significant body of data which supports evolutionary data" (Item 8) as opposed to 41.4% who disagreed and were undecided respectively. When the item was negatively phrased as "The available data are ambiguous as to whether evolution actually occurs" (Item 6), 5.1%, 53.4% and 41.4% of the pre-service biology teachers correspondingly agreed, were undecided, and disagreed with the statement.

Table 19.5 Prospective biology teachers' understanding of evolutionary theory at UMS (N = 58)

Item	Statements	M (SD)	Agree % (#)	Undecided % (#)	Disagree % (#)
1	Organisms existing today are the result of evolutionary processes that have occurred over millions of years	2.21 (1.12)	18.9 (11)	6.9 (4)	74.2 (43)
2[a]	The theory of evolution is incapable of being scientifically tested	2.48 (0.978)	15.5 (9)	36.2 (21)	48.3 (28)
3	Modern humans are the product of evolutionary processes which have occurred over millions of years	3.36 (1.321)	51.7 (30)	19.0 (11)	29.3 (17)
4[a]	The theory of evolution is based on speculation and not valid scientific observation and testing	2.86 (1.115)	31.0 (18)	27.6 (16)	41.3 (24)
5	Most scientists accept evolutionary theory to be a scientifically valid theory	2.36 (0.950)	12.0 (7)	27.6 (16)	60.3 (35)
6[a]	The available data are ambiguous as to whether evolution actually occurs	2.62 (0.768)	5.1 (3)	53.4 (31)	41.4 (24)
7[a]	The age of the earth is less than 20,000 years	3.22 (1.027)	34.5 (20)	48.3 (28)	17.2 (10)
8	There is a significant body of data which supports evolutionary theory	2.69 (0.902)	17.2 (10)	41.4 (24)	41.4 (24)
9[a]	Organisms exist today is essentially the same form in which they always have	2.95 (1.191)	41.4 (24)	22.4 (13)	36.2 (21)
10[a]	Evolution is not a scientifically valid theory	2.95 (1.016)	34.4 (20)	29.3 (17)	36.2 (21)
11	The age of the earth is at least 4 billion years	2.83 (1.028)	13.8 (8)	50.0 (29)	36.2 (21)
12	Current evolutionary theory is the result of sound scientific research and methodology	2.52 (0.800)	6.9 (4)	44.8 (26)	48.3 (28)
13	Evolutionary theory generates testable predictions with respect to the characteristics of life	2.34 (0.762)	6.9 (4)	31.0 (18)	62.0 (36)
14[a]	The theory of evolution cannot be correct since it disagrees with the Biblical account of creation	2.05 (1.234)	12.1 (7)	20.7 (12)	67.3 (39)
15[a]	Humans exist today is essentially the same form in which they always have	2.26 (1.163)	17.2 (10)	25.9 (15)	56.9 (33)

(continued)

Table 19.5 (continued)

Item	Statements	M (SD)	Agree % (#)	Undecided % (#)	Disagree % (#)
16	Evolutionary theory is supported by factual, historical, and laboratory data	2.72 (1.022)	19.0 (11)	41.4 (24)	39.7 (23)
17[a]	Much of the scientific community doubts if evolution occurs	2.43 (0.901)	13.8 (8)	29.3 (17)	56.9 (33)
18	The theory of evolution brings meaning to the diverse characteristics and behaviors observed in living forms	2.22 (0.727)	3.4 (2)	29.3 (17)	67.2 (39)
19[a]	With few exceptions, organisms on earth came into existence at about the same time	3.03 (1.075)	32.7 (19)	37.9 (22)	29.3 (17)
20	Evolution is a scientifically valid theory	2.86 (1.034)	29.3 (17)	34.5 (20)	36.2 (21)
		59.26 (0.983)			

Note 1 = strongly disagree, 2 = disagree, 3 = undecided, 4 = agree, 5 = strongly agree; [a]negatively-worded items; Cronbach's Alpha = 0.663

19.7 Prospective Biology Teachers' Acceptance of Evolutionary Theory: A Glimpse at Sultan Idris Education University (UPSI)

Table 19.6 shows the means and standard deviations of MATE scores obtained from a group of prospective biology teachers at Sultan Idris Education University (UPSI).

Table 19.6 shows the descriptive statistics of the evolution acceptance among 41 pre-service biology teachers at Sultan Idris Education University. It appears that there are also consistencies in the responses to certain complementary items. For example, it was found that 75.6% of the pre-service biology teachers agreed, 12.2% were undecided, and 12.2% disagreed with the statement that "Evolution is a scientifically valid theory" (Item 20). However, when the item was reworded negatively as "Evolution is not a scientifically valid theory" (Item 10), only 31.7% of the pre-service biology teachers agreed as the majority of participants (53.7%) felt undecided or disagreed (14.6%). The majority of the pre-service biology teachers (53.7%) agreed that "Current evolutionary theory is the result of sound scientific research and methodology" (Item 12) while less than the majority (46.3%) agreed that "The theory of evolution cannot be correct since it disagrees with the religious account of creation" (Item 14). A great majority of the pre-service biology teachers (82.9%) agreed that "Organisms existing today are the result of evolutionary processes that have occurred over millions of years" (Item 1) while less than the

Table 19.6 Prospective biology teachers' understanding of evolutionary theory at UPSI (N = 41)

Item	Statements	M	Agree % (#)	Undecided % (#)	Disagree % (#)
1	Organisms existing today are the result of evolutionary processes that have occurred over millions of years	3.73 (0.775)	82.9 (34)	2.4 (1)	14.6 (6)
2[a]	The theory of evolution is incapable of being scientifically tested	3.66 (0.728)	68.3 (28)	26.8 (11)	4.8 (2)
3	Modern humans are the product of evolutionary processes which have occurred over millions of years	3.56 (0.673)	51.2 (21)	46.3 (19)	2.4 (1)
4[a]	The theory of evolution is based on speculation and not valid scientific observation and testing	3.37 (0.733)	48.8 (20)	41.5 (17)	9.7 (4)
5	Most scientists accept evolutionary theory to be a scientifically valid theory	3.61 (0.628)	63.4 (26)	31.7 (13)	4.9 (2)
6[a]	The available data are ambiguous as to whether evolution actually occurs	3.34 (0.911)	53.6 (22)	29.3 (12)	17.1 (7)
7[a]	The age of the earth is less than 20,000 years	2.07 (1.149)	9.8 (4)	39.0 (16)	51.2 (21)
8	There is a significant body of data which supports evolutionary theory	3.56 (0.867)	73.1 (30)	9.8 (4)	17.0 (7)
9[a]	Organisms exist today is essentially the same form in which they always have	3.49 (0.898)	68.3 (28)	19.5 (8)	12.2 (5)
10[a]	Evolution is not a scientifically valid theory	3.12 (0.872)	31.7 (13)	53.7 (22)	14.6 (6)
11	The age of the earth is at least 4 billion years	2.20 (1.229)	14.6 (6)	34.1 (14)	51.2 (21)
12	Current evolutionary theory is the result of sound scientific research and methodology	3.49 (0.597)	53.7 (22)	41.5 (17)	4.9 (2)
13	Evolutionary theory generates testable predictions with respect to the characteristics of life	3.49 (0.675)	58.5 (24)	31.7 (13)	9.8 (4)
14[a]	The theory of evolution cannot be correct since it disagrees with the Biblical account of creation	3.39 (0.666)	46.3 (19)	48.8 (20)	4.8 (2)
15[a]	Humans exist today is essentially the same form in which they always have	3.34 (0.693)	41.4 (17)	48.8 (20)	9.8 (4)

(continued)

Table 19.6 (continued)

Item	Statements	M	Agree % (#)	Undecided % (#)	Disagree % (#)
16	Evolutionary theory is supported by factual, historical, and laboratory data	3.29 (0.929)	46.3 (19)	29.3 (12)	24.4 (10)
17[a]	Much of the scientific community doubts if evolution occurs	3.41 (0.499)	41.5 (17)	58.5 (24)	0.0 (0)
18	The theory of evolution brings meaning to the diverse characteristics and behaviors observed in living forms	3.66 (0.825)	68.3 (28)	19.5 (8)	12.2 (5)
19[a]	With few exceptions, organisms on earth came into existence at about the same time	3.46 (0.552)	43.9 (18)	56.1 (23)	0.0 (0)
20	Evolution is a scientifically valid theory	4.37 (1.113)	75.6 (31)	12.2 (5)	12.2 (5)
		62.29 (1.136)			

Note 1 = strongly disagree, 2 = disagree, 3 = undecided, 4 = agree, 5 = strongly agree; [a]negatively-worded items; Cronbach's Alpha = 0.783

majority (43.9%) agreed that "With few exceptions, organisms on earth came into existence at about the same time" (Item 19). Equally, the majority of the pre-service biology teachers (63.4%) agreed that "Most scientists accept evolutionary theory to be a scientifically valid theory" (Item 5) but less than the majority (41.5%) agreed that "Much of the scientific community doubts if evolution occurs" (Item 17). Similarly, 73.1% of the pre-service biology teachers agreed that "There is a significant body of data which supports evolutionary data" (Item 8) as opposed to 17.0% and 9.8% of them disagreed and undecided respectively. When the item was negatively phrased as "The available data are ambiguous as to whether evolution actually occurs" (Item 6), 53.6%, 29.3% and 17.1% of the pre-service biology teachers agreed, were undecided, and disagreed with the statement respectively.

The inconsistencies of responses among pre-service biology teachers for certain pairs of positively and negatively-worded complementary MATE items indicate incongruent patterns for the acceptance of evolutionary theory. These misconceptions prevent students from fully understanding the real impact of evolution in biological diversity (Campos & Sá-Pinto, 2013). According to Rusell (2011), when people understand this new, more integrated theory of evolution, they will see that evolution makes more sense and thus more easily embrace its ability to explain the origins of the great diversity of life on Earth. In addition, controversy and misconceptions exist among the public who sometimes reject the validity of evolutionary theory because biological evolution is not readily observable or testable. Other than that, perceived conflict between evolutionary theory and students' religious beliefs, particularly the conflicting religious and scientific explanations about human origins, lead to lower acceptance of evolutionary theory among religious students.

19.8 Suggestions to Improve Evolution Education in Malaysia

The introduction of courses on biological evolution that cover its most fundamental principles is crucially needed in the biology teacher education programs in the Malaysian context. This will enable both educators and students to fully understand the real impact of evolution in biological sciences and integrate this knowledge with other fields such as human health and agriculture. In relation to that, more research is crucially needed especially in developing instrumental and effective biology curricula that can increase students' knowledge on evolutionary theory while addressing misconceptions, especially in the Malaysian context. Borgerding et al. (2015) highlighted four challenges faced when teaching evolution such as limited content knowledge, teachers' own evolution rejection, resistance to instruction, and concerns about religion. With this, preparing future educators who have complete science knowledge is necessary to cope with these challenges, avoid science misconceptions and provide better conceptual development of scientific explanations. Despite the controversy and misconceptions, biological evolution is the unifying idea in biology, and emphasis should be given to avoid science misconceptions. It is important to utilize empirically supported instruction and curriculum to effectively address teachers' evolution misconceptions and promote the development of accurate evolution knowledge prior to their entering service (Nadelson, 2009). Without doubt, if educators want to promote acceptance of biological evolution, they must effectively promote a deep understanding of its most fundamental principles (Rice et al., 2015).

19.9 Conclusion

In general, the level of acceptance of evolutionary theory among Malaysian prospective biology teachers was low. The evolutionary mechanisms are not fully explored and understood as it was only taught in biology-related courses like Zoology, Botany, and Biodiversity courses. Hence, the introduction of specific courses on biological evolution that cover its most fundamental principles is crucially needed in the biology teacher education programs in the Malaysian context to enhance prospective biology teachers' acceptance and understanding of evolutionary theory.

References

Asghar, A., Hameed, S., & Farahani, N. K. (2014). Evolution in biology textbooks: A comparative analysis of 5 muslim countries. *Religion and Education, 41*(1), 1–15. https://doi.org/10.1080/15507394.2014.855081.

Ayala, F. J. (2008). Science, evolution and creationism. *Proceedings of the National Academy of Sciences USA, 105*(1), 3–4.

Barnes, R. M., Keilholtz, L. E., & Alberstadt, A. L. (2009). Creationism and evolution beliefs among college students. *Skeptic, 14*(3), 13–16.

Borgerding, L. A., Klein, V. A., Ghosh, R., & Eibel, A. (2015). Student teachers' approaches to teaching biological evolution. *Journal of Science Teacher Education, 26*, 371–392.

Brem, S. K., Ranney, R., & Schindel, J. (2003). Perceived consequences of evolution: College students perceive negative personal and social impact in evolutionary theory. *Science Education, 87*(2), 181–206.

Campos, R., & Sá-Pinto, A. (2013). Early evolution of evolutionary thinking: teaching biological evolution in elementary schools. *Evolution: Education and Outreach, 6*(25). https://doi.org/10.1186/1936-6434-6-25.

Cavallo, A. M., & McCall, D. (2008a). Seeing may not mean believing: Examining students' understandings and beliefs in evolution. *The American Biology Teacher, 70*(9), 522–530.

Cavallo, A. M. L., & McCall, D. (2008b). Seeing may not mean believing: Examining students' understandings and beliefs in evolution. *The American Biology Teacher, 70*(9), 522–530.

Chinsamy, A., & Plagányi, É. (2008). Accepting evolution. *Evolution, 62*(1), 248–254.

Colburn, A., & Henriques, L. (2006). Clergy views on evolution, creationism, science, and religion. *Journal of Research in Science Teaching, 43*(4), 419–442.

Cotner, S., Brooks, D. C., & Moore, R. (2010). Is the age of the earth one of our "sorest troubles?" student's perceptions about deep time affect their acceptance of evolutionary. *Evolution, 64*(3), 858–864. Retrieved October 2016, from https://www.cbs.umn.edu/sites/default/files/public/downloads/age-0f-earth-evo-march2010.pdf.

Darwin, C. (1859). *On the origin of species by means of natural selection, or the preservation of favoured races in the struggle for life*. London: John Murray.

Dobzhansky, T. (1973). Nothing in biology makes sense except in light of evolution. *JSTOR, 35*(3), 125–129. https://doi.org/10.1016/j.shpsc.2013.06.006.

Faculty of Science and Mathematics. (2016). *Academic Guidelines. Bachelor Programmes*. 2016/2017 Session. Tanjung Malim: Universiti Pendidikan Sultan Idris.

Hameed, S. (2008). Bracing for islamic creationism. *Science, 322*(3), 1637–1638. https://doi.org/10.1126/science.1163672.

Hassan, R. (2007). On being religious: Patterns of religious commitment in muslim societies. *Muslim World, 97*(3), 437–478. https://doi.org/10.1111/j.1478-1913.2007.00190.x.

Hermann, R. S. (2008). Evolution as a controversial issue: A review of instructional approaches. *Science and Education, 17*(8–9), 1011–1032.

Hildebrand, D., Billica, K., & Capps, J. (2008). Addressing controversies in science education: A pragmatic approach to evolution education. *Science and Education, 17*(8–9), 1033–1052.

Ingram, E. L., & Nelson, C. E. (2006). Relationship between achievement and students' acceptance of evolution or creation in an upper-level evolution course. *Journal of Research in Science Teaching, 43*(1), 7–24.

Losh, S. C., & Nzekwe, B. (2011). Creatures in the classroom: Pre-service teacher beliefs about fantastic beasts, magic, extra-terrestrials, evolution and creationism. *Science and Education, 20*(5), 473–489. https://doi.org/10.1007/s11191-010-9268-5.

Malaysia Department of Statistics. (2012). *Malaysia population projections*. Kuala Lumpur: Government of Malaysia.

Malaysian Examinations Council. (2012). *Malaysia higher school certificate examination biology syllabus and specimen papers*. Kuala Lumpur: Malaysian Examinations Council.

Miller, J. D., Scott, E. C., & Okamoto, S. (2006). Public acceptance of evolution. *Science, 313* (5788), 765–766.

Miller, K. R. (1999). *Finding Darwin's god: A scientist's search for common ground between god and evolution*. New York: Cliff Street Books, HarperCollins.

Ministry of Education Malaysia. (2006). *Integrated curriculum for secondary schools curriculum specifications biology form 5*. Putrajaya: Ministry of Education Malaysia.

Ministry of Education Malaysia. (2017a). Educational pathway in Malaysia. Retrieved April 21, 2017, from http://www.moe.gov.my/index.php/en/dasar/laluan-pendidikan-di-malaysia.

Ministry of Education Malaysia. (2017b). Educational pathway in Malaysia. Retrieved April 21, 2017, from http://www.moe.gov.my/index.php/en/dasar/falsafah-pendidikan-kebangsaan.

Moore, R., & Cotner, S. (2009a). Rejecting Darwin: The occurrence and impact of creationism in high school biology classrooms. *The American Biology Teacher, 71*(2), e1–e4.

Moore, R., & Cotner, S. (2009b). The creationist down the hall: Does it matter when teachers teach creationism? *BioScience, 59*(5), 429–435.

Nadelson, L. S. (2009). Pre-service teacher understanding and vision of how to teach biological evolution. *Evolution Education and Outreach, 2*(3), 490–504. https://doi.org/10.1007/s12052-008-0106-z.

National Academy of Sciences and Institute of Medicine. (2008). *Science, evolution, and creationism*. Washington. DC: The National Academies Press. Retrieved from http://www.csun.edu/sites/default/files/ScienceEvolution%26Creationism_NAS2008.pdf.

Nehm, R. H., & Schonfeld, I. S. (2007). Does increasing biology teacher knowledge of evolution and the nature of science lead to greater preference for the teaching of evolution in schools? *Journal of Science Teacher Education, 18,* 699–723. https://doi.org/10.1007/s10972-007-9062-7.

Paz-y-Miño, C. G, & Espinosa, A. (2011). New England faculty and college students differ in their views about evolution, creationism, intelligent design, and religiosity. *Evolution: Education and Outreach, 4,* 323–342.

Paz-y-Miño, C. G, & Espinosa, A. (2012). Educators of prospective teachers hesitate to embrace evolution due to deficient understanding of science/evolution and high religiosity. *Evolution: Education and Outreach, 5,* 139–162.

Rice J. W., Clough, M. P., Olson, K. J., Adams, D. C., & Colbert, J. T. (2015). University faculty and their knowledge and acceptance of biological evolution. *Evolution. Education and Outreach, 8,* 8. https://doi.org/10.1186/s12052-015-0036-5.

Rusell, C. G. (2011). *Evolution of evolutionary theory. Epic of Evolution*. Retrieved from http://epicofevolution.com/dialog/evolution-of-evolution.html.

Rutledge, M. L., & Sadler, K. C. (2011). University students' acceptance of biological theories-is evolution really different? *Journal of College Science Teaching, 41*(2), 38–43.

Schauer, A., Cotner, S., & Moore, R. (2014). Teaching evolution to students with compromised backgrounds & lack of confidence about evolution-is it possible? *The American Biology Teacher, 76*(2), 93–98. Retrieved November, 2016, from https://www.cbs.umn.edu/sites/default/files/public/downloads/schauer-abt2014.pdf.

Smith, M. U., & Siegel, H. (2004). Knowing, believing and understanding: what goals for science education? *Science and Education, 13*(6), 553–582.

Tidon, R., & Lewontin, R. C. (2004). Teaching evolutionary biology. *Genetics and Molecular Biology, 27,* 124–131. https://doi.org/10.1590/S1415-475720054000100021.

Universiti Malaysia Sabah. (2017). *Undergraduate prospectus 2016/2017*. Retrieved January, 2017, from http://www.ums.edu.my/v5/files/Prospectus2016-2017_BI.pdf.

Verhey, S. D. (2005). The effect of engaging prior learning on student attitudes toward creationism and evolution. *BioScience, 55*(11), 996–1000.

Wiles, J. R. (2010). Overwhelming scientific confidence in evolution and its centrality in science education-and the public disconnect. *The Science Education Review, 9*(1), 18–27.
Wiles, J. R. (2014). Gifted students' perceptions of their acceptance of evolution, changes in acceptance, and factors involved there. *Evolution: Education and Outreach.* https://doi.org/10.1186/s12052-014-0004-5.

Yoon-Fah Lay is an Associate Professor of Science Education at the Faculty of Psychology and Education (FPE), Universiti Malaysia Sabah (UMS), Kota Kinabalu, Malaysia. Currently, he is a Research Fellow of the Unit for Rural Education Research (URER), FPE, UMS and an Associate of the Centre for Research in International and Comparative Education (CRICE), Universiti Malaya. His primary research interests are in the areas of scientific skills, logical thinking abilities, attitude toward science, environmental education, professional development of science teachers, and international studies e.g., TIMSS and PISA. He is the Editor-in-Chief of Journal of Educational Thinkers (Jurnal Pemikir Pendidikan), UMS; the Associate Editor for the special issues of EURASIA Journal of Mathematics, Science, and Technology Education, the Editorial Board Member of international journals, and manuscript reviewer of high impact international journals.

Eng-Tek Ong is a Professor of Science Education at Sultan Idris Education University (UPSI). He teaches undergraduate, masters, and doctoral level courses in science education program at UPSI. His research interests include science process skills, inquiry-based science education, cooperative learning, concept cartoon, and the integration of Education for Sustainable Development (ESD), Science, Technology, Engineering and Mathematics (STEM) Education, and Higher Order Thinking Skills (HOTS) in science across levels and disciplines. He has conducted numerous continued professional development courses on science pedagogy for teachers in SEAMEO member countries and beyond.

Crispina Gregory K. Han is a senior lecturer of Science Education at Universiti Malaysia Sabah (UMS). She teaches undergraduate and masters level courses at UMS. Her research scopes include science and mathematics education in urban and rural schools, evaluation and management in education. She is recently the Head of Unit for Rural Education Research at the Faculty of Psychology and Education, UMS. The UREdR conducts and supports research pertaining to rural education which will contribute to the development of the society and the nation.

Sane-Hwui Chan is a high school biology teacher at Kian Kok Middle School, Kota Kinabalu, Sabah, Malaysia. She obtained her bachelor's degree and master's degree in science education at Universiti Malaysia Sabah (UMS). She is currently pursuing her doctoral degree on the effect of affective domains on behavioural intention in teaching science among pre-service science teachers using Partial Least Squares-Structural Equation Modeling approach.

Chapter 20
Biological Evolution Education in Malaysia; Where We Are Now

Kamisah Osman, Rezzuana Razali and Nurnadiah Mohamed Bahri

Abstract This paper presents the public acceptance of evolutionary theory on the current state of biological evolution education in Malaysia. The public acceptance of evolutionary theory amongst Malaysians differ according to religion but not according to educational background and economic statuses. Like other Muslim dominant countries, Islamic resurgence that took place in the 1980s has influenced the Malaysian acceptance towards evolution through (a) education reform and Islamization of Science, (b) impact of prominent Islamic scholars and their publications and (c) publications of anti-evolution by Malaysian academicians. Academic reform and Islamization of Science has caused the withdrawal of evolution from the Biology syllabus. Today's Biology subject does not contain any chapter on the theory of evolution. Generally, 70% of Malaysian teachers were shown to be most radically creationist and their attitude and conception were significantly shaped by their religious beliefs and values. With regards to the Malaysian scenario on the evolution education, no emphasis could be discussed in detail concerning the context of Malaysia's biology teachers' evolutionary theory education programs. However, evolution theory education could be a set of knowledges that the citizen should be aware of so that they will be equipped with and exposed to comprehensive and meaningful scientific knowledge supported with religious explanation. This is in order to build a literate nation and produce individuals who have stances based on their pure understanding and knowledge.

Keywords Evolution education · Malaysia

K. Osman (✉) · R. Razali · N. M. Bahri
Faculty of Education, Universiti Kebangsaan Malaysia, Bangi, Malaysia
e-mail: kamisah@ukm.edu.my

20.1 Biological Evolution Education in Malaysia; Where We Are Now

Malaysia is a complex multi-ethnic and multi-religious country located in the heart of South-eastern Asia. It consists of the Peninsular Malaysia (131,800 ft^2) and the states of Sarawak (124,400 ft^2) and Sabah (73,700 ft^2) which are separated by the South China Sea. According to the 2010 population census, the total population of Malaysia is 28.3 million and it is comprised of diverse ethnic groups, namely the Malays, Chinese, Indians, Indigenous, Kadazan, Muruts, Dayaks, Bajaus and Melanaus. Among the Malaysian citizens, the Malays represent the predominant ethnic group. Since most Malays are Muslim, Islam is the most widely practised religion in the country with 61.3% of its citizens being Muslim. On the other hand, other major religions being embraced by Malaysians are Christianity, Buddhism, Hinduism, Sikhs, Confucianism, Taoism, other traditional Chinese religions, and various folk religions. Even though Islam has been the official religion of this country since it gained its independence from British in 1957, Malaysia is neither an Islamic state nor a secular state. Non-Muslims are given the freedom to practise their own religions and are governed by a civil code based largely on English Law. Muslims, however are subjected to a mixture of civil and Islamic Law (Loo, 1999). The *Bahasa Melayu* (Malay Language) is the official language of Malaysia while English is the second language.

As a land of diverse races and religions, unity has always been a priority in developing the nation. In an attempt to establish national unity among all its citizens regardless of ethnic origins or religious affiliations, especially in reaction to the tragic racial riot in 1969, the Malaysian government has introduced the principles of the National ideology, or the *Rukunegara* in 1970 (Khalid & Saad, 2010). The *Rukunegara* is the Malaysian declaration of national principles which is represented by five national principles that guide Malaysians namely: (i) belief in God; (ii) loyalty to the King and country; (iii) upholding the constitution; (iv) rule of law, and (v) good behaviour and morality. Apart from acting as a tool to foster national unity and harmony among disparate ethnic and religious communities amongst Malaysians, the introduction of *Rukunegara* had also given a major impact to the educational development in Malaysia. Its first pillar i.e. belief in God and religion is one of the main principles embodied in the National Education Philosophy, thus playing a significant role in the development and formation of curriculum implemented in Malaysian schools.

20.2 Public Acceptance of Evolutionary Theory in Malaysia: Overall Scenario

The publication of Darwin's most controversial book, *On the Origin of Species Through Natural Selection* in 1859 has created immense debates between the scholars which lead to a split in views between science and religion. Like other Muslim countries, biological evolution is a relatively new concept in Malaysia. Evolution itself is a wide-ranging topic, and the acceptance of evolution relies heavily on the definition of evolution as understood by individual respondents. This is especially a problem when many perhaps most people in the Muslim world confuse evolution with atheism and consider it as inherently against religion (Hameed, 2008). In the context of Malaysia, it could be argued that serious debates on and opposition to evolution are often focused on the origin of man rather than the issue of common ancestry.

In 2008, the Malaysian Science and Technology Information Centre (MASTIC) conducted a survey to assess the public awareness and understanding of, interest in, and attitudes towards science and technology. Malaysians who are in the 12–60 years old age group were selected as respondents for the survey. One of the questions in the questionnaire contains statement that says, "Human beings as we know them today developed from earlier species of animals". Analysis of findings revealed that only 17.0% of Malaysians who responded to the survey agreed to the statement which suggests that the majority of Malaysians rejected the notion that man originated from apes. The inclination to agree with Darwin's evolution theory was noticeably less among Muslims than respondents of other faiths. In the survey, only 13.7% of the Muslims support the theory, as opposed to 26.5% of the Buddhists, 33.8% of the Hindus, 26.4% of the Christians, 42.1% of the Confucians, 26.5% of the Taoists and 19.4% of the other believers. However, it was later reported that the theory of evolution was not included in computing the level of the Malaysian public's understanding of science and technology issues. This is because the statement "Human beings as we know them today developed from earlier species of animals" was argued as an expression of beliefs and philosophies rather than scientific facts (Ministry of Science Technology and Innovation Malaysia, 2010).

As stated earlier, serious debates on and opposition towards evolution in Malaysia are often focused on the origin of man. And throughout the years, it is important to note that the public acceptance towards the theory of evolution in Malaysia remains unchanged. A recent study by Kasmo, Usman, Hassan, Yunos, and Mohamad (2015b) has also showed that Malaysians in general rejected the theory that human beings originated from apes. It has been demonstrated from the study that the strongest rejection on the theory of evolution are from the Muslim community with 77% of the Muslim respondents rejected that humans originated from apes. On the other hand, the Christian community had shown the highest level of acceptance of evolutionary theory followed by the Hindus and Buddhists.

During the 7th World Conference of Science Journalist which was held in Qatar, Salman Hameed, Director of the Centre for the Study of Science in Muslim Societies (SSiMS) at Hampshire College, United States presented initial results from a survey that examined the attempts of educated Muslims to reconcile their religion with the evolutionary science. Respondents of the survey were mainly doctors and medical students in five Muslim countries namely Egypt, Indonesia, Malaysia, Pakistan and Turkey and in three countries hosting Muslim diaspora i.e. Turkish doctors in Germany; Pakistani doctors in the United Kingdom; and Arab, Pakistani and Turkish doctors in the United States. Based on the preliminary results, most Malaysian doctors in Malaysia, despite being scientifically literate, rejected the theory of evolution, especially with regard to humans (Padma, 2011).

Once again, it should be noted that the Malaysians responses to the evolution theory do not indicate familiarity with or ignorance of scientific concepts and issues. Instead, they are statements of a person's belief or philosophical stand more than his or her knowledge of the concepts (Ministry of Science Technology and Innovation Malaysia, 2010). Thus, it is not surprising to see that the public acceptance amongst Malaysians towards this theory does not differ according to educational backgrounds and social economic statuses (Padma, 2011).

20.3 Reasons for Strong Rejection Towards Evolution by Malaysian Muslims

Over the past three decades, Malaysia has undergone radical development and transformations caused mainly by the country's rapid economic growth. Besides gigantic economic transformation programmes, another significant change that can be seen is the Islamization process that has been occurring since the early 1980s and is still giving a huge impact on the country many aspects of life including the education system. Muzaffar (1987) in his books Islamic Resurgence in Malaysia states that:

> Islamic resurgence is a description of the endeavour to re-establish Islamic values, Islamic practices, Islamic institutions, Islamic laws, indeed Islam in its entirety, in the lives of Muslims everywhere. It is an attempt to re-create an Islamic ethos, an Islamic order at the vortex of which is the Islamic human being, guided by the Qur'an and the Sunnah (p. 2)

He further describes that, the signs of Islamic resurgence in Malaysia can be seen through the Islamic attire, decline in social communication between the sexes, the popular acceptance of Islamic greetings, the increase in concern about the Muslim dietary rules, hobbies, tastes and even values. Besides, the rise of Islamic consciousness amongst Malaysian has also increased the popularity of audio tapes and variety of publications dealing with Islam. In Muslim bookstores and little roadside stalls in small towns and big cities, tapes and booklets which discuss personal morality, religious rituals, duties to God, the Day of Judgement and the Hereafter are making brisk business.

From our perspective, there are various factors that may lead to strong rejection of theory of evolution particularly by Malaysian Muslims. In this section, we will discuss how the Islamic resurgence that took place in Malaysia has influenced the Malaysian acceptance of the theory of evolution through (a) educational reform and Islamization of science (b) impacts of prominent Islamic scholars and their publications, and (c) anti-evolution publications by Malaysian academicians.

20.3.1 Education Reform and Islamization of Science Education

As a former British Colony country, the National Education System of Malaysia, was inherited from the British colonial government. From 1968 until 1981, the Malaysian school science curriculum was adopted from three British science curricula. The Scottish Integrated Science Syllabus was adopted for lower secondary schools, replacing the old general science curriculum. The Nuffield Secondary School Science Curriculum was adopted for the Arts Stream of upper secondary schools and renamed Modern General Science; and the Nuffield "O" Levels Pure Science Syllabi were adopted for the Science Stream of upper secondary school and renamed Modern Biology, Modern Chemistry, and Modern Physics (Tan, 1991). Nearly all these curricula were under strong Western influence and did not produce the results expected (Lee, 1999).

As mentioned in the previous section, the occurrence of the Islamic resurgence had witnessed the increase of Islamic consciousness among Muslims in Malaysia which subsequently contributed to educational reform (Hashim & Langgulung, 2008). In April 1977, the First World Conference on Muslim Education was held in Mecca, Saudi Arabia. The conference participants believed that the absence of genuine Islamic education is the cause of growing westernization and secularization in Muslim societies. Such a phenomenon is regarded as endangering the distinctive Islamic identities of Muslims. They also believed that the Muslim world can preserve its identity and save the Ummah (Muslim society) from confusion and the erosion of Islamic values through a true Islamic education (Hassan, 2007). At the same time, at the end of the 1980s, the need for the teaching of values was formally acknowledged in Malaysia. The Cabinet Committee Report recommended that the Ministry of Education develop a curriculum that integrates the teaching of values in the form of moral education (as a subject) for the non-Muslim students, and for it to be made mandatory as well as examinable. It was to be taught at the same time when the Muslim students were taught the Islamic Education subject (Ahmad, 1998).

Therefore, in response to the demand for Islamization as described in the preceding section and as well as the failure of the adopted curricula to produce expected results, thus starting from 1982, the Malaysian school curriculum was totally revamped right from Primary 1 up to Form 5 (Lee, 1999). The Ministry of Education had launched the New Primary School Curriculum, which was introduced with much emphasis given to the learning of the 3Rs (reading, writing and arithmetic) and the inculcation of moral values (Lee, 1992). In 1988, the new curriculum was extended to the lower secondary level and the reform was known as the Integrated Secondary School Curriculum. This curriculum was guided by the recommendation of the first World Conference on Muslim Education 1977, in Mecca (Langgulung, 1993) and in line with the National Philosophy of Education that was launched in 1988 (later was replaced with a more comprehensive version in 1996). The philosophy emphasized a firm belief in and devotion to God as an essential element in shaping a holistic Malaysian citizen. The National Philosophy of Education stated that:

> Education in Malaysia is an on-going effort towards further developing the potential of individuals in a holistic and integrated manner, so as to produce individuals who are intellectually, spiritually, emotionally and physically balanced and harmonious, based on a firm belief in and devotion to God. Such an effort is designed to produce Malaysian citizens who are knowledgeable and competent, who possess high moral standards, and who are responsible and capable of achieving high level of personal well-being as well as being able to contribute to the harmony and betterment of the family, the society and the nation at large (Ministry of Education, 1996).

Every phrase in the National Philosophy of Education unequivocally shows the strong influence of Islamic values in the Malaysian education system. This is why the population of Malaysia is now witnessing and experiencing an Islamic resurgence in almost every dimension of its everyday life including the field of Science Education. It could, therefore, be argued that Malaysia has also responded positively to support the effort towards the "institutionalization" of Islamic Science.

This kind of approach is perceived by Lee (1992) as an attempt to reconcile the western institution of science and technology with the moral and cultural values of Islam. The acceptance of an "Islamic" approach to science is based upon the notion that the scientific methods are not the sole way to knowledge acquisition, but places equal importance on other way of knowing such as by intuition and revelation. And more importantly, the insertion of Islamic and moral values in teaching science is to make the students recognize that science is not only a way of gaining new knowledge but also a means of appreciating and realising the presence and greatness of the Creator.

The Malaysian education system had once again undergone a major curriculum reform in 2011 through the introduction of Primary School Curriculum Standard for Standard One students. The new curriculum was implemented in stages to ensure a gradual transition from primary school to secondary school. Six years after the implementation of the Primary School Curriculum Standard (2011–2016), the Secondary School Curriculum Standard has been introduced to Form One students starting this year (2017) as a continuation of the Primary School Curriculum Standard. The Primary School Curriculum Standard (Standard 1–Standard 6) and Secondary School Curriculum Standard (Form 1–Form 5) is the current syllabus that is being used by all government and private schools in Malaysia. Once again, a firm belief in and devotion to God remains as an essential element in the formation of the science curriculum in Malaysia.

20.3.2 Influence of Prominent Islamic Scholars and Their Publications

Strong rejections to evolution theory by Muslims in Malaysia are caused by the widespread of Islamic Science concept that was influenced by prominent international Muslim Creationists who reject the modern theory of evolution based on Darwinism. Maurice Bucaille and Adnan Oktar are two scholars who are not only popular but also very prominent in Malaysia. Their views on evolution have influenced public perceptions and acceptance towards the theory.

Maurice Bucaille (19 July 1920–17 February 1998). Maurice Bucaille was a French medical doctor and a renowned scholar who supported the idea that the Darwinian Theory was against the teachings of the Qur'an. He rose to prominence throughout Islamic world following the publication of 'The Bible, The Qur'an and Modern Science', a book that is a best seller in the popular literature in Muslim countries, particularly Turkey and Malaysia. His book, 'The Bible, the Qur'an and Modern science' and 'What is the Origin of Man' were translated to the Malay language. The information on the relation between the Qur'an and science has shaped the general opinion of the people towards the Qur'an and science (Kasmo, et al., 2015a). Islamic establishments worldwide and Malaysia are using this book up until today (Abdullah, 2015).

In 1989, Maurice Bucaille was invited by Universiti Sains Malaysia, the second oldest university in Malaysia, to present a paper under the university's Public Lecture Series (National and International Category). This is an honour reserved only for those regarded by the said university as having achieved the highest level of scholarly excellence (Loo, 2001).

Harun Yahya (2 February 1956–present). Adnan Oktar, who goes by the pen name of Harun Yahya is a Muslim Creationist from Turkey. He is a strong critique of the Darwin Theory and propagates his ideas through publications. In Malaysia, the Saba Islamic Media Sdn Bhd which is one of the largest Islamic book stores in Malaysia is responsible in publishing and distributing books by Harun Yahya. Apart from books, they also distributed almost all lecture series by Harun Yahya including his best seller, *Sign of the Creator* public lecture which also has been translated to Malay language. Since 2010, Saba Islamic Media in collaboration with the Harun Yahya Foundation, are also very active in organizing public lectures and seminars in Malaysia. According to the Harun Yahya's website, two of his most successful conferences were the "The Collapse of the Evolution Theory" and "The Creation Facts". Participants of the conferences were academicians, university students and professionals. On 6th May 2015 Saba Islamic Media and Faculty of Dentistry, University of Malaya organised an open public lecture and two representatives from Harun Yahya Foundation were invited as speakers.

Adnan Oktar is also an avid columnist for Malaysian newspapers. His articles are often published in Malaysian daily newspapers like *The Malaysian Insider*,

Harakah, *Malaysia Today* and the *New Straits Times*. One of his latest articles pertaining the latest scientific findings that disproved Darwinism was published in the *News Straits Time* on September 2016 (Yahya, 2016).

20.3.3 Publications of Anti-evolution by Malaysian Scholars

One of the books that were published during the occurrence of Islamic Resurgence in Malaysia is a book with the title of *Teori Evolusi: Suatu Fakta atau Asas Ideologi? (Evolution Theory: Facts or Basic Ideologies)* by Sulaiman Noordin and Suzanah Abdullah. The book was published by Malaysia Islamic Science Academy in 1978. The book consists of articles that were presented during an Islamic Science Seminar that was held in December 1977. Through the publication of the book, the authors argue that the concept of evolution is nothing more than just a hypothesis that does not have any scientific value. The book also emphasizes how this concept is against Islamic principles and teachings.

20.4 The Place of the Evolutionary Theory in the Malaysian Science Curriculum

The formal public school system in Malaysia involves six years of primary education (Primary 1–6), three years of lower secondary education (Form 1–3), two years of upper secondary education (Form 4–5) and 2 years of post-secondary education (Form 6). There are three types of public examinations in Malaysia i.e. Primary School Achievement Test (*Ujian Penilaian Sekolah Rendah, UPSR*) which is compulsory for all Primary 6 students, Malaysian Certificate of Education (*Sijil Pelajaran Malaysia, SPM*) which is compulsory for all Form 5 students and Malaysian Higher School Certificate (*Sijil Tinggi Pelajaran Malaysia, STPM*) or Malaysian Higher Religious Education Certificate (*Sijil Tinggi Agama Malaysia, STAM*) compulsory for all Form 6 students. In Malaysia, the Biology subject is offered as one of the elective subjects for upper secondary level (Form 4–5) and post-secondary level (Form 6) for science stream students.

Although evolutionary theory is included in the high-school curriculum of many Muslim countries, this is not the case in Malaysia. Theory of evolution is not included in the biology subject that is currently taught in upper secondary level (Form 4–5) and post-secondary level biology curriculum (Form 6) and subsequently, the Biology textbook in Malaysia does not contain any chapter on evolution. To understand the underlying reasons for the non-existence of theory of evolution in the current Biology syllabus, we must first understand the history of the National Curriculum of Malaysia. As indicated earlier, Malaysian science education for upper secondary level (Form 4–5) was adopted from three British science curricula. But, the adoption of British science curricula has raised concerns regarding the teaching of the evolution theory in the Modern Biology subject. Before the introduction of the National Integrated Secondary Curriculum in 1988, the theory of evolution, although was not explicit part of the school biology curriculum due to its sensitivity to conservative Muslims who believed that the teaching of evolution was contradictory to religious teachings (Kasmo, Usman, Hassan, Yunos, and Mohamad (2015b), was taught on a voluntary basis by many teachers, especially in urban schools. Teachers were not barred from teaching the subject and evolutionary theory was at least presented in biology textbooks and reference books (Loo, 1999).

As previously highlighted, Islamic Science manifested itself in the Malaysian science curriculum at the beginning of 1980s (Lee, 1999) with the introduction of the New Primary School Curriculum (KBSR) in 1983 and the Integrated Secondary School Curriculum (KBSM) in 1989. Since then, the theory of evolution was withdrawn from the curriculum and Malaysian students have not learned Darwin's theory of evolution in schools because it is considered to be contradictory to the Islamic belief that Allah is the Creator of the universe. Hence, unlike Indonesia, another Muslim Country in Southeast Asia, Malaysia did not sign a statement by the Interacademy Panel (IAP), in support of the teaching of evolution, including human evolution (IAP, 2006). Therefore, it is not surprising to see that there is no formal teaching of the theory of evolution in Malaysian's upper secondary education (Form 4 and 5) as well as post-secondary level (Form 6).

Although the theory of evolution was not presented as a fact, and the word 'evolution' itself appears to be non-existent in the current upper secondary biology text book, the concept of evolution and processes that result in the evolution of organisms, i.e. natural selection, inbreeding, hybridization, and mutations, are thoroughly discussed in the current Biology syllabus. For example, in Form 5 Malaysian Biology, students are taught about variations and mutations by using the peppered moth *Biston betularia* as an example. In learning this topic, mutations are discussed as a source of genetic variations. Students are also taught on the importance of variations for the survival of organisms (Ministry of Education Malaysia, 2006b). Meanwhile, inbreeding and hybridization are discussed in Form 4 Malaysian Biology curriculum focusing on cloning techniques and breeding in the agricultural field (Ministry of Education Malaysia, 2006a). However, all these processes are not used as evidence of evolution and the word natural selection or "evolution" is not included in the upper secondary Biology textbook.

Beginning from 2017, the Integrated Curriculum for Secondary School has been replaced with the new Secondary School Curriculum Standard which is implemented on all Form 1 students. This situation has resulted in major changes in science curriculum in Malaysia and may bring some light in evolutionary theory education in Malaysia. Based on the new syllabus, new topics have been introduced to Form 1 students beginning 2017. In the new syllabus, evolutionary concepts will be introduced to Form 1 students indirectly through the introduction of Palaeontology study. Students are introduced to the geologic time scale and how it is uses to determine the Earth's age. Students are also introduced to the importance of fossils in providing information about the species that have disappeared from the Earth as well as how animals and plant species on earth have changed over billions of years (Ministry of Education Malaysia, 2015a, b).

The idea of ancient earth is not controversial among Muslims, and young-Earth creationism is wholly absent in the Muslim world. Muslims around the world commonly accept that the universe is billions of years old (Hameed, 2008). Therefore, the inclusion of the Earth's age in the new secondary science syllabus is not surprising and will not lead to controversial debates among the Muslims in Malaysia. Similar to other evolutionary concepts included in the previously mentioned Biology syllabi, the introduction of palaeontology to all Form 1 students will not be discussed as an evidence supporting evolution. However, it will still be a positive development in evolution education in Malaysia as these topics were never introduced in either biology or science subjects in the former syllabus.

Similar situation can be seen in the Malaysian post-secondary level biology (Form 6) where theory of evolution that was once existed in the previous syllabus was withdrawn from the current syllabus through syllabus revision. In the previous Form 6 Biology syllabus, that was implemented since 2001, the theory of evolution was formally introduced to all science stream students. With reference to the previous syllabus, Form 6 students were introduced to Lamarck's theory, Darwin-Wallace's theory, and evidences supporting theory of evolution were thoroughly discussed. Various evolutionary mechanisms, including natural and artificial selection and speciation process were also discussed. Relationships between population genetics and evolution were also discussed through the introduction of the concept of gene pool. However, in 2012, the Form 6 Biology syllabus that had been in use since 2001 was revised. Lamarck's Theory, Darwin-Wallace's Theory and evidences supporting the theory of evolution were withdrawn from the current syllabus. But evolution is still being taught indirectly through discussion on natural selection, speciation and natural selection (Malaysian Examination Council, 2002).

20.5 Biology Teacher's Attitudes Towards Teaching Evolutionary Theory

It is almost axiomatic that teachers are the key players in any curriculum reform and implementation. Several studies have revealed that teachers' instructional and content knowledge, acceptance and attitude about the subject matter impact their instructional decisions. With regards to evolution education in Malaysia, generally 70% of Malaysian teachers were shown to be most radically creationist and their attitude and conception were significantly shaped by their religious beliefs (Yok, Clément, Leong, Shing, & Ragem, 2015). Malaysia is well known as a multi-religious and multi-racial country. Based on the 2010 Malaysia population census, 61.3% of its population is composed of Muslims, 19.8% Buddhists, 9.2%

Christians, and 6.3% Hindus. Even though Muslim religion in Malaysia has the reputation of being moderate and tolerant, Muslim teachers were recorded to be the most creationist, while their fellow Buddhist colleagues being the most evolutionists. Whilst Christian teachers' attitudes and conceptions were recorded to be intermediate between these two poles.

Studies done by Kose (2010), Yok et al. (2015) and Downie and Barron (2000) had shown that rejection of evolution among Muslim teachers is correlated and can be explained from the point of view religious beliefs and values. The majority of Malaysian Muslim teachers are rooted by their firm belief that God created life. This belief is consistent with the Quran teachings which state that Allah is Al Khaliq (the Creator) and is also Al Bari' (The Evolver). Based on description of Islamic science in Malaysian science curriculum earlier, for Muslims, religion and science cannot be separated. The Quran has explained in specific details regarding the creation of mankind as well as the creation and various functions of the universe and all its contents. Furthermore, the direction of education in Malaysia is structured by its National Education Philosophy as discussed earlier. This philosophy is a document that contains the policy statement as well as guidelines for policy implementation. It firmly holds on to the concept of national identity, unity and the principals of national ideology consistent with the education aims and goals. It is clearly highlighted in the National Education Philosophy that Malaysian education is closely tied to the belief and devotion to God. This National Education Philosophy forms a vertical relationship between individuals and God; acknowledgement of the existence of God; the acceptance of God the Almighty as the creator of the universe and its contents; as well as the awareness that all natural phenomena are by the regulation of God.

In addition to religious beliefs and values, most of the teachers are perceived as radically creationist due to their relatively low knowledge about the evolutionary theory itself (Kose, 2010). Many students graduated from university and college with a poor understanding of this issue. Compounding the problem is the fact that many of these graduates become teachers and inevitably contribute to the low level of attitude and conception regarding teaching evolution in schools. Kose (2010) further argue that teachers who lack an understanding of evolution and the nature of science may be incapable of making informed decisions of acceptance and rejection of the evolutionary theory. As a result for this scenario, Malaysian teachers would probably face difficulties to teach the biological evolution, because creationism is not scientific and hence cannot be taught in any formal biology class (Yok et al., 2015).

20.6 Emphasis Given to Evolutionary Theory in Biology Teacher Education Programs

As what had been discussed earlier, the Malaysian educational system does not offer nor provide theory of evolution education in the biology subject that is currently taught in upper secondary level and post-secondary level biology curriculum.

Subsequently, the Biology textbook in Malaysia also does not contain any chapter on evolution. In conjunction to this scenario, no emphasis could be discussed in detail regarding the context of Malaysia's biology teachers' evolutionary theory education programs. The Malaysian orientation as an Islamic country and also the philosophy used in the application of all its educational strategy implementation and policy regard God as the sole creator of the universe and its contents. Thus, this fact has established a creationism belief and stance among the majority of Malaysian citizens.

However, the authors do believe that evolution theory education should be taken into account in the national education as the basic and foremost important purpose of science education to enable us to understand nature. Progress in this knowledge field has produced numerous scientific theories that provide better explanation of the natural phenomena and the causes and effects of them. However, with reference to (Tao, 2002), the scientific theories constructed do not necessarily represent the reality. The theory of evolution is one of these explanations. Due to that, teaching about the nature of science should be integrated with teaching about evolution (Kose, 2010). Inquiry and the nature of science are not entities to be separated from the development of scientific theories. Teachers teaching the subject of evolution should connect the relationship between the scientific processes and the development of a theory (Bybee, 2001).

In science education in general, and in the evolution education in particular, science teachers are signified as an important 'missing link' between scientists' understanding and the general public ignorance of or resistance to the idea (Nehm & Schonfeld, 2007). Science education in schools is one of the few possible mediums for the evolution learning to take place. As for that, teachers should be fully equipped with the relevant and comprehensive knowledge to disseminate it to the students. Teachers' conception and knowledge structures have been found to powerfully impact the students' understandings (Rutledge & Mitchel, 2002). Arguably, research also found that many biology teachers avoid teaching evolution because they have low level of evolution content knowledge and understanding (Nehm & Schonfeld, 2007).

One of the possible ways that Nehm and Schonfeld (2007) suggested in order to ensure a solid and firm content knowledge and understandings to be possessed by those teachers is by establishing a smart collaboration between science teacher educators and scientists. One obvious approach to foster this link between scientists and public is to require a college course in evolution as part of all science teachers' certification program in biology. Such course could provide content knowledge on evolution and the nature of science, employ conceptual-change strategies, address well-documented misconceptions, and model pedagogical content knowledge necessary for the teaching and learning of evolution. The fundamental goal of this evolution instruction is to increase teachers' knowledge of evolution and the nature of science, thus providing one of the factors that enables the teachers to make informed and professionally responsible instructional and curricular decisions (Rutledge & Warden, 2000).

20.7 Evolution Education in Malaysia: A Way Forward

As discussed earlier, Malaysia is a country that does not instil evolution theory in its education due to the fact that Malaysia's official religion is Islam. Islam is a religion that adheres to the principle that God is the ultimate Creator of the universe and all its content. In addition, the National Education Policy also strongly emphasises beliefs and devotion to God in all its educational policies and strategy implementation. Due to this fact, research on evolution theory education research in Malaysia is scarce. However, there is no harm in commencing studies in order to look into the students', teachers' and education administrators' acceptance, understanding, knowledge, attitude and belief of evolution. This is to determine whether their stances are based on their in-depth understanding and knowledge, or just merely following others. This is in conjunction with the research done by Kose (2010) which highlighted that there are numerous factors that shape people's attitude and acceptance towards evolutionary theory. The most frequently mentioned factor is religion, followed by personal relationships with parents, teachers, friends and school itself. These entities and environment that are closely related to one's personal life may contribute and affect the values, characters and attitudes of an individual. Other supporting factors mentioned are the media, evidence for evolutionary theory, and flaws or lack of proof for evolution.

Besides, in this globalised world and Malaysia's multi-cultural atmosphere, evolution education could be a set of knowledges that the students should be aware of, similar as the knowledge of religious comparisons that the Malaysian students are allowed to learn. In addition, the students should also be given comprehensive and meaningful scientific knowledge supported with religious explanation for every scientific phenomenon taught in the class. This also includes the evolution-creationism aspect so that we could become a more literate nation. Nehm and Schonfeld's (2007) research states that the maturing field of global evolution education faces three challenges. One of them is related to the understanding of the interrelationship among cognitive, epistemological and religious aspects related to the evolution-creationism subject. The current instructional strategy on this subject from around the world showed that concepts and processes related to the nature of science are not explicitly discussed, particularly in relation to the construction of knowledge in the area of evolutionary science (Asghar, Hameed, & Farahani, 2014). Although there are some of these concepts embedded in the discussion of the mechanisms and evidences for evolution, students may not develop a deeper understanding of them without learning the explicit connections between scientific epistemology and biological evolution. In relation to that, Rutledge and Warden (2000) suggested that a total structure of a subject should be composed of both substantive and syntactic elements. The substantive structure refers to the concepts and proposition of a domain and their organizational framework, whilst the syntactic structure consists of the means by which knowledge is generated within a given domain, which is the nature of science. Thus, to overcome this limitation, developing a comprehensive and meaningful understanding of the nature of science and the inclusion of the nature of science-related concepts specifically in the context of evolution-creationism (e.g., the role of physical evidence in constructing scientific knowledge, various methods used by the scientists to test the validity of their claims, and the role of inference in connecting data to theory) in the science-biology curricula as well as teacher education programs would potentially enhance teachers' and students' understanding.

20.8 Conclusion

Through the detailed elaboration discussed in this chapter, it is clearly known that Malaysia is firmly guided by the national ideology principles of *Rukunegara* and also National Education Philosophy which mainly embody the principles of beliefs and devotion to God and religion. These key national principles in Malaysia have directly and indirectly shaped the entire nation which consists of various religions. Besides, the Islamic resurgence in Malaysia has contributed to the education reform and Islamization of science education, influence of prominent Islamic scholars and publications, as well as publication of anti-evolution by Malaysian scholars. Thus, these overall factors and scenarios greatly contribute to the low acceptance and attitudes of Malaysians in the public communities as well as the school and education communities towards the subject and idea of the evolution theory. In particular, Yok et al. (2015) highlighted that this rejection and low level of acceptance and attitudes are clearly translated across the religions irrespective of educational backgrounds and social economic statuses. This trend was reported to be significantly unchanged throughout the years. Besides, Malaysia has opposed the teaching of evolution, including human evolution in its entire national science curriculum.

However, in this 21st century world, where information is widely and freely disseminated via numerous technological platforms, everyone needs to be equipped with and exposed to comprehensive and meaningful scientific knowledge supported with religious explanation for every scientific phenomenon occurred. This also includes the evolution-creationism aspect. Plus, we believe that adequate knowledge regarding this subject is important to develop well-informed individuals who have stances based on their pure understanding and knowledge—not merely following others. It is not the aim of this research to produce citizens that can be labelled as pro-evolution or anti-evolution. However, it is the basic purpose of education to produce a literate nation that is able to think critically, justify rationally and make informed decisions based on their knowledge in science and religion—not solely via other people's information and beliefs. In addition, they will also be able to informatively debate their belief and stances. Furthermore, it is also one of the main 21st century education agendas to produce students who are capable of facing the challenges of the globalized world.

References

Abdullah, A. (2015). Can religion have a place in modern science? In *International Conference on Aqidah, Dakwah and Syariah (IRSYAD)* (pp. 1–15). Kuala Lumpur, Malaysia.

Ahmad, R. (1998). Educational development and reformation in Malaysia: Past, present and future. *Journal of Educational Administration, 36*(5), 462–475. https://doi.org/10.1108/09578239810238456.

Asghar, A., Hameed, S., & Farahani, N. K. (2014). Evolution in biology textbooks: A comparative analysis of 5 Muslim countries. *Religion and Education, 41*(1), 1–15. https://doi.org/10.1080/15507394.2014.855081.

Bybee, R. W. (2001). Teaching about evolution: Old controversy, new challenges. *BioScience, 51*(4), 309–312.

Downie, J. R., & Barron, N. J. (2000). Evolution and religion: Attitudes of Scottish first year biology and medical students to the teaching of evolutionary biology. *Journal of Biological Education, 34*(3), 139–146.

Hameed, S. (2008). Bracing for islamic creationism. *Science, 322,* 1637–1638. https://doi.org/10.1126/science.1163672.

Hashim, C. N., & Langgulung, H. (2008). Islamic religious curriculum in Muslim countries: The experiences of Indonesia and Malaysia. *Bulletin of Education and Research, 30*(1), 1–19.

Hassan, R. (2007). On being religious: Patterns of religious commitment in muslim societies. *The Muslim World, 97,* 437–478.

IAP, IAP statement on the teaching of evolution. (2006). Retrieved from www.interacademies.net/id=6159.

Kasmo, M. A., Usman, A. H., Haron, H., Yusuf, A. S., Idris, F., Yunos, N., et al. (2015a). The compatibility between the Quran and modern science: A comparative study among Malaysian. *Asian Social Science, 11*(10), 299–306. https://doi.org/10.5539/ass.v11n10p299.

Kasmo, M. A., Usman, A. H., Hassan, W. Z. W., Yunos, N., & Mohamad, Z. (2015b). The perception on the theory of evolution, man from ape: A cross ethnic, education and religious backgrounds study. *Mediterranean Journal of Social Sciences, 6*(1), 338–344. https://doi.org/10.5901/mjss.2015.v6n1s1p338.

Khalid, K. A. T., & Saad, S. (2010). Power sharing and unity: The policies of alliance in plural society. In Ho Khek Hua, Sivamurugan Pandian, Hilal Hj. Othman and Ivanpal Singh Grewel (Eds.), *Managing success in unity* (pp. 112–147). Putrajaya: Department of National Unity and Integration.

Langgulung, H. (1993). *Curriculum development and textbook production in lower and upper secondary level*. Paper presented at international seminar on islamic education in Makkah (23–26 May).

Lee, M. N. N. (1992). School science curriculum reforms in Malaysia: World influences and national context. *International Journal of Science Education, 14*(3), 249–263. https://doi.org/10.1080/0950069920140302.

Lee, M. N. N. (1999). Education in Malaysia: Towards vision 2020. *School Effectiveness and School Improvement: An International Journal of Research, Policy and Practice, 10*(1), 86–98. https://doi.org/10.1076/sesi.10.1.86.3514.

Loo, S. P. (1999). Scientific understanding, control of the environment and science education. *Science and Education, 8,* 79–87.

Loo, S. P. (2001). Islam, science and science education: Conflict or concord? *Studies in Science Education, 36,* 45–77.

Muzaffar, C. (1987). *Islamic resurgence in Malaysia*. Petaling Jaya: Penerbit Fajar Bakti Sdn. Bhd.

Malaysian Examination Council. (2002). *Biology syllabus* (2nd ed.).

Ministry of Education Malaysia. (2006a). Integrated curriculum for secondary schools curriculum specifications biology form 4.

Ministry of Education Malaysia. (2006b). Integrated curriculum for secondary schools curriculum specifications biology form 5.
Ministry of Education Malaysia. (2015a). Sains Tingkatan 1 Dokumen Standard Kurikulum dan Pentaksiran.
Ministry of Education Malaysia (2015b). Pertimbangan Dasar, *60*, 40.
Ministry of Science Technology and Innovation Malaysia. (2010). the public awareness of science and technology Malaysia 2008.
Nehm, R. H., & Schonfeld, I. S. (2007). Does increasing biology teacher knowledge of evolution and the nature of science lead to greater preference for the teaching of evolution in schools? *Journal of Science Teacher Education, 18*(5), 699–723. https://doi.org/10.1007/s10972-007-9062-7.
Ozay Kose, E. (2010). Biology students' and teachers' religious beliefs and attitudes towards theory of evolution. *Hacettepe University Journal of Education, 38*, 189–200. Retrieved from http://search.proquest.com.proxy.bc.edu/docview/757172652?accountid=9673.
Padma, T. (2011). *Complex Islamic response to evolution emerges from study-scidev.net*. Retrieved January 18, 2017, from http://www.scidev.net/global/health/news/complex-islamic-response-to-evolution-emerges-from-study-1.html.
Rutledge, M. L., & Mitchel, M. A. (2002). High school biology teachers' knowledge structure, acceptance and teaching evolution. *The American Biology Teacher, 64*(1), 21–28.
Rutledge, M. L., & Warden, M. A. (2000). Evolutionary theory, the nature of science and high school biology teachers: Critical relationships. *Evolution, Science and High School Teachers, 62*(1), 1–9.
Tan, S.-B. (1991). The development of secondary school science curriculum in Malaysia. *Science Education, 75*(2), 243–250.
Tao, P.-K. (2002). A study of students' focal awareness when studying science stories designed for fostering understanding of the nature of science. *Research in Science Education, 32*, 97–120. https://doi.org/10.1023/a:1015010221353.
Yahya, H. (2016). Darwinism disproved once again|New Straits Times|Malaysia General Business Sports and Lifestyle News. Retrieved January 18, 2017, from http://www.nst.com.my/news/2016/09/175385/darwinism-disproved-once-again.
Yok, M. C. K., Clément, P., Leong, L. K., Shing, C. L., & Ragem, P. A. (2015). Preliminary results on Malaysian teachers' conception of evolution. *Procedia—Social and Behavioral Sciences, 167*, 250–255. https://doi.org/10.1016/j.sbspro.2014.12.670.

Kamisah Osman is a Professor from Universiti Kebangsaan Malaysia in the Department of Teaching and Learning Innovation, Faculty of Education. Her expertise is STEM education specializing in the assessment of problem-solving and higher order thinking as well as innovative pedagogical approaches in STEM learning. She is one of the prominent key players of STEM education not only at the national but also at the international levels. Kamisah is also an active member in securing the Quality Assurance and Programme Accreditation processes, not only at the university level, but also at the national and international levels and is currently holding administrative responsibilities as the Deputy Director (Audit and Benchmarking) at the university Centre of Quality Assurance.

Rezzuana Razali is a Phd student at the Faculty of Education, Universiti Kebangsaan Malaysia. Her research focuses on the development of innovative biotechnology teaching and learning modules in order to stimulate students' interest towards the subject.

Nurnadiah Mohamed Bahri is currently a Phd student at the Faculty of Education, Universiti Kebangsaan Malaysia. Her background is biotechnology and her research focuses in measurement and evaluation particularly in revealing factors which affect students' attitudes towards STEM related subjects.

Chapter 21
Evolution Education in the Philippines: A Preliminary Investigation

Jocelyn D. Partosa

Abstract Although the literature on evolutionary theory points to its central role in biology, conversations regarding its place in the curriculum remain scanty and vague in terms of content and pedagogy, particularly in the Philippines. In fact, research on evolutionary theory mainly reports on students' beliefs and concepts including their conceptual understanding. Whether evolutionary theory is getting the attention it deserves is uncertain, even to these days. It is thus relevant and timely to determine the place of evolutionary theory in the curriculum in the context of the Philippines. This paper aims to answer the following questions: 1. What specific legal provisions refer to the inclusion of evolution in the curriculum? 2. In terms of content, what concepts of the evolutionary theory are emphasized? 3. In terms of research on evolutionary theory, what has been the focus? The significance of this chapter is two-fold. First, the current literature on evolutionary theory in the Philippines is fragmentary. This chapter aims to address this gap by attempting to corroborate available data. Second, this chapter hopes to serve as a basis for a more focused and streamlined research agenda on teaching and learning evolutionary theory in the Philippines.

Keywords Evolutionary theory · Global evolution education · Science education

21.1 Introduction

> Nothing in Biology Makes Sense Except in the Light of Evolution.
>
> —Theodosius Dobzhansky (1973)

I have come to appreciate the evolutionary theory only in recent years, beginning in 2012 when I was asked to teach evolutionary biology, an elective in the curriculum for Bachelor of Science in Biology. It was then my first time to teach evolution after

J. D. Partosa (✉)
Ateneo de Zamboanga University, Zamboanga City, Philippines
e-mail: ojdpartosa@yahoo.com; partosajocd@adzu.edu.ph

© Springer International Publishing AG, part of Springer Nature 2018
H. Deniz and L. A. Borgerding (eds.), *Evolution Education Around the Globe*, https://doi.org/10.1007/978-3-319-90939-4_21

nineteen years of being in the teaching profession. I remember wondering what to teach and how to go about teaching the same to my students. I asked colleagues in my department if they had a syllabus on evolutionary biology. Using one colleague's previously prepared syllabus and another I found in literature elsewhere, I began to write my own.

Since evolutionary biology is a three-unit course in our BS in Biology program, I prepared a syllabus whose focus was three-fold. In terms of content, it centers on basic principles behind evolution and its mechanisms, particularly natural selection and adaptation. As the attempt was integration, key concepts in cell and molecular biology, genetics, ecology, classification and phylogeny were also built-in. In view of developing skills in reading, writing and critical thinking, students were required to turn in article reviews and summaries on key evolutionary issues, in addition to major examinations.

My background in evolution was almost non-existent. It was barely discussed when I was in college. It was taught quite extensively in one of my classes in the graduate school though. Now, I am writing a chapter on global evolution education relative to curriculum, content and research. Initially the intent was to review current curricula from across the country at various levels of education and do a systematic review of literature. Regrettably, these barely were the case. Therefore, it would be presumptuous to say that this narrative is a collective view of the Philippines.

Essentially, this chapter will address aspects of evolution education particularly: public acceptance of evolutionary theory within the social, political, and cultural context of the Philippines, existence and extent of influence of anti-evolution movements in the country, place of evolutionary theory in the curriculum, emphasis given to evolutionary theory in biology teacher education programs, biology teachers' attitudes toward teaching evolutionary theory and suggestions to improve evolution education in the country. Occasional references to my colleagues' experience in teaching and doing research in evolution within the social and cultural context of the Philippines appear in some parts of the chapter as well.

21.2 General Information of Country

As of July 2016, the Philippines' population was estimated to be 102, 624,209, majority of which (36.86%) belong to the 25–54 age group (Philippines Demographics Profile, 2016). In terms of religious affiliations, the Philippines is mainly Catholic (82.9%) followed by Islam (5%), Evangelical (2.8%) and Iglesia ni Kristo (2.3%). Some 4.5% of the population belong to other Christian denominations, 1.8% to others, 0.6% unspecified, and 0.1% none (2000 census, cited in Philippines Demographics Profile, 2016). Such diversity is further convoluted by ethnicity resulting from a unique geography and rich history of enculturation. Thus, the Philippines has a semblance of its European, American, and Asian colonizers.

Education in the Philippines has been a confluence of influence of its natives, migrants and settlers. Durban and Catalan (2012) describe the evolution of Philippine education in a timeline covering the Spanish, American, and Japanese occupations, through the EDSA revolution, and up to the present. Education during the Spanish Colonial Era (1521–1898) was mainly selective and elitist. Learning was more a privilege than a right. The curriculum heavily focused on the teachings of Christian Doctrines, Spanish, Latin, and the Filipino language. Math and science were either neglected or non-existent especially for girls. During the American occupation (1898–1946), schools were established across the country whose curriculum gave primacy to reading and writing American literature (that is, geography, history, lives of heroes and English). Education became accessible to all during this time. However, the Japanese destroyed the public school system within their three-year occupation (1942–1945). After the war, education in the Philippines underwent transition—from massive rebuilding of infrastructures to restoring Filipino values that were eclipsed by colonial mentality. The ensuing years in the 1950s and 1960s were marked by rapid economic growth and student activism. Even up to these days, education in the Philippines is yet to address issues and concerns both emergent and pressing. One such concern points to '*curriculum as not responsive to the basic needs of the country.*' And the growing demands for globalization of education in view of program alignment to meet international standards. This is utterly overwhelming for a country that has been struggling amid political divide, corruption of sorts, poverty, and a growing indifference among the mass to improve their lot.

21.3 Public Acceptance of Evolutionary Theory Within the Social, Political and Cultural Context of the Country

Where does evolution stand amid a very diverse setting then? Why should evolution deserve attention anyway? Is it generally accepted? To what extent is such acceptance palpable and in what terms? Whether the public accepts the evolutionary theory and its related constructs remains contentious. Despite provisions in the curriculum, conversations about the subject tend to be spurious and divisive. Although there are no known anti-evolutionist movement in the Philippines, the divide between adherents of creationism and evolution is common. The following are a couple of excerpts available online: One said, "I was raised catholic and now an Atheist. I think my Philippine education of the late 80s and 90s did not really do a good job in communicating to me what the 'randomness' in evolution is and what natural selection is all about (gmvancity, 2014)." Amparo (2012) said, "evolution is still treated, even within academic circles, as a scientific principle that can be reasonably doubted. Grossly unscientific ideas like creationism continue to permeate the academic, and that people who are products of our country's so-called 'premiere university' keep on spouting nonsense against evolution, makes me seriously doubt the effectiveness of our system of science education." Interestingly, both excerpts alluded to Philippine science education as ineffective.

Apparently, evolutionary theory has yet to gain public acceptance even to these days. And even in academic discourse, it remains in the peripheries. People tend to overlook it, dismiss it or treat it superficially to the point of furthering deep-seated misconceptions. Survey and case studies about biological evolution among students for example (Clores & Limjap, 2006; Clores & Bernardo, 2007) show that attitudes and beliefs both affect acceptance and understanding of evolutionary theory. Students accept, reject or doubt evolutionary theory in the foregoing survey and ensuing case analyses on beliefs and concepts of evolutionary theory among 37 freshman students in a general college biology class at a Catholic University in Philippines (Clores & Limjap, 2006). Those who accept the theory were further categorized based on their strong scientific inclination, preference for evidence, and misconceptions on evolution. Those who reject evolution remain adherent to creation and refuse to change position, even after four weeks of constructivist-inspired instruction.

Similarly, Yasri and Mancy (2016) surveyed Buddhist and Christian high school students in a course on evolution at a Christian school in Thailand using a tool they developed consisting eight positions and a question as to reasons for change in position. They investigated student changes in position on the relationship between evolution and creation. Several students changed their position towards increasing acceptance of evolution which was noticeable among Christian students. Participants averred that such changes were influenced by their understanding of the evidence for evolution and of ways of relating evolution and their religious belief (Yasri & Mancy, 2016).

Both studies show that faulty prior knowledge, deep-seated beliefs and predispositions often lead to misunderstanding evolution and impede learning. It has been 20 years since studies on conceptual change and movement towards deep understanding in science education became popular following reform movements at all levels and in all disciplines (Tanner & Allen, 2005). This has been the scenario elsewhere in the globe and quite recently in Philippine research on conceptual change across disciplines—biology, mathematics, chemistry and statistics—(Clores & Limjap, 2006; Clores & Bernardo, 2007; Halili & Trillanes, 2012; Jugar, 2013). Sadly, conceptual change studies on evolutionary theory remain scarce.

21.4 Place of Evolutionary Theory in the Curriculum

This lack of research on evolution education possibly stems from a general disinterest on the subject despite the explicit mandate on the inclusion of evolution in the BS Biology core program curriculum along with other basic concepts in biology like structure/function; regulation; growth; and development (Sample Outcomes-Based Curriculum for the Bachelor of Science in Biology as per

Commission en Banc Resolution No. 085-2015). However, there is no reference to evolution in the suggested outcomes-based curriculum for the Bachelor of Secondary Education major in natural sciences or biological sciences. Only one private and Catholic school among 22 schools offering Bachelor in Secondary Education major in Biological Sciences has a 4-unit subject on evolution and genetics combined.

21.5 Emphasis Given to Evolutionary Theory in Biology Teacher Education Programs

For higher education institutions in the Philippines, there is an explicit legal provision for placing evolutionary theory in the curriculum. This, however, is limited to Bachelor of Science in Biology. And both public and private higher education institutions (HEIs) are advised to align their curricula according to the foregoing mandate. Currently, the Philippine Education Reform (Enhanced 12-year curriculum) is in transition following its implementation in 2012. In the revised secondary education curriculum (RSEC), evolution is moved from grade 8 to grade 9. Whereas evolution is taught in primary grades, nothing in the provision points to its inclusion in teacher education program for elementary teachers—a blatant gap that has to be addressed.

Even with the foregoing provisions in the Bachelor of Science in Biology curriculum for evolution, implementation varies from one school to another and from one teacher to another, as evidenced by two biology teachers whose thoughts on evolution are presented in Table 21.1. Both teachers are teaching in their respective Catholic universities. I collaborated with one of these teachers on an educational research project about misconceptions with regard to natural selection and photosynthesis. They are referred to as T1 and T2, respectively, hereafter.

T1 first taught evolution in 2006 and T2 in 2013. Both prepared their own syllabus using existing syllabi, books and online sources as guide. T2 further said that she based her syllabus on the practical application and relevance of evolutionary concepts to taxonomy, genetics and biodiversity.

21.6 Biology Teachers' Attitudes Toward Teaching Evolutionary Theory

Both find the evolutionary theory acceptable. According to T1 there is "growing evidence for evolution in many scientific disciplines and the theory is relevant and an intelligible explanation for many natural phenomena." T2 said something akin relative to studies and experiments supporting evolution. She further commented that "evolution is evidently taking place now as seen in how organisms differ then

Table 21.1 Some thoughts on teaching evolutionary theory in terms of focus

Thoughts on evolution when asked:	Teacher 1	Teacher 2
What are the basic tenets of evolutionary theory?	Referred to Darwin's five major theories related to variational evolution: *Theory of evolution as such* *Theory of common descent* *Theory of multiplication of species* *Theory of gradualism* *Theory of natural selection*	Said that organisms change overtime. And whose changes are shown not only in their morphology, but chemistry and genetic composition as well. "These changes are driven by environmental factors
What concepts of the evolutionary theory have you been emphasizing?	• Darwin's basic ideas on evolution • Evidence for evolution • Misconceptions about evolutionary theory	• Physico-chemical theory • Darwin's theory of evolution • Dobzhansky's modern evolution synthesis
Why do you think these concepts should be emphasized?	"Without clear and accurate understanding of these concepts, students might not be able to understand the theory of evolution in particular and biological issues in general"	The physico-chemical theory answers where and how questions. Darwin's natural selection explains why organisms change and how they do so "Dobzhansky's modern evolutionary synthesis points to the significance of chromosome recombination and gene mutation on the evolution of organisms"

and now; and how individuals differ within the same species." As to how they feel towards teaching evolutionary theory, T2 said she feels fortunate since it makes her cognizant of its concepts and that it helps her in teaching other biology subjects.

As can be gleaned from Table 21.1, the two participants appear to have reasonable parallel focus. For example, both point to the theory of evolution as such (organisms change over time); theory of natural selection (changes driven by environmental forces and hinted on genetic composition as the source of variation); and scientific evidence (morphology, genetics, phylogeny) for evolution. T1 sees the relevance of addressing misconceptions about evolutionary theory as well.

Their rationale for emphasizing the foregoing concepts has to do with either ensuring accurate understanding of evolution, its relationship to biology in general (T1); or showing the connections among concepts like the physico-chemical theory, Darwin's natural selection, and Dobzhansky's modern evolutionary synthesis. Both teachers recognize concept integration as primary in understanding evolution.

21.7 Suggestions to Improve Evolution Education in the Country

When asked how evolution education may be improved in the country, T1 said to "engage teachers in helping students fully understand biological evolution." Ergo, research in teaching and learning evolution must be done vis-à-vis students' prior conceptions and or predispositions. Likewise, T2 said that it is essential to "strengthen evolution education in the country owing to its importance in explaining changes in the ecosystem resulting from climate change, pollution, and other organisms" for example. According to her, the "Philippines' rich plant and animal biodiversity can be used as basis for studying patterns of growth, life cycle and behaviour in response to changing environment"—which actually reflect and form part evolutionary processes. T2 further said that to address this "the Commission on Higher Education (CHED) should provide schools with solid and strong guidelines on what concepts should be taught, focusing on Philippine flora and fauna, and how it should be taught."

21.8 An Attempt to Reconcile Curriculum, Focus, Pedagogy and Research in Evolutionary Theory: Further Suggestions to Improve Evolution Education in the Country

The Enhanced K-12 Basic Education Program which was launched on April 24, 2012 resulted from a long history of studies on the inadequacy of the basic education curriculum and whether adding or restoring 7th grade would assuage this enduring problem (DepED Discussion Paper, 2010). The Enhanced K-12 Basic Education aimed to: enhance the quality of basic education owing to its poor quality as evidenced by low achievement in the National Achievement Test (NAT) for basic education and high school and in Trends in International Mathematics and Science Study (TIMSS) in 2003 and 2008, respectively; decongest the curriculum; better prepare high school graduates either for work or higher education, consequently making them more emotionally prepared for entrepreneurship, employment or higher education—here or abroad (DepEd Discussion, 2010).

The K-12 basic education program consists of kindergarten and the 12 years of elementary (6 years) and secondary education (4 years junior high school and 2 years senior high school). Students in senior high school can choose from among specializations in science and technology, music and arts, agriculture and fisheries, sports, business and entrepreneurship. In a nutshell, the Philippines envisions to produce students and graduates who have sound educational principles, are lifelong learners, are competent and productive, coexist in fruitful harmony with local and global communities, are critical thinkers, and are capable of transforming others and self (DepEd Discussion, 2010).

In view of the foregoing, what then should constitute a science curriculum framework for basic education? Problems on quality of teachers, the teaching-learning process, the school curriculum, and instructional materials and administrative support have been identified in many education and graduate student researches (DOST-SEI, 2006 cited in SEI-DOST & UP-NISMED, 2011; Durban & Catalan, 2012). As part of its threefold function, the University of the Philippines National Institute for Science and Mathematics Education Development directed its efforts at creating a science curriculum aimed at improving the quality of education at the elementary and secondary levels. In consultation with key stakeholders from the industry, university, scientists, parents, teachers, school administrators, community leaders, media and students in 2006, the institution resolved to form a *"coherent, comprehensive science curriculum framework for basic education with development of scientific inquiry as its overarching emphasis and the promotion of core science concepts and skills to enable students to 'learn how to learn'"* (SEI-DOST & UP NISMED, 2011).

The framework gives an overall structure for organizing learning and teaching three interlocking components: inquiry skills, scientific attitudes and content and connections. It is non-prescriptive but should provide a common curriculum direction for educators, curriculum developers, and textbook writers in making learning activities and experiences coherent in view of preparing students to become scientifically literate amid a dynamic, ever changing and increasingly technological society (SEI-DOST & UP NISMED, 2011).

Genetics, evolution and biodiversity (under life science) are offered in grades 9 and 10 (Junior High School) whose focus questions are outlined in Table 21.2 (excerpt).

Table 21.2 Grades 9 and 10 focus questions and science ideas for evolution and biodiversity

Focus	Science ideas
Why are there different kinds of organisms? How did each kind come to be?	When changes in the genetic material (mutations) result in individuals that can no longer reproduce with members of the original population of organisms, a different kind of organism (species) evolves
Why are there more kinds of organisms in some areas than others?	There are more kinds of organisms in the tropics than in temperate regions. Scientists propose varied reasons for this observation
Why is high biodiversity important?	Biodiversity promotes stability in a constantly changing environment Biodiversity provides a wider range of resources for food, medicine, fuel and other essential needs of human and other living organisms Evolution and biodiversity are the results of genetic changes Extinction of species may occur when the adaptive characteristics of a species are insufficient to permit its survival in a changing environment

Adopted with permission from SEI-DOST & UPNISMED (2011)

In Junior High School, therefore, evolution is taught relative to natural selection as its mechanism, mutations as sources of variation, speciation, environmental pressures and biodiversity. With the science curriculum framework in place, the gap evidently points to non-inclusion of evolution in the elementary and secondary teacher education program. Possibly it is time to rethink our teacher education program and make it more consistent with the signs of the times. Despite suggested revisions for the Bachelor in Elementary Education and Bachelor in Secondary Education programs in the framework for Philippine Science Teacher Education (SEI-DOST & UP NISMED, 2011), these changes are yet to take effect. Again, the framework gives a broad description of science course for general education and specialization in science. The new teacher curriculum sets 12 and 60 units of science subjects for pre-service teachers in the elementary and secondary levels respectively.

In the past, pedagogy mainly involved recall of information and the teacher with the central role in the educative process. Assessment usually meant having to repeat the same information and with little opportunities for cohesion and integration. Recently, studies in constructivism and conceptual change have slowly permeated education in the Philippines. In effect, approaches that promote constructivism and conceptual change like integration, reflection, collaboration, and inquiry-based problem solving are highly advocated in the Enhanced K-12 Basic Education primer.

As described early on in this chapter, research in evolutionary theory in the Philippines is very limited. Two fairly recent works by Clores and Limjap (2006) and Clores and Bernardo (2007) mostly dealt with beliefs about evolution among students in one Catholic school. The most recent dealt with understanding of natural selection among pre-service and in-service secondary biology teachers (Clores et al., 2014). Here teachers from various public and private secondary high schools in Regions V (Bikol) Region IX (Zamboanga City as representative) generally had low understanding of natural selection. Of 20 items in the Conceptual Inventory of Natural Selection questionnaire, concepts like *origin of variation, variation inherited, change in population and limited survival* were especially difficult for several teachers. When asked about their understanding of natural selection, most teachers referred to *survival of the fittest* and *adaptation*, whereas other important concepts like *overpopulation, migration, dominance and reversibility of evolution* were referred to once only. Elsewhere in the world, studies in evolution education dealt with perceptions (Woods & Scharmann, 2001), scientific views and religious beliefs (Dagher & Boujaoude, 1997), and teachers' conceptions and knowledge structures and acceptance (Rutledge & Mitchell, 2002).

21.9 A Synthesis: What Now?

21.9.1 Of Irreconcilable Thoughts or Imagined Divisions

Apparently, whether it is here or abroad, research on evolution education largely points to students and teachers' misconceptions (Clores & Limjap, 2006; Clores & Bernardo, 2007; Yates & Marek, 2014) and opposing views and associated beliefs and attitudes (Dagher & Boujaoude, 1997). And there is growing evidence relating misconceptions with beliefs and a certain predisposition or religious inclinations (Woods & Scharmann, 2001; Yasri & Mancy, 2016). Interestingly, misconceptions cut across students and teachers irrespective of position (for or against) evolution. What is the root cause of such confusion? Perhaps, the growing dissent even among evolutionists themselves contributes to the furtherance of misconceptions and ill-constructed understandings. The divide among researchers as to which processes should be considered vital is discussed in the paper of Laland et al. (2014). In a nutshell, proponents of the extended evolutionary synthesis (EES) aver that processes like *phenotypic plasticity, niche construction, inclusive (extragenetic) inheritance and developmental bias* control evolution and not solely by genes, as opposed to the "gene-centric" view among advocates of the standard evolutionary theory (SET). According to advocates of the SET, the said processes are add-ons and have long been integrated in discussions of evolutionary theory and the goal has always been towards a more collective, cohesive theory (Laland et al., 2014). The SET asserts that both groups recognize the foregoing processes; yet the genes remain central along with natural selection, drift, mutation, recombination, and gene flow. So, how then do we address misconceptions in the classroom? Or how do we ensure that students are getting the right information? It is unlikely that anything will ever be removed from bias.

21.9.2 Problematic Science Education

Science education in the Philippines has yet to gain momentum and become globally competitive. The implementation of the enhanced K-12 basic education in 2012 was a huge step towards the said direction. The teacher education program though has yet to align with the science framework for basic education chiefly on the inclusion of evolution which is lacking in the current teacher education program. In terms of explicit focus on natural selection and mutations as sources of variation; speciation; and environmental pressures; teacher education program has yet to work on mastery of content to preclude them from spreading ill-constructed concepts in the future. As advanced in the science framework, the skills and attitudes like *critical thinking, curiosity, creativity, intellectual honesty, accuracy, objectivity, independent thinking, active listening, assuming responsibility, taking initiative and perseverance* must be developed and strengthened in science classrooms. Therefore, the learning opportunities must be one where students discuss issues, postpone judgement pending availability of acceptable data, and maintain a tolerant disposition towards diverse ideas, opinions including belief systems. As neatly offered by one teacher in evolution, the Philippines being one of the biodiversity hotspots is a potential material for discussion in science classes—an excellent platform to discuss a frequently undermined concept like evolution.

21.9.3 Research and Teaching Must Inform Each Other

There is no denying the role of research in the classroom. While research is supposed to inform teaching, issues surrounding the teaching and learning process provide a plethora of impetus for research. Much has to be done in the areas on misconceptions; research along this line must be long term and extensive. It is one thing to identify the misconceptions; the work has to move towards correcting those misconceptions (conceptual change studies). Local studies involving intervention are usually short term and for an effect to be truly attributed to the intervention, they require longer exposure and practice. Research on the effects of belief systems on students' acceptance or rejection of evolution inarguably remains challenging. As a science educator, the goal is not to annihilate those belief systems; rather focus on redirecting students' attention to recognizing the relevance of other perspectives, such as those offered by science. I think the gap lies in habitually presenting science and religion as opposing views relative to life, its various forms, and origin, with hardly an opportunity for interaction or connection. Teachers often approach science in a fragmented, disparate and absolute fashion, losing sight of its revisionary nature. Whereas science is essentially systematic, rigorous, controlled, empirical, critical, valid and verifiable, it is never absolute. This paradigm shift in thinking was championed in Kuhn's The Structure of Scientific Revolutions (1970). The history of science is proof of the temporal and revisionary nature of science. Kuhn (1970) referred to critical points in scientific development like those of Copernicus, Newton, Lavoisier, and Einstein. Revolutionary means having to reject one time-honored scientific theory in favor of another incongruous to it. The new paradigm stems from an apparent '*malfunction*' (Roberts, 2000) in the existing paradigm as it ceases to address problems, thereby creating a *crisis* and ensuing revolution. The new paradigm then takes on one of three ways: the community manages the crisis and keeps its paradigm; or on occasions, the community relegates the paradigm for future query; or usually, the new paradigm emerges and the community struggles with its acceptance. Eventually, the new supersedes the old in overall perspective, methods, and goals (Roberts, 2000). Again, our roles as science educators is neither to present science in absolute terms nor simply present it as a collection of facts which is often fragmentary and incoherent. Although there is no undermining the importance of facts, we will do well with integrating the nature of science in class. McComas (2004) describes nine keys to teaching the nature of science in attempts to assuage the problems in science education. The core NOS ideas are:

science demands and relies on empirical evidence; knowledge production in science includes many common features and shared habits of mind – there is no single step-by-step scientific method though by which all science is done; scientific knowledge is tentative but durable; laws and theories are related but distinct kinds of scientific knowledge; science is a highly creative endeavour; science has a subjective element; there are historical, cultural and social influences on science; science and technology impact each other, but they are not the same; and science and its methods cannot answer all questions. (McComas, 2004, pp. 24-27)

Integrating the core NOS ideas in the Philippines basic education and considering its long term effects are rich potential research areas as well. Another area of research that is worth exploring is the teachers' attitude towards teaching evolutionary theory. If teachers were to successfully integrate the core NOS in science teaching, it is imperative that they keep an impartial and clear perspective about what they are getting into. Finally, the absence of evolution in the teacher education program is a huge gap in the science curriculum. Clearly, in terms of emphasis given to teaching evolutionary theory, this is an opportunity for extensive research yet in the Philippines. Elsewhere, Yates and Marek (2014) surveyed 35 students and their respective 536 students in one of 32 public high schools in Oklahoma. They identified types and prevalence of biological evolution-related misconceptions held by high school biology teachers and their students. Furthermore, they identified factors that contribute to the acquisition of misconceptions among students, particularly emphasizing the teachers' role. One factor they explored was number of hours spent in teaching evolution. In the survey they used, teachers were selected from among teachers who spent 0, 1–5, 6–10, 11–15 and >15 h on teaching evolution. Accordingly, the significant difference ($p < 0.01$) in the mean difference between students' numbers of pre-and-post-instruction misconceptions were related to the number of hours teachers spent in teaching biological evolution concepts (Yates & Marek, 2014). Additionally, the optimum duration is 6–10 h. This number of hours (6–10) though neither reduced the occurrence of misconceptions nor added to the existing ones. Possibly, a similar and more extensive study is needed for the Philippines.

21.9.4 Challenges, Issues and Concerns

It would be a misnomer to end this chapter with a conclusion. Since this is a preliminary look at the Philippine scenario, I believe it is fitting to end by recalling the challenges, issues and concerns that continue to haunt and daunt science education in the Philippines.

Overall the problem with ill-equipped classrooms, inadequate equipment, facilities and even infrastructures continue to overwhelm our teachers and administrators across all levels from basic education to higher education particularly in public schools. Equally pressing and dismal is the lack of competent teachers both in science content and process and pedagogy, particularly those in basic education and even secondary education, mostly in the outskirts. This is further convoluted by a growing number of teachers who are inarticulate either in the oral or written forms. Sadly, the better teachers are either abroad or are not evenly distributed. In the Philippines, English is one of the media of instructions. Even with the recent move to use the mother tongue in basic education, language facility remains challenging.

The inclusion of evolution in the curriculum much less its acceptability is no longer an issue at least as evidenced in the new science framework previously discussed. However, evolution must be part of the teacher education programs for both our elementary and secondary teachers. At present, it is only in the Bachelor of Science in Biology that evolution is offered as an elective. The concern, therefore, is one of aligning teacher education programs with the Enhanced K-12 Basic Education. There are opposing views; again, these are mainly spurious and divisive, though such debates have not led to any known anti-evolution movement. Filipinos tend to be dismissive and frequently choose to be tolerant of others' views. Others would rather stay silent about the topic many times either for lack of familiarity or understanding. Still others are more vocal and aggressive in their belief in evolution. Yet again both sides of the fence are marked with bias and prejudice stemming from a stiff outlook or rigid view of evolutionary theory. Perhaps we will never be impartial as everybody is situated in particular contexts, culture, experiences and belief systems that serve as our filters and lenses. The history of science is replete with stories of disunity within the church, within the scientific community and between church and science. Even to these days such disunity is palpable in various forms and shapes.

So how then do we envision science education classrooms to be? Research on evolution-related misconceptions shows that the way out is to focus on conceptual change. Conceptual change studies in the Philippines continue to be sparse and fragmented though. Moreover, attitudes and beliefs of teachers and students regarding evolution are yet to be extensively explored. Specifically, research that attempts to show how attitudes and beliefs interconnect with students' understanding and emphasis (time spent) on teaching evolution are critical. Because attitudes and beliefs are often ingrained in students, a call for integration of the nature of science (NOS) in science education can potentially appease a long-standing divide between those who subscribe to creation theory and evolutionary theory.

Science has thrived because of faith as well—faith in its assumptions, theories, and laws in view of attempts to explain the world and how it works. Again, history is replete with stories of discoveries and scientific breakthroughs championed by Catholic scientists. The list includes Rene Descartes, who came up with analytic geometry and the laws of refraction; Blaise Pascal, who invented the adding machine, hydraulic press, including the mathematical theory of probabilities; Augustinian priest Gregor Mendel, father of modern genetics; Louis Pasteur, for microbiology and inventor of the first vaccine for rabies and anthrax; and Nicolaus Copernicus, for the heliocentric model of the solar system (Kaczor, 2012). Additionally, the "Big Bang Theory" was proposed by Georges Lemaitre, a Belgian physicist and Roman Catholic priest; and there are several Nobel Laureates in Physics, medicine, and physiology who are Catholics, such as Erwin Schrodinger, John Eccles and Alexis Carrel, to name a few (Kaczor, 2012).

Our inclinations and all of our faculties—mental, physical and spiritual—should not be divisive. They are meant to be integrated, complementing and supplementing each other. In ending, I would like to reiterate Kaczor (2012) quoting Pope John Paul II in his 1988 letter addressing the Director of the Vatican Astronomical Observatory saying "*Science can purify religion from error and superstition; religion can purity science from idolatry and false absolutes.*"

References

Amparo. (2012, April 30). Re: Evolution is not a religious issue [Blog post]. Retrieved from http://filipinofreethinkers.org/2012/04/30/evolution-is-not-a-religious-issue/.
Clores, M. A., & Bernardo, A. B. (2007). *Proceedings of the redesigning pedagogy: Culture, knowledge and understanding conference in Singapore.*
Clores, M. A., & Limjap, A. A. (2006). Diversity of students' beliefs about biological evolution. *Asia Pacific Journal of Education, 26*(1), 65–77.
Clores, M., Partosa, J. D., Conde, A. A, Prudente, M. M., Goingo, L., & Reganit, A. (2014). Secondary biology teachers' understanding of natural selection. *The Philippine BIOTA, 47*, 42–68.
Dagher, Z. R., & Boujaoude, S. (1997). Scientific views and religious beliefs of college students: The case of biological evolution. *Journal of Research in Science Teaching, 34*(5), 429–445.
DepEd Discussion Paper. (2010). Discussion paper on the enhanced K+12 basic education program.
Dobzhansky, T. (1973). Nothing in biology makes sense except in the light of evolution. *The American Biology Teacher, 35*(3), 125–129.
Durban, J. M., & Catalan, R. D. (2012). Issues and concerns of Philippine education through the years. *Asian Journal of Social Sciences & Humanities, 1*(2), 61–69.
Halili, L. M., & Trillanes, M. O. (2012). Qualitative reasoning approach in understanding mathematical concept in statistics. *Educational Research, 3*(3), 284–289.
Jugar, R. R. (2013). Promoting conceptual change of students' understanding of gases using instructional materials derived from conflict maps. *The International Journal of Social Sciences, 8*(1), 11–20.
Kaczor, C. (2012). "The church opposes science: the myth of Catholic irrationality" chapter 1 in the seven big myths about the Catholic church: Distinguishing facts from fiction about Catholicism (pp. 19–25). San Francisco, CA: Ignatius Press. Reprinted by permission of Ignatius Press. Available Online: Catholic Education Resource Center (CERC) www.catholiceducation.org/en/science/Catholic-contributions/the-church-opposes-science-themythsof-catholic-irrationality.html.
Kuhn, T. S. (1970). *The structure of scientific revolutions* (2nd ed.). Chicago: University of Chicago Press.
Laland, K., Uller, T., Feldman, M., Sterelny, K., Müller, G. B., Moczek, A., ... Strassmann, J. E. (2014). Does evolutionary theory need a rethink? *Nature, 514*, 162–165.
McComas, W. F. (2004). Keys to teaching the nature of science. *The Science Teacher, 71*(9), 24–27.
Gmvancity. (2014). On how well has your HS and college taught evolution? Retrieved from https://www.reddit.com/r/Philippines/comments/20rdf0/how_well_has_your_high_school_and_university.

Philippines Demographics Profile. (2016). Retrieved January 23, 2017, from http://www.indexmundi.com/philippines/demographics_profile.html.
Roberts, L. J. (2000). Thomas Kuhn's the structure of scientific revolutions. *ETC: A Review of General Semantics, 57*(1).
Rutledge, M.L. & Mitchell, M.A. (2002). High school biology teachers' knowledge structure, acceptance & teaching of evolution. *The American Biology Teacher, 64*(1), 21–28.
SEI-DOST & UP NISMED, (2011). Science framework for Philippine basic education. Manila: SEI-DOST & UP NISMED. Retrieved February 14, 2017, from http://www.sei.dost.gov.ph/images/downloads/publ/sei_scibasic.pdf.
Tanner, K., & Allen, D. (2005). Approaches to biology teaching and learning: understanding the wrong answers—Teaching toward conceptual change. *Cell Biology Education, 4*, 112–117.
Woods, S. C., & Scharmann, L. C. (2001). High school students' perceptions of evolutionary theory. *Electronic Journal of Science Education, 6*(2).
Yasri, P., & Mancy, R. (2016). Student positions on the relationship between evolution and creation: What kinds of changes occur and for what reasons? *Journal of Research in Science Teaching, 53*(3), 384–399.
Yates, T. B., & Marek, E. A. (2014). Teachers teaching misconceptions: A study of factors contributing to high school biology students' acquisition of biological evolution-related misconceptions. *Evolution: Education and Outreach, 7*(7).

Jocelyn D. Partosa is a Professor of the Natural Sciences Department of Ateneo de Zamboanga University. She was department chair for three years (June 2006–May 2009) and associate dean of the school of arts and sciences for two years (April 2013–Mar 2015). She teaches in the undergraduate and graduate levels and has been doing thesis / dissertation advising for several years now. Her research and publications include the works in metacognition, knowledge structures and vegetation analysis; and of these, two won international awards. She is currently finalizing the remaining two chapters of her first book on research.

Part VI
Africa

Chapter 22
The Unusual Case of Evolution Education in South Africa

Martie Sanders

Abstract This chapter provides an account of the history of evolution in the South African school curriculum, against a background of social, political and scientific influences in a country with the most prolific hominin fossil finds in the world. After being excluded from the school curriculum for almost 50 years for political and religious reasons, evolution was introduced into the school curriculum only about 10 years ago, at both junior and senior levels. This unusual situation provided a fertile field for new directions in research about the teaching and learning of evolution, as discussed in this chapter. Although there were initially sporadic objections from religious groups and individuals, arguments for its inclusion in the last three years of schooling for *Life Sciences* students prevailed. However, the inclusion of natural selection in the *Natural Sciences* (grades 7–9) was short-lived, and the topic disappeared without explanation after one of the frequent curriculum revisions of the last two decades, just as teachers and students were starting to accept it.

Whilst many challenges involved in the teaching and learning of evolution in South Africa are similar to those in other countries, three factors make South Africa a unique case to study. Firstly, South Africa has only incorporated the topic of evolution into the school curriculum during the last decade, leaving students schooled in the previous 50 years ignorant of the explanatory framework necessary to understand so many biological phenomena and technological innovations. This recent inclusion in the curriculum provides a new lens for viewing the teaching and learning of evolution. Secondly, South Africa has a rich fossil record, making the study of evolution a more authentic, interesting and relevant experience for students in the country. Thirdly, the introduction of evolution into the school curriculum initiated new research focus areas that provide a different perspective from which to view the challenges that accompany the teaching of evolution at school level. Each of these factors is explained in greater detail in this chapter.

M. Sanders (✉)
Animal, Plant and Environmental Sciences, University of the Witwatersrand, Johannesburg, South Africa
e-mail: Martie.Sanders@wits.ac.za

© Springer International Publishing AG, part of Springer Nature 2018
H. Deniz and L. A. Borgerding (eds.), *Evolution Education Around the Globe*, https://doi.org/10.1007/978-3-319-90939-4_22

22.1 Relevant Demographics, to Set the Scene

South Africa is the most southerly country on the African continent, and the 25th largest country out of 195 in the world, covering 1.2 million km^2 (471,000 square miles). It has a population of 55.65 million (Statistics South Africa, 2016), which has expanded rapidly in the last two decades as millions of people flee the political malfeasance, poverty, and civil war in countries to the north, seeking a better life (Facchini, Mayda, & Mendola, 2013). The country comprises nine provinces, and has 11 official languages, typically associated with the dominant tribal composition of different regions. Although English is the mother tongue of only 9.6% of the population, and only the fourth most prevalent mother-tongue language in the country, it is the main medium of instruction from Grade 4. In spite of the language difficulties this poses in the early years of schooling, and particularly the problems identified in the learning of science subjects, this policy is supported by most parents because it permits universal communication, economic access, wider opportunities for further education and jobs, is the language of science and technology, and opens doors to the Western world.

South Africa is a secular country in terms of government, but because religion is a prominent factor affecting the attitudes of students, parents, and teachers worldwide when evolution is taught, it is relevant to note that South Africans tend to be strongly religious. The most recent census (2011) did not ask people about their religious affiliations, but in the 2001 census only 15% professed to follow no religious faith. The country is predominantly Christian (80%, with 8% of the population belonging to Pentecostal or Charismatic churches typified by fundamentalist beliefs about creation). The two other religions that arose in the middle East (Islam and Judaism), and also based on a six-day creation story during which a supreme being creates the Earth and all living things, have relatively small groups of adherents: 1.5% Muslims, and 0.2% of the Jewish faith (Statistics South Africa, 2004). Hinduism and Buddhism, which do not have problems with the concept of evolution (Reddy, 2012), also have small numbers of adherents.

In 2014 there were 12,655,436 pupils in the school system, taught by 425,090 teachers, in 25,741 schools (Department of Basic Education, 2016). Pupils usually start school in the year they turn five, and follow a 13-year curriculum (a reception year followed by grades 1 to 12). No entrance tests are written at any stage of their schooling. Schooling is divided into four phases: *Foundation Phase* (grades R to 3), *Intermediate Phase* (grades 4 to 6); *Senior Phase* (grades 7 to 9); and the *Further Education and Training Phase* (grades 10 to 12). Secondary schools start at the Grade 8 level, and have only one school-leaving examination, at the end of Grade 12. Schooling is compulsory up to the age of 16.

Aspects of evolution are taught at both junior and senior levels. At the more junior level (grades 7–9, where 23% of the school-going population is found) two compulsory subjects, *Natural Sciences*, and the *History* component of *Social Sciences*, include aspects of evolution (currently predominantly only adaptation), so close to three million students should be learning about some evolution-related topics. However, it is only in *Life Sciences* that evolution by natural selection is covered in detail. *Life Sciences* is the most popular of the optional subjects offered at the Further Education and Training level (grades 10–12) that includes 21% of the school-going population. *Life Sciences* is a three-year course, taken by 53.4% of the students writing the final-year school exit examination (Department of Basic Education, 2016). This means that almost 300,000 students a year take *Life Sciences,* and learn about evolution in some detail.

22.2 The History of the Introduction of Evolution into the School System

The unusual history of the teaching of evolution in South African schools is the first aspect distinguishing the country from others in the world. It has been strongly influenced by the social, political, and cultural contexts of two distinct periods in the country's recent history.

22.2.1 *The Barren Years*

For almost 50 years (1948–1994) the National Party governed South Africa, and wielded hegemonic control over the education system. They were strongly religious Afrikaans-speaking Calvinists whose ideological, historical, social, political, and cultural circumstances led to their strong anti-evolution stance (van den Heever, 2009). Their ideological, religious and political resolves were fueled by historical circumstances that induced a hatred for the British who governed South Africa in the 1800s. By 1820 Afrikaners were migrating north from Cape Town in large numbers, in an effort to escape the British yoke. The Anglo-Boer[1] war (1899–1902) aggravated matters, because the British implemented a scorched earth policy, destroying the boers' farms and putting their wives and children in concentration camps under appalling conditions which resulted in thousands of deaths. When the war was over they struggled in poverty-stricken conditions to re-establish their razed farms. In 1918 a powerful organisation, the Broederbond,[2] was formed to foster the Afrikaner culture and their economic and political aspirations. The Broederbond became so powerful that "blatant falsehoods could eventually be stated without fear of contradiction" (van den Heever, 2009). All members of the National Party government, which came into power in 1948 after the Second World War, were members of the Broederbond, and they were responsible for the design and implementation of the apartheid system of separate development based on race.

---

[1]'Boer' translates literally as 'farmer'.
[2]Translates as 'brotherhood'.

In 1967 they implemented a 'Christian National Education' policy, based on earlier writings of their ideals. One document from 1948 stated that "the spirit and direction of every subject taught must correspond to the Christian and National life- and world-view … and … in no subject may anti-Christian, unchristian or anti-national or un-national propaganda be conveyed" (Lever, 2002, 34). However, Lever claims it was more a case of non-Darwinism than anti-Darwinism, as Darwin was mentioned in a section on leading biological figures, in the early years of National Party rule, and evolution was included in some textbooks of the time. van den Heever (2009) reports that a committee of academics was set up to check new textbooks; he claims there was evidence that at least one textbook was rejected in the 1970s for various spurious reasons, although a government official later admitted it had been rejected because of its pro-evolution leanings.

22.2.2 A Time of Radical Curriculum Changes for the "Rainbow Nation"

When the African National Congress came into power in 1994, they wanted to use the education system to achieve social transformation. They initiated a curriculum (known as Curriculum 2005) that was introduced progressively in different grades from 1998. Based on ideological rhetoric, and attempting to introduce every curriculum innovation being tried in other countries around the world, the outcomes-based initiative resulted in an outcry from teachers at all levels. They were expected to apply radical new methods, based on a jargon-ridden curriculum document that few people understood, and without any content being spelled out. The mostly underqualified teachers simply could not cope with the radically different requirements: these included designing their own materials, using activity-based and learner-centred methods, and providing formative assessment. A government-initiated review of the situation, released in 2002, resulted in significant changes being recommended in the 'Revised National Curriculum Statement'. As shown in Fig. 22.1, this was the first of three new curricula introduced in quick succession during the decade that followed. Teachers were very angry, because no sooner did they become familiar with a curriculum statement than it was changed.

Figure 22.1 indicates the evolution-related topics for the two main subjects that included aspects of evolution: *Natural Sciences* (grades 7 to 9) and *Life Sciences* (grades 10 to 12).

Natural Sciences. The Revised National Curriculum Statement was implemented at the Grade 7 level only in 2006. A major change was that content was spelled out, albeit superficially, and new content areas added. One new topic was natural selection, although the word 'evolution' was never used, probably to avoid contention. Strangely, the Grade 7 *History* component of *Social Sciences* included human evolution and hominid discoveries in South and East Africa, but these were

22 The Unusual Case of Evolution Education in South Africa

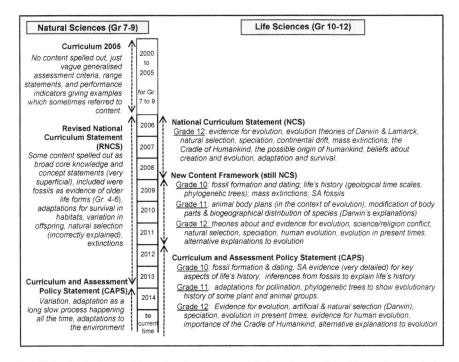

Fig. 22.1 Curriculum revisions showing content and timing changes for subjects where evolution was included

not explained from a biological point of view. A third new curriculum, outlined in the Curriculum and Assessment Policy Statement (CAPS), was introduced in progressive grades from 2012, reaching grades 7–9 in 2014. Natural selection had inexplicably been removed at the junior secondary level, although the topic of adaptation remained. No-one will admit to who did this, or why, but it left *Life Sciences* students in later grades underprepared to cope with evolution.

Life Sciences. Because the new curriculum was implemented progressively by grade, the first new *Life Sciences* curriculum, the National Curriculum Statement (NCS), was only implemented from 2006. Doidge, Dempster, Crowe, and Naidoo (2008) report that the curriculum committee was required to focus more on skills and less on content, so the content specified was very vague. 'Learning programme guidelines' had to be hastily supplied to specify additional content detail. Natural selection and some other evolution-related topics were included in the Grade 12 year, where the NCS was implemented for the first time in 2008 when the 2006 cohort reached this level (see Fig. 22.1). Evolution was intended to be taught for seven weeks of the 30-week academic year, and made up almost 25% of the *Life Sciences* content in the final public examination. Before the NCS was even fully implemented, a new version of the curriculum, called the "new Content Framework', was issued, which provided further content detail. It was introduced progressively from 2009 with the next Grade 10 group. A radical change was that the evolution content had been split, with some topics moving down to grades 10 and 11, leaving evolution by natural selection in Grade 12. Within three years the new CAPS curriculum was implemented, starting with Grade 10 in 2012, reaching Grade 12 in 2014. The evolution content was little changed, except that the contentious 'alternatives to evolution' section included more detail about the content that had to be taught ("different cultural and religious explanations for the origin and development of life on Earth; Creationism; Intelligent Design; Literalism; and Theistic evolution"). Teaching religious beliefs in a science subject may have been an effort to appease religious groups, but it did not sit well with scientists as it was worded as 'alternatives to evolution' which seemed to legitimise religious beliefs as an acceptable scientific alternative.

Teacher training programs. In 2001 colleges of education were either closed or they merged with local universities, and all teachers since then have had to obtain a degree in order to teach. Primary-school teachers and some secondary-school teachers follow a four-year combined academic and professional BEd degree. However, a three-year Bachelors degree, majoring in two teaching subjects, followed by a one-year postgraduate teaching diploma, is an alternative qualification for secondary school teachers. It tends to have more stringent entry requirements.

Ten years ago practising teachers commonly claimed not to have learned about evolution when they trained. For example, Coleman, Stears, and Dempster (2015) report that teachers were not taught evolution before 2006. However, this seems to have depended on which tertiary institution was involved. Even prior to 2001 many teacher training colleges were linked to a local university, and were required to cover in two years of academic work what the universities did in one, so that after four years teachers graduated with the equivalent of two second-year university sub-majors. Based on these regulations, I taught a detailed evolution course in the 1970s to second-year student-teachers at the Johannesburg College of Education.

22.3 Responses from Various Stakeholders, to the Addition of Evolution to the Curriculum

The introduction of evolution at school level was strongly supported by the scientific community (*Academy of Science of South Africa*, undated) and science educationists (Dempster & Hugo, 2006). As with scientific and professional organisations elsewhere in the world, the *Academy of Science of South Africa* released a statement about the teaching of evolution in schools. Whilst acknowledging the sensitivity of teaching evolution, especially for some religious groups, it stated that no-one had the right to prevent students access to learning about an explanatory framework that is crucial for understanding aspects of the modern world (e.g. HIV/aids, disease pandemics, agriculture, and environmental sustainability), so that people can make informed decisions about matters affecting their lives. They also pointed out that without such knowledge students would be prevented from entering some professions.

Drafts of the curriculum documents were widely circulated for comment, and drew condemnation from certain quarters. van den Heever (2009) described the reaction as an anti-evolution outcry "in the media, in hastily written books and ill-prepared talks". Chisholm (2002, 53) said "it provoked a storm of controversy" from individuals, university student associations, and conservative Christian organisations (which she said were by far the loudest and most vocal) and their members, who signed and submitted a form letter, jamming the Education Department's fax machines. The complaints were documented by academics such as Doidge et al. (2008), and Chisholm (2002), representing the Education Department. Objections from religious communities were voiced in the newspapers and in magazines like the creationist Christian magazine *Joy*. The religious arguments used tend to be based on common erroneous thinking about what evolution is about, and an insistence that it is anti-Christian, unscientific, lacking evidence, an unproven theory, an infringement on parent's rights, and something that would corrupt children (Chisholm, 2002; *Joy* magazine articles between 2007 and 2009). *Joy* magazine articles made various incorrect overgeneralised claims such as "All the major religions, Christianity, Judaism and Islam, by definition reject Evolution" and chastised those lured away from the church by theistic evolution ideas, labelling all those who supported evolution as 'atheists', and asking readers to pray for them. However, religious academics such as Tucker (2012) voiced their views through well researched academic papers. Tucker claimed that banning the teaching of evolution would not sort out the objections, recommending instead teaching it alongside the story of creation, so students could apply a critical thinking approach to dealing with the matter of evolution. He added that it would also provide a platform for Christian teachers to think through the challenges, reinforce their spirituality, and allow them to teach about the interface between religion and science. He said this should be facilitated by theologians and ministers with a knowledge of the science behind evolution.

The backlash against the teaching of evolution seemed to be sporadic but prominent in that it was publicised through articles and letters in the media, via newspapers and magazine articles. It inevitably came from religious-text literalists who would be classified by Scott (2000) as Young Earth Creationists, who believe that all living things were created by a supreme being in a short period of time (usually six twenty-four hour days) about six thousand years ago, and have not changed since. Numerous websites appeared condemning evolution, but most soon became defunct. One creationist organisation handed out pamphlets and CDs at an in-service session that had been organised to help teachers to teach evolution effectively. Chisholm (2002) reported that ultimately those supporting the inclusion of evolution (academics and science educators) became the dominant voices heeded by the Department of Education. Teachers expected to teach the new evolution topics in Grade 12 voiced a long list of concerns (Sanders & Ngxola, 2009). The most frequent comments in that study (49%) were about the teachers' own lack of confidence in their content knowledge about teaching evolution, or how to teach it. About a fifth of the comments (21%) were concerns about a potential religious clash.

22.4 Professional Development to Prepare Teachers to Teach Evolution

Numerous groups offered in-service courses for teachers, because the week-long training offered by the Department of Education was judged by teachers to be unhelpful as the departmental presenters had spent the week trying to justify their ideological stance and explain the vast vocabulary of confusing new jargon in the curriculum statements, and spent no time on the new content to be taught. The biggest teacher union (the South African Democratic Teachers Union) with its militant political agenda and regular strike action (Pattillo, 2012; Tangwe, Tanga, & Tanyi, 2015) and about a quarter of a million members, did nothing to help their teachers cope with the teaching of evolution. The more educationally-oriented but smaller National Professional Teachers' Organisation of South Africa (Tangwe et al., 2015), with about 55,000 members and a policy emphasising students' right to learn, tried to assist teachers by publishing a number of helpful articles in their in-house magazine, *Naptosa Insight*. They also approached academics involved in teaching and/or research about the teaching of evolution, to give talks for their teachers. Also, many of the universities, and some individual academics, offered courses for those teachers who asked for help, although these were sporadic and isolated, and had no way of systematically reaching all teachers. They did, however, prove very helpful for teachers who attended, particularly workshops offering strategies to address the perceived religion/ evolution conflict (Sanders & Ngxola, 2009; Sanders, 2010a). Over the years the Department of Education has also approached individual academics to run workshops for teachers and for subject advisors, but these, too, were sporadic.

22.5 The Rich History of Fossil Finds in South Africa, and Their Influence on Evolution Education

The second way in which the teaching of evolution may differ from that in other countries is associated with the fossil-rich context in the country. South Africa is a world-renowned source of fossil evidence, which makes learning the topic more relevant, fascinating, and motivating for South African students, especially those living near fossil sites and museums they can visit. Although some students (mainly religious Christians and Muslims) say evolution is boring (Naude & de Beer, 2014) others find it interesting and want to learn more (Chinsamy & Plagányi, 2008).

Furthermore, new findings are constantly being announced, so the media are full of relevant current reports that teachers can use as resources for their teaching, and which students see as applicable to what they are learning about. Evolution is taught in a dynamic and ever-changing real world of science. The history of fossil finds, described next, provides an enthralling narrative for teachers and students, particularly because some claims about the finds were initially disputed by international scientists, who later had to recant in the face of evidence to support the assertions. The use of narratives in teaching is important. van der Mark (2011) found that training South African teachers to explain evolution through narratives resulted in significant increases in the teachers' use of higher-order thinking skills.

22.5.1 A Brief Overview of Major Fossil Discoveries in South Africa

A historical review of the narrative begins with Darwin's visit to Cape Town (31 May to 18 June 1836) during the homeward voyage of the Beagle. This helps students to see Darwin as a real person, not just a figure from history. The South African visit is well documented in Darwin's diaries (Darwin, 1839), which formed part of his first book, published in 1839, and now known as 'The Voyages of the Beagle'.

Next came the riveting fairy-tale discovery in 1938 of a coelacanth, a lobe-finned fish thought to have become extinct about 66 million years ago in the Cretaceous period (Smith, 1956). The story involves Marjorie Courtenay-Latimer, a young woman curator of the museum in a small coastal town in South Africa (East London) summoned by a local fisherman she knew to view a strange catch in his nets. She later described it as "the most beautiful fish I had ever seen, five feet long, and a pale mauve blue with iridescent silver markings" (South African Agency for Science and Technology Advancement, 2013). She transported it to the museum in a taxi (much against the will of the taxi driver, who objected to its smell and size). Her research led her to observe its similarity to the long-extinct lobe-finned coelacanth fishes. This suggestion was scoffed at by the museum director, and later by incredulous international scientists, but was confirmed by a renowned ichthyologist (J. L. B. Smith) from a nearby university. This 'Lazarus fossil' or 'living fossil' was "considered to be one of the most notable zoological finds of the twentieth century" (Amemiya et al., 2013, 311), as the lobed fins suggested a possible transitional form to the tetrapod limbs of terrestrial animals.

Other 'missing links' found in South Africa include extinct mammal-like reptiles found in the semi-arid Karoo desert in the hinterland of South Africa (e.g. Rubidge, 1991; Rubidge & Day, 2015), and a range of hominin fossils which are filling in gaps in our knowledge of the possible ancestry of humans. The Karoo is internationally acknowledged as one of the biggest and most important fossil deposits in the world. The hominin fossils were found further north, and some of them were also originally rejected by international scientists, before later being verified. The stories of their discovery also provide motivating narratives for pupils. The first, named as a new type species, *Australopithecus africanus,* by Raymond Dart in 1925, was the skull and brain cast of what turned out to be a very small child (estimated to be a three-year old). Dart recognised its potential importance in the early ancestry of humans because the position of the foramen magnum was at the base of the skull (showing it had been bipedal) and its short canine teeth that were more similar to those of humans than to the ape-like common ancestors. Marks on the skull appear to show that the child was killed and eaten by a carnivore, now thought to have been a large bird of prey. Because of its tiny size, and because it was found during blasting for limestone in caves near the town of Taung in 1924, Dart nicknamed it the 'Taung baby'. This established a long practice of giving nicknames to important fossil finds. Although initially repudiated by international scientists, who were sceptical because they did not believe humans could have originated in Africa as Darwin had predicted, further evidence resulted in an acceptance, in the late 1940s, of its classification and importance as a hominin ancestor.

A number of other hominin finds came from an area close to Johannesburg, declared a UNESCO World Heritage Site in 1999 because of its rich yield of hominin fossils. This area is somewhat misleadingly named the *Cradle of Humankind* (the oldest hominin fossils were, in fact, found in East Africa). The Sterkfontein Caves in this area, a 45-min drive from Johannesburg and Pretoria, had, up to 2010, yielded almost a third of the world's hominin fossils. These include two iconic fossils dated to be between 3.3 and 2 million years old (Berger & Hilton Barber, 2006). The first was a 2.3 million year-old skull discovered in 1947. It was identified as *Australopithecus africanus,* and nicknamed Mrs Ples. Although later thought to be a 'master Ples', 'she' retains 'her' original name. This skull provided the evidence Dart needed to verify his identification of the Taung skull, found more than 20 years earlier, and sparked an interest in the locality from world palaeontologists and archaeologists. The second fossil from the Sterkfontein Caves, *A. prometheus* (Clarke, 2008; Dart, 1948; Granger et al., 2015), nicknamed Little Foot because of its small size, provides an intriguing mystery story full of coincidences. It starts with Ron Clark's discovery, as he rummaged through some old boxes of bones from Sterkfontein, of what he realised were hominin foot bones, probably from the same individual (Clarke & Tobias, 1995). Several years later he found the broken off distal part of a tibia in another box. Believing that the rest of the skeleton must still be in the caves he gave the broken tibia to two of the people working on fossils at Sterkfontein (Nkwane Molefe and Stephen Motsumi) and asked them to search for the rest of the skeleton (a task Clarke described as like looking for a needle in a haystack). Yet after only a two-day search in 1997 they found the perfect match, the other half of a broken off tibia embedded in the wall of the blasted cave (Clarke, 1998). After 13 years of careful excavation Little Foot was extracted, one of the most complete fossil hominins to have been found at that date. Recent dating with new techniques has placed the age of Little Foot at 3.67 mya, contemporaneous with East African hominins of *A. afarensis* (Granger et al., 2015). This started a search for more fossils in the area, involving post-graduate students from the University of the Witwatersrand (known colloquially as Wits), and later by international scientists.

Three more discoveries of iconic fossils occurred near Sterkfontein. The narratives about their discovery have also proved riveting for South African pupils learning about evolution because the discoveries occurred in the last 10 years, while senior students were learning about evolution at school. The 2008 discovery of *A. sediba* was announced in 2010 (Berger et al., 2010). A single bone found by a nine-year old boy and his dog, walking with his palaeontologist father in the Cradle of Humankind, led to more extensive explorations of the area. Bones from two individuals, a young woman and a boy, were eventually found. The woman and child are assumed to have fallen through a vertical chute into an underground cave from which they could not climb out. Following the habit of creating a common 'nickname' for important fossils, a competition was launched for South African school children to find a common name for *A. sediba*. The name selected, suggested by Johannesburg school-girl, is 'Karabo' (meaning 'the answer', in the local language of the area). The second find, in the Dinaledi Chamber cave in the Cradle, was announced with great fanfare to the world press by National Geographic in September 2015 (Shreeve, 2015).

Although there was some controversy in terms of whether it should be classified as *Homo* or *Australopithecus,* the species was named *Homo naledi* (Berger et al., 2015). Because the cave was accessible only by a long tunnel with a very narrow opening deep underground, an international search was launched for small-bodied archaeologists who would be able to fit through the opening. Four were recruited to go into the cave to film and recover the fossils, while the rest of the team waited outside. The remains of about 15 skeletons of all ages were found. In May 2017 the discovery of another cave close to the Dinaledi Chamber was announced, containing more *H. naledi* bones, including parts from two adults and a child (Hawks et al., 2017). Rigorous recent dating of the 2015 specimens, using six independent methods, now suggests that *H. naledi* lived as recently as only 250,000 years ago, raising the possibility that they existed alongside *H. sapiens*.

These three most recent discoveries have been characterised by a new atypical approach to researching new fossil finds, which traditionally involved decades of slow, painstaking work by just one or a few scientists. The discoveries of *A. sediba* and *H. naledi* were typified by very rapid processing by multi-disciplinary teams of international scientists, using modern tools and techniques such as scanning electron microscopes, micro-CT scanners, 3D printers, and virtual image processing electron microscopes. X-ray topography allows virtual sections to be viewed, modelled, and printed in 3D. This provides information about internal structures without having to remove the fossils from embedding rock, or sectioning them (an irrevocable step). Scientific publications in prestigious journals have been swift, accompanied by prominent media coverage for the public. There is an almost 'open access' policy to researchers, and replicas of the fossils are readily available. Funding and support is now obtained from prominent organisations like National Geographic.

22.5.2 Implications of the Fossil Finds for Evolution Education

The public accessibility of numerous South African museums and fossil sites has the potential to both motivate and educate visitors, although the research findings regarding the educational impact are ambivalent, as discussed later. Easiest to access is the UNESCO Cradle of Humankind World Heritage Site, which includes the Maropeng museum and exhibition centre, and the Sterkfontein Caves museum and fossil site where visitors can climb through the caves. Both places conduct or facilitate school visits and have numerous exhibits and activities that tie in well with the curriculum. Because the sites are in the populous province of Gauteng, close to the border with North West province, almost two million pupils from Gauteng and three-quarters of a million from the North West have ready access to these facilities, and other provinces stay in touch through the media and internet. The University of the Witwatersrand (Wits) curates Sterkfontein and owns all intellectual property rights. The Evolutionary Studies Institute at Wits houses a rich collection of original fossils (including the Taung skull, Little Foot, *A. sediba,* and *H. naledi*). The Institute is associated with the Origins Centre museum on the campus, has a rigorous research programme, and actively engages with schools and the public through outreach programmes.

Three other provinces have important archaeological cave sites: Makapansgat[3] in Limpopo province; Sibudu Cave, near Durban in Kwazulu-Natal; and Blombos Cave in the Western Cape, where evidence of the evolution of modern human behaviour has been found. Investigated since 1991, Blombos is the subject of international discussion on the origins of modern human behavior and the evolution of complex language (Henshilwood et al., 2001). Thus five provinces have major fossil sites in caves, and three more (the Western, Northern, and Eastern Cape) are covered by the Karoo system which is also a very rich source of fossils, although not in caves.

22.6 Lessons Learned from the South African Research, and Implications for the Future

Research into the teaching and learning of evolution in South African schools proliferated as evolution started to be taught. There had been sporadic research into university students' ideas about evolution (e.g. Moore et al., 2002), even prior to the introduction of evolution in schools, but such research followed the international thrust of investigating the understanding, misconceptions, or acceptance of evolution, some looking at pre-and post-teaching data to focus on conceptual and

---

[3] 'Gat' translates literally from Afrikaans as 'hole'.

acceptance changes. However, the third difference noted in the South African context was the shift in the focus of the research on evolution education. This may have been because evolution was a new content area, being taught for the first time in decades, and there was a sense of urgency about finding a solution to some of the problems that had emerged. The first new focus was the identification of teachers' concerns and needs when first expected to teach this new content area (Sanders & Ngxola, 2009; White, de Beer, & Ramnarain, 2014). A second new area of research was a critical analysis of the curriculum documents, both initially (Dempster & Hugo, 2006) and later as the various revised curriculum documents appeared (e.g. Johnson, Dempster, & Hugo 2011, 2015). Thirdly, there was a concerted effort to investigate possible sources of unscientific ideas, a necessary first step if the 'misconceptions' problem was to be addressed. The factors investigated included teachers (Molefe & Sanders, 2009; Ngxola & Sanders, 2009), textbooks (Robertson, 2015; Sanders, 2014; Sanders & Makotsa, 2016; Tshuma & Sanders, 2015), and examination papers and memoranda (Reddy & Sanders, 2014).

The research led to number of valuable lessons that provide direction for future efforts to address the problems identified.

1. **Identification of teachers' concerns and needs is important if appropriate support is to be offered**. White et al. (2014) found teachers from different backgrounds and school types express very different needs. Thus it is important when designing professional development interventions to start by identifying teachers' concerns and needs, and to be aware that 'one-size-fits-all' training may be ineffective (White et al., 2014). Keke (2014) established that the content area 147 KwaZulu-Natal teachers most needed help with was evolution. Sanders and Ngxola (2009) found that the main concerns (discussed earlier) of four groups of teachers attending professional development workshops before they started teaching evolution for the first time were mainly self-concerns, which Hall and Hord (2006) consider to be 'early stage' concerns. They point out that if teachers are to move to the more important 'impact concerns', self-concerns must first be addressed. A number of postgraduate reports investigating teachers and learners from specific religions contributed information about the concerns of students and teachers when learning or teaching about evolution: Christians (Mpeta, 2013; Naude & de Beer, 2014; Pillay, 2011), Muslims (Yalvac, 2011), and Hindus (Reddy, 2012).
2. **Knowledge is power**. Numerous researchers found that even attending a conventional course on evolution, i.e. with no specific intervention, reduces both learners' (Chinsamy & Plagányi, 2008; Schröder & Dempster, 2014) and student teachers' (Coleman et al., 2015) concerns and levels of misconceptions, and promotes more positive attitudes to learning about or teaching evolution (Stears, 2012). Chinsamy and Plagányi (2008) noted that students' understanding of evolution improved when presented with "facts" in the form of evidence supporting evolution, even if their levels of acceptance did not change. Sanders (2010a) found that attending an in-service course designed to develop teachers' pedagogical content knowledge in five ways (see Sanders, 2008),

particularly emphasising activity-based ways to deal with potential conflict, and simply going through the experience of teaching evolution made teachers realise that either their fears had been unwarranted, or had been resolved. Not only that, but many teachers had moved away from a dogmatic religious-text literalist approach to a more accepting religious viewpoint on Scott's beliefs continuum (Scott, 2000). Coleman et al. (2015) noted that student teachers who had attended courses on evolution had far fewer worries than the teachers who were about to teach it for the first time in 2007 and 2008. Mokgobanama (2011) observed that some concerns expressed by teachers attending an in-service course were reduced after they had visited museums dealing with hominin evolution.

3. **Do not stereotype people based on their religion**. Teachers in two studies came to realise that there is a wide range of beliefs about evolution, both between and within congregations of a religious denomination (Sanders, 2010a; Stears, Clément, James, & Dempster, 2016) and that views expressed are often those of the pastor rather than the church (Sanders, 2010a). Stears et al. (2016) found that their expectation based on a stereotyped view that African traditional churches would be biblical literalists was inaccurate, and that members of the Zionist and Shembe churches unexpectedly had more evolutionistic views than all other religious groups except Hindus, who were outmatched only by agnostics. Mpeta, de Villiers, and Fraser (2015) noted that learners' ideas were not linked to their level of religiosity. Because of the wide range of ideas within a religious group, it is not appropriate to make generalisations about a particular group of people based on a particular shared characteristic such as religion. Workshops I run show that making teachers aware of this helps reduce their resistance to learning about evolution, and the same strategy has proved effective for pupils.

4. **Overgeneralised results in the literature create inaccurate expectations**. The impression gained from reading the global and South African literature on the teaching and learning of evolution is that lack of acceptance of evolutionary ideas is very high and affects the majority of students, but this is not always the case. This is illustrated by three studies conducted in Kwazulu-Natal, South Africa: Stears (2006) found that more students accepted than rejected statements about evolution, and Coleman et al. (2015) established that first- second- and third-year student-teachers all had greater than 70% levels of acceptance. Schröder and Dempster (2014) found that only 27% of the Grade 12 s in their sample did not find natural selection credible, and this dropped to 16% once they had covered evolution in class. Kyriacou, de Beer, and Ramnarain (2015) also found unexpectedly high levels of acceptance of human evolution among practising teachers from Johannesburg: 45% agreed that humans evolve and more than half said that religion and evolution do not conflict. Furthermore, even if students do not find evolution to be credible, they are nevertheless interested in learning about it (Chinsamy & Plagányi, 2008). It was also unexpected that as many as 41% of the university students in one study conducted before evolution was taught at school level, could correctly explain the

cause of changes in pepper moths (Moore et al., 2002). Researchers need to be careful to look at the evidence before accepting generalised claims.

5. **The research design and methods used can affect the results obtained**, so both researchers and those reading the research need to be cautious about inferences made from the research data. Studies asking teachers to rate their own knowledge (e.g. Abrie, 2010; Ngxola & Sanders, 2009) may well yield inaccurate overestimates, revealed if checked against the more objective view of others, or the results of pencil-and-paper diagnostic instruments (Molefe & Sanders, 2009; Ngxola & Sanders, 2009; Stears et al., 2016). Furthermore, the decision of what design to use for developmental studies used to track changing trends in misconceptions as students learn more about evolution each year should not be taken without due consideration, because longitudinal and cross-sectional studies can yield very different results (Lawrence & Sanders, 2014).

6. **Results of different studies do not always agree and may be context-dependent, so generalisations should be made with caution**. Naude and de Beer (2014) and Yalvac (2011) found students thought evolution was 'boring', while Chinsamy and Plagányi (2008) found students became more interested after having learned something about evolution. And while numerous studies found students' acceptance levels were affected by their religiosity, Mpeta et al. (2015) found this not to be the case. Furthermore, Kyriacou et al. (2015) established that a surprisingly high number of teachers did not see any conflict between religion and evolution. Acceptance levels could be affected by the strategies teachers use in the classroom, some of which can cause conflict, as found by Sanders (2010a) when investigating which of Barbour's four strategies teachers used when teaching evolution.

7. **Anyone wanting to alleviate negative attitudes to the teaching or learning of evolution needs first to identify the factors that aggravate such attitudes**. Working with a group of very religious Jewish students, Kagan was surprised to find that many accepted evolution, contrary to what the literature depicts. Based on her interviews she developed a model of factors affecting acceptance of evolution by this ultra-religious group (Kagan & Sanders, 2012). This model suggests areas which can be targeted by teachers wanting to improve students' attitudes in order to make learning about evolution less stressful for religious students needing to "border-cross" (Phelan, Davidson, & Cao, 1991) when they move from home to the science classroom. Open-mindedness; critical thinking; open discussions with parents, moderate rabbis, and peers; and a desire to learn more, all led to these students being unexpectedly accepting of evolutionary theory.

8. **The impact of museums on the understanding of evolution is ambiguous**. South African teachers assume that visiting museums will improve the knowledge of their students, but this may not always be the case. Based on the assumption that teachers need to know how to run effective field visits, Sanders (2010b) had run an extensive professional development workshop stretching over several weekends. One component introduced how to conduct fieldtrips to

promote meaningful learning, which included designing effective worksheets. The teachers then spent a day at Maropeng and Sterkfontein so they could plan a visit. Yet when she interviewed 27 teachers eighteen months later, and analysed six worksheets they had designed, the results were disappointing for a number of reasons. These were used to develop a model of factors affecting the success of such visits (Sanders, 2010b). Mokgobanama (2011) found that while many of the Life Sciences teachers visiting the same museums improved their knowledge, attitudes, and confidence levels, others were confused by the museum exhibits, and the misconception that humans evolve from apes increased. Lelliott (2016) surveyed more than 800 members of the public after they had visited either Sterkfontein or Maropeng, and found that many had misconceptions about what 'The Cradle of Humankind' meant.

22.7 Improving Evolution Education in the Country

Many lessons arising from the existing research suggest areas which could be targeted for improvement. For example, whilst museums and visitors' centres play a crucial role in both formal and informal learning some visitors still have misconceptions. Pillay (2010) provides a critical analysis of the museums, and suggests ways in which they could be improved. A fruitful area for development would be targeting teachers and their field-trip planning so that school visits result in meaningful learning of evolutionary concepts.

It is clear that including evolution in the school curriculum improves knowledge and attitudes, even without specifically-designed interventions. Pre-service and in-service education initiatives also have an impact. But in spite of this some teachers and students still have negative attitudes, and some misconceptions persist. Sanders (2010a) found that activities based on Scott's beliefs continuum, and on Barbour's strategies for teaching evolution, change teachers' attitudes and address many of their concerns, and Lawrence (work ongoing) has found Scott's continuum to be very effective in reducing Grade 12 learners' concerns about a religion-evolution clash. This avenue needs to be investigated further.

References

Abrie, A. (2010). Student teachers' attitudes towards and willingness to teach evolution in a changing South African environment. *Journal of Biological Education, 43*(3), 102–107.

Academy of Science of South Africa (ASSA). (undated). Teaching evolution in South African schools. https://www.assaf.org.za/files/statements/Teaching%20Evolution.pdf.

Amemiya, C., et al. (2013). The African coelacanth genome provides insights into tetrapod evolution. *Nature, 496*, 311–316.

Berger, L., & Hilton-Barber, B. (2006). *A guide to Sterkfontein: The Cradle of Humankind*. Cape Town: Struik Publishers.

Berger, L., de Ruiter, D., Churchill, S., Schmid, P., Carlson, J., Dirks, P., et al. (2010). *Australopithecus sediba:* A new species of homo-like Australopith from South Africa. *Science, 328*(5975), 195–204.

Berger, L., Hawks, J., de Ruiter, D., Churchill, S., Schmid, P., & Delezene, L. (2015). *Homo naledi*, a new species of the genus *Homo* from the Dinaledi Chamber, South Africa. *eLife, 4*. https://doi.org/10.7554/elife.09560.

Chinsamy, A., & Plagányi, E. (2008). Accepting evolution. *Evolution, 62*(1), 248–254.

Chisholm, L. (2002). Religion, science and evolution in South Africa: The politics and construction of the Revised National Curriculum Statement for Schools (R-9). In W. James & L. Wilson (Eds.), *The architect and the scaffold: Evolution and education in South Africa* (pp. 51–59). Cape Town: Human Sciences Research Council Press.

Clarke, R. (1998). First ever discovery of a well-preserved skull and associated skeleton of *Australopithecus*. *South African Journal of Science, 94*(10), 460–463.

Clarke, R. (2008). Latest information on Sterkfontein's *Australopithecus* skeleton and a new look at *Australopithecus*. *South African Journal of Science, 104*(11/12), 443–449.

Clarke, R., & Tobias, P. (1995). Sterkfontein member 2 foot bones of the oldest South African hominid. *Science, 269*(5223), 521–524.

Coleman, J., Stears, M., & Dempster, E. (2015). Student teachers' understanding and acceptance of evolution and the nature of science. *South African Journal of Education, 35*(2). https://doi.org/10.15700/saje.v35n2a1079.

Dart, R. (1948). The Makapansgat proto-human *Australopithecus prometheus*. *American Journal of Physical Anthropology, 6*(3), 259–284. https://doi.org/10.1002/ajpa.1330060304.

Darwin, C. (1839). Narrative of the surveying voyages of His Majesty's ships Adventure and Beagle between the years 1826 and 1836, describing their examination of the southern shores of South America, and the Beagle's circumnavigation of the globe. In *Journal and remarks. 1832–1836* (Vol. III). London: Henry Colburn. Retrieved December 2016, from http://darwin-online.org.uk/content/frameset?itemID=F10.3&viewtype=text&pageseq=1.

Dempster, E., & Hugo, W. (2006). Introducing the concept of evolution into South African schools. *South African Journal of Science, 102*(3/4), 106–112.

Department of Basic Education. (2016). *Education statistics in South Africa 2014*. Department of Basic Education, Pretoria.

Doidge, M., Dempster, E., Crowe, A., & Naidoo, K. (2008). The Life Sciences content framework: A story of curriculum change. In *Proceedings of the 4th Biennial Conference of the South African Association for Science and Technology Educators* (pp. 9–14). University of the Witwatersrand, Johannesburg.

Facchini, G., Mayda, A., & Mendola, M. (2013). What drives individual attitudes towards immigration in South Africa? *Review of International Economics, 21*(2), 326–341.

Granger, D., Gibbon, R., Kuman, K., Clarke, R., Bruxelles, L., & Caffee, M. (2015). New cosmogenic burial ages for Sterkfontein Member 2 *Australopithecus* and Member 5 Oldowan. *Nature, 522*(7554), 85–88. https://doi.org/10.1038/nature14268.

Hall, G., & Hord, S. (2006). *Implementing change: Patterns, principles, and potholes* (2nd ed.). Boston: Pearson.

Hawks, J., Elliott, M., Schmid, P., Churchhill, S., de Ruiter, D., Roberts, D., et al. (2017). New fossil remains of *Homo naledi* from the Lesedi Chamber, South Africa. *eLife, 6*, e24232. https://doi.org/10.7554/eLife.24232.

Henshilwood, C., D'errico, F., Curtis, W., Marean, C., Milo, R., Yates, R., et al. (2001). An early bone tool industry from the Middle Stone Age at Blombos Cave, South Africa: Implications for the origins of modern human behaviour, symbolism and language. *Journal of Human Evolution, 41*(6), 631–678. https://doi.org/10.1006/jhev.2001.0515.

Johnson, K., Dempster, E., & Hugo, W. (2011). Exploring the recontextualisation of biology in the South African Life Sciences curriculum, 1996–2009. *Journal of Education, 52*, 27–57.

Johnson, K., Dempster, E., & Hugo, W. (2015). Exploring the recontextualisation of biology in the CAPS for Life Sciences. *Journal of Education, 60,* 101–121.

Joy Magazine. 2007–2009. http://www.joymag.co.za/past-issue.php May 2009: Shocking revelation! For Christians in western Cape schools. November 2009: The evolution debate continued. Darwin's impact. December/January 2008: The bloodstained legacy of evolution.

Kagan, T., & Sanders, M. (2012). Learners' ideas about evolution, and factors affecting such ideas in a religious Jewish school. In *Proceedings of the 20th Annual Conference of the Southern African Association for Research in Mathematics, Science, and Technology Education.* Lilongwe, Malawi.

Keke, B. (2014). Understanding Life Sciences teachers' engagement with ongoing learning through continuous professional development programmes. Unpublished Ph.D. thesis, University of KwaZulu-Natal, Durban.

Kyriacou, X., de Beer, J., & Ramnarain, U. (2015). Evolutionary ideas held by experienced South African biology teachers. *African Journal of Research in Mathematics, Science, and Technology Education, 19*(2), 118–130. https://doi.org/10.1080/10288457.2015.1014231.

Lawrence, S., & Sanders, M. (2014). Potential artefacts of cross-sectional and longitudinal surveys for tracking developmental trends in student misconceptions. In *Proceedings of the 22nd Annual Conference of the Southern African Association for Research in Mathematics, Science, and Technology Education* (pp. 67–73). Nelson Mandela Metropolitan University, Port Elizabeth.

Lelliott, A. (2016). Visitors' views of human origins after visiting the Cradle of Humankind World Heritage Site. *South African Journal of Science, 112*(1/2), 132–139.

Lever, J. (2002). Science, education and schooling in South Africa. In W. James & L. Wilson (Eds.), *The architect and the scaffold: Evolution and education in South Africa* (pp. 10–44). Cape Town: Human Sciences Research Council.

Mokgobanama, M. (2011). Learning about evolution: The influence of an educational visit and intervention on teachers' knowledge and attitudes. Unpublished Masters research report. University of the Witwatersrand, Johannesburg, South Africa.

Molefe, L., & Sanders, M. (2009). The pedagogical content knowledge of Life Sciences teachers having to teach evolution for the first time. In *Proceedings of the Seventeenth Annual Conference of the Southern African Association for Research in Mathematics, Science and Technology Education* (Vol. 2, pp. 337–346). Rhodes University, Grahamstown, South Africa.

Moore, R., Mitchell, G., Bally, R., Inglis, M., Day, J., & Jacobs, D. (2002). Undergraduates' understanding of evolution: Ascriptions of agency as a problem for student learning. *Journal of Biological Education, 36*(2), 65–71.

Mpeta, M. (2013). The influence of the beliefs of teachers and learners on the teaching and learning of evolution. Unpublished Ph.D. thesis. University of Pretoria, South Africa.

Mpeta, M., de Villiers, J., & Fraser, W. (2015). Secondary school learners' response to the teaching of evolution in Limpopo province, South Africa. *Journal of Biological Education, 49* (2), 150–164.

Naude, F., & de Beer, J. (2014). The theory of 'evilution': Christian teachers' and learners' perspectives on evolution. In *Proceedings of the International Conference on Mathematics, Science and Technology Education (ISTE)* (pp. 263–274). Mopani Camp in Kruger National Park, Limpopo, South Africa. Produced by UNISA.

Ngxola, N., & Sanders, M. (2009). Teachers' content knowledge for teaching evolution for the first time: Perceptions and reality. In *Proceedings of the Seventeenth Annual Conference of the Southern African Association for Research in Mathematics, Science and Technology Education* (Vol. 2, pp. 417–426). Rhodes University. Grahamstown, South Africa.

Pattillo, K. (2012). Quiet corruption: Teachers unions and leadership in South African township schools. Unpublished Honors thesis. Wesleyan University, Middletown, Connecticut, USA.

Pillay, C. (2011). The difficulties faced by some teachers with strong religious beliefs when they teach evolution. Unpublished Masters research report, University of the Witwatersrand, Johannesburg, South Africa.

Pillay, M. (2010). A critical evaluation of representations of hominin evolution in the museums of the Cradle of Humankind World Heritage Site, South Africa. Unpublished Masters research report. University of the Witwatersrand, Johannesburg, South Africa.

Phelan, P., Davidson, A., & Cao, H. (1991). Students' multiple worlds: Negotiating the boundaries of family, peer, and school cultures. *Anthropology & Education Quarterly, 22*(3), 224–250.

Reddy, C. (2012). The lived experiences of Hindu teachers and learners in the teaching and learning of evolution in Life Sciences in the FET phase. Unpublished M.Ed. research report. University of Johannesburg, South Africa.

Reddy, L., & Sanders, M. (2014). The potential influence of Life Sciences examinations on misconceptions about evolution. In *Proceedings of the 22nd Annual Conference of the Southern African Association for Research in Mathematics, Science, and Technology Education* (pp. 146–154). Nelson Mandela Metropolitan University, Port Elizabeth.

Robertson, S. (2015). Grade 8 learners' interpretations of problematically worded textbook statements about evolutionary adaptation. Unpublished Honours research report, University of the Witwatersrand, Johannesburg, South Africa.

Rubidge, B. (1991). A new primitive dinocephalian mammal-like reptile from the Permian of southern Africa. *Palaeontology, 34*(3), 547–559.

Rubidge, B., & Day, M. (2015). Why South Africa's Karoo is a palaeontological wonderland. *The Conversation Africa*, June 2015. http://theconversation.com/why-south-africas-karoo-is-a-palaeontological-wonderland.

Sanders, M. (2008). Teaching about 'evolution': More than just knowing the content. In *Proceedings of the Fourth Biennial Conference of South African Association for Science and Technology Educators* (pp. 91–103). University of the Witwatersrand, Johannesburg.

Sanders, M. (2010a). Meeting the curriculum requirement of 'learner-centredness' when teaching evolution: Giving learners a fair deal. In *Proceedings of the Eighteenth Annual Conference of the Southern African Association for Research in Mathematics, Science and Technology Education* (Vol. 3, pp. 22–31). University of KwaZulu-Natal.

Sanders, M. (2010b). Planning a fieldtrip to the Cradle of Humankind: A model of factors affecting the success of museum visits. In *Proceedings of the Eighteenth Annual Conference of the Southern African Association for Research in Mathematics, Science and Technology Education* (Vol. 3, pp. 32–40). University of KwaZulu-Natal.

Sanders, M. (2014). Teleological and anthropomorphic thinking, and misconceptions about biological adaptations in Natural Sciences textbooks. In *Proceedings of the 22nd Annual Conference of the Southern African Association for Research in Mathematics, Science, and Technology Education* (pp. 139–145). Nelson Mandela Metropolitan University, Port Elizabeth.

Sanders, M., & Makotsa, D. (2016). The possible influence of curriculum statements and textbooks on misconceptions: The case of evolution. *Education as Change, 20*(1), 216–238.

Sanders, M., & Ngxola, N. (2009). Identifying teachers' concerns about teaching evolution. *Journal of Biological Education, 43*(3), 121–128.

Schröder, D., & Dempster, E. (2014). Factors influencing conceptual change when South African learners encounter evolution. In T. Tal & A. Yarden (Eds.), *Proceedings of the 10th Conference of European Researchers in Didactics of Biology (ERIDOB)*. Haifa, Israel.

Scott, E. (2000). The creation/evolution continuum. National Centre for Science Education: Defending the teaching of evolution in the public schools. http://ncselegacy.org/creationism/general/creationevolution-continuum.

Shreeve, J. (2015). This face changes the human story. But how? *National Geographic, 228*(4), 30–57.

Smith, J. (1956). *Old Fourlegs: The story of the coelacanth*. London: Longmans, Green.

South African Agency for Science and Technology Advancement. (2013). Celebrating 75 years of coelacanth research. *GetSetGo* April issue. http://www.saasta.ac.za/getsetgo/issues/201304/03.php.

Statistics South Africa. (2004). *Census 2001: Primary tables South Africa: Census '96 and 2001 compared*. Statistics South Africa Report No. 03-02-04.

Statistics South Africa. (2016). *Community survey 2016*. Statistical release PO301, Statistics South Africa. Pretoria.

Stears, M. (2006). The FET curriculum and the teaching of evolution—What are the challenges facing teachers? In *Proceedings of the 3rd Biennial Conference of the South African Association for Science and Technology Educators* (pp. 182–189). University of KwaZulu-Natal, Durban, South Africa.

Stears, M. (2012). Exploring biology education students' responses to a course in evolution at a South African university: Implications for their roles as future teachers. *Journal of Biological Education, 46*(1), 12–19.

Stears, M., Clément, P., James, A., & Dempster, E. (2016). Creationist and evolutionist views of South African teachers with different religious affiliations. *South African Journal of Science, 112*(5/6), 76–85.

Tangwe, M., Tanga, P., & Tanyi, P. (2015). Teachers' strikes and the right of learners to education in South Africa: A critical literature review. *International Journal of Educational Sciences, 11*(3), 234–243.

Tshuma, T., & Sanders, M. (2015). Textbooks as a possible influence on unscientific ideas about evolution. *Journal of Biological Education, 49*(4), 354–369.

Tucker, R. (2012). Practical theological research into education and evolution in South African high schools—Teaching learners to think! *The Dutch Reformed Theological Journal, 53*(1/2), 219–231. http://ngtt.journals.ac.za.

van den Heever, J. (2009). Creationism in the colonies: Science, religion and the legacy of apartheid in South Africa. In *Proceedings of the Fourteenth conference of the South African Science and Religion Forum (SASRF) of the Research Institute for Theology and Religion*. University of South Africa, Pretoria. http://hdl.handle.net/10500/4291.

van der Mark, M. (2011). The use of narratives and concept cartoons in the professional development of teachers to achieve higher-order thinking skills and deep learning about the evolution of life and geological time. Unpublished Ph.D. thesis, Johannesburg University, South Africa.

White, L, de Beer, J., & Ramnarain, U. (2014). "One size does not fit all": Curriculum support groups as structured support for teachers' professional development. In *Proceedings of the International Conference on Mathematics, Science and Technology Education (ISTE)* (pp. 221–230). Kruger National Park, South Africa.

Yalvac, G. (2011). Barriers in the teaching and learning of evolutionary biology amongst Muslim teachers and learners in South African Muslim schools. Unpublished M.Ed. research report. University of Johannesburg, South Africa.

Since her retirement in December 2017, **Martie Sanders** holds an Honorary Associate Professorship in the School of Animal, Plant and Environmental Sciences at the University of the Witwatersrand, Johannesburg (Wits), where she taught for 35 years. She started her career as a physical science and biology teacher at senior secondary level. After four years, she was recruited into teacher education, and for forty years has been involved in both pre-service and in-service education, mainly for biology teachers. She offered the first "Science Education" Masters topic at the university in 1986, and was active in developing their Masters in Science Education degree first offered in 1988. Her teaching has involved all levels of postgraduate science education students. Since evolution was introduced into the life sciences curriculum for South African schools about 11 years ago, all her research, and that of her postgrads, has focussed on various aspects of evolution education. She is currently supervising postgrads and writing papers.

Part VII
New Zealand

Chapter 23
Evolution Education in New Zealand

Alison Campbell

Abstract The teaching of evolution in New Zealand has followed its own 'evolutionary' trajectory, from being taught only on the university stage in the late 19th and early 20th centuries, to inclusion in the senior curriculum only in secondary schools, to the point where the subject is now a unifying theme throughout the national school science curriculum. The move to this thematic structure of the curriculum was effectively paralleled by changes in how student learning was assessed, which in turn has had impacts on teacher education and the need for resources and ongoing professional development. This series of curriculum modifications was not achieved without some resistance from those opposed to teaching evolutionary biology on both religious and cultural grounds. Even today, while the majority of New Zealand students gain an understanding of the subject, those educated at 'special character' schools rather than within the state school system can still be taught a curriculum based on a creationist world-view. It is also possible, in a system where schools are sensitive to the needs of their local communities, for the relevant sections of the senior curriculum to become 'the part we don't teach'. Thus, in education as in actual biological systems, there is always the potential for further change and adaptation

23.1 Demographic Context

New Zealand's current population is approximately 4.6 million people. It has three official languages: English, Māori, and New Zealand Sign Language, although a number of other languages are also spoken in the home. While the majority of those who identify as having a religious faith are Christian, there are also small proportions of other religions, including Hinduism, Buddhism, and Islam (Statistics NZ, 2013). However, New Zealand is a secular state and the great majority of

---

A. Campbell (✉)
Faculty of Science and Engineering, University of Waikato,
Hamilton, New Zealand
e-mail: alison.campbell@waikato.ac.nz

students attend secular public (state) schools, which they are legally required to attend between six and 16 years of age. In practice the majority of students enter primary school close to their fifth birthday.

23.2 An Historical Perspective on the Place of Evolution in the New Zealand Curriculum

The New Zealand education sector has itself evolved over time, with its first schools run by churches and private secular organisations. When the country was divided into provinces in 1852, each province took on responsibility for its own education services. While these were often church-based, 1871 saw the establishment of provincial boards of education and the first 'state' (public) schools (McLintock, 1966). In 1875 state-funded education became the responsibility of central government: the 1877 Education Act delineated the national system of primary education, followed in 1903 by free state-funded secondary schools; a system of 'private' schooling remained alongside the state schools. In this system, children usually enter primary school at age 5 (year 1, new entrants), where they remain until year 8 unless intermediate schools are available, in which case students move on to this level of schooling for years 7 and 8, and then to secondary school for years 9–13 of their education.

Darwin's great book, *The Origin of Species*, was published in 1859, but while it attracted a significant amount of attention and was very popular with the general public, the scientific response at first was more muted. In fact, Darwin was not the first person to propose some form of biological evolution, and it was not until the development of genetics and the 'new synthesis' in the 1930s that the theory of evolution by means of natural selection achieved consensus recognition. Thus, in the 1890s the textbooks used by New Zealand schoolchildren were mostly from the UK and tended to provide biblical explanations for the natural world. However, the primary school curriculum had its first major revision in 1904 and this resulted in a range of new texts, some of which explicitly referred to Darwinian evolution (McGeorge, 1992), but which did not have evolution as an organising theme. Indeed, "references to evolution occurred when the author had been more or less cornered by his material" (McGeorge, 1992: 206). In addition, both educators and religious leaders seem to have taken the 'non-overlapping magisteria' view, promoted more recently by evolutionary biologist Gould (1997): the scientific view of origins was felt to be compatible with religious belief (Stenhouse, 1984), although—*contra* Gould—natural selection was seen as providing a means to God's ends.

New Zealand has not experienced campaigns against teaching evolution in its schools—or, perhaps more accurately, campaigns to include creationism in the national curriculum—in quite the same way as the various attempts to do so in the USA. Nevertheless, the suggestion, in the 1928 revision of the school curriculum, that older primary school children "should be given some definite ideas of the principle of evolution" (NZ Department of Education, 1928; cited in McGeorge, 1992) was met with significant opposition from creationists. Despite this, the syllabus was retained, and was revised in the mid-1940s to propose a core curriculum for senior primary students. The list of texts to accompany this curriculum included two that discussed evolution, which led to meetings and letter writing campaigns (both ineffectual) by those concerned that the teaching of evolution would lead to moral decay in the country. However, in 1947 complaints from members of the Evolution Protest Movement and others resulted in the cancellation of a series of radio programs that included references to evolution, and despite widespread concern from scientists and educators, the series was not reinstated. Interestingly, the then-Minister of Education did hold and express the view that evolution should be presented as a manifestation of God's will for the Universe, and McGeorge (1992) suggests that the government of the time, which had a shaky majority in the House, was wary of alienating the considerable bloc of Catholic voters in New Zealand.

While all this might sound like a (temporary) win for creationists, McGeorge states that it was, in fact, a loss. The leader of much of the protest, Dr. Milne, had wanted the opportunity to present the "case against evolution" to the nation's schoolchildren—an opportunity that was never offered. Nonetheless, the 1950 'nature study' syllabus contained no mention of evolution at all, and the next (1960s) revision of the national curricula for primary and secondary schools saw first, a shift from nature study to an emphasis on physical sciences; and second, the removal of any specific mention of evolution to year 13 (Form 7) biology classes (McGeorge, 1992). This meant that relatively few students were exposed to this sensitive topic, something that continued even after a further curriculum revision in 1993, which saw evolutionary concepts explicitly taught only to biology students in years 12 and 13. In fact, the 'Achievement Aims' (learning objectives) for the "Living World" (Biology) strand of this curriculum made no specific reference to evolution at all. Even for year 13 students the curriculum stated only that they would be able to "investigate and describe the diversity of scientific thought on the origins of humans". Students could learn about this by

> holding a debate about evolution and critically evaluating the theories relating to this biological issue" and "presenting a seminar about the discovery and suggested lifestyle of Australopithecus africanis [sic] to develop an awareness of its significance in current theories about human evolution, (Ministry of Education [MoE], 1993: 68)

which suggests that the writers may have been confused about the nature of the word 'theory' in science. Indeed, in the early 2000s, a Ministry representative wrote to an evangelical organisation that

> It is not the intention of the science curriculum that the theory of evolution should be taught as the only way of explaining the complexity and diversity of life on Earth. ... [the curriculum] does not require evolution to be taught as an uncontested fact at any level. (Listener, 2006)

In some cases, creationism gained even more of a foothold: in 1982 schools could use a creationist textbook distributed by the then Auckland College of Education's Science Resource Centre (Numbers & Stenhouse, 2000). Furthermore, comments by a spokesman for the New Zealand Education Department (cited by Numbers and Stenhouse) make it clear that this was viewed as acceptable, as long as creationism was not presented as the *only* explanation for life's diversity.

In 1992 McGeorge was able to comment that "there is no strong pressure to deal more fully with evolution either, although that might arise from increasing concern for the environment and for New Zealand's endangered species" (McGeorge, 1992: 217). However, the implementation of the 2007 version of the curriculum (Ministry of Education [MoE], 2007) positions evolution as one of the key organising themes in biological sciences, to be presented (in age-appropriate ways) in all classes from new entrants to year 13. The document states that students should "develop an understanding of the diversity of life and life processes, of where and how life has evolved, and of evolution as the link between life processes and ecology" (MoE, 2007: 28).

Thus, students in their earliest years of primary schooling are expected to learn that the wide range of living things they see can be grouped together, and to be able to explain how we know about extinction. By the time they reach year 7 or 8 they will be learning how groups of living things have changed over geological time, and that some of the plants and animals found in New Zealand differ from those found elsewhere. And in year 13 the curriculum includes an exploration of the evolutionary processes driving life's diversity, including how these processes include our own evolutionary trajectory.

Because the science disciplines are not separated into individual subjects until year 11 at the earliest, this means that all students in schools following the national curriculum document should receive some exposure to basic evolutionary concepts. It is worth noting that a significant majority (75%) of New Zealanders now feel that evolution is probably or definitely 'true' (UMR, 2007); in contrast, Miller, Scott, and Okamoto (2006) found that in the USA only 40% of those surveyed felt that evolution was 'true', lower than any other European country in the survey.

However, this does not mean that all students in primary and secondary schools in New Zealand *do* receive this exposure. For example, Peddie (1995) described how cultural perspectives can influence parents' and students' attitudes to curriculum content. In his experience, and that of other teachers he spoke with during his research, at least some Māori and Pacific Island families objected to their students learning about evolutionary biology as this conflicted with their own culture's views on human origins. (The majority of Māori students are taught in state schools, although a small proportion attend the Māori-language immersion schools known as kura kaupapa). Private and 'special character' schools (described later) are also free to deliver their own curricula, as are home-schoolers. In 2016 29,344 students were enrolled in private schools in New Zealand (Education Counts, 2017).

23.3 The Presence of Anti-evolution Movements, and Education, in New Zealand

Writing in 1986, Hewitt reviewed the status of creationist groups in New Zealand. He noted that arguments about evolution went back to the 1870s, but at this time they involved scientists who also expressed religious orthodoxy. As noted earlier, this was not a conflict of science and religion. While the Evolution Protest Movement distributed creationist literature, actual support for creationist scientists had dwindled away by the early 1900s (Hewitt, 1986). However, a number of US-based organisations—the Institute for Creation Research (ICR), Christian Heritage Colleges (CHS), and the Creation Research Society—have been and are active in New Zealand. Both Henry Morris (who founded both the ICR and CHS) and Duane Gish (then the ICR Deputy Director) visited this country in the 1980s to speak at schools, universities, and schools, including a push for creationism to be taught alongside evolution in science classes. More recently, the group Scientists Anonymous (NZ) sent a thinly disguised 'Intelligent Design' resource, provided via the ICR website, out to the Heads of secondary school Science departments around the country (Campbell, 2010). Two years later (Campbell, 2012), the same group provided schools with a further 'resource' promoting Intelligent Design, this time with an attempt to link their material to one of the four themes of the current national curriculum, i.e. The Nature of Science. The number of schools that actually made use of this material in their science classes is unknown.

The conservative lobby group, Focus on the Family, has also been involved in promoting creationism in New Zealand. The organisation has hosted workshops that included among their speakers William Dembski of the Discovery Institute—events that were advertised in the *Education Gazette* as presenting "an excellent learning opportunity that offers both a professional development opportunity and a fresh look at some knotty problems in science and biology" (Ministry of Education [MoE], 2003). In 2008 Focus on the Family sent the creationist CD-ROM *Icons of Evolution* to every secondary school in the country, which provoked concern from universities and the NZ Royal Society. However, a Ministry of Education representative responded that parents "had a right to withdraw children from religious instruction" (Dominion Post, 2008). This, of course, begged the question of whether parents would know in advance that this material would be presented in a science class. Furthermore, the MoE spokesperson added that "[the] science curriculum does not require evolution to be taught as an uncontested fact at any level. The theory of evolution cannot be replicated in a laboratory and there are some phenomena that aren't well explained by it" (P. Spratt, cited in Campbell & Otrel-Cass, 2011).

When the draft version of New Zealand's current national curriculum document went out for public consultation, the conservative group Creation Ministries International (previously the Creation Science Foundation) asked its members to lobby for the retention of the status quo, with its limited coverage of evolution.

CMI does not suggest evolutionists be forced to teach about creation. What we do suggest is that freedom be retained for the presenting of both evolution-based and Creation-based frameworks of science. We support the teaching of evolution provided it is done accurately, 'warts and all', i.e. with open discussion of its many scientific problems included. (Campbell, 2007: para. 12)

There were a number of other similar submissions (Watson, Bowen, Tao & Earle, 2006), but fortunately these failed to sway those developing the curriculum documents.

Nonetheless, it remains possible for publicly funded schools to minimise students' exposure to concepts related to evolution: a 2013 survey indicates that a number of state schools (92 in 2013) were not teaching the relevant parts of the science curriculum (Robins, 2013). At the senior school level, this is made possible by the nature of current assessment practices under the National Certificate of Educational Achievement (NCEA), which means that schools can simply choose not to offer the Assessment Standards relating to evolutionary biology, reflecting Peddie's (1995) observations. This applies particularly to those year 13 standards that address human evolution and patterns and processes of evolution,

In addition, as previously mentioned, New Zealand has long had both a state school system, which is free to students and where there is a state-mandated core curriculum, and a 'private' system. Many but not all private schools have a history as faith-based, predominantly Catholic, schools. In 1975 the introduction of the Private Schools Conditional Integration Act offered those schools the opportunity to move to 'integrated' status: they then received state funding and largely adopted the state-mandated school curriculum. By mid-2012, there were 332 state-integrated schools in New Zealand; their 86,500 students comprised about 14% of the total number of students in the nation's schools (Lynch, 2012). But even now, such schools can avoid teaching some content, due to flexibility around curriculum delivery and the requirements for entry to tertiary-level study. The legislation allows them to "[teach], develop and implement programmes reflecting the faith and or educational philosophy articulated in the special character, and to establish customs and traditions that authentically reflect these" (Association of Integrated Schools NZ [AISNZ], n.d., para. 2).

Special character schools are designated by the Minister of Education as schools that "have a character that is in some specific way or ways different from the character of ordinary State schools" (Sect. 156 of the 1989 Education Act, NZ Government, 1989) and are free to set their own curriculum, as are fully-private establishments. Currently there are 72 of these schools, which provide students with an education that "differs significantly from the education they would get at an ordinary State school; … is not available at any other State school that children … can conveniently attend" (NZ Government, 1989); and is seen as desirable by the parents involved. In 2010 the then-Associate Minister for Education made this clear in comments reported in the national press: "[I]ndependent schools are free to set their own curricula and to have their own distinct ethos. Parents choose to send their children to these schools" (NZ Herald, 2010). Charter, or 'partnership', schools have similar flexibility in the curriculum (Radio New Zealand (RNZ), 2013; Ministry of Education [MoE], 2016).

In some instances, special-character schools teach a 'science' curriculum based on young-Earth creationism, which explicitly denies evolution and as a corollary provides very poor understanding of the nature of science (also an organising theme in the National Curriculum). For example, one school says in its mission statement (Jireh School, n.d.: Statement of Faith, clause e) that "We believe God's acts of creation, and not evolution, are responsible for the origin of all things, and that creation according to the Biblical record is an historical event."

Another offers a curriculum "founded on a Biblical World View" (Westminster Christian School, 2015), while a third chooses which areas of "commonly accepted science" to believe in, and which to reject (Ponatahi School, n.d.). In addition to rejecting evolutionary biology, this approach will not give students at these schools a good understanding of the nature of science. Similarly the Accelerated Christian Education (ACE) curriculum (2014), used by at least some home-schooling families, takes a strong Biblical slant on science in general and does not teach evolution. Nonetheless, Universities New Zealand (2016) recognises this curriculum as appropriate preparation for university study.

23.4 The Place of Evolutionary Theory in the New Zealand School Science Curriculum

Prior to 2002 students' learning in their final years of school (years 11–13, Forms 5–7) was assessed by a series of norm-referenced national examinations; more specifically, School Certificate in year 11, and Bursary examinations in year 13. As previously described, teaching and assessment of students' knowledge of evolutionary biology was limited to senior examinations. However, in 2002 the country introduced the standards-based National Certificate of Educational Achievement (NCEA), which uses Achievement and Unit Standards that describe how an individual's performance compares to a range of indicators (Campbell & Otrel-Cass, 2011; New Zealand Qualifications Authority [NZQA], n.d., a). There are a total of eight biology Achievement Standards (ASs) available at year 12 and seven at year 13 (Ministry of Education [MoE], n.d), although many schools do not offer the full suite of ASs due to time pressures and the demands of other, competing, subjects, or because they offer tailored combinations of subjects (Hipkins & Bolstad, 2005). And some schools may simply avoid those ASs with an explicit evolutionary focus (e.g. Robins, 2013).

Several of these Achievement Standards assess students' mastery of evolutionary concepts and related material in genetics: Year 12 students may be assessed on their "understanding of genetic variation and change"; in Year 13 they may complete assessments of their understanding of "trends in human evolution" and "evolutionary processes leading to speciation", via short- and extended-answer questions (MoE, n.d). For example, in 2016 one question for year 13 students sitting the examination paper on evolutionary processes provided some contextual information and then asked them to

> Compare and contrast the impact of disruptive and stabilising selection on genetic diversity AND discuss how speciation could occur in the Mexican spadefoot toad.
>
> In your answer you should: describe genetic variation; describe the terms disruptive and stabilising selection, and describe which population(s) of Mexican spadefoot toad tadpole is associated with each type of selection; explain the selection pressures that promote disruptive selection, AND the selection pressures that promote stabilising selection in the Mexican spadefoot toad tadpole. (New Zealand Qualifications Authority [NZQA], n.d., b)

Students able to answer this particular question successfully would demonstrate a good understanding of natural selection. Similarly, the 2015 examination on human evolution included a question that provided information on skeletal comparisons between modern humans and the great apes, and then asked students to

> Discuss the importance of bipedalism in the development of hominins by linking the skeletal features to their adaptive significance. In your answer, describe what is meant by the terms quadruped and biped; explain how any three of the skeletal features [illustrated in the exam paper] provide evidence for the form of locomotion changing to bipedalism; justify why bipedalism was so significant to the evolution of hominins. (NZQA, n.d., b)

However, each Standard assesses a portion of the wider curriculum, and there has been an understandable tendency to teach to the standards, thus fragmenting the curriculum and potentially affecting students' ability to see evolution as an integrating principle in biology. In my personal experience, examiners cannot use material from one standard (e.g. human evolution) to provide context for another (e.g. patterns and processes of evolution), because students may not have studied both—this also works against integration of concepts across the curriculum.

23.5 Emphasis Given to Evolutionary Theory in Biology Teacher Education Programs

New Zealand has a system of professional teacher education, delivered by training colleges, polytechnics, and Universities and wananga (Māori tertiary institutions), with the Education Council of Aotearoa NZ providing accreditation and quality assurance.

However, there is no requirement for demonstrating content knowledge of concepts related to evolution in professional teacher examinations or accreditation requirements. Those students intending to become secondary school biology teachers would normally have studied for a B.Sc. (or higher) in biology, although unpublished data suggest that their understanding and acceptance of evolution may not be much affected by this (Campbell & King, n.d.). In at least some institutions graduates taking their post-graduate professional training year will have a class or classes on addressing contentious issues in biology, and so evolution is discussed at this point. However, students aspiring to teach other science disciplines (chemistry, physics, earth sciences) do not necessarily study biological sciences during their university years. This is potentially a problem as, while they will almost certainly be required to teach science—including biology—to their junior classes, they may not have been exposed to any biology at all beyond their own junior years in high school.

Intending primary school teachers usually study towards a three-year Bachelor of Teaching degree, a professional qualification accredited at a national level (thus, outcomes are consistent across institutions). They are expected to teach across the curriculum, although some may specialise in a particular subject area. In their first year of study students take a range of papers, including one on science teaching, which largely takes a pedagogical and curriculum-development focus: for example, at one institution (University of Waikato [UoW], 2017) student teachers can

> explore the nature of science; strengthen their own science knowledge base; recognise some of the variety of science ideas that children already have when they come to their science learning; and find ways to interact with children's existing ideas to challenge and extend their science understanding.

Student teachers may take additional elective papers in a range of subjects, but for those who are focused elsewhere this first-year paper, augmented by learning while on practicum (practical experience in school classrooms), may be their sole exposure to science. Those who study for their Bachelor of Teaching as a conjoint with a Bachelor's degree in some other discipline will obviously gain a much deeper understanding of that discipline, but the number of intending primary school teachers who chose to take a science degree is small. The immediate consequence of this is that many primary teachers have very little background in delivering the science part of the curriculum, a fact highlighted by the Prime Minister's Chief Science Advisor in a paper advocating for 'science champions' in the nation's primary schools (Gluckman, 2011), alongside other suggestions for strengthening primary school children's engagement with science. This is similar to the finding by Coleman, Stears, and Dempster (2015) that intending teachers enrolling in a graduate training program in South Africa had a higher understanding of both evolution and the nature of science than those embarking on a Bachelor of Education degree. Thus, while the intention of the 2007 national science curriculum was that children would be exposed to evolutionary concepts from the start of their formal education, the reality may well be quite different.

23.6 Biology Teachers' Attitudes Toward Teaching Evolutionary Theory

There is a large body of data in the international literature on the attitudes of biology teachers towards teaching evolutionary theory (e.g. Athanasiou, Katakos, & Papadopoulou, 2016 [Greece]; Nadelson & Sinatra, 2010 [USA]; Rutledge & Warden, 2000 [USA]), which shows regional variation. Campbell & Cooke (unpublished data) found that New Zealand secondary school teachers had a reasonably high acceptance of evolutionary theory (an average Likert score of 84.55 from a possible 100), albeit with some uncertainty in relation to concepts such as the duration of life on earth, the testability of evolutionary theory, and the evidence in support of the theory. The New Zealand teachers also outperformed the US cohort sampled by Rutledge and Warden (2000) in their understanding of evolutionary concepts (scoring 76% on a series of multiple-choice items, compared to 71% for the US cohort), perhaps because the New Zealand curriculum expects that senior biology classes will cover Darwinian theory and mechanisms of evolution in some detail. However, the teachers' understanding of the nature of science was only moderate, and this is of concern given the role and impact of science on everyday life. Less than 70% of teachers surveyed were able to give correct answers on the goals and scope of science, the relationship between evolution and religious faith, the tentativeness and limits of science, and the nature of scientific theories. (However, only a very small proportion were openly creationist in their opinions.) These findings highlight a need for ongoing professional development for New Zealand science teachers that focuses on the nature of science. In addition, as already noted, there is no requirement for pre-service teachers to receive instruction that focuses on evolution, so that both primary school teachers and those teaching other disciplines at secondary school may have different attitudes to the subject.

23.7 Enhancing Evolution Education in New Zealand

The New Zealand government coordinates professional learning development (PLD) for practising teachers and provides some funding to allow schools to access accredited facilitators; teachers may also self-fund their PLD, in which case they can choose from a wider list of providers. Once a school (or individual) has selected a facilitator, they collaborate to develop the program that they require, with provision ranging from one-off sessions to programs lasting up to six months (Bull, 2016). Significantly, Bull (2016) identifies the widespread use of short-term PLD— only 12% of providers offer programs of up to six months' duration—as a potential problem in terms of their limited opportunities for deep learning by teachers.

In addition, organisations like BEANZ (Biology Educators Association NZ), the Ministry of Education via its website *Te Kete Irirangi*, and journals such as the *New Zealand Science Teacher* provide resources. The Sir Paul Callaghan Science Academy (http://www.scienceacademy.co.nz/content/scienceAcademies.php) runs an annual academy for primary and intermediate school teachers, and there is also Primary Science Week. This annual event not only enhances links between primary schools and the wider science community, but also encourages networking between teachers, who come together to share resources and discuss the classroom challenges they face. In my experience, evolution is a popular Science Week topic with both teachers and students.

At the secondary school level, Campbell and Otrel-Cass (2011) suggest tailored professional development for practising teachers. This should focus not only on maintaining currency of knowledge in a rapidly-moving field, but also encouraging and supporting teachers to move away from a didactic approach to knowledge transmission, and towards an approach that recognises the uncertainty of science. Particularly in teaching about evolution, it is important to have the pedagogical skills to enable any discussion of what questions science can and cannot answer; similarly, teachers need the skills to address students' misconceptions around evolutionary topics including adaptation, natural selection, and fitness (misconceptions identified on a regular basis in national examiners' annual reports, which are published annually on the NZQA website: New Zealand Qualifications Authority [NZQA], n.d., c).

However, "[c]entrally funded PLD will be focussed on building equity and excellence in a small number of priority areas" (Education Services, n.d.); PLD around teaching evolution is not likely to be one of these and indeed, the pressing need is to increase science literacy per se. This is especially the case for teachers in primary schools, where literacy and numeracy are key foci of professional development opportunities, particularly since the 2010 inception of National Standards that set expectations for student performance in these areas (NZ Curriculum On Line [NZC], n.d.). Anecdotally, the National Standards have led to a focus on literacy and numeracy at the expense of other curriculum areas, including science, and so it will be interesting to observe the effects of their imminent removal, following a change of government in 2017. Bull (2016: 1) also describes science as being 'marginalised' in the primary curriculum, and cites research indicating that "many primary teachers do not feel confident about either teaching science or being able to access the support they need", with little in-school support and limited access to appropriate professional development opportunities.

In fact, Bull found that 51% of PLD providers focused on "developing teachers' confidence to teach science"; only 5% saw "developing teachers' knowledge of science" as important. She argues that there is a need for a greater focus on content knowledge as this can also enhance confidence around delivery, citing other research concluding that primary teachers' lower confidence in teaching science reflected their lesser degree of content knowledge. Built into this, however, must be recognition that understanding the *nature* of science is particularly important in understanding and accepting the theory of evolution (e.g. Akyol, Tekkaya, & Sungur, 2010; Fensham, 2001). This means that PLD provision focusing on the nature of science could have wide positive impact on science teaching as a whole, not just delivery of evolution-related content, and so is highly desirable. Bull, however, is somewhat pessimistic about the current ability of the New Zealand system to deliver this:

> Facilitating PLD is a complex job, requiring a range of different sorts of expertise. Facilitators need a deep knowledge of curriculum and science education, and need to know about adult learning and development, be familiar with how both schools and the science community operate, and be well networked so that they can draw on relevant resources. This sort of expertise requires time and support to develop, and the current insecurities

around the employment of many providers seem unlikely to encourage a commitment to the necessary growth and development. (Bull, 2016: 6)

This makes the need for 'science champions' in primary schools even more pressing (Gluckman, 2011)—either as individual teachers, identified and trained via initiatives such as the Sir Paul Callaghan Science Academy, or through provision via clusters of schools. Recently the New Zealand government moved to set up and fund "Communities of Schools" (Ministry of Education [MoE], 2015), which can cooperate to improve their pupils' achievement. This may also enhance schools' delivery of science, especially because these clusters of schools cover all ages of schooling and so primary schools (with arguably the greatest need for support around science teaching and learning) can benefit from the expertise of subject teachers in secondary schools and the mentoring available to primary school teachers:

> Each Community of Schools will work with parents, whānau [family] and the community to identify achievement challenges, reflecting the specific needs of their students. (MoE, 2015)

However, while STEM subjects should surely receive consideration, evolution—as one theme, albeit the unifying thread, in *one* of the science disciplines—is unlikely to receive much specific attention. Thus, any evolution-focused PLD will likely be accessed piecemeal by individual schools and teachers, perhaps via discipline-based education conferences (such as the biennial BioLive conference run for and by New Zealand secondary school biology teachers) or relationships with members of the local scientific community (again, recommended by Gluckman, 2011).

A further option for delivery of learning opportunities in science in general and evolutionary biology in particular is the development of on-line resources that teachers can access whenever convenient and at no cost, and which allow them the opportunity to resolve personal misconceptions. Nadelson and Sinatra (2010) discuss one such website, which appears to have had a "modest success" in increasing acceptance of evolution, and reiterate the importance of teachers' understanding of the nature of science if they are in turn to have a good understanding of the nature and workings of evolution.

In similar vein, the Science Learning Hub (SLH) is a New Zealand website, which provides a range of classroom resources to primary and secondary teachers, including a number specifically targeted to understanding evolutionary concepts (https://www.sciencelearn.org.nz/concepts/evolution). However, beginning teachers in particularly may need support in learning to use this resource to their best advantage. Concerned at the evidence that primary science programs are unappetising and failing to deliver students with either science understanding or critical thinking skills, Hume and Buntting (2014) developed an initiative—intended to develop teachers' pedagogical and content knowledge—focused on the SLH resources. They worked with primary teacher trainees, encouraging them to "consider pedagogical prompts about what science ideas to teach their students and why, when and how" while using those resources. The intention of the project was to provide these pre-service teachers with the knowledge, skills and confidence to become the 'champions of science' envisaged by Gluckman (2011). Their findings were very positive:

> Even student teachers who had previously felt very apprehensive about teaching science reported feeling far more confident about the prospect after completing the CoRe assignment. (Hume & Buntting, 2014)

However, there does need to be a commitment to provide suitable and ongoing mentoring to new teachers who have experienced this program, in order to see them reach their full potential as 'science champions' in our primary schools (Hume & Buntting, 2014). By itself, this will not necessarily address teachers' ability to deliver on evolutionary concepts. However, as others have noted (e.g. Fensham, 2001; Nadelson & Sinatra, 2010), once teachers have a good base of knowledge of content and pedagogy and an equally good appreciation of the nature of science, their ability to address evolutionary content and concepts also improves. In an ideal world, pre-service teacher education would increase the attention given to science, especially in light of the current government's recognition of the importance of the STEM disciplines.

23.8 Conclusions

It is apparent that little has changed since Campbell and Otrel-Cass (2011) noted that "pressure on teaching time remains a significant issue" (2011). A two-fold approach may be the best, pragmatic, option for improving students' understanding and acceptance of evolution: enhancing primary school teachers' knowledge of the nature of science and their ability to deliver a science curriculum per se, and providing tailored professional development for secondary school teachers. For both, learning opportunities that model a historically-rich curriculum would support delivery around the nature of science (Campbell & Otrel-Cass, 2011), enhancing students' learning opportunities and their recognition of the place that science in general, and evolution in particular, play in our 21st-century lives.

References

Accelerated Christian Education. (2014). A.C.E. Curriculum: Scope and sequence. Retrieved from https://www.aceministries.com/homeschool/pdf/Home_Educators_Scope_and_Sequence.pdf.
Association of Integrated Schools NZ (AISNZ). (n.d.). Special Character. Retrieved from http://www.aisnz.org.nz/special-character.
Akyol, G., Tekkaya, C., & Sungur, S. (2010). The contribution of understandings of evolutionary theory and nature of science to pre-service science teachers' acceptance of evolutionary theory. *Procedia—Social and Behavioural Sciences, 9,* 1889–1893.

Athanasiou, K., Katakos, E., & Papadopoulou, P. (2016). Acceptance of evolution as one of the factors structuring the conceptual ecology of the evolution theory of Greek secondary school teachers. *Evolution: Education and Outreach 9*, 7. https://doi.org/10.1186/s12052-016-0058-7.

Bull, A. (2016). *Developing primary science teacher expertise: Thinking about the system.* New Zealand Council for Education Research, Wellington, NZ.

Campbell, A. (2007). Evolution in the NZ school curriculum. *NZ Skeptics.* Retrieved from http://skeptics.nz/journal/issues/85/evolution-in-the-nz-school-curriculum.

Campbell, A. (2010). So who are these 'scientists anonymous'? Retrieved from http://sci.waikato.ac.nz/bioblog/2010/11/so-who-are-these-scientists-an.shtml.

Campbell, A. (2012). 'Scientists Anonymous (NZ)' write again. Retrieved from http://sci.waikato.ac.nz/bioblog/2012/03/post-2.shtml.

Campbell, A., & King, C. (n.d.). The effect of an evolution-focused third-year paper on biology students' attitudes to and understanding of evolution. (Unpublished data).

Campbell, A., & Otrel-Cass, K. (2011). Teaching evolution in New Zealand's schools—Reviewing changes in the New Zealand science curriculum. *Research in Science Education, 41,* 441. https://doi.org/10.1007/s11165-010-9173-6.

Coleman, J., Stears, M., & Dempster, E. (2015). Student teachers' understanding and acceptance of the nature of science. *South African Journal of Education, 35*(2), 1–9. https://doi.org/10.15700/saje.v35n2a1079.

Dominion Post. (2008). Christians challenge teaching of evolution. Retrieved from http://www.stuff.co.nz/national/education/509003/Christians-challenge-teaching-of-evolution.

Education Counts. (2017). http://www.educationcounts.govt.nz/data-services/directories/private-schools.

Education Services. (n.d.). Professional Learning and Development. Retrieved from http://services.education.govt.nz/pld/news/what-does-the-redesign-of-pld-mean-for-my-school/.

Fensham, P. (2001). Science content as problematic—Issues for research. In H. Behrendt, H. Dahncke, R. Duit, W. Graber, M. Komored, A. Kross, & A. Reiska (Eds.), *Research in science education—Past, present, and future* (pp. 27–42). New York: Kluwer Academic.

Gluckman, P. (2011). *Looking ahead: Science education for the twenty-first century: A report from the Prime Minister's Chief Science Advisor.* Office of the Prime Minister's Science Advisory Committee.

Gould, S. J. (1997). Non-overlapping magisteria. *Natural History, 106,* 16–22.

Hewitt, G. (1986). Creationism: Gospel, heresy, or science? *Tuatara, 28*(2), 71–82.

Hipkins, R., & Bolstad, R. (2005). *Staying in science: Students' participation in secondary education and on transition to tertiary studies.* New Zealand Council for Educational Research, Wellington.

Hume, A., & Buntting, C. (2014). Creating 'science champions' in teaching training. *New Zealand Science Teacher,* 6 August 2014. Retrieved from http://nzscienceteacher.co.nz/teacher-education-in-science/primary/creating-science-champions-in-teaching-training/#.WIxSGht97b0.

Jireh School. (n.d). Mission statement. Retrieved from http://www.jireh.school.nz/about-jireh/mission-statement.

Listener. (2006). By accident or design. *Listener Archive,* 15 April 2006. Retrieved from http://www.noted.co.nz/archive/listener-nz-2006/by-accident-or-design/.

Lynch, P. (2012). Affidavit presented to the High Court, Wellington, on behalf of the New Zealand Catholic Education Office—Historical overview of the history of integrated schools in New Zealand. Retrieved from http://www.nzceo.catholic.org.nz/media/resources/brief-history-of-integration.pdf.

McGeorge, C. (1992). Evolution in the primary school curriculum. *History of Education, 21*(2), 205–218. https://doi.org/10.1080/0046760920210207.

McLintock, A. H. (Ed.). (1966). Education—Evolution of present system. In *An encyclopaedia of New Zealand.* Originally published in 1966. Te Ara—The Encyclopedia of New Zealand. Retrieved from http://www.TeArra.govt.nz/en/1966/education-evolution-of-present-system.

Miller, J. D., Scott, E. C., & Okamoto, S. (2006). Public acceptance of evolution. *Science, 313,* 765–766.

Ministry of Education (MoE). (n.d). Level 3 Biology Assessment Resources, Biology Matrix. Retrieved from http://ncea.tki.org.nz/Resources-for-Internally-Assessed-Achievement-Standards/Science/Biology/Level-3-Biology.
Ministry of Education (MoE). (1993). *Science in the New Zealand Curriculum.* Learning Media, Wellington. Retrieved from http://nzcurriculum.tki.org.nz/content/download/63077/504797/file/ScienceInTheNewZealandCurriculum.pdf.
Ministry of Education (MoE). (2003). *Education Gazette,* 22 August 2003, Wellington, NZ.
Ministry of Education (MoE). (2007). *The New Zealand Curriculum for English-medium teaching and learning in years 1–13.* Learning Media Ltd, Wellington, NZ.
Ministry of Education (MoE). (2015). Communities of Schools underway. *NZ Education Gazette,* 09 February 2015. Retrieved from http://www.edgazette.govt.nz/articles/article.aspx?articleid=9057.
Ministry of Education (MoE). (2016). Partnership Schools/Kura Hourua. Retrieved from http://partnershipschools.education.govt.nz/pskh/what-are-partnership-schools.
Nadelson, L. S., & Sinatra, G. M. (2010). Shifting acceptance of the Understanding Evolution website. *The Researcher, 23*(1), 13–29.
New Zealand Government. (1989). Education Act. Retrieved from http://www.legislation.govt.nz/act/public/1989/0080/latest/whole.html.
New Zealand Qualifications Authority (NZQA). (n.d., a). NCEA. Retrieved from http://www.nzqa.govt.nz/qualifications-standards/qualifications/ncea.
New Zealand Qualifications Authority (NZQA). (n.d., b). Standards and assessment. Retrieved from http://www.nzqa.govt.nz/ncea/assessment/search.do?query=Biology&view=exams&level=03.
New Zealand Qualifications Authority (NZQA). (n.d., c). Examiners reports. Retrieved from http://www.nzqa.govt.nz/ncea/assessment/search.do?query=Biology&view=reports&level=03.
NZ Curriculum Online (NZC). (n.d.). National Standards. Retrieved from http://nzcurriculum.tki.org.nz/National-Standards.
NZ Herald. (2010). Sacked for teaching King Lear. Retrieved from http://m.nzherald.co.nz/nz/news/article.cfm?c_id=1&objectid=10642209&pnum=0.
Numbers, R. L., & Stenhouse, J. (2000). Antievolutionism in the Antipodes: From protesting evolution to promoting creationism in New Zealand. *The British Journal for the History of Science, 33,* 335–350.
Peddie, W. S. (1995). *Alienated by Evolution: The educational implications of creationist and social Darwinist reactions in New Zealand to the Darwinian theory of evolution.* Unpublished Ph.D. thesis, University of Auckland.
Ponatahi School. (n.d.). Content outline. Retrieved from http://www.ponatahi.school.nz/creation-evolution.html#Science.
Radio New Zealand (RNZ). (2013). Creationism will be taught in two charter schools. Retrieved from http://www.radionz.co.nz/news/political/222211/creationism-will-be-taught-in-two-charter-schools.
Robins, J. (2013). Christianity in classrooms: Survey reveals religious instruction in schools. Retrieved from https://nz.news.yahoo.com/top-stories/a/18215257/christianity-in-classrooms-survey-reveals-religious-instruction-in-schools/#page1.
Rutledge, M. L., & Warden, M. A. (2000). Evolutionary theory, the nature of science and high school biology teachers: Critical relationships. *The American Biology Teacher, 62*(1), 23–31.
Statistics NZ Tatauranga Aotearoa. (2013). 2013 Census totals by topic. Retrieved from http://www.stats.govt.nz/Census/2013-census/data-tables/total-by-topic.aspx.
Stenhouse, J. (1984). The wretched gorilla damnification of humanity: The "battle" between science and religion over evolution in nineteenth century New Zealand. *New Zealand Journal of History, 18,* 143–162.
UMR Research. (2007). *Morality, Religion and Evolution: A Comparison of New Zealand and the United States* [Omnibus Results—August 2007]. Wellington. Retrieved from https://openparachute.files.wordpress.com/2007/10/finalmorality-religion-evolution-nz_uscomparison-sep07.pdf.

Universities New Zealand. (2016). *Admission requirements for domestic students*. Retrieved from http://www.universitiesnz.ac.nz/studying-in-nz/domestic.
University of Waikato (UoW). (2017). *Learning and teaching science—Paper information*. Retrieved from https://papers.waikato.ac.nz/papers/TEMS121.
Watson, S., Bowen, E., Tao, L., & Earle, K. (2006). *The New Zealand Curriculum draft for consultation: Analysis of long submissions draft Supplementary Report*. Lift Education.
Westminster Christian School. (2015). *Our school*. Retrieved from http://www.westminster.school.nz/our-school.

Alison Campbell is a Senior Lecturer in Biology, and Associate Dean (Teaching & Learning) in the Faculty of Science & Engineering, The University of Waikato, New Zealand. She is currently teaching an undergraduate course in science education and co-developing a course about the interplay between science and traditional Maori knowledge. She also writes a blog with a focus on communicating about science and pseudoscience, another about science education, co-developed the Evolution for Teaching website, and does outreach in schools. Alison is a member of the Ako Aotearoa Academy, an organisation for the top tertiary teachers in New Zealand, and is a mentor for colleagues.

Part VIII
Conclusion

Chapter 24
Evolution Education Around the Globe: Conclusions and Future Directions

Lisa A. Borgerding and Hasan Deniz

Abstract The twenty-two chapter contributions in this book paint a portrait of evolution education across six continents. The range of religions, geography, relationship to evolutionary theory, education systems, and teacher education systems is broad, and yet some consistent issues and opportunities unite evolution education efforts across these diverse regions. This concluding chapter is organized around the six requested elements by directly comparing and contrasting the information presented in each chapter. We conclude this chapter with some general comments about the place of evolution education within the broader discourse of these countries and offer some future directions for the evolution education community.

The twenty-two chapter contributions in this book paint a portrait of evolution education across six continents. The range of religions, geography, relationship to evolutionary theory, education systems, and teacher education systems is broad, and yet some consistent issues and opportunities unite evolution education efforts across these diverse regions.

Chapter authors were invited to share information around the six suggested elements so that comparisons could be made. These suggested elements included the public acceptance of evolutionary theory within the social, political, and cultural context of the country; the existence and extent of influence of anti-evolution movements in the country; the place of evolutionary theory in the curriculum, the emphasis given to evolutionary theory in biology teacher education programs, biology teachers' attitudes toward teaching evolutionary theory; and suggestions to improve evolution education in the country.

Eleven chapter authors also included new data regarding the teaching of evolution in their countries. These new findings portray preservice and inservice

---
L. A. Borgerding (✉)
College of Education, Health, and Human Services, Kent State University, Kent, OH, USA
e-mail: ldonnell@kent.edu

H. Deniz
College of Education, University of Nevada Las Vegas, Las Vegas, NV, USA
e-mail: hasan.deniz@unlv.edu

teachers' attitudes toward evolution, knowledge, acceptance and their evolution teaching experiences (Malaysia (Fay et al.), Mexico, Galapagos, Indonesia, Philippines, Scotland, U.S. Southwest), preservice teachers' views of evolution and its instruction (U.S. Southeast), innovative preservice teacher education programs for evolution (U.S. Central), textbook analyses of evolution coverage (Brazil, Hong Kong), and students' conceptions and views of evolution (France; Scotland).

This concluding chapter is organized around the six requested elements by directly comparing and contrasting the information presented in each chapter. Particular aspects of evolution education recur through many of the chapters, and other aspects are unique to particular countries or regions. We conclude this chapter with some general comments about the place of evolution education within the broader discourse of these countries and offer some future directions for the evolution education community.

24.1 Public Acceptance of Evolutionary Theory Within the Social, Political, and Cultural Contexts Across Countries

As an indicator of the global contention surrounding the teaching of evolution, eight chapter authors chose to reference Dobzhansky's claim (1973) that "Nothing in biology makes sense except in light of evolution" to defend the high scientific status of evolution in the midst of societal refutation. This justification was used by authors in countries very accepting of evolution as well as countries in which the teaching evolution is banned. Authors reported a wide range of acceptance across the countries included in this book. Comparisons of these findings are difficult because of different sampling, instruments, and timing of data collection for each of these studies, but the overall findings reveal a yet unfocused image of how different places conceive of this foundational bedrock of modern biology. Taking acceptance of human evolution as an example, countries represented high acceptance (80% in France and 75% in New Zealand), medium acceptance (69% in the United Kingdom, 67% in Palestine, 60% in German-speaking countries, and 60% in Mexico) and lower acceptance (55% in Greece, 54% in Missouri, U.S., 52% in Jordan, 50% in Ecuador; and 17% in Malaysia). In other countries such as Iran and Hong Kong, there have been no published assessments of evolution acceptance. This wide range of acceptance is related to religious views in some cases (but not others), cultural traditions, connections to Darwin, and geographical diversity.

Across the majority of these countries, the acceptance of evolution and views of evolution teaching are related to the influence of religion. Around 20% of samples in research studies in Austria, Switzerland, and Germany hold young-earth Creation ideas, and evolution rejection is most widespread among Muslims and evangelical Christians in Germany. In newly presented data from Scottish college biology students, all rejecters stated religious reasons for rejection, and Islamic students were disproportionately more rejecting than their non-Islamic peers. In other countries, the tension between science and religion drives evolution rejection. When religion is viewed as harmonious with science (some Arab countries) or when harmonious Islamic hadiths are accepted (American Muslims), evolution acceptance is more possible. When evolution is equated with atheism (Malaysia), evolution rejection is the norm. This strong religious rejection of evolution has consequences for evolution education. For example, there was large public outcry, especially from conservative Christian organizations, when evolution was introduced to the South African curriculum recently. Similarly, several authors noted significant proportions of the public that support the teaching of Creationism (Austria, 21%; Mexico, 47%; Brazil, 89%).

Yet, in other countries, religion is less important for evolution positions. For example, French evangelicals tend to be more accepting of evolution than American counterparts, possibly explaining in part why evolution education is not contentious in France. Similarly, Stasinakis and Kampourakis report studies of Greek teachers in which religiosity does not seem to be a predictor of evolution acceptance. Instead, the authors suggest a more important factor is likely the paltry ways evolution is taught in Greek education. Similarly, in Hong Kong, the greatest threat to sound evolution education is simply that biology is not a required secondary course, and fewer students are choosing to take it as an elective. From Germany, Eder, Seidl, Lange, and Graf cite a study in which religiosity was not as important as attitudes toward science for evolution acceptance. Thus, availability of evolution education and attitudes toward science also frame some countries' evolution acceptance.

Authors cite several important cultural traditions that impact evolution acceptance and evolution instruction. Maori and Pacific Island families in New Zealand object to evolution specifically because it conflicts with their cultural views about human origins in particular. In the Central U.S., the cultural traditions of the Ozarks (religion, distrust of government, emphasis on privacy) lend support to creationism and state/local rights to determine educational practices. Similarly, in the Southeastern U.S., the Southern world view that embraces religion, treats discussions of evolution as taboo, predominance of small communities with little anonymity, and the church as important for social as well as religious life creates negative consequences for those who appear to reject these tenets and accept evolution. Fisher (U.S. Southwest) also reminds us to attend to often-forgotten cultural identifiers such as ethnicity and non-mainstream religions when understanding evolution education in different contexts. BouJaoude notes that while evolution rejection in Arab countries is often rooted in views of the compatibility of science and religion, simplistic NOS views that demand certainty, experimentation, reproducibility, and direct evidence are also cited for evolution rejection. In a related way, Iran's commitment to advance science and technology and treatment of evolution as a separate field of knowledge from religion are cultural traditions supportive of evolution education. Across these diverse contexts, cultural traditions, commitment to science, and the demarcation of science and non-science are important for evolution education.

In many chapters, the importance of Darwin himself emerged as important for evolution education in various country contexts. Several countries can take pride in the role their scientists and biota played in shaping the development of Darwin's theory. England proudly claims Charles Darwin as a native son, and *On the Origin of Species* was first published here. Furthermore, when Darwin was a medical student in Scotland, contemporary scientists influenced his thinking about the unity of human species, non-superiority of Europeans, an older age of the earth than previously thought, and the need to have copious evidence before publicly making evolutionary assertions. Similarly, several French scientists such as Lamarck and Cuvier were critical to the early development of Darwin's theories, and their influences still shape French evolution learning. Certainly, Darwin's visit to the Galapagos archipelago in 1835 and his study of native mockingbirds were critically important for his speciation ideas culminating in *On the Origin of Species*, and the history of Darwin and evolution are now economically-significant tourist attractions in the Galapagos. Finally, Darwin documented his visit to Cape Town, South Africa on the return voyage of the Beagle, and this history helps South African students see Darwin as a real person.

Initial reactions to Darwin's work also had important consequences on evolution education in various countries. In some countries, these initial reactions cleared the way for early adoption of evolution in biology curricula. For example, in Scotland and England, *On the Origin of Species* was initially met with skepticism but was soon accepted widely even by the church. Consequently, evolutionary topics were incorporated into curricula in the 1860s and 1870s. Similarly, *On the Origin of Species* was not translated into Greek until 1915. While there was a mildly-negative initial reaction in Greece, The Church of Greece never officially condemned it, and religious rejection of evolution is not a widespread problem in Greece today. In France, the inclusion of prehistoric humans in early syllabi probably led to the political decision to secularize French public schools in 1880. On the contrary, Ataturk's secularization and replacement of the Creation story with evolution of the history curriculum in the 1930s were quickly reversed soon after his death in 1938. Early reactions to Darwin's ideas took a different path in Germany. The first German translation of *On the Origin of Species* occurred within two months of its publication, and Ernst Haeckel was the first German biologist to embrace evolution. Haeckel's interpretation of evolution was more directional (finalist with humans at the pinnacle) and more anti-religious than Darwin's writings. His influence is still perceived today as many regard evolution as contrary to religion.

Several places have important geographic and historical connections to evolution. For example, Mexico's immense biodiversity and conservation efforts necessitate an evolutionary perspective. Similarly, Partosa's Filipino teachers emphasized the relevance of evolution in a place with such rich plant and animal biodiversity. Additionally, South Africa is the home of iconic fossil finds and especially of hominin fossils. Local African evolution fossil discoveries engage South African students and makes evolution learning relevant. Consequently, many of these sites and associated universities have outreach programs for schools.

Evolution has a particularly dark history in Germany. Evolution became situated in a Social Darwinist perspective in German-speaking countries. Many German evolutionary scientists had eugenic views. Under Hitler's direction, evolutionary biology was used to promote eugenics, racism, and ultimately mass murder. Present-day Creationists and clergymen still refer to this history when discussing evolution, and this historical shadow still likely influences public acceptance of evolution and evolution instruction.

24.2 The Existence and Extent of Influence of Anti-evolution Movements in Countries

Across the countries represented in this book, there exists a wide range of anti-evolution movements. Although some countries (France, Greece, Hong Kong, Iran, Mexico, and Philippines) appear to have very few present-day anti-evolution movements, many countries have faced historical organized opposition to evolution education. Still, many others face current and emerging threats to evolution education. These anti-evolution efforts have had a range of impacts as well.

Historically, several countries faced anti-evolution movements that are no longer influential today. France endured waves of finalist thinking in the 1950s and Lamarckianism in the 1970s, but there are no active movements today. In the U.S. Southwest, Phoenix Arizona was a testing ground for the U.S. National Science Foundation Biological Sciences Curriculum Study in the 1960s and several parents petitioned to have their children excused from the evolution instruction portion of this curriculum. In 1882, biology teaching was completely forbidden in Germany because of evolution. In Austria, church leaders spoke negatively of evolution but have softened their criticism. In Switzerland, Creationist teaching materials were developed but met with public protest and revised. In South Africa, Conservative Christian groups organized in early 2000s to protest the introduction of evolution into the national curriculum, but scientific and academic theologians reacted and quelled much vocal resistance. In New Zealand where the primary school curriculum had addressed evolution since 1904, an Evolution Protest Movement briefly curtailed radio broadcasts referencing evolution, evolution was moved to secondary grades, and creationist textbooks circulated in the 1980s before a robust, integrated evolution curriculum was adopted in 2007.

The majority of countries represented in this book face present-day antievolution forces. England claims the oldest anti-evolution organization, the Creation Science Movement (formerly, the Evolution Protest Movement), which has been active since 1932. In Germany, Creationist ideology again rose during the 1980s with the advancement of the WuW Young Earth Creationist organization which continues to distribute materials and create textbooks. In Brazil, the Brazilian Creation Society and Brazilian Institute of Intelligent Design shape public opinion about evolution. Because these efforts are recent, pro-science groups have not yet organized consistently to refute these groups. In the United States, anti-evolution organizations such as the Discovery Institute and Answers in Genesis develop and distribute materials while several anti-evolution groups advocate for Academic Freedom Laws designed to weaken the teaching of evolution and open the door to alternatives. American Muslims, the United Kingdom, Turkey, Malaysia, and Arab countries also face anti-evolution messages from influential theologians such as Harun Yahya, and American Muslims also encounter anti-evolution rhetoric from organizations such as the Nation of Islam. While Indonesia has not experienced a specific anti-evolution movement, anti-evolutionary ideas from theologians such as Harun Yahya and Michael Denton have been incorporated into the national biology curriculum. In Scotland, Glasgow's Center for Intelligent Design and Highland Theological College have been recent active opponents of evolution instruction. In France, fundamentalist Muslims openly criticized evolution and distributed antievolution materials, but the French Ministry reacted quickly. Recently, the German-speaking countries have also seen a rise in Islamic Creationism's antievolution rhetoric. Turkey recently made international news as its ministry of education removed evolution from its curricula, citing that students were too young to understand controversial topics. Thus, antievolution activities are not obstacles faced by a few countries or by those practicing particular religious practices—this anti-science work is pervasive throughout the globe.

The sources of antievolution movements range widely. In some cases, the antievolution messages are heralded from the heads of state, such as the efforts of Brazil's Ministry of Education and politicians or the conservative political party in Turkey. In other cases, organized Creationist groups (Brazil, German-speaking countries, U.S.) with religious and political ties are the source of these movements. In some cases, U.S.-based antievolution organizations have disseminated to other countries such as Scotland and New Zealand. Other authors described less organized anti-evolution efforts led by religious scholars (Malaysia), church leaders (American Muslims, Austria), Seventh-Day Adventist churches (Galapagos), journalists (Malaysia), and concerned parents (US Southwest). In Turkey, anti-evolution proponents have become active even within the Turkish scientific community, TUBITAK.

Antievolution movements wield their influence in a variety of ways. In Brazil, members of the Ministry of Education pressure teachers to teach Creation. In several countries antievolution groups distribute Creationist materials (France, German-speaking countries, Scotland, New Zealand, United Kingdom), Creationist teaching materials (Austria, Scotland, New Zealand), and Creationist textbooks (Germany). Some anti-evolution groups have a very influential web presence (American Muslims). In Scotland, groups like People with a Mission Ministry develop extracurricular materials disseminated via buses that visit schools. In Lebanon and Turkey, religious antievolution forces successfully pressured for the removal of evolution from the curriculum. In the U.S., there have been government bills have been created (U.S. Central; U.S. Southeast; U.S. Southwest) and sometimes passed (U.S. Southeast) to weaken and eliminate evolution instruction. A similar situation occurred in Hong Kong in 2009 when the biology curriculum was changed to encourage students to explore non-evolutionary explanations for life. In other countries, antievolution groups are involved in teacher preparation and hiring such as the Creationist ministers and school chaplains who sit on Local Authority Education Committees that hire teachers in Scotland. In other countries, antievolution groups wage a public campaign to promote Creationism such as Creationist billboards in the Galapagos or anti-evolution newspaper articles in Malaysia. In response to these efforts, some countries have formally decreed or through court decisions that Creationism/Intelligent Design should not be taught (Germany, Scotland, U.S.).

Even in countries faced by staunch anti-evolution forces, evolution education and research persist. For example, across the Gulf states where the teaching of evolution is sometimes specifically banned, international collaborations sustain the funding and proliferation of biological evolution research.

24.3 Place of Evolutionary Theory in Curriculum Across Countries

Given the diversity in public opinion on evolution, the range of the place and status of evolution within science curricula. An important issue arose in comparing different countries' and regions' inclusion of evolution in curricula in terms of the question, "What counts as inclusion?" Some countries include related concepts (adaptation, variation, geologic time) at various ages but explicitly exclude the word "evolution" and make no reference to Darwin or natural selection. Table 24.1 compares where evolution first appears in various countries' curricula with respect to grade band, according to the interpretations of "inclusion" of each chapter author or author team.

Countries range in the extent to which evolution is present in national curricula. In France, Greece, and New Zealand curricula, evolution is treated as a unifying theme. For example, France's comprehensive coverage of evolution begins when children aged six to eight learn about biodiversity, children aged nine to eleven classify organisms according to an evolutionary perspective, and children aged 12–14 learn about fact versus belief, human evolution, and evolutionary mechanisms. In Iran, fifth graders learn about the history of the Earth and the evolutionary emergence of life; eighth-graders learn about evidence for evolution, adaptation,

Table 24.1 Range of treatment of evolution by grade band

Grade band of first evolution instruction	Countries
Primary (before age 10)	England, France, New Zealand, Philippines, U.S. (some states)
Middle (age 10–13)	Austria, Germany, Greece, Iran, Kuwait, Luxembourg, Mexico, Scotland, South Africa, South Tyrol, Switzerland, U.S. (some states)
High School (above 14)	Brazil, Egypt, Greece, Hong Kong, Indonesia, Malaysia (Fah et al.), Syria, Tunisia, U.S. (some states)
Not required/Omitted	Galapagos, Lebanon, Malaysia (Osman), Turkey (2017)
Banned	Algeria, Morocco, Oman, Saudi Arabia

and natural selection; and high school students explore these topics in more depth with a population genetics focus. Whereas in other countries such as Mexico, evolutionary topics are taught but in disjointed ways, reducing evolution to adaptation, for example. Evolution instruction has been included for a century and a half in some countries, while South Africa only recently (2006) began including evolution in their national curriculum. Evolution has been inserted and removed from national curricula as well, as in Turkey.

Even when evolution is included in curricula, some evolution content tends to be emphasized more than others. Adaptation is often taught (for example, in Mexico and South Africa) non-contentiously, and the focus of evolution tends to be on micro rather than macroevolution (U.S. Southwest). As such, natural selection is frequently included as a high school topic (German-speaking countries, South Africa). Evolutionary history is frequently included (German-speaking countries) and is a central focus of the Brazilian curriculum. Some countries avoid using the "e-word" altogether or only in high school curricula (South Africa, U.S. Southwest). In many curricula, more contentious topics are minimized or avoided. Although deep time is addressed in middle school in German-speaking countries, it is avoided elsewhere (U.S. Southwest). Variation has limited coverage in Mexico, and human evolution is often excluded in Hong Kong, Iran, and the U.S. In Jordan, evolution is taught from a religious perspective consistent with the Quran. While alternatives to evolution are explicitly taught in South Africa and Brazil, the national curriculum requires Indonesian teachers to connect evolution instruction to religious ideas and to include antievolutionary ideas. A critical examination of Creation from a scientific perspective is built into the curriculum of some German-speaking countries. Some evolutionary content is included in History/Social Studies curricula including prehistory in France and hominid evolution in South Africa. Clearly, evolutionary content is parsed across courses and grades in ways particular to each unique context.

When the "e-word" evolution is not explicitly included in curricula, principles and tenets of evolution are often embedded in national biology curricula. For example, in Malaysia where evolution is excluded, concepts such as natural selection, inbreeding, hybridization, mutation, genetic variation, and the ancient earth are taught but not used as evidence for or components of evolutionary theory. For Malaysian students preparing to attend a university, students prepare for a Malaysian Higher School Certificate Examination by exploring Lamarck's theory of inheritance of acquired characteristics, fossil record, and some reference to Darwin-Wallace's natural selection.

Evolution is present in several informal education contexts in some countries as well. In South Africa, the rich fossil sites of hominid evolution serve as a means to educate the public about evolution. In Iran, museums and even a dinosaur-themed amusement park offer opportunities to explore evolutionary content.

Evolution's treatment in science textbooks is often a matter of concern. Up until 2012, a major textbook used in Scotland omitted evolution, and it is frequently relegated to the end of the textbook where it can be easily omitted in countries like Greece. Occasionally, textbooks reference nature of science (Brazil). Creationist influences are present in some textbooks in Brazil and Scotland (although these have been discontinued). In textbooks in the U.S. Southeast, students face disclaimers of evolutionary content when they examine their biology books. Other textbooks include non-Darwinian ideas such as teleology, anthropomorphism, and finalism (German-speaking countries). On the contrary, Hong Kong textbooks have been increasing their coverage, depth, learning activities, and nature of science connections related to evolution over the past three decades.

Evolution appears on some college entrance exams and compulsory school completion exams but not others. This evolutionary content is present in college entrance exams in Brazil, Greece (although relatively few people take the latter exam), Hong Kong (although relatively few students take biology as an elective), and New Zealand. Although evolution is present on New Zealand examinations at the year 13 level, publicly funded schools can opt out of the portion of standards pertaining to evolution. While present in compulsory school completion exams in the U.S. Southwest, fewer than 10% of the state exam questions address evolution. Elsewhere in the U.S. compulsory school completion exams do not address evolutionary content at all (U.S. Central).

Even when the national curriculum includes evolution, students in private schools in different countries may never receive evolution education. For example, about 14% of New Zealand's students attend private "special character" schools that reflect the chartering groups' faiths and philosophies, possibly circumventing evolutionary content.

24.4 Emphasis Given to Evolutionary Theory in Biology Teacher Education Programs

Many countries reported that teacher education programs do not emphasize evolution education. A rare exception to this lack of emphasis is France where secondary life and earth science teachers master phylogenetic classification in their coursework. Sometimes, evolution content is integrated throughout all biology courses regardless of whether evolution-specific courses are offered (German-speaking countries, Malaysia (Fay et al.) In some places, evolution coursework was not previously required but now is required (South Africa, U.S. Central). More often, evolution coursework is presented sporadically, treated as an elective, or is not available at all (Arab countries, Philippines, U.S. Southeast). Clearly, teacher expertise in evolutionary content is an important issue across the globe. Even when evolution content is included in university coursework for science teachers, these science content courses sometimes explicitly endorse Creationist views (Brazil).

Given this sporadic coursework in evolution, preservice science teachers across countries often have misconceptions about evolution and reject it. Authors cited common evolution misconceptions like Lamarckian ideas (Brazil, France) and teleological reasoning (Indonesia) as common among preservice teachers. Teacher rejection of evolution or parts of evolution is an important challenge in many contexts (Galapagos, Malaysia, U.S. Southeast).

Several authors mention specific kinds of pedagogical strategies that should be embedded in science teacher preparation programs to better teach evolution. These include nature of science (Hong Kong), connecting evolutionary content to the everyday lives of students (Mexico), and strategies for dealing with Creationist and antievolution arguments (German-speaking countries). Unfortunately, many authors also report how science methods courses are often overly broad by including all aspects of science pedagogy in a single course (U.S. Southeast) or by encompassing all sciences instead of biology alone (U.S. Central) so that evolution education receives little attention. To attend to these challenges, multiple authors (Greece, U.S. Central) advocated for new efforts to develop teachers' evolution-specific pedagogical content knowledge, and the U.S. Central chapter specifically details a combined methods/content course designed for biology preservice teachers.

Several authors described larger issues affecting biology and science teacher preparation that impact the way teachers are prepared to teach about evolution. In some countries, the availability of teacher preparation programs is limited. For examples, there is only a single Escuela Normal that is bilingual and indigenous in Mexico, and Greece lacks a systematic preservice and inservice teacher training program so scientists like biologists and physicists become teachers with insufficient didactics support. In Greece and New Zealand, another larger issue is out of area teaching. Preservice teachers preparing for non-biology science disciplines at the secondary level may end up teaching biological science (and evolution) without much evolution-specific content or pedagogical preparation. Access to sustained professional development in general and about evolution in particular is sporadic in Greece and South Africa. One final larger issue impacting teacher preparation is the rapidly changing curriculum in places like Turkey.

Another teacher preparation issue that surfaced across countries was the need to support primary grades teachers for teaching about evolution. Some countries (France, Mexico, and the United Kingdom) already include evolution content and teaching strategies in their preservice teacher education programs. Chapter authors from other countries (New Zealand, Philippines, and Scotland) suggest that this type of teacher preparation is much needed in their contexts. These findings make sense in light of Table 24.1 that shows the countries that address evolution at the primary level.

24.5 Biology Teachers' Attitudes Toward Teaching Evolutionary Theory

When science teachers in various countries have been surveyed with respect to their evolution acceptance, several interesting results have emerged. Not surprisingly, in countries with high public acceptance of evolution, teacher acceptance of evolution is also high. For example, Quessada and Clément report findings that less than 2%

French teachers accepted Creation only with 98% accepting evolution. In France, most of the religious teachers also accepted evolution. In Scotland, Downie, Southcott, Braterman, and Barron shared that 78% Scottish biology teachers endorsed secular evolution; 16% indicated that God had a role in evolution, and 6% were young-earth Creationists. Similarly, in New Zealand, science teachers have high evolution acceptance and understanding. However, in Mexico where public acceptance of evolution is mid-range, the vast majority of Monterrey teachers showed high evolution acceptance. Conflicting results were reported by Cotner and Moore where 30% Galapagos biology teachers indicated that evolution cannot be correct since it disagrees with the Bible, but 86% also agreed that evolution is the result of sound scientific research and methodology. In Brazil where Creationist views are very common, Brazilian teachers showed a higher acceptance of Creation concepts than other countries, but biology teachers showed less, in a study cited by Oliveira and Cook. In Arab countries, teachers are much more rejecting of evolution, often maintaining that God alone created living things. In Malaysia where evolution is widely rejected, 70% of teachers are considered radically creationist.

Across these countries, chapter authors also reported a range of teachers' attitudes toward teaching evolution. At the positive end, the majority of Monterrey Mexico biology teachers had positive attitudes toward teaching evolution, and some U.S. Southwest teachers revealed a passion for teaching evolution. In Hong Kong, teachers consistently teach evolution because of its emphasis on public examinations. In terms of self-perceived content knowledge, 92% of Scottish teachers and 97.6% of Galapagos teachers said they were confident in their evolution content knowledge. However, teachers' evolution teaching practices are often less than ideal. For example, Greek teachers often intend to teach evolution but many times do not as the topic is at the end of the textbook. Similarly, while the majority of Galapagos teachers claims to enjoy teaching evolution, 79–82% indicated they are uncomfortable teaching about evolution and 82% believe it presents a conflict between science and religion. In the Central U.S., Friedrichsen, Brown, and Schul report that beginning teachers often faced resistance to evolution instruction and did not want evolution taught while more experienced teachers sought to build trust in the community in order to teach about evolution. Biology teachers in the Southeast U.S. described evolution teaching as frustrating and challenging as they felt public pressure to avoid teaching the subject. Likewise, Kuwaiti teachers resisted the teaching of evolution because they perceived it to be incongruous with their culture and religion. Similarly, some Arab state and Turkish biology teachers and professors thought evolution should not be taught, thought evolution and creation should be given equal time, or taught about God's role in creation during evolution instruction. Finally, some Southwest U.S. teachers were simply not comfortable teaching about evolution.

Science teachers from many countries have reported similar concerns about teaching evolution. For example, Greek teachers are concerned about their content knowledge, lack of teaching skills, low self-efficacy for teaching biology relative to other biology concepts. Similarly, South African biology teachers expressed concerns about their limited content knowledge about evolution, how to teach it, and concerns about religious conflicts. Turkish teachers cited concerns about student/parent resistance, student misconceptions, negative influence of media, administration pressures, and their own lack of content knowledge. Similar concerns were reported for Southeast and Southwest U.S. teachers. These concerns often stem from a lack of pedagogical content knowledge for teaching evolution as mentioned by chapter authors from Greece, Turkey, and U.S. Central. This pedagogical content knowledge could include understandings of common student misconceptions about evolution, knowledge of good labs and activities, and awareness of evolution teaching resources for engaging students and teaching about NOS and science/religion distinctions.

24.6 Suggestions to Improve Evolution Education Across Countries

Despite the wide range of evolution acceptance, place of evolution in the curriculum, and teacher preparation and attitudes toward teaching evolution, several chapter authors offered similar suggestions for how to improve evolution education in their countries. These suggestions included curricular changes, textbook changes, a focus on informal science education, better preservice and inservice teacher education, policy changes, cautions about assumptions, and calls for more research. Table 24.2 outlines these suggestions and countries in which such changes are advocated.

Several chapter authors called for specific curricular changes regarding evolution education. The most commonly sought change is a more integrated approach for curriculum. For example, treating evolution as an organizing principle was recommended for Mexico, German-speaking countries, and Hong Kong. In a related

Table 24.2 Suggestions for improving evolution education across countries

Suggestions for improvement	Countries advocating such change
Curricular and textbook changes	American Muslims, Brazil, France, German-speaking countries, Greece, Hong Kong, Malaysia (Osman), Mexico, Turkey, United Kingdom
Informal science education	United Kingdom
Improved preservice teacher preparation	Galapagos, German-speaking countries, Greece, Hong, Kong, Malaysia (Osman), Mexico, New Zealand, Philippines, Scotland, Turkey, U.S. Central, U.S. Southeast, U.S. Southwest
Improved inservice teacher professional development	Brazil, France, Galapagos, Hong Kong, Mexico, New Zealand, South Africa, U.S. Southeast, U.S. Southwest
Policy changes about evolution education	Greece, Scotland
Cautions about assumptions	Brazil, South Africa
More research about evolution education	Indonesia, Iran, Malaysia (Osman), Mexico, American Muslims, Philippines

way, Quessada and Clément call for evolution education that represents an even broader interdisciplinary approach melding science and philosophy. Other countries such as the German-speaking countries, authors suggest including evolution in primary grades rather than only at the secondary levels. More inclusion of NOS was recommended for American Muslims, Greek, Hong Kong, Malaysian, New Zealand, and Philippine evolution curricula. Reiss recommended curricular approaches that allow for genuine discussions of students' doubts about evolution as these are opportunities to highlight scientific and non-scientific worldviews. Similarly, Fouad recommended careful sequencing of evolution instruction beginning with microevolution, then macroevolution, and offering different ways to think about science/religion relationships beyond Gould's non-overlapping magisteria approach. At the other extreme, chapter authors like Mugaloglu suggested the need to restore evolution back into the Turkish curriculum. One final set of curricular changes includes revisions to biology and science textbooks to mitigate Creation references (Brazil) and update content (Greece).

Reiss offered the suggestion that U.K. evolution educators draw upon the strengths of informal science education opportunities to teach about evolution. Museums and galleries often have rare specimens, and patrons have the benefit of being motivated to learn about such content when they voluntarily seek out these informal opportunities. Yet, Reiss cautions that misconceptions about evolution (e.g. progress narratives) that arise in formal science education contexts also loom in these informal contexts.

The most commonly offered suggestions for improving evolution education centered upon preservice science teacher preparation for teaching about evolution. Some of these suggestions focused on better pedagogical preparation of science teachers (Greece) in general or calls for making evolution more of an emphasis in science teacher preparation (German-speaking countries, Mexico, U.S. Southeast). Other authors called for more and specific evolution content courses (Greece, Malaysia, Philippines, U.S. Central) and opportunities to reflect on common evolution misconceptions (Galapagos). Several authors suggested attending to preservice secondary biology teachers pedagogical content knowledge for evolution teaching (Hong Kong, U.S. Central, U.S. Southwest) and including specific strategies for addressing students' antievolution resistance (Scotland). Authors (e.g. New Zealand, Philippines, Scotland) also suggested making evolution more of an emphasis for primary grades preservice teacher education. The need for more culturally-responsive evolution teaching strategies for biology preservice teachers was also noted (U.S. Southeast, U.S. Southwest), especially calling for opportunities for preservice science teachers to explore and define their own worldviews (US Southeast). Some suggestions for improving preservice teacher education with respect to evolution education included programmatic changes such as using evolution acceptance as a prerequisite for admission into biology teaching programs (Scotland) or developing strong mentor programs for evolution teaching during preservice teacher clinical experiences (US Southwest).

Several chapter authors suggested specific types of evolution education professional development for teachers. These suggestions included professional development to improve content knowledge (Galapagos) and pedagogical content knowledge (U.S. Southwest). Several authors also suggested multicultural and culturally relevant pedagogy professional development for teachers. For example, Oliveira and Cook (Brazil), Glaze and Goldston (U.S. Southeast), and BouJaoude (Arab countries) suggested culturally-sensitive approaches that take into account students' worldviews. Similarly, Fisher (U.S. Southwest) and Galindo, Franco, Ramos, Pérez, and Frias (Mexico) advocate culturally relevant pedagogy that attends to students' racial, ethnic, and religious heritages. Other authors suggested specific evolution education topics that should be addressed in science teacher professional development programs: guidance for dealing with student resistance (Brazil), Eugenie Scott's evolution-creation continuum (South Africa), strategies for facilitating student debates (France), teaching evolution in a historically-rich curriculum focusing on nature of science (New Zealand), and strategies for improving evolution-related field trips (South Africa).

Some chapter authors suggested specific policy changes to improve evolution education. For example, Stasinakis and Kampourakis suggested making the teaching of evolution mandatory to overcome evolution omission excuses in Greece. Furthermore, Downie, Southcott, Braterman, and Barron suggested changing Scotland's Local Authority Education Committees' composition given the current undemocratic role of faith-based representatives.

Several authors emphasized how evolution educators at all levels must be aware of the worldviews and unique cultural/religious milieus that exist across countries. As BouJaoude noted for Arab countries, meaningful evolution education changes necessitate changes outside scientific and educational spheres alone.

In their review of their countries' evolution education, some authors noted the need for evolution educators and researchers to be careful of assumptions about evolution education. For example, Oliveira and Cook cautioned science teachers not to blankly assume that teaching Creationism will appease evolution-deniers and create a friendly classroom atmosphere. In the South African context, Sanders advised evolution researchers not to make the assumption that all religious students have a particular evolution view or learning experience. In a similar way, BouJaoude points out how Druze Muslim students in Lebanon are very accepting of evolution even as many other Lebanese Muslims were more firmly rejecting of evolution.

A final suggestion for improving evolution education is to conduct more evolution education research. This suggestion was offered by Galindo, Franco, Ramos, Pérez, and Frias who emphasized the need for evolution education research across the diverse contexts within a country such as Mexico. Similarly, Fouad suggested that very little research has been done regarding American Muslims' evolution learning experiences, and this must be remedied. Furthermore, Partosa called for more Filipino evolution education research, especially pertaining to conceptual change and how attitudes and beliefs are connected with student understanding. Likewise, Kazempour and Amirshokoohi called for more Iranian evolution education research, especially focusing on students' and teachers' understanding and acceptance of evolution. Rachmatullah et al. emphasized the need for more research about how students and teachers make sense of the religious connections to science content in the Indonesian context where the two ways of knowing are so intertwined. Osman also called for evolution education research in Malaysia to assess students', teachers', and administrators' acceptance, understanding and attitudes toward evolution. While much of this research would address gaps in knowledge about evolution education in countries and regions for which little is known, this research is also essential for understanding fundamental questions about the teaching and learning of evolution. How can evolution educators best prepare teachers? How can evolution instruction be culturally sensitive? How can common misunderstandings about evolution and the nature of science be addressed?

24.7 Conclusions and Directions Forward

As biological and molecular technologies flourish and re-shape agriculture, medicine, and conservation efforts, scientific literacy about the most foundational concept in modern biology is essential. Yet, evolution education is contentious in many regions and places around the globe. While anti-evolution forces manifest differently depending on the extent to which religious and civic lives are intertwined, very few countries represented in this volume are unaffected by these efforts. Across contexts (Christian, Muslim, secular, etc.), authors called for stronger efforts to help students, teachers, and the public address misconceptions about evolution and distinctions between science and religion. Evolution education is not a "Muslim problem" or a "Christian problem"—it is a global issue.

While global, the challenges of evolution education manifest differently in different contexts. In countries with tight autocratic leadership, antievolution forces and policies can sweep through ministries of education, obliterating evolution education. Most recently, several Arab countries and especially Turkey have experienced such injuries to their science education curricula. In other countries where curricula are not as tightly controlled by governments, antievolution forces work through informal means to erode public understanding of science. One sentiment we heard from many chapter authors was that evolution education was once never under siege in their countries, but now antievolution forces have heightened in recent years.

Evolution teaching serves as "a canary in a coalmine" as a cultural indicator of many other issues of authoritarianism and public understanding of and relationship to science. As authoritarian forces arise and simmer across the globe, evolution education wanes and waxes like phases of the moon. Efforts to address the challenges of teaching and learning evolution alone, without recognition of the larger social, political, and cultural forces at play, will likely be ineffective.

And yet, addressing the prevalence of so many misconceptions about evolution and the nature of science across societies is surely within the purview of science educators. These efforts to education the public across societies must address how evolution is foundational to modern biology, how it is widely embraced within the scientific community, and how it embodies solid scientific practices and yet remains tentative as does all science. Additionally, the public must understand how applications of evolution already permeate our lives in terms of medicine and agriculture. When a person rejects evolution, they are often not aware that they are rejecting much of the biotechnologies already pervasive in our worlds. Part of this public understanding of evolution challenge is the disconnect between the scientific community and the public. While the scientific societies of so many of the countries represented in the book signed the Interacademy Panel (IAP) statement in support of the teaching of evolution in 2006, this recognition of the status of evolution within this international scientific community is not often apparent in primary and secondary curricula and the preparation of teachers.

In terms of future research, many voices from various countries and regions are not present in this volume. Specifically absent are Russia, China, India, most African nations, and most South American nations. Future efforts to explore and communicate evolution teaching practices and challenges must be made to inform this global alliance of evolution educators.

Chapter authors contributed many ideas for moving evolution education forward. Clearly, curriculum development, preservice and inservice teacher preparation efforts, policy reform, and research are clearly needed across diverse contexts. Specifically, earlier evolution education, resources for teaching about science and religion relationships, and culturally- and geographically-responsive teaching strategies and resources are needed. We also call for possibilities of cross-pollinating global education initiatives. We have already seen how anti-evolution forces from one nation can impact another, but more needs to be done to develop a more concerted alliance of evolution educators to mobilize.

Reference

Dobzhansky, T. (1973). Nothing in biology makes sense except in the light of evolution. *American Biology Teacher, 35,* 125–129.

Lisa A. Borgerding is an Associate Professor of Science Education at Kent State University. She teaches undergraduate, masters, and doctoral level courses in science education and research methodology. Her research centers upon the teaching and learning of biological evolution from early childhood through college, nature of science instruction, preservice and inservice teacher development, and service learning in teacher education.

Hasan Deniz is an Associate Professor of Science Education at University of Nevada Las Vegas (UNLV). He teaches undergraduate, masters, and doctoral level courses in science education program at UNLV. His research agenda includes students' and teachers' epistemological beliefs about science (nature of science) and evolution education. He is recently engaged in professional development activities supported by several grants targeting to increase elementary teachers' knowledge and skills to integrate science, language arts, and engineering education within the context of Next Generation Science Standards.

Printed by Printforce, the Netherlands